SHARKS and RAYS
OF AUSTRALIA

This guide was funded by a research grant from the
Australian Fisheries Research and Development Corporation.

SHARKS and RAYS
OF AUSTRALIA

P.R. LAST & J.D. STEVENS
CSIRO Division of Fisheries

COLOUR ILLUSTRATIONS
R. SWAINSTON
LINE ILLUSTRATIONS
G. DAVIS

CSIRO
AUSTRALIA

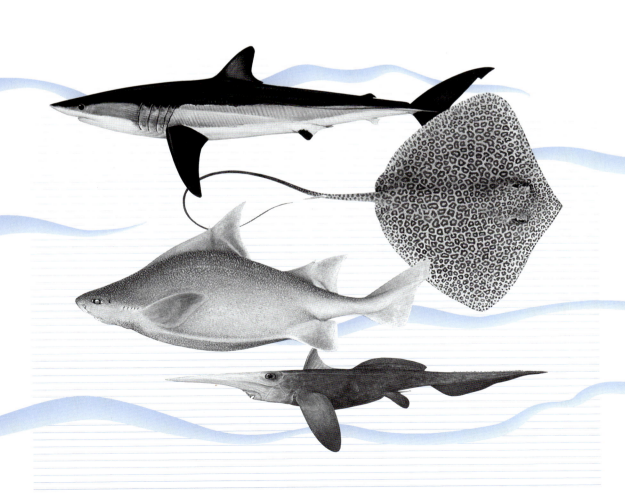

National Library of Australia Cataloguing-in-Publication entry

Last, P.R. (Peter Robert).
 Sharks and rays of Australia.

 Bibliography.
 Includes index.
 ISBN 0 643 05143 0

 1. Sharks – Australia – Identification.
 2. Rays (Fishes) – Australia – Identification.
 I. Stevens, J.D. (John Donald).
 II. Title.

597.310994

© CSIRO Australia, 1994

Managing Editor: Kevin Jeans
Editor: Marta Veroni
Designers: Linda Kemp, Kerry Slaven and Gordon Yearsley
Production Manager: Jim Quinlan
Layout: Linda Kemp

FOREWORD

The shark and ray fauna is of considerable commercial interest worldwide. For example, up to 60 000 tonnes of stingrays are caught annually off India, and the dogfish fishery of the North Sea sustains the traditional 'fish and chips' of the British Isles. In the Australian region, fisheries for school and gummy sharks are well established in the south of the continent, and a fishery for blacktip sharks is developing in the north. There is the potential for development of fisheries for blue shark, skates, and deepwater dogfishes for both food and fish oils.

Apart from this commercial significance, the Australian shark and ray fauna is unique in its diversity and scientific importance: it is almost 50 per cent more diverse than other comparable regions of the world. However, despite an almost universal popular interest in sharks and rays, it is over 50 years since their identification and ecology in Australia have been examined comprehensively. During that time the number of identified species in the Australian region rose from about 160 to 300.

This book is an invaluable source of information for anyone wanting to know more about this ancient, unique, and commercially important group of animals. Within its pages you will find colour plates painted with exquisite care to record accurately details of form and colour. There are taxonomic keys, technical descriptions for the professional scientist, and over 1400 black and white illustrations to aid in the identification of every species. There are maps of their geographical distributions, and information on the feeding, biology, reproduction and other aspects of the natural history of each species.

The production of the book was financed by CSIRO, with a significant contribution from the Fishing Industry Research and Development Council. It could not have been written without the presence of the taxonomic collection of fishes at the CSIRO Division of Fisheries. The I.S.R. Munro collection was built up over many years, and owes its existence to the lodging of specimens collected during the Division's research cruises, donations from other research organisations and from specimens brought in by fishermen curious about the amazing fauna that inhabits our waters. The collection is now one of the two most significant taxonomic collections of sharks and rays in the Southern Hemisphere, and is a significant record of Australia's marine vertebrate biodiversity.

The book is a fascinating source of information to the general-interest reader, an indispensable compendium to the research taxonomist, a catalogue of information pertaining to Australia's biodiversity, and a baseline reference for the fishing industry in its search for sustainable development of our marine fisheries resources. I congratulate the authors for producing this significant contribution to knowledge, and am sure that it will find a place wherever books on our Australian animals are found — from fishing boats to academic libraries.

Peter C. Young
Chief
CSIRO Division of Fisheries
Australia

ACKNOWLEDGEMENTS

While many people have helped us in one way or another in the preparation of this book, we would particularly like to thank Mr G. Yearsley (CSIRO Division of Fisheries, Hobart) who prepared a mockup of the text, co-authored the urolophid chapter and assisted in X-raying and measuring specimens. We are also particularly grateful to Dr M. Stehmann (Institut für Seefischerei, Hamburg) for providing data on a new species of skate (*Bathyraja* sp.), and to Mr K. Graham (Fisheries Research Institute, Sydney) for providing many valuable specimens and detailed distributional information on fishes off the New South Wales coast.

We are deeply indebted to the fine efforts of the illustrators, Mr R. Swainston and Ms G. Davis, for their high quality work which so strongly enhances the guide. Most of the colour illustrations were drawn from excellent colour transparencies taken by CSIRO photographers Mr T. Carter and Mr G. Johnson.

Mr N. Kemp (Tasmanian Museum and Art Gallery, Hobart) and Dr G. Edgar (Melbourne University) kindly reviewed the manuscript and provided constructive advice. Dr V. Mawson (CSIRO Division of Fisheries, Hobart) edited the text and made many useful suggestions on style and design.

During the prepartion of this guide we received technical and administrative assistance from many colleagues at the CSIRO Division of Fisheries. In particular we would like to thank Mr A. Graham, Ms S. Kalish, Mr M. Gregory, Mr M. Edmunds, Ms J. O'Regan, Ms J. Burdon, Ms C. Bulman, Ms S. Davenport, Mr J. Salini and Mr G. West. We also thank Mr T. Fountain, Mr M. Reed, Ms F. Klok, Mr J. Palmer, Ms M. Bresic, Dr A. Williams, Mr W. Whitelaw, Mr D. Abbott, Mr G. Forbes, Mrs T. Cracknell, Ms A. Constantine, Ms K. Meyer, Ms F. Barbour, Mr D. Taylor, Ms B. Hansen, Mr A. Rees, Mr K. McLoughlin, Dr A. Koslow, Dr K. Sainsbury, Dr S. Blaber, Mr D. Evans, Mr D. Wright, Dr B. Wallner, Ms B. Potter, Mr G. Thill, Mr T. Mangan and Dr B. Hill.

Many of the specimens used to describe species in the guide were collected by the CSIRO research vessels *Courageous*, *Soela* and *Southern Surveyor*. We thank the masters and crew of these vessels, who over the years have indirectly contributed so much to our knowledge of the Australian shark and ray fauna.

We also received assistance from colleagues at other research organisations and we are particularly grateful to Dr L. Compagno (South African Museum, Capetown), Dr J. McEachran (Texas A & M University, College Station), Dr G. Burgess (Florida Museum, Gainesville), Dr T. Taniuchi and Dr H. Ishihara (University of Tokyo), Dr K. Nakaya (Hokkaido University), Mr L. Paul (Ministry of Agriculture and Fisheries, Wellington), Mr A. Stewart (National Museum of New Zealand, Wellington), Dr B. Seret (ORSTOM, Paris), Dr G. Dingerkus (Museum d'Histoire naturelle, Paris), Dr N. Parin (Russian Academy of Sciences, Moscow). We also thank Dr M. Dunning (Queensland Department of Primary Industries, Brisbane), Mr G. McPherson (Queensland Department of Primary Industries, Cairns), Dr R. Lenanton and Mr M. Cliff (Department of Fisheries, Perth), Mr I. Wittington (University of Queensland, Brisbane), Dr D. Ramm and Dr D. Grey (Department of Primary Industry and Fisheries, Darwin), Dr J. Lyle (Division of Sea Fisheries, Hobart), Mr J. Garrick (New Zealand) and Mr J. West (Taronga Zoo, Sydney).

Curators and staff of Australasian museums made available their specimens for examination and loan. We thank Dr J. Paxton, Dr D. Hoese and Mr M. McGrouther (Australian Museum, Sydney), Dr B. Russell and Dr H. Larson (Northern Territory Museum, Darwin), Dr B. Hutchins and Mr N. Haigh (Western Australian Museum, Perth), Dr J. Glover and Mr T. Sim (South Australian Museum, Adelaide), Dr R. Mackay and

Mr J. Johnson (Queensland Museum, Brisbane), Dr M. Gomon and Ms R. Poole (National Museum Victoria, Melbourne), and Dr C. Paulin and Dr C. Roberts (National Museum of New Zealand, Wellington) for their assistance.

We are very grateful to the Commonwealth foreign fishing observers and a group of enthusiastic commercial and recreational fishermen for collecting important specimens, providing distributional information, and allocating cabin space on their boats. We particularly thank Mr M. Baron, Mr R. Cropp, Mr K. Staich and Mr A. Grice (Commonwealth Observers); Mr N. Mooney and Mr N. Brothers (Parks, Wildlife and Heritage, Hobart); Mr G. and Mrs B. Lane, Mr P. Lewis, Mr R. Lowden, Mr N. Soulos, Mr B. Scimone, Mr P. Osborne, Mr K. Ranger and Mr R. Johnston (commercial fishermen). The efforts of Mr N. McLaughlin (Department of Fisheries, Perth), Mr P. Johnston (Department of Fisheries, Broome), Mr C. Done and Mr R. Gaho (Conservation and Land Management, Kununurra), Mr I. and Mrs S. Sinnamon, Mr G. Cooke, Mr W. Robb, Mr R. Cooley, Mr R. Exten, Mr B. Davison, Mr G. Stables, Mr Murray Walton and Mr Mervin Walton were greatly appreciated.

Technical advice and assistance with artwork and layout were provided by Dr K. Marlowe, Ms S. McDonald (Tasmanian State Library, Hobart), Ms C. Clark (Queen Victoria Museum, Launceston), Mr M. Averice and Mr P. Kilbourne (CSIRO, Melbourne), and Mr K. Jeans, Ms K. Slaven, Mr P. Reekie, Mr J. Quinlan, Ms H. Kinniburgh, Ms M. Veroni and Ms L. Kemp (CSIRO, Information Services).

This guide could not have been compiled without the financial support of the Fishing Industry Research Development Trust and the CSIRO Division of Fisheries. We thank chiefs Dr F. Harden-Jones and Dr P. Young, and program leader Dr R. Thresher for making Divisional resources available.

Finally, we would like to express our gratitude to Claire, Jayne and Josephine for their help, patience and tolerance during the prepartion of this tome.

CONTENTS

	Foreword	v
	Acknowledgements	vi
1	Introduction	1
2	How to Use This Book	12
3	Glossary	16
4	Key to Families	27
5	Chlamydoselachidae (Frilled Sharks)	36
6	Hexanchidae (Sixgill and Sevengill Sharks)	38
7	Echinorhinidae (Bramble Sharks)	44
8	Squalidae (Dogfishes)	47
9	Oxynotidae (Prickly Dogfishes)	104
10	Pristiophoridae (Sawsharks)	106
11	Heterodontidae (Horn Sharks)	112
12	Parascylliidae (Collared Carpet Sharks)	117
13	Brachaeluridae (Blind Sharks)	122
14	Orectolobidae (Wobbegongs)	125
15	Hemiscylliidae (Longtail Carpet Sharks)	134
16	Stegastomatidae (Zebra Sharks)	138
17	Ginglymostomatidae (Nurse Sharks)	140
18	Rhincodontidae (Whale Sharks)	142
19	Odontaspididae (Grey Nurse Sharks)	144
20	Mitsukurinidae (Goblin Sharks)	148
21	Pseudocarchariidae (Crocodile Sharks)	150
22	Megachasmidae (Megamouth Sharks)	152
23	Alopiidae (Thresher Sharks)	154
24	Cetorhinidae (Basking Sharks)	159
25	Lamnidae (Mackerel Sharks)	161

26	Scyliorhinidae (Catsharks)	167
27	Triakidae (Hound Sharks)	205
28	Hemigaleidae (Weasel Sharks)	217
29	Carcharhinidae (Whaler Sharks)	221
30	Sphyrnidae (Hammerhead Sharks)	270
31	Squatinidae (Angel Sharks)	277
32	Rhinobatidae (Shovelnose Rays)	283
33	Rhynchobatidae (Sharkfin Guitarfishes)	295
34	Rajidae (Skates)	299
35	Anacanthobatidae (Leg Skates)	357
36	Pristidae (Sawfishes)	360
37	Torpedinidae (Torpedo Rays)	368
38	Hypnidae (Coffin Rays)	372
39	Narcinidae (Numbfishes)	374
40	Dasyatididae (Stingrays)	380
41	Urolophidae (Stingarees)	416
42	Gymnuridae (Butterfly Rays)	443
43	Hexatrygonidae (Sixgill Stingrays)	445
44	Myliobatididae (Eagle Rays)	447
45	Rhinopteridae (Cownose Rays)	456
46	Mobulidae (Devilrays)	458
47	Callorhinchidae (Elephant Fishes)	465
48	Chimaeridae (Shortnose Chimaeras)	467
49	Rhinochimaeridae (Spookfishes)	479
	References	484
	Checklist of Australian Sharks, Rays and Chimaeras	495
	Index of Scientific Names	501
	Index of Common Names	508
	List of Plates	
	Plates	

1 INTRODUCTION

Chondrichthyan fishes *(sharks, rays and chimaeras)*

The sharks, rays and chimaeras (ghost sharks), collectively known as the 'cartilaginous fishes', comprise one of the two major taxonomic groups of contemporary fishes, the Class Chondrichthyes. The earliest known chondrichthyans evolved in the late Silurian period more than 400 million years ago. Today, more than 950 species live in the seas and freshwater systems of the world. The other major group, the Osteichthyes or bony fishes, makes up about 95% of the modern fish fauna and includes forms such as the tunas, mackerels and cods.

The main difference between the two major groups is that the bony fishes have a skeleton made of bone, whereas chondrichthyans have a cartilaginous skeleton without true bone, although it was apparently present in ancestral cartilaginous fishes. Bone was probably lost in the evolution of contemporary groups to reduce body weight and facilitate buoyancy. Other features that, in combination, distinguish the chondrichthyans from bony fishes include the presence of small, structurally tooth-like scales (dermal denticles), a skull without sutures, mouths and nostrils that are usually on the underside of the head, pelvic claspers on males, and teeth that are either continuously replaced and embedded in the gums (rather than attached to the jaws) or that are fused into plates that grow with the animal.

All sharks and rays fertilise ova internally and produce large yolked eggs (about 57% are also viviparous). They retain urea in their blood to control osmosis; the urea on breaking down after death gives them their characteristic smell of ammonia. Cartilaginous fishes are predators and they have evolved a battery of finely tuned sensory systems to locate their prey. In addition to the more common senses of smell, hearing, vision, touch and taste, they are also able to detect vibrations through the lateral line system and are sensitive to electrical and magnetic fields. They have a relatively large brain to handle and process this diverse array of sensory information.

Cartilaginous fishes occupy a variety of habitats. They are found in all oceans of the world, from near the shoreline to the deep abyss. They are most numerous above 2000 m in tropical and warm temperate continental marine habitats, but a few species are found in freshwater and hypersaline habitats.

The taxonomic relationships of the chondrichthyan fishes are not fully understood and there is some disagreement between taxonomists as to their phylogeny. Cartilaginous fishes comprise two readily identifiable taxonomic subgroups: the Elasmobranchii (sharks and rays) and Holocephali (chimaeras or ghost sharks). Elasmobranchs are generally further subdivided into four superorders, of which three make up the 'true sharks' and the remaining superorder, the 'rays'. Some shark groups, however, appear to be more closely related to rays than to other sharks. Evolutionary issues, however, are beyond the scope of this book and for our purposes sharks, rays and chimaeras are conveniently treated as separate groups. The term 'sharks and rays' has been used broadly in many texts, including the title of this guide, to refer to all cartilaginous fishes.

The 'true sharks' are mainly fusiform in shape (a few are ray-like) with 5–7 gill openings. They have one or two dorsal fins with or without spines, usually an anal fin (absent in some species), and most have a well-developed caudal fin for swimming. The 30 or so families are represented by over 370 living species of sharks world-wide, only one of which is regularly found in freshwater well away from marine influences. The features which are important for identifying sharks are colour, body shape, fin size and position, and tooth shape and number. Fully grown sharks can be as small as 20 cm or as large as 1200 cm, but the average length is about 150 cm. A few species feed on plankton, but most are predators at the top of the food chain, with some taking such large prey as marine mammals. About 70% of sharks bear live young, while the remainder lay eggs protected by horny cases. A number of species are commercially important, mainly for their meat and fins.

'Rays' are thought to have evolved from sharks. Most are dorsoventrally flattened as an adaption for life on the bottom, however, some are rather shark-like in form and a few live near the surface. They differ from sharks in having greatly enlarged (often wing-like) pectoral fins attached to the head in front of 5–6 ventrally located gill openings. The pectoral fins and the body together often form a large structure which is referred to as the 'disc'. Living forms usually have one or two dorsal fins (occasionally none) without fin-spines, a thin (often whip-like) tail, and no anal fin. There are 16 or so families and more than 500 living species of rays world-wide; most are marine, although a few sawfishes and stingrays live in freshwater. The important features for identifying ray species include colour, disc and tail shape, structure of the oronasal region, dorsal-fin position, and the distribution and shape of dermal thorns and denticles. When fully grown, rays can be as short as 25 cm long or as large as 880 cm long with a disc width of 670 cm (length measurements are not always useful in describing size because the long, whip-like tails of some species are often damaged). Most species prey largely on invertebrates or small fish, but many scavenge for food, and some pelagic forms are plankton feeders. Apart from skates, which are oviparous, rays bear live young, some with a placenta-like connection between the female and the embryo. In many parts of the world, stingrays and skates are of considerable commercial importance for food.

Chimaeras probably evolved from an ancient shark group. They differ from true sharks and rays in having the upper jaw fused to the skull, only one external gill opening (compared with 5–7 in sharks and rays), and a largely naked skin. They have additional claspers on the head and in front of the pelvic fin of adult males, and their teeth are fused into plates which are often beak-like. They have a strong spine in front of the first dorsal fin, large pectoral fins, a weak (often partly filamentous) caudal fin, and may have a small anal fin that is barely separable from the caudal fin. There are three families and more than 38 living species of chimaeras in marine waters world-wide; they are found mainly in deepwater on the continental slopes, although some commercially important species occur inshore in relatively shallow water. The features important for identifying chimaeras include the head shape, fin positions and shape, the relative sizes of the first dorsal fin and spine, tooth plate structure, and colour. Fully grown, they vary in length from about 50–200 cm. All living species are oviparous and feed mainly on bottom-dwelling invertebrates.

Studies of sharks and rays

We know less about the taxonomy and biology of chondrichthyan fishes than we know of the more commercially important bony fishes. However, a recent world-wide resurgence in chondrichthyan systematics has resulted in some extensive reviews and revisions of major groups. Compagno's (1984) excellent two-part FAO series *Sharks of the World* includes much new information as well as a review of our biological knowledge of the world's shark fauna; it is rightly considered the modern shark taxonomists 'Bible'. The order Carcharhiniformes has been revised and reviewed by Compagno (1988) and the diverse family Carcharhinidae by Garrick (1982, 1985). The scyliorhinids were revised by Springer (1979), the sphyrnids by Gilbert (1967), and the squalids by Bigelow and Schroeder (1957). Studies by Springer (1964), Garrick (1967), Heemstra (1973) and Dingerkus and DeFino (1983) led to revisions of several smaller shark groups over the last 30 years.

The taxonomy of most ray and chimaera families is less well known. Major species-level taxonomic problems exist in the Indo–Pacific region in particular and very few of the genera have been revised on a word-wide basis. Only one recent study, a review of the mobulid rays by Notarbartolo-Di-Sciara (1987), attempts to review an entire family.

Some of the shark and ray faunas of regions outside Australia have been described. The sharks of the east coast of southern Africa were covered in an excellent series of reports by Bass *et al.* (1973, 1975a–d, 1976). For the North Atlantic chondrichthyan fauna, the western component was treated by Bigelow and Schroeder (1948, 1953) and the eastern component by Cadenat and Blache (1981). Regional coverages have been provided for the eastern tropical Pacific (Kato *et al.* 1967), Hawaii (Tester 1969; Tinker and DeLuca 1973), Taiwan (Chen 1963), Thailand (Monkolprasit 1984), Philippines (Fowler 1941), Red Sea (Gohar and Mazhar 1964), south-west Indian Ocean (Wheeler 1953, 1959, 1960, 1962), southern Africa (Wallace 1967a,b,c), Madagascar (Fourmanoir 1961), Seychelles (Smith and Smith 1963), New Caledonia (Fourmanoir 1975–1979), Japan (Masuda *et al.* 1984; Nakaya 1975) and New Zealand (Garrick 1954–1960; Paulin *et al.* 1989). Popular books on the chondrichthyan faunas (mainly dealing with sharks) of different regions have been produced for South Africa (Compagno *et al.* 1989), Arabia (Randall 1986), North America (Castro 1983) and Polynesia (Johnson 1978). Several comprehensive regional fish guides have updated local information on sharks and rays.

Australian shark and ray fauna

The cartilaginous fish fauna of Australia has received little specialised attention since the work of Gilbert Whitley between 1930 and 1950. His classic 1940 review, *Sharks, rays, devilfish, and other primitive fishes of Australia and New Zealand,* which covers 162 chondrichthyan species, was reprinted and revised by Lincoln Smith (1981). Recent Australian ichthyologists such as Munro (1956), Stead (1963) and Grant (1978) dealt with aspects of the Australian shark and ray fauna but did not make major original contributions. Several regional guides to Australian fishes have provided new information on Australian sharks and rays: Scott *et al.* 1980; Maxwell 1980; Last *et al.* 1983; Hutchins and Thompson 1983; Gloerfelt-Tarp and Kailola 1984; Sainsbury *et al.* 1985; May and Maxwell 1986; Allen and Swainston 1988. In the first half of a catalogue of Australian fishes, Paxton *et al.* (1989) reviewed the scientific literature and listed 177 sharks and rays from the region.

Earlier faunal studies were disadvantaged because of a lack of specimens and information about the communities in the more remote parts of the 200 mile Australian Fishing Zone. Only within the last decade has much of the upper continental slope of Australia been explored. Deepwater trawl-fish surveys of these areas have provided a more complete sampling coverage of our coastline and yielded many first records for the region. After examining this material and the collections of sharks, rays and chimaeras held by museums in this country, we estimate that at least 296 species (166 sharks, 117 rays and 13 chimaeras) live in Australian waters. This list includes 97 species not identifiable from the current literature, many of which appear to be new to science. The absence of many of these species in previous guides, and an increasing interest by industry and the general public in these fishes, provided the impetus for a revision of the Australian fauna.

Habitats

Australia has a long coastline (more than 36 000 km) covering a broad range of climates from equatorial zones at about 10°S to cool temperate zones at about 45°S. The 200 nautical mile Australian Fishing Zone, which includes some offshore reefs, banks and seamonts, as well as Lord Howe and Norfolk Islands, is large in total area. The proportion of continental shelf and slope, however, is relatively small compared with other regions of similar total area. The continental shelf ranges from about 5 to 400 km wide but, apart from the northern sector, it is very narrow and the adjacent continental slope is steep. Australia is also an arid continent and, compared with other regions, has only a small network of rivers. A major topographic feature of the coastline is the Great Barrier Reef, which stretches for almost 2000 km along the Queensland coast from about 10° to 23°S and covers over 200 000 km^2.

The Australian continent is bordered by three oceans: the Pacific, Indian and Southern Oceans. The fauna is influenced by two major warm currents and one cold current, the West Australian Current, which is driven by the West Wind Drift and moves in a northerly direction up the western seaboard. The warm currents seasonally bring tropical species into more southerly latitudes. The warm East Australian Current flows down the New South Wales coast, extending to at least 160 km offshore and to a depth of 1100 m.

Off Sydney, surface velocities are greatest in February and least in June–July, with average net southerly drifts of about 32 and 16 km per day, respectively. Effects of the East Australian Current may be felt as far south as Tasmania. The Leeuwin Current flows down the coast of Western Australia and around the south-west corner of the continent into the Great Australian Bight; this current is strongest in winter.

Sharks and rays are widely distributed throughout most habitats of the region. They are found more than 200 km up rivers and in the ocean below 2000 m deep; the fauna below 1500 m, however, has not been well sampled and is virtually unknown. Some species occupy a range of niches, others have narrow distributions and habitat preferences. The coastal region, which extends from the intertidal zone to about 30 m depth, contains a broad suite of species, including transients normally found well offshore. The continental shelf fauna, which includes the fishes that live primarily between the 30 and 200 m depth contours, is highly diverse and incorporates both pelagic and demersal habitats. Some fishes beyond the shelf live in close association with the continental slope (from 200 to about 2000 m depth), while others wander freely through the open ocean.

Zoogeography

The distribution patterns of Australian sharks and rays vary from restricted to Australia (endemic) to widespread throughout the world (cosmopolitan), widespread in tropical regions (circumtropical) or widespread in temperate regions (circumtemperate). Between these distributional extremes are those with Indo–Pacific (found in both the Indian and Pacific Oceans) or Pacific distributions, some which occur in the Australasian region (Australia, New Zealand and Papua New Guinea), and a few which have very patchy or unusual distributions.

The extreme richness of the Australian shark and ray fauna (296 species or more) is evident from comparisons with faunal lists from other areas: 182 species have been recorded from southern Africa, 174 from the Japanese Archipelago, 130 from the eastern North Atlantic and Mediterranean Sea, and 83 from New Zealand. Of these regions, Japan and South Africa have the most similar faunas to those of Australia; there is considerably less faunal overlap between Australia and the eastern North Atlantic or, surprisingly, with New Zealand. Although New Zealand is geographically close to Australia, it is less subject to extremes of climate and has more uniform sea conditions and less complex habitats. Thirteen families (43%), 24 genera (28%) but only 18 species (6%), are shared by all five regions.

More than half of the entire chondrichthyan fauna (54%) is endemic to Australia. Of the shark fauna, 48% of the species are endemic, 29% are widespread, 21% are Indo–Pacific or Pacific, and 2% have Australasian distributions. Only one species is commonly found in freshwater far from the sea. The endemic species include several deepwater sharks from the continental slope that are likely to have more widespread distributions. Almost a half of Australian sharks are demersal on the continental slope; of the rest, 20% are demersal on the continental shelf, 15% are pelagic on the continental shelf, and 8% are oceanic. Most of the endemic sharks are demersal, either on the continental slope (65%) or shelf (24%); a small proportion of the others are pelagic (3%) and, not surprisingly, none is oceanic. Most of the Australian endemics are found in tropical or warm temperate areas; only a few occur in southern waters.

Most Australian rays (73%) are endemic to the region; the rest are part of a wider Indo–Pacific fauna (20%), but only a small proportion is more widespread (7%). Two species are probably restricted to freshwater, or at least spend most of their lives in rivers; two others are estuarine. Typically, most of the rays are demersal on the continental shelf or slope (90%); about 8% are coastal pelagic or oceanic species. Almost all of the endemic species are demersal (98%). The chimaeras are demersal, mainly on the continental slope, and slightly more than half appear to be regional endemics.

Biology of sharks and rays

Reproduction and development

Cartilaginous fishes have a very different reproductive strategy to bony fishes, which generally produce millions of small eggs that are liberated into the water and fertilised externally. This strategy is relatively wasteful because the unprotected eggs and larvae are subject to high mortality from predation and unfavourable environmental conditions. In contrast, sharks, rays and chimaeras produce fewer, larger eggs or young, fertilisation is internal and the survival rate of offspring is much higher. Internal fertilisation in cartilaginous fishes is accomplished by the paired claspers of the male, which are modifications of the pelvic fin used for transferring sperm to the female. They have a range of reproductive methods varying from simple oviparity (egg-laying) to advanced viviparity (live-bearing), where the embryos may be nourished through a placenta analogous to that of humans. About 57% of sharks and rays are viviparous; viviparity may be favoured in this group because their young are larger and better protected, and because wide-ranging pelagic species are not restricted to certain habitats for egg laying. In oviparous species, such as chimaeras, skates and some sharks, the eggs are protected in horny cases ('mermaid's purses'), which are usually more or less rectangular in shape (although there is considerable variation) with tendrils at each corner to anchor them to the bottom on weed or rocks. The young embryos are nourished by their yolk supply and usually take from several months to a year to hatch.

In viviparous species the embryos are nourished by a variety of methods. In some, the young hatch out in the uterus and obtain all their nourishment from the yolk sac, which is completely resorbed just before birth. In others, the yolk is supplemented by nutrients secreted by the mother. In the most advanced forms, such as the whaler sharks (Carcharhinidae) and hammerheads (Sphyrnidae), the embryos initially obtain nourishment from the yolk sac, but later in pregnancy the sac becomes modified into a placenta and nutrients are then obtained directly from the mother. In the mackerel sharks and some of their allies (Order Lamniformes), the female continues to ovulate after the young have hatched and they feed in the uterus on the supply of unfertilised eggs (which are much smaller and more numerous than in other chondrichthyans). This form of reproduction is taken to a bizarre extreme in the grey nurse shark (*Carcharias taurus*) which practises intra-uterine cannibalism; the strongest embryos actively hunt and eat their weaker siblings in the uterus.

The litters of viviparous sharks are normally between 2 and 40, but over 100 have been recorded in some species. The size at birth ranges from about 6 cm in the smallest species to possibly 150 cm in the basking shark (*Cetorhinus maximus*) and white shark (*Carcharodon carcharias*). Gestation usually lasts about 12 months but varies between species from 6–22 months. Some sharks and rays give birth throughout the year, while in other species birth is distinctly seasonal; young may be produced every year, in alternate years or in a few cases twice a year. To protect the young from predation by other sharks of the same species, and to reduce competition for food, most sharks and rays tend to segregate by size or sex (or both). The young may stay in nursery areas that are well away from adult populations.

Age and growth

Sharks and rays generally grow more slowly than bony fishes. Growth can be measured directly in captivity (although this may well be different from growth in the wild) or from tagging experiments in the wild. Alternatively, growth can be calculated by estimating the animal's age from growth rings in hard skeletal parts such as the vertebrae or dorsal-fin spines.

Most species of sharks whose age and growth have been estimated seem to have life spans of 20–30 years and to take six or seven years to attain sexual maturity. Some gummy (*Mustelus*) and whaler (*Carcharhinus*) sharks reach sexual maturity in 2–3 years and live for 10–15 years. At the other extreme, the white-spotted spurdog (*Squalus acanthias*) grows about 4 cm per year up to sexual maturity (which it reaches at about 20 years) and lives for at least 70 years. School sharks (*Galeorhinus galeus*) tagged in Australia have been recaptured after 36 years, indicating that this species lives for 50 years or more. Some of the large pelagic sharks, such as the blue whaler (*Prionace glauca*), shortfin mako (*Isurus oxyrinchus*) and white shark, show the fastest absolute growth rates of about 30 cm per year to maturity.

Functional morphology

The body form of sharks and rays mostly reflects their way of life. 'Typical sharks' have streamlined, fusiform bodies, a longish snout and pectoral fins, and a tail fin with the upper lobe longer than the lower lobe. Sharks control their position in the water by balancing the action of the upper tail lobe (which drives the shark downwards) against lift produced by the pectoral fins and the flattened underside of the snout. Buoyancy control with minimum energy expenditure is important; while bony fishes evolved gas-filled swim bladders, sharks reduced their density by replacing heavy bone with lighter cartilage and by acquiring large oily livers. Bottom-dwelling sharks and rays are more dense than their pelagic counterparts. 'Typical sharks' can cruise at around 2–5 km per hour with short sprints of at least 30 km per hour. The dermal denticles covering the skin of sharks, particularly pelagic species, may channel the water to produce laminar flow and reduce friction over the body — making the shark 'hydrodynamically quiet' and perhaps providing some advantage in stalking prey.

There are many variations in chondrichthyan body form. Almost all rays and some sharks have become dorsoventrally flattened as an adaptation for life on the bottom. In most rays, which are normally slow moving, locomotion is achieved by vertical muscle waves passing along the enlarged pectoral fins. Some of the more active pelagic rays with this general body form, such as the devilrays (Mobulidae), have strong, rather firm, wing-like pectoral fins with pointed apices (rather than rounded) to suite this mode of existence.

The mackerel sharks (Lamnidae) have become highly specialised for a pelagic existence. They have a conico-cylindrical body close to the perfect hydrodynamic shape, a caudal fin that provides maximum thrust with minimum drag, and large gills. A heat exchanger system maintains their body temperatures at 5–11°C above ambient water temperatures, making muscle operation more efficient. Mako sharks (*Isurus*), which are the fastest chondrichthyan fishes, can leap repeatedly out of the water; to do this requires a starting velocity of about 35 km per hour.

Many deepsea sharks achieve near-neutral buoyancy with large livers full of oil with a low specific gravity. This allows them to hang nearly motionless in the water or to ascend or descend after prey more quickly than fish with gas-filled swim bladders. The largest chondrichthyans, the whale shark (*Rhincodon typus*), basking shark and manta ray (*Manta birostris*), feed on plankton by cruising slowly near the surface; these species also have large oily livers and are close to neutral buoyancy.

Some extremes of body form are evident in the group. The head of hammerheads (Sphyrnidae) acts as a hydrofoil, giving these sharks great manoeuverability which enables them to catch agile prey such as squid. The enormously long tail of thresher sharks (Alopiidae), like the 'bill' on marlins and swordfish, is used rather like a baseball bat to stun fish. The long, flexible, eel-like body of the frill shark (*Chlamydoselachus anguineus*) may be an adaptation for hunting prey hidden in caves and crevices. The sharp, saw-like snout of sawsharks (Pristiophoridae) is equipped with tactile barbels and electroreceptors. It is probably used for flushing prey out of the substrate and, as in sawfishes (Pristidae), as a weapon for disabling prey.

Teeth and diet

The teeth of sharks and rays are replaced continuously through life. Replacement overcomes the problem of broken and blunt teeth and allows the teeth to increase in proportion to the size of the fish. In whaler sharks (Carcharhindae), each tooth is replaced every 8–15 days. The teeth of chimaeras, which feed mainly on molluscs and other invertebrates, grow continuously, as they are worn down with use, rather like the incisors of rodents. The shape of the teeth is an important taxonomic character. It also varies between the sexes in some groups: males of many rays have longer, sharper teeth than females, and some catsharks (Scyliorhinidae) have extra cusps on the teeth of females. Such differences in tooth structure may also reflect a subtle difference in diet between the sexes.

There is considerable variation in tooth shape and function, although most are variants of three basic types. Most rays and skates, as well as some sharks, have grinding, molariform teeth adapted for an invertebrate diet. Many sharks have essentially triangular or blade-like cutting teeth for feeding on fish and cephalopods, and occasionally larger prey such as turtles and marine mammals. Some predators of fish and cephalopods have long fang-like teeth for seizing and holding the prey, which may then be swallowed whole.

The popular belief that all sharks are scavengers is a misconception. Many species are very selective feeders, although they may opportunistically exploit other prey. For example, whaler sharks feed mainly on small fish and squid, Port Jackson sharks (*Heterodontus*) feed largely on sea urchins, gummy sharks mainly on crustaceans, and some weasel sharks (*Hemigaleus*) on cephalopods. Some shark species such as the tiger shark (*Galeocerdo cuvier*) and bull shark (*Carcharhinus leucas*) are omnivorous. Rays tend to be less specialised feeders than sharks, but still may have preferred groups of prey. Research has shown that co-existing species of stingarees (Urolophidae) have preferences for either invertebrates living on the bottom (mainly crustaceans) or animals buried in the sediments (mainly polychaete worms). Some chondrichthyans are very specialised feeders: a few sharks and rays eat only plankton; nurse sharks (*Nebrius ferrugineus*) feed by suction; some dogfishes (Squalidae) are ectoparasitic ambushing and biting the unsuspecting host, then rolling on the bite to remove an almost circular patch of flesh ('cookie-cutting').

A battery of sense organs is used in prey detection. Both sharks and rays use electroreception to detect the small electric fields emanating from their prey at close range. Their lateral line system, which is sensitive to vibrations and pressure changes, is used mainly for detecting prey over middle distances. The sense of smell is very acute in some sharks, which can detect fish oil and blood in concentrations of about one part per million over a long distance. The more common senses of vision, taste and touch are also used in prey detection at close range.

Ecological role

Chondrichthyan fishes, particularly sharks, contain many species at or near the top of the marine food chain and thus play an important role in the ecosystem. Large predators, whether on land or in the sea, help keep the population sizes of their prey in check; they also maintain the 'genetic fitness' of their prey by weeding out the sick and weak. Ecological balance within fish communities is such a complex process that we cannot accurately predict what the effects of interference may be. In the absence of more precise information, however, the roles of these fishes should not be underestimated. Indiscriminate removal of apex predators from marine habitats could disastrously upset the balance within the sea's ecosystems.

Impact of sharks and rays on humans

Shark attack

The incidence of shark attack on humans has in the past been exaggerated out of all proportion to the facts, largely by sensational media reporting and popular films such as *Jaws*. While there is still a general misconception of the dangers from sharks, a number of more recent factual accounts and more responsible reporting by some elements of the media are helping to redress the situation. The reality that shark attack is not a major killer of Australians is slowly becoming appreciated. In fact, humans are a far greater threat to sharks than sharks are to humans.

Over 1000 shark attacks in the period 1940–70 are recorded in an International Shark Attack File (currently administered by the American Elasmobranch Society): an average of about 30 a year. Because attacks in Third World countries are often unreported, the true figure is likely to be much higher: closer to 100, of which about 30 are fatal. Until the 1940s, Australia had the worst shark attack record in the world. Even so, since the first recorded attack in 1791, fewer than 200 people have been killed by sharks in this region. Most attacks have occurred on the more populated eastern coast, particularly around

Sydney. According to the Australian Shark Attack File, which is currently administered by scientists at the Taronga Zoo, since 1791 a total of 76 fatal attacks has occurred in New South Wales, 69 in Queensland, 15 in South Australia, eight in Tasmania, seven in Victoria and Western Australia and three in the Northern Territory. Since shark nets were installed off New South Wales beaches in 1937 and Queensland beaches in 1962, fatalities at these highly populated localities have been virtually eliminated. Most recent attacks have occurred in more remote areas. In the last 10 years (to 1992), six fatalities resulting from shark attack have occurred in Queensland, four in South Australia, one in New South Wales and Tasmania and none in Victoria, Western Australia and the Northern Territory. When compared with other water-related deaths (22 drowning fatalities in coastal New South Wales in 1984 alone), the incidence of shark attack is insignificant.

Why sharks attack people is still not fully understood, although it is clear that not all attacks are directly related to feeding. The pattern of bite wounds on victims and the reported behaviour of the shark suggests that many attacks may be a reaction to an invasion of the shark's personal space or to a perceived threat, in which case a victim may be mouthed (causing minor lacerations) and then released. Some attacks, particularly by white sharks, are probably cases of mistaken identity: a diver in a black wet suit may resemble a seal, which is a food item for adult white sharks. Similarly, a surfer paddling a board at the surface may resemble a seal to a white shark, or a slow-moving tuna to other active sharks such as the whalers. After heavy rain, which may increase coastal turbidity and bring prey species close inshore, sharks may become over-stimulated to bite at almost anything. Very occasionally, when a victim is repeatedly attacked, the attacks may be direct feeding responses, usually from one of the few larger, omnivorous species. While there has been considerable research into the development of chemical, electrical and mechanical shark repellants, none is currently effective against all species under all conditions.

Only a few shark species are potentially dangerous to humans. There are about 370 species world-wide, but nearly all fatal attacks can be attributed to only four species. The white shark is responsible for most cool-water attacks, particularly on divers. The tiger and bull sharks, which are often found inshore in shallow water, are large, warm-temperate and tropical species with omnivorous diets. Both have attacked swimmers, particularly in murky waters near shore or in estuaries. The oceanic whitetip (*Carcharhinus longimanus*), while not normally found close to land, has probably been responsible for many open-ocean attacks, particularly after air or shipping disasters. Some of the larger members of the whaler family are abundant inshore in some regions and may become aggressive to swimmers. The dusky shark (*Carcharhinus obscurus*) and bronze whaler (*Carcharhinus brachyurus*) have menaced divers and surfers off southern Australia, but serious injury from these species is rare.

The bad reputations of hammerheads and the grey nurse shark are ill-founded. Large hammerheads are capable of causing serious injury, but few fatalities can be attributed to them. They are usually timid towards divers, but can be a nuisance to spearfishermen. Shark attacks off eastern Australia have often been attributed to the grey nurse. This unfortunate, fearsome-looking but docile, shark has been hunted ruthlessly by spearfishermen after being wrongly blamed for these attacks. Other sharks, even bottom-dwelling species such as the wobbegongs (Orectolobidae), may attack if provoked or disturbed.

Control of sharks

Shark nets have been associated with a reduction in the number of shark attacks off popular beaches in New South Wales and Queensland, but their use is controversial. The nets catch local sharks, reducing their numbers and the statistical chances of attack, but they do not prevent the sharks from entering an area. Not all areas can be netted (ideal locations are gradually shelving, sandy beaches) because of the coastal topography, rough sea conditions or remoteness. As well as nets, some Queensland beaches also use setlines (baited hooks suspended from drums anchored to the bottom).

There are negative aspects to the shark netting program that need addressing. The current operation costs more than half a million dollars a year yet no serious effort is made to utilise the sharks for scientific research. On average, over 1500 sharks are caught each year in the program. Some basic catch details are recorded before the sharks are dumped at sea, but the species identification is generally so poor that the data are of little use. In

South Africa, which has a similar netting program, valuable information on the distribution and biology of the captured sharks has been obtained. In both the United States and South Africa, the shark attack problem has been the stimulus for considerable research into sharks. In Australia, little research has been carried out on the larger, potentially dangerous species of shark. Furthermore, the removal of large numbers of sharks may have serious effects on the ecosystem. In South Africa, shark netting is thought to have caused a large increase in the population of small sharks by removing the larger species that normally feed on them. These small sharks are now affecting the important bony fish resources in the area. Nets also kill large numbers of harmless species; for example, from 1962–1978 a staggering 10 889 rays, 2654 turtles, 468 dugongs and 317 dolphins were caught in the Queensland program alone.

Other potentially harmful cartilaginous fishes

Some other chondrichthyans can cause minor injury to humans, but these are rarely serious. Several ray groups have a barbed, venomous stinging spine on the tail. Confirmed reports of large stingrays killing swimmers and divers are extremely rare, although a few such fatalities have been reported from Australia. These animals, which are often very large, are not naturally aggressive but their long, sharp sting is capable of inflicting deep wounds; the relative effect of the toxin compared to the physical injury is unknown. Smaller stingarees, common inshore in southern Australian waters, are capable of inflicting a shallow but painful wound. The toxin of all rays is a large, water-soluble protein that is destroyed by heat. The pain can be alleviated by immersing the wound in hot water (about 50°C). Dogfishes, Port Jackson sharks, and chimaeras have venom glands associated with the dorsal spines. Wounds from these can produce mild to severe pain which is generally milder than that from stingrays. Treatment of the victim is the same for both types of sting.

Electric rays are capable of producing a strong shock from paired organs arranged horizontally on each side of the disc. The organs contain a 'honeycomb' of cells filled with a jelly-like substance that functions like a storage battery; the dorsal surface is positive and the ventral surface negative. These rays are reputed to produce shocks of up to 220 volts, although the voltage strengths produced by Australian species are unknown. Accounts of fishermen being thrown across the deck of a boat after touching large torpedo rays (*Torpedo*) or numbfishes (*Narcine*) are not uncommon.

Impact of humans on sharks and rays

Traditional fisheries

Sharks and rays have been a very important food for the Australian Aborigines for many centuries. They are caught seasonally off northern Australia (October to April) and usually prepared as buunhdhaarr: the liver and flesh are boiled separately, minced and mixed together. Various species of stingrays are considered to be in season after the first thunderstorms of the wet season, the actual time varying between different species. On capture, the liver is checked and if it is pinky white and oily the animal is considered suitable to eat. Similar criteria are used for evaluating small species of shark. Stingrays are speared by fishermen operating either from the bow of a small boat or by wading into chest-deep water off a beach. Rays are speared at distances of up to 15 m by experienced throwers. The stinging spine is sometimes used as a spear tip. Stingrays with two spines are considered inedible, as are the manta rays.

Commercial fisheries

Sharks and rays are important primary produce in many parts of the world. Between 1947 and 1986 more than 20 million tonnes of these fishes were taken by target fisheries throughout the world. These figures do not include large but unknown quantities taken as bycatch of other fisheries; a significant part of the bycatch is also discarded. While these quantities are small in comparison with the total bony fish catch, landings of cartilaginous fishes are becoming increasingly important.

Chondrichthyan fisheries require careful and specialised management as they are more susceptible to overfishing than most other fisheries. Sharks and rays are generally slow growing and late maturing; many species do not reach sexual maturity until they are 10–12 years old. Their reproductive strategy of producing relatively few well-developed young means that the number of young produced is fairly closely linked to the size of the adult population. Most bony fishes, on the other hand, produce millions of eggs very few of which survive; it is possible to fish the adult population to a very low level and the survivors will still produce enough eggs to replenish the population.

Cartilaginous fishes have been exploited for a range of commercial purposes. The flesh is used for food in many parts of the world, including Australia. The fins of sharks (and some rays) fetch high prices (currently up to A$100 per kilogram) on the oriental market for shark fin soup and as an aphrodisiac. During the 1940s shark liver oil was in demand as a major source of vitamin A, but this market collapsed when synthetic products became available. There is still a demand for squalene oil, which is abundant in the livers of some sharks and dogfishes; squalene is used in the cosmetic industry and as a high-grade machine oil.

Interest in cartilaginous fishes for medical purposes is growing. Extracts from the cartilage have been found to suppress the development of tumours and may play a role in cancer treatment. Chondroiten, derived from cartilage, has been used as artificial skin for burn victims, extracts of shark bile have been used for treating acne, shark corneas have been used in human transplants, and anticoagulant bloodclotting agents have been extracted from some sharks.

Small markets exist for the skin of some species for leather and shagreen (used by cabinet makers instead of sandpaper), and for jaws and teeth as curios and in the jewellery trade. Large sharks and rays are important exhibits in public aquariums, and dogfishes are often used for training anatomy students and as experimental animals.

In the Australian region, sharks and rays are principally used for food: about 7000 tonnes of shark are landed annually. The southern shark fishery (mainly school and gummy shark) accounted for a yearly catch of about 5000 tonnes off Victoria, Tasmania and South Australia until recent management restrictions were imposed. Western Australian vessels land about 1200 tonnes of whiskery (*Furgaleus macki*), gummy and dusky sharks, and in the tropics Queensland and Northern Territorian fishermen catch up to 500 tonnes consisting mainly of Australian blacktip (*Carcharhinus tilstoni*) and spot-tail sharks (*Carcharhinus sorrah*). From the early 1970s until 1986, a licensed Taiwanese fishery also exploited these tropical species, taking about 7000 tonnes a year.

Almost all shark landed in Australia is used for domestic consumption, much of it sold in Victoria under the marketing name of 'flake'. The southern shark fishery was first exploited in the 1920s, but large increases in fishing effort in recent times have led to overexploitation, and there is now considerable concern over the state of the stocks. The Western Australian fishery for dusky shark is based on newly born fish of about 100 cm length but nothing is known about the state of the stocks of this species. There is also some concern about the stocks of whiskery shark landed in Western Australia.

Rays and skates are important food fish in Europe and parts of Asia. Some are among the most highly priced fishes in these regions, and catches are large in some areas; the inshore ray fishery of India, for example, has exceeded 60 000 tonnes annually. Although similar rays occur off Australia, there is currently only a small market for them locally. This situation may change, however, with market promotion and better consumer education. Skates (Rajidae) are abundant on the continental shelf and upper slope off southern Australia, and stingrays are one of the most common large fishes in northern coastal waters. The potential value of the skin, fins and stomach bag of these fishes has been highlighted in recent fisheries publications.

Chimaeras are taken in the trawl and Danish seine bycatch and sold in markets of southern Australia. The elephantfish (*Callorhinchus milii*) is targeted and sold as whitefish fillets. Other members of the group, which are quite abundant on the continental slope, are currently underexploited, but are likely to be used for food more often in the future.

Recreational fisheries

Many sharks and some rays are important sport fish in various parts of the world. In Australia, large numbers of sharks are caught by anglers near populated areas, particularly near Sydney. At present, many of these sharks are killed and dumped back at sea. These catches could be a potentially valuable source of information for shark biologists. Since little is known about the movements of larger species, such as whalers and mackerel sharks (particularly in Australia), tag-release programs should be encouraged. Many game fishing clubs keep meticulous records of their catches, which are important data for scientists. When collecting data on any fish, however, it is important to accurately identify the species.

Aims of this book

The main aim of this book is to provide an identification guide to Australian sharks, rays and chimaeras for the use of ichthyologists, fishermen, divers and the general public. The fishing industry subsidised the book to make it more affordable. The guide is intended to be functional in the field; users can simply match illustrations to specimens or follow the keys to the descriptions of each species. Where possible, we have used mainly external characters that are easy to see, particularly in the field. Apart from vertebral counts, we have not used specialised internal characters such as the skull structure, fin ray counts or intestinal valve counts, even though they may sometimes provide a better means of distinguishing between closely related species.

Preparation of this book entailed a major revision of the Australian shark and ray fauna. In the process, 97 species that are either new to science or unable to be identified to a known species were examined. We did not attempt to formally name them in this guide, as more detailed descriptions will be published later in technical papers; they are denoted in the text as 'sp. A', 'sp. B', etc. The introductory account given in this chapter of general chondrichthyan biology, faunal reviews, zoogeography, interaction with humans, and shark attack, is intended only as a brief overview as other authors have treated these subjects in more detail.

2 HOW TO USE THIS BOOK

This book is designed to assist in the identification of Australian sharks, rays and chimaeras and has been organised so that users can select their own approach to achieving this aim. Those wishing to systematically identify an animal can start at the family key to find the correct chapter, then work through the species key and appropriate species description to confirm their identification. If the correct family is already known, the reader can move straight to the species key and/or description. Alternatively, users can go directly to the colour plates at the back of the book, which are laid out so that similar or related species are on the same or adjacent plate. Plates are arranged more or less according to the order of chapters, but the sequence is sometimes modified to assist the reader; for example, unrelated groups or species that have a similar body form, such as the saw sharks and sawfish (which are rays), are grouped near each other.

Descriptions of species have been kept as simple as possible, but the frequent use of some technical ichthyological terms is unavoidable without the repetitive use of long phrases. The general reader should consult the illustrated glossary in Chapter 3 to clarify these terms.

Classification

Classification is a means of cataloguing the millions of plants and animals on the planet based on similarities in external and internal structure. Originally the system was used to group superficially similar organisms, but following the development of evolutionary science it has been modified to reflect the evolutionary history of the organisms within a group in a hierarchical sense. This hierarchical system classifies organisms into major groups called phyla, which are then further subdivided, in decreasing order of status, into classes, orders, families, genera and species. Species, which form the basic rank of classification, are named according to an internationally accepted binomial system. Species names are used in preference to common names which can vary from place to place or from person to person. The scientific species name consists of two italicised words: the first is the name of the genus to which the animal belongs and starts with a capital letter; the second is the unique species name, and is not capitalized. Some species have more than one scientific name (synonyms), but only one of these (usually the oldest) is valid. The classification used in this guide follows classifications adopted by Compagno (1984) for sharks, and McEachran (1982) for rays and chimaeras.

Illustrations

Colour illustrations of each fish at the back of the book are also reproduced in black and white within the appropriate chapter to make identification easier and eliminate the need for cross-referencing. In almost all cases, specimens and photographs of Australian fish were used to obtain details of shape and colour during the preparation of plates and illustrations. Additional line illustrations are used in the keys, species sections and glossary to demonstrate special characters or particular features. Where possible, the undersurface of the head and teeth are represented for at least one member of each genus. In some families, where many of the species are previously undescribed or where particular special anatomical features distinguish closely related species (e.g. teeth, oronasal area, cutaneous tail folds or claspers), additional characters are sometimes illustrated.

Species treatments

Aspects of identification, size and distribution are discussed for each species under the following subheadings. Where no information is available for a particular subsection, it is omitted.

Common and scientific names

As for fishes in general, there is little consistency in the common names for sharks and rays. The most widely used Australian names or universally accepted names are recommended for general use. The common names of many sharks follow those adopted in the FAO guide (Compagno, 1984). Some names are used here for the first time to represent genera or groups of closely related species, and some new names are proposed for species that are currently unnamed. Other names used in Australia are given under 'Alternative names'.

The scientific name of each species consists of genus and species names, the name of the author who named it, and the year in which it was named. Parentheses around the author and date indicate that the author originally placed the species in a different genus. New species or species of uncertain identity are referred to by a generic name and an alphabetic second name.

Species codes

The official codes for Australian Aquatic Biota are provided for each species. This system, which is presently being updated (Yearsley, Last and Morris, in preparation), is used by the fishing industry to record commercial catch information and by research organisations to obtain ecological data. The first two of the eight digits refer to the main animal or plant group; the next three digits refer to the family; the last three digits to the species. The major group 00 refers to an old archival list that will eventually be replaced with a specific code for fish.

Field characters

These are the main characters by which a species can be most easily identified in the field. This section should be referred to in conjunction with the relevant family description. Easy-to-use characters, usually based on external morphology, are selected where possible. Characters considered to be very important for identification may be repeated in the 'Distinctive features' section.

Distinctive features

This section, which is written in an abbreviated sentence style for concise presentation, provides more detailed information on each species, including such features as the body shape, teeth and denticle shapes, relative fin positions and dimensions, and tooth and vertebral counts. Extra details are given for species that are difficult to separate, or where information on the species is new or scant. The use of technical terms in this section is unavoidable, so the reader should be prepared to use the glossary when necessary. Counts are usually provided where known; in many instances these will provide supplementary information for specialists rather than being important for separating species in the book. Most counts were taken from Australasian specimens; those in square brackets are from specimens collected elsewhere. When fewer than five specimens were examined, the values for counts and measurements are marked with an asterisk. Proportions may be expressed as either percentages or as ratios (rounded off to the nearest 0.05).

Vertebral counts were obtained by dissection, from radiographs or published descriptions. They may be made in several ways. The vertebral column is made up of two main types of vertebrae: monospondylous (relatively large centra of the trunk) and diplospondylous (smaller centra supporting the tail and caudal fin). Counts of one or other of these types may be useful in distinguishing between species. Total counts include all vertebrae from the back of the cranium to the extremity of the caudal fin (tip of upper lobe in most sharks). In sharks, precaudal counts refer to all vertebrae from the back of the cranium to the posterior edge of the precaudal pit or origin of the dorsal lobe of the caudal fin.

Precaudal counts in some literature accounts that we used were sometimes taken to the anterior edge of the precaudal pit, which makes a difference of about one vertebra; these are reported as given. The vertebrae of the caudal fin (caudal counts), are often difficult to count but the information is useful for some groups. In rays, the caudal fin is absent or the tail tip may be damaged, so monospondylous counts or predorsal diplospondylous counts (tail vertebrae to the first dorsal-fin origin) are generally used. In rhinobatids, the nerve pores at the front of the vertebral column are not counted as monospondylous centra.

Colour

Some chondrichthyans, like many other fishes, change colour soon after death. For example, many species of whaler sharks are a metallic bronze colour in life, but a drab brown or grey after death. Further colour change may occur in preservative; spots and colour patterns are sometimes barely detectable after long periods of preservation. Where possible, we have tried to describe the living colours of each species.

Size

Unless stated, all size measurements refer to the total length of a species (see p. 26). In the case of sharks, this is measured as a straight line from the tip of the snout to the tip of the extended upper caudal-fin lobe. Ray sizes are also given as total lengths except in the families Dasyatididae, Gymnuridae, Myliobatididae, Rhinopteridae and Mobulidae, in which the tail is frequently damaged. For these groups the main size measurement used is the disc width; total length is also generally provided to give an indication of the attainable length. The caudal filament at the tail tip of chimaeras is not included in total length.

In addition to maximum size, the sizes at birth or hatching, and at sexual maturity are given for each species when known. As size may vary from one region to another, data relate where possible to Australian specimens. Substantial differences in size between populations from Australia and elsewhere are noted.

Distribution

This section covers geographic and depth distributions and includes basic information on the habitats (pelagic and oceanic, demersal on the continental shelf, etc.) of species. Distributional ranges shown on maps, which were compiled from local data and from the literature, are reasonably comprehensive. Questionable distributions are demarcated accordingly.

Remarks

This rather informal section includes relevant or interesting features of a species, such as its biology or distribution, relationship to close relatives, interest to fisheries, and potential danger to humans.

Local synonymy

Synonyms are other scientific names for the same species. This section lists invalid scientific names from the Australasian region, together with once used combinations of the species name with other generic names. Misidentifications, demarcated with a colon between the species name and author (misidentifier) are given when considered to be important. Compagno (1984) should be consulted for shark synonyms from other regions; no equivalent source is available for ray synonymies.

References

References to species in the Australasian region that contribute significant additional information or qualify taxonomic authorship are listed within the appropriate species treatment. Readers should also consult Whitley (1940) and Compagno (1984), which because of their relevance to many of the species, are not cited individually.

Preservation and storage

Although the cartilaginous fishes of the Australian region are now reasonably well known, rare or unusual species, particularly those that cannot be identified using this book, should be taken to a museum or fisheries department. If this cannot be done immediately, specimens should be either frozen or preserved in 5% formalin solution (1 part formaldehyde concentrate to 19 parts water). Full strength formalin, which is used in bulk as sheep dip, can be obtained from pharmacies, chemical distributors or most agricultural suppliers (as 'formol'). Formalin is a poisonous liquid with a pungent odour; it should not be inhaled or touch the skin. It must be handled with extreme care while wearing a face mask and gloves. For sharks that are too large to preserve whole, good photographs should be taken of the entire fish in lateral view and of the underside of the head, and the head (or at least the jaws) retained. For large rays, the jaws, claspers and tail should be retained, and photographs taken of the whole animal in dorsal and ventral view. Specimens should be accompanied by basic collection data which includes the date, site, captor and method of capture. Museums or other appropriate scientific organisations should be contacted regarding transport of frozen or preserved specimens.

3 GLOSSARY

abdomen (*adj.* abdominal) – the part of the body that contains the digestive and reproductive organs; the lower part of the body in front of the cloaca.

aberrant – unusual in form or behaviour, abnormal.

abyssal plain – the ocean bottom from about 2000 to 6000 m depth; the upper abyssal plain (2000–4000 m) is also often referred to as the continental rise.

acuminate – tapering to a point.

acute – sharp or pointed.

adipose – fatty.

adpressed – pressed flat against the body.

advance (in advance) – in front of.

alar thorns – paired patches of thorns on the outer disc of most mature male skates (fig. 3.8).

alimentary canal – the passage through which food passes and is digested and absorbed; includes the oesophagus, stomach and intestine.

allopatric – populations or species occupying mutually exclusive geographic areas.

anal fin – the unpaired fin placed ventrally behind the anus (figs 3.1, 11).

angular – forming a distinct angle.

anterior (*adv.* anteriorly) – relating to the front of or head end of an object (fig. 3.15).

anterolateral – pertaining to the direction or position between the front and side of an object.

antitropical – found in both hemispheres but not in equatorial regions.

antrorse – turned forward or upward.

apex (*adj.* apical) – the tip, pointed end or extremity (figs 3.4, 5).

appendage – a major projection from the body of an animal.

articulating – united by means of a moveable joint.

asymmetrical – not symmetrical; one side is not the mirror image of the other.

axil – the angle formed by the inner edge of a fin and the body at the point of attachment.

backward (backward of) – behind.

bar – *see* saddle.

barb – *see* stinging spine.

barbel – a slender, tentacle-like sensory structure on the head (fig. 3.3).

basal – at or towards the base.

base – the part of a projection (often a fin) connected to the body (figs 3.5, 16).

bathyal – benthic habitats from 200 to 4000 m depth.

bathypelagic – living above the bottom in the deep sea at depths of between 1000 and 4000 m.

behind – refers to the posterior placement of one part of a fish relative to another (i.e. along the horizontal axis); should not be confused with beneath.

bell-shaped – resembling a bell, refers to the the shape of the internasal flap or snout undersurface.

beneath – refers to the placement of one part of a fish relative to another (i.e. one underneath the other).

benthic – living on the bottom of the ocean.

benthopelagic – occurring near or just above the bottom.

bi – prefix meaning two.

bicuspid – with two cusps or projections.

bifid – having two ends; split into two parts.

bifurcated – split or divided into two parts; bifid.

bilobate – having two lobes.

blotch – an enlarged area or patch (often irregular in shape) and different in colour to that adjacent.

body – the portion of a chondrichthyan fish other than its head; bordered anteriorly by the last gill opening.

bone (*adj.* bony) – hard calcareous substance that makes up the skeleton of some fish.

border – margin, edge.

brackish – having a salt concentration between that of freshwater and seawater (usually 0.5–30 parts per thousand of salt).

buccal – pertaining to the mouth cavity.

bulbous – shape swollen or bulging.

buoyancy – the ability to float, rise or sink in water.

bycatch – the component of the catch (often discarded) excluding the targeted commercial species.

canine tooth – an enlarged jaw tooth, conspicuously longer than others nearby; adapted for holding prey.

carinate – having a keel or ridge.

carnivorous (*n.* carnivore) – preying on other animals.

cartilage – a skeletal material consisting of a matrix of soft, white or translucent chondrin.

caudal – pertaining to the tail region.

caudal filament – fine, flexible, filamentous extension of the caudal-fin tip of chimaeras (fig. 3.11).

caudal fin – the tail fin (figs 3.1, 6, 7, 10, 11).

caudal keel – a longitudinal fleshy ridge along the side of the caudal peduncle (fig. 3.1).

caudal peduncle – the posterior part of the body supporting the caudal fin; measured from the insertion of the anal fin to the lower lobe of the caudal fin.

caudal pit – a small groove or depression on the caudal peduncle of some sharks; *see* precaudal pit.

caudal vertebra (*pl.* –vertebrae) – centrum of the caudal fin.

centrum (*pl.* – centra) – a bony segment of the backbone.

cephalic – pertaining to the head.

cephalic lobe – broad lobe on the forehead of some rays.

cephalopod – animal group including the cuttlefishes, squids and octopi.

cetaceans – group of aquatic mammals including whales and dolphins.

chest – the front, lower portion of the body containing the heart.

chin – the anterior part below or immediately behind the lower jaws.

chondrichthyan – member of a major group of fishes including sharks, rays and chimaeras.

ciguatoxin – a toxic substance accumulated up the food chain in the flesh and viscera of some fish.

circumglobal – distributed around the world within a range of latitudes.

circumnarial fold – skin fold around the nostrils (fig. 3.3).

circumnarial grooves – grooves around the nostrils (fig. 3.3).

circumtropical – distributed throughout the tropics.

claspers – modified portions of the pelvic fins in male sharks, rays and chimaeras used for transferring sperm to the female; *see also* head and prepelvic claspers (figs 3.1, 7, 11, 12).

cloaca – a common opening for digestive, urinary, and reproductive tracts in many fishes (fig. 3.7).

common name – the informal vernacular name for a fish (or other organism), which may vary from place to place.

compressed – flattened laterally, from side to side.

concave – hollowed out, curved inwards (opposite of convex).

confluent – joined together.

conical teeth – teeth shaped like a cone.

conspecific – individuals or populations of the same species.

contiguous – touching at edges but not actually joined.

continental shelf – the shelf-like part of the seabed adjacent to the coast extending into a depth of about 200 m.

continental slope – the often steep, slope-like part of the seabed bordering the continental shelf and extending to a depth of about 2000 m.

continuous – not interrupted; a fin not divided into two portions.

convex – arched, curved outwards (opposite of concave).

copepod – a major group of small crustaceans.

corrugation – alternating furrows and ridges.

cosmopolitan – having a worldwide distribution.

cranial – pertaining to the skull.

cranium – the part of the skeleton containing the brain.

crenate – having a margin shaped into small rounded scallops.

crenulate – *see* crenate.

crescentic – shaped like the new moon.

crustaceans – major group of animals, including crabs, shrimps, prawns, lobsters and crayfish.

cryptic – applied to fishes that live amongst sheltering and concealing cover, or that have protective coloration.

cusp – a projection on a tooth.

cusplet – small cusp.

cutaneous – pertaining to the skin; dermal.

Danish seine – a method of boat seining with a large net, but landing the catch on the vessel.

deciduous – easily shed or rubbed off; usually referring to denticles.

demersal – living on or near the bottom of the ocean.

dentate – bearing teeth or tooth-like projections.

denticle – a small, tooth-like structure; placoid scale of cartilaginous fish.

denticulate – with small, tooth-like projections.

depressed – dorsoventrally flattened; flattened from top to bottom.

depth (*adj.* deep) – height of body or head from dorsal to ventral surface (excluding fins) (fig. 3.15); also refers to the distance below the sea surface in which the fish lives.

dermal – pertaining to the skin.

dermal flaps – skin outgrowths.

dichromatic (*n.* dichromatism) – having different colour patterns within a species; usually related to sexual or growth differences.

dimorphic (*n.* dimorphism) – existing in two forms; usually refers to differences between the sexes in body shape and/or colouring.

diphycercal – caudal fin shape which is primitively symmetrical and pointed.

diplospondylous – having two vertebrae in each body segment, as in the tail region of certain fishes; centra smaller than monospondylous centra.

direct length – shortest distance between two points.

disc – the combined head, trunk and enlarged pectoral fins of some cartilaginous fishes with depressed bodies (fig. 3.17).

distal – region, border or point remote from the site of attachment (opposite of proximal).

dorsal – pertaining to the upper part or surface of back (figs 3.15, 17).

dorsal fin – an unpaired fin on the back or upper tail (figs 3.1, 4, 7, 10, 11).

dorsolateral – positioned or orientated between the dorsal and lateral surfaces.

dorsoventral – referring to the direction from top to bottom.

dredge – equipment for collecting and bringing up objects from the seabed by dragging.

dropline – a deepwater fishing method involving the use of a vertical line bearing rows of baited hooks.

duct – small tube or canal through which some material (e.g. a secretion) is conveyed.

dusky – slightly dark or greyish in colour.

elasmobranch – member of a major group of fishes including sharks and rays.

electric organ – organ capable of delivering an electric shock.

element – a ray or spine of a fin.

elevated – higher.

elevated fin – some part of a fin higher than the adjacent parts of the fin or body.

elliptical – shaped like an ellipse, oval.

elongate – drawn out or extended in length relative to some other criterion (usually depth).

emarginate – with the margin slightly hollowed.

encapsulated egg – egg contained in either a thick horny, or thin membranous case.

endemic – native and restricted to a defined area.

entire – with a continuous margin.

epipelagic – the upper part of the oceanic zone from the surface to about 200 m depth.

erectile – capable of being raised or erected.

estuarine – living mainly in estuaries.

euphausids – small, pelagic, shrimp-like crustaceans.

euryhaline – able to live in a wide range of salinities.

excised – with the margin cut out, or concave.

eyelid – moveable, muscular fold of skin capable of covering all or part of the exposed portion of the eyeball (fig. 3.2).

falcate – curved like a sickle.

family – one of the categories in animal and plant classification; contains one or more closely related genera.

fauna – the communities of animals in an area.

fertilisation – the union of male and female cells to form a new individual.

filament (*adj.* filamentous) – a thread-like process or appendage.

filter feeding – filtering suspended food particles from a water current by means of the gill rakers.

fimbriate – with a fringed margin.

fork length – length of a fish measured from the snout tip to the centre of the caudal fin.

forked caudal fin – a caudal fin with a deeply concave or excavated hind margin.

fossa – a groove or pit.

free rear tip – posterior tip of a fin closest to the fin insertion (figs 3.4, 5).

fringe – edge adorned with fine tassels (e.g. posterior margin of internasal flap of some rays) (fig. 3.9).

front – anterior position.

fusiform – spindle-shaped, tapering at both ends.

gape – the expanse of the open mouth.

gelatinous – like jelly.

genus (*pl.* genera) – a group term used in classifying organisms; contains one or more related species.

gill arch – a cartilaginous arch bearing the gills.

gillnet – a net used to tangle or snare fishes.

gill opening – an opening behind the head that connects the gill chamber to the exterior; usually slit-like in cartilaginous fishes.

gill – organ for breathing or extracting oxygen contained in water.

gill slit – a long, narrow gill opening (figs 3.1, 7, 11).

gonad – the organ containing the reproductive tissues; ovaries in females, testes in males, both in hermaphrodites.

graball net – gillnet.

granular – a rough or grainy surface.

granulations – fine denticles (fig. 3.10).

gregarious – tending to live in groups.

habitat – the locality with its own particular environment in which an organism lives.

hammer-shaped – shaped with paired lateral expansions; resembling the head of a mallet.

head – specialised anterior part of an animal on which the mouth and major sensory organs are located; part other than the body (snout to the posterior gill opening in fish) (figs 3.14, 15, 17, 18).

head clasper – small appendage on the forehead of mature male chimaeroid fishes (fig. 3.11).

heterocercal – caudal fin shape with unequal lobes, the upper lobe being larger than the lower.

holotype – a single specimen designated as the 'type' (i.e. name bearer) of a new species by the author of the original description.

horny – hard or solid form.

horizontal length – distance between two points measured parallel to the longitudinal axis of the fish.

hyaline – transparent.

hyomandibular pores – line of enlarged pores extending posteriorly from the mouth corners.

hypocercal – caudal fin shape in which the lower lobe is larger and often more posteriorly directed than the upper lobe.

imbricated – overlapping, like shingles.

incised membrane – membrane having a notch or with a concavity between supports.

indented – refers to a structure with a small notch in the middle.

indigenous – native to, but not necessarily limited to, an area.

inferior – lower (opposite of superior).

inflatable – capable of expanding in volume.

infra – prefix meaning below.

infraorbital – the area below the eye.

inner corner – corner or angle of pectoral fin closest to body; *see also* free rear tip.

insertion (of fin) – posterior point of attachment of a fin to its base (figs 3.4, 5, 7).

integument – covering, skin.

inter – prefix meaning between.

interbreeding – breeding between groups or populations of animals.

interdorsal – the space on the dorsal surface between the first and second dorsal fins; measured from the point of insertion of the first to the origin of the second.

interdorsal ridge – ridge of skin between first and second dorsal fins.

internarial space – distance between the nostrils; area between the nostrils.

internasal flap – a fleshy flap extending between the nostrils and partly covering the mouth of some rays and sharks (figs 3.7, 9).

interorbital space – the area on top of the head between the eyes.

interorbital distance – the shortest distance between the eyes.

interspace – the area between two given features.

interspecific – between separate species.

intraspecific – within one species.

iridescent – displaying a wide range of changing and often brilliant colours.

jaws – part of the mouth supporting the teeth.

jugular – related to the throat.

juvenile – young fish, mostly similar in form to adult but not yet sexually mature.

keel – a fleshy ridge; usually relates to a skin fold on the caudal peduncle (fig. 3.7).

labial – pertaining to the lips.

labial furrows – shallow grooves around the lips (figs 3.1, 3).

lanceolate – broad at base and tapering to a point; spear-shaped or lance-shaped.

lateral – referring to the sides (fig. 3.15).

lateral keel – prominent ridge along the side of the body.

lateral line – a canal or row of sensory pores in the skin along the side of the body (fig. 3.11).

lateral skin fold – fine fold of skin along the side of the tail of some rays (fig. 3.10).

laterosensory pores – small pores on the head and body forming part of the lateral line system.

life cycle – the generalised history of a species from birth to death.

linear – in a line.

lip – fleshy outer portion of jaws (fig. 3.14).

lip groove – *see* labial furrows.

lobate – divided into lobes.

lobe – a rounded outgrowth (figs 3.4, 6, 9, 10, 11).

longitudinal – lengthwise (opposite of 'transverse').

longline – a line of considerable length bearing numerous baited hooks that is usually set horizontally in the water column.

lozenge-shaped – shaped like a rhombus.

lunate – shaped like a crescent moon.

malar thorns – patches of thorns beside the eyes of many mature male skates (fig. 3.8).

margin – edge, rim.

matrix – an embedding or enclosing substance.

medial – on or towards the middle of the body.

medial cartilage – support cartilage of the snout of some rays (often obvious on the midline of the snout as a ridge) (fig. 3.7).

median – pertaining to the middle.

membrane – the thin layer of tissue covering a part of an animal or connecting the fin elements.

meristics – countable features (e.g. tooth rows, fin radials or vertebrae).

mesopelagic – living in the open ocean at depths of between 200 and 1000 m.

migrating – moving from one area of inhabitation to an another.

molar – blunt and rounded grinding tooth.

molariform – resembling a molar.

monospondylous – having a single vertebra in each body segment, as in the trunk of certain fishes; centra larger than diplospondylous centra.

monotypic – including only a single species.

morphology (*adj.* morphological) – pertaining to the physical form and structure of an animal.

mouth – the opening through which food enters the alimentary canal (figs 3.1, 3, 7, 9, 11, 14).

muciferous – producing or containing mucus or slime.

mucin – one of a group of nitrogenous substances secreted by a mucous gland.

mucous canal – part of the sensory network of the head of chimaeras; appearing as a distinct (often wavy) line on the head (figs 3.11, 14).

mucous gland – a gland secreting mucus.

mucous membrane – a membrane secreting mucus.

mucus (*adj.* mucous) – a slimy solution of mucin or other viscous substances.

multicuspid – with multiple tooth cusps.

multiserial – arranged in several rows.

myomere – a muscle segment of the body, separated from adjacent segments by connective tissue.

naked – skin smooth, without denticles or thorns.

nape – the region of the head above and behind the eyes; the back of the neck.

narial – *see* nostril.

nasal capsule – cartilaginous envelope containing the nasal organs.

nasal curtain – *see* internasal flap.

nasal flap – *see* internasal flap.

nasal lobe – prominent skin fold beside the nostril.

nasal organs – sensory structures for detecting smell; usually appearing externally as one or two pores or slits on each side of the fish.

nasal tentacle – fleshy protrusion near the nasal pores or nostrils (*see* barbel).

nasoral grooves – *see* oronasal grooves.

neritic – the shallow pelagic zone over the continental shelf.
neutral buoyancy – capable of maintaining fixed depth in the sea.
niche – the role or specialised position of an organism in its environment.
nictitating eyelid – a transparent, moveable membrane or inner eyelid that protects and helps keep the eye clean (fig. 3.2).
nomenclature – the systematic naming of animals and plants.
nostril (*adj.* nasal, narial) – external opening of the nasal organs (figs 3.1, 3, 9, 14).
notched fin – a groove or dip in the profile of a fin.
nuchal – pertaining to the nape (see fig. 3.8).

obsolete – a taxonomic character that is disappearing or scarcely evident.
obtuse – broadly rounded, having a blunt end.
oceanic – living in the open ocean.
ocellus (*pl.* ocelli) – an eye-like spot or marking with a marginal ring.
ocular – related to the eye.
oesophagus – beginning of the digestive tract, between the mouth and the stomach.
oral – pertaining to the mouth.
orbit – bony cavity in skull where eyeball is housed.
organism – an organised body consisting of mutually connected and dependent parts constituted to share a common life.
origin (of a fin) – the most anterior point of a fin base (figs 3.4, 5, 6).
oronasal curtain – *see* internasal flap.
oronasal groove – furrow in some sharks and rays connecting the mouth to the nasal organs; usually concealed beneath internasal flap (fig. 3.3).
oviduct – a tube leading from the ovary to the cloaca or external genital opening along which the ova pass during spawning.
oviparous – producing eggs that hatch after being ejected from the body of the parent female.
oviphagous – method of embryonic nutrition where the embryo feeds on unfertilised eggs or other embryos within the uterus.
ovoid – egg-shaped.
ovoviviparous – producing eggs that hatch within the body of the parent female, but where there is no placental connection.

palate – the roof of the mouth.
papilla (*pl.* papillae) – a small fleshy projection (fig. 3.9).
papillose – covered with papillae.
papillate – *see* papillose.
parasitic – living and feeding in or on another organism to the detriment of that organism.
paratype – a specimen, other than the holotype, on which the description of a new species is based.

pearl thorn – thorn resembling half of a pearl in shape and colour.
pectoral – pertaining to the breast.
pectoral–pelvic interspace – distance or area between pectoral-fin insertion and pelvic-fin origin.
pectoral fin – paired fins just behind or below the gill opening; united to form disc in most rays (figs 3.1, 5, 7, 11).
pectoral girdle – the cartilaginous skeletal arch supporting the pectoral fins.
pelagic – free swimming in the seas, oceans or open water, not in association with the bottom.
pelvic–anal interspace – distance or area between pelvic-fin insertion and anal-fin origin.
pelvic fins – paired fins (rarely joined) positioned on the ventral surface between the head and vent; also referred to as ventral fins (figs 3.1, 7, 11).
perinasal – around the margin of the nostril.
pharynx – the part of the throat into which the gill slits open; the space between the mouth and the oesophagus.
pigmented – coloured.
placoid – plate-like; applied to modified scales in some extinct fishes, and in sharks and rays.
plain – uniformly coloured, without a contrasting colour pattern.
planktivorous – feeding on plankton.
plankton – small animals or plants that float or drift in open water.
plica (*pl.* plicae) – fold of skin.
plicate – when folds of skin are arranged to form a fan-shaped structure.
population – a biological unit; representing the individuals of a species living in a particular area.
pore – a small secretory or sensory opening or pit.
postdorsal ridge – prominent cutaneous ridge behind a dorsal fin.
posterolateral – pertaining to the direction or position between the rear and side of an object.
postorbital – the region behind the eye.
posterior (*adv.* posteriorly) – relating to the hind or rear portion; situated farther back than something else (fig. 3.15).
pre – prefix meaning in front of.
precaudal pit – in sharks, a transverse or longitudinal notch on the caudal peduncle just in front of the caudal fin (fig. 3.1).
precaudal vertebrae – centra other than those of the caudal fin.
predator (*adj.* predatory) – feeding on other animals.
predorsal – area or distance anterior to first dorsal fin.
pre-oral – before the mouth.
pre-oral cleft – a deep groove extending forward from the lateral border of the mouth.
prepelvic claspers – small, retractable appendages on each side just forward of the pelvic fins of mature male chimaeroid fishes; concealed within a slit when retracted; also known as prepelvic tentacula (fig. 3.11).

proboscis – the elongated mouth parts or snout of some animals (fig. 3.11).

process – a natural outgrowth or projection of part of an organism.

procumbent – lying down and pointing forwards.

produced – elongated or projecting.

projecting – extending beyond something else.

projection – a part that juts out.

protractile – capable of being drawn out or extended forwards.

protrusible – a condition of the jaws in which the mouth projects forward as a tube when the mouth is opened.

protruberance – an outward bulge.

proximal – region, border or point adjacent to the place of attachment of a projection or appendage (opposite of distal).

purse seine – a fishing net used to encircle surface-dwelling fish; it is usually landed aboard a boat rather than beached.

quadradiate – with four radiating arms or extensions.

quadrangular – shaped with four distinct edges or margins.

recurved – curved backwards.

reflexed – bent or turned backwards.

respiratory – associated with breathing.

reticulated – divided into a network.

reticulations – markings in the general form of a net.

retrorse – pointing or curved backwards.

rhomboid – diamond-shaped.

rostral cartilage – a gristly structure supporting the snout.

rostral teeth – tooth-like projections on the sides of the snout of sawfishes and sawsharks.

rostrum (*adj.* rostral) – a projecting snout; protracted anterior part of the skull in sharks and rays.

rounded – margin evenly convex.

rugose – rough.

saddle – a blotch extending across the dorsal surface from one side to another.

salinity – the concentration of salt in water.

scale (*adj.* scaly) – a small membranous or horny modification of the skin of many fishes.

scapular – the shoulder region (see fig. 3.8).

school – a close aggregation of fish that swim in association with each other.

scientific name – the formal binomial name of an organism consisting of the genus and specific names; a species has only one valid scientific name.

secretory – involved in producing a secretion, or exuding a substance.

seine – a fishing net designed to hang vertically in the water, the ends being drawn together to encircle fish; to fish with a seine net (*see also* purse and Danish seine).

sensory – relating to the reception and transmission of a sense impression (e.g. sight, smell, touch, taste or hearing).

sensory canal – a tube beneath the skin connecting a series of sensory pores.

serrate – saw-like.

sexual dichromatism – difference of colour between the sexes.

sexual dimorphism – difference of physical form (shape) between the sexes.

shagreen – hard, granular skin, especially that of a shark.

simple – singular, not divided into one or more branches.

skeleton – a structure whose main function is to strengthen and maintain the shape of an animal.

skin fold – an area where skin is bent over upon itself, forming a fleshy ridge (fig. 3.10).

skirt-shaped – broadening distally in the shape of a short skirt (e.g. the internasal flap of some rays).

snout – that part of the head in front of the eyes; distance from the eye to the anterior tip of the head above the upper jaw (figs 3.1, 15, 18).

solitary – used in reference to a fish that occurs alone, not in schools or shoals.

spatulate – broad and greatly flattened.

spawning ground – geographic area where shedding and fertilisation of eggs takes place.

species – actually or potentially inter-breeding populations that are reproductively isolated from other populations; the basic rank of biological nomenclature.

speciose – rich in number of species.

spine (*adj.* spinous) – a sharp projecting point; a stiff unsegmented, undivided and unbranched element supporting a fin (figs 3.1, 4, 10, 11).

spiracle – a respiratory opening behind the eye in sharks and rays (figs 3.1, 7).

spiral valve – a spiral structure in the intestines of some fish.

spot – a regularly shaped or rounded area (usually small in size) of a colour different from that of the area adjacent.

squalene – oil produced in the liver of some sharks.

stellate – star-shaped; with radial form.

striated – marked with narrow lines or grooves, usually parallel.

stinging spine – the large, serrated, dagger-like bony structure on the tail of some rays; sometimes abbreviated to 'sting' (figs 3.7, 10).

stripe – a contrasting longitudinal pattern in the form of a line.

sub – prefix meaning below.

subcutaneous – positioned beneath the skin.

subequal – not quite equal, almost equal.

subgenus – group taxon below a genus.

submarginal – nearly to or slightly inside of margin.

suborbital – area beneath the eye.

substrate – the substance forming the bottom of the sea or ocean floor.

subterminal – positioned near but not at the end of something.

subterminal notch – a notch in the caudal fin created by the subterminal lobe (fig. 3.6).

superior – upper (opposite of 'inferior').

supra – prefix meaning above.

supra-orbital crests – enlarged ridges above the eyes.

suture – line of juncture of two parts.
symmetrical – capable of being divided into two equal halves.
sympatric – living together in the same spatial or geographic area.
symphysial groove – longitudinal furrow emanating from behind the symphysis of the lower jaw in some sharks (fig. 3.3).
symphysis (*adj.* symphysial) – the junction of two bones; particularly relating to the medial junction of either the upper or lower jaw.
synonym (*adj.* synonymous) – each of two or more scientific names of the same rank used to denote the same taxon.

tail – the part of the fish between the vent and the origin of the caudal fin (figs 3.10, 15, 16, 17).
taxon – any formal taxonomic unit or category of organisms (genus, species, family, etc.).
taxonomy – the science of classification of animals and plants.
teeth – hard outgrowths on the jaws, roof of the mouth or pharynx; used for biting and masticating food.
teleost – a large group containing most bony fishes.
tendril – a slender, curling barbel.
terminal – situated at or forming the end of something.
terminal lobe – posterior upper lobe of the caudal fin of some cartilaginous fishes (fig. 3.6).
thoracic – pertaining to the chest.
thorn – large denticle on the surface of a ray or skate (figs 3.7, 8, 10).
tip – the extremity of part of a fish.
tooth rows – horizontal rows of teeth in the jaws; usually distinct in sharks but sometimes difficult to count in rays.
total length – longest length of the fish, measured from the snout tip to the upper caudal-fin tip or tail tip (excluding caudal filaments) (figs 3.15, 18).
toxin (*adj.* toxic) – any poisonous substance of microbial, mineral, vegetable or animal origin.
translucent object – semi-transparent; an object that may transmit light but through which objects are not clearly visible.
transparent object – clear, an object not impairing light transmission or sight.
transverse – directed crosswise, across the width (opposite of longitudinal).
trawl – a fishing net that is dragged behind a boat; to fish with a trawl net.
tricuspid tooth – tooth with three cusps.
triradiate – with three radiating arms or extensions.
tritors – knobs on the teeth of chimaeroid fishes.
truncate – terminating abruptly, as if cut off square.
trunk – that part of a fish (other than the fins) between the head and the tail; the region between the last gill opening and vent (figs 3.15, 17, 18).
tubercles (*adj.* tuberculate) – either soft or hardened projections on the surface of the skin.
tubule – a small hollow, cylindrical structure.

uniserial – arranged in a single row.
united – joined together.

venomous – capable of producing a poisonous fluid that is transmitted by a bite or sting.
vent – the terminal external opening of the alimentary canal.
ventral – pertaining to the lower part or surface (figs 3.6, 10, 15, 17).
ventral fins – *see* pelvic fins.
ventrolateral – positioned or orientated between the ventral and lateral surfaces.
vermiculations – a pattern of fine, wavy, worm-like lines or streaks of colour.
vertebra (*pl.* vertebrae) – a bony segment of the backbone; centrum.
vertebral column – the backbone; consisting of monospondylous and diplospondylous centra.
vertebrate – animal having a vertebral column or backbone.
vestigial – pertaining to the remaining part or indication of a structure that earlier was developed and functional.
villiform teeth – small slender teeth that form velvety bands.
viviparous – producing live young from within the body of the parent female.

Shark terminology

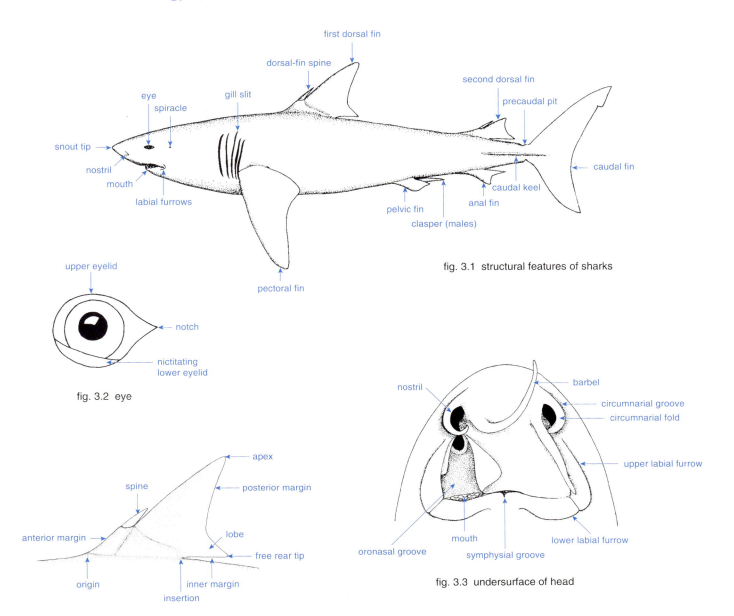

fig. 3.1 structural features of sharks

fig. 3.2 eye

fig. 3.3 undersurface of head

fig. 3.4 dorsal fin

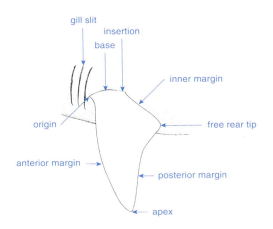

fig. 3.5 pectoral fin

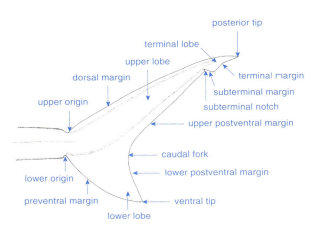

fig. 3.6 caudal fin

Ray terminology

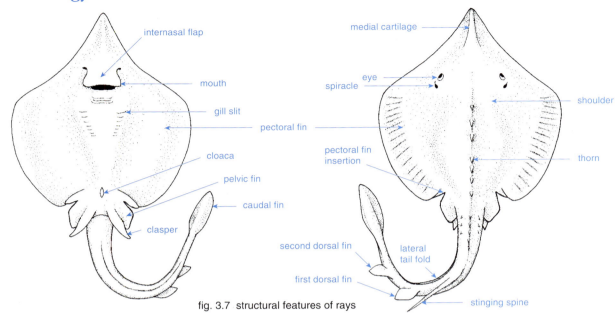

fig. 3.7 structural features of rays

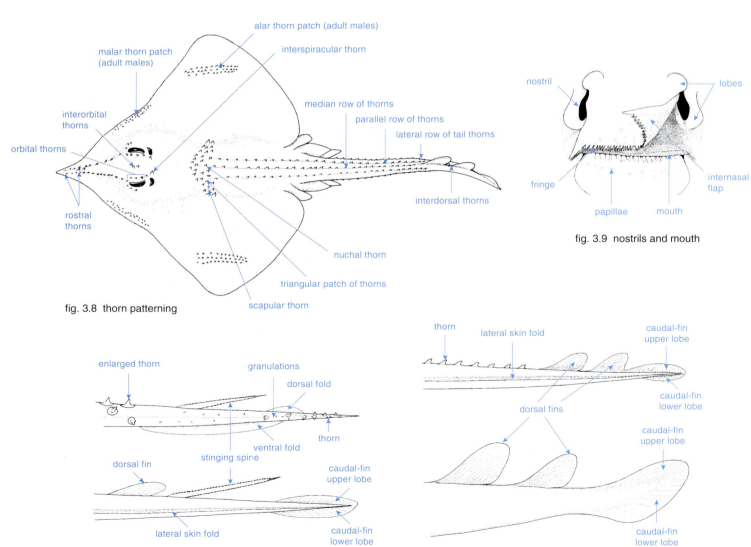

fig. 3.8 thorn patterning

fig. 3.9 nostrils and mouth

fig. 3.10 tail variations in rays

Chimaera terminology

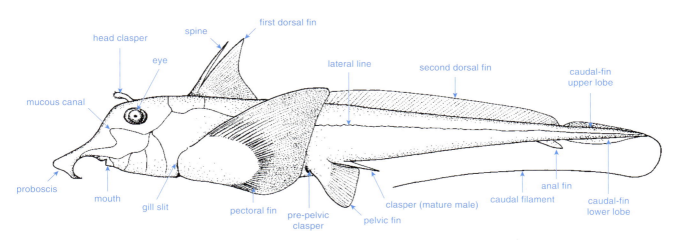

fig. 3.11 structural features of chimaerids

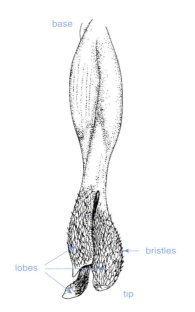

fig. 3.12 trilobate clasper

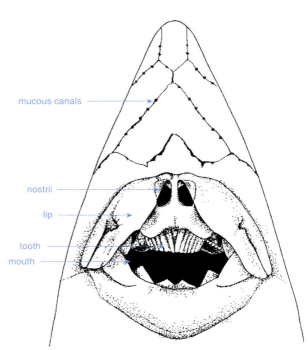

fig. 3.14 undersurface of head

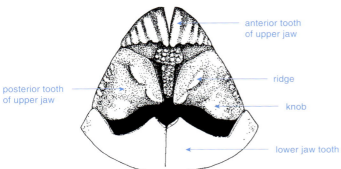

fig. 3.13 mouth

Dimensions

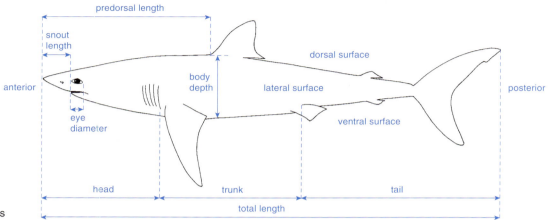

fig. 3.15 shark dimensions

fig. 3.16 fin measurements

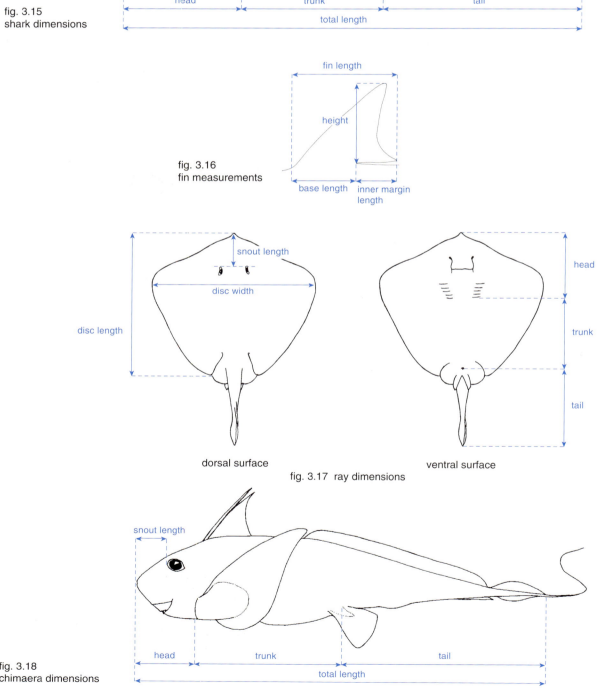

fig. 3.17 ray dimensions

fig. 3.18 chimaera dimensions

4 KEY TO SHARK AND RAY FAMILIES FOUND IN AUSTRALIAN WATERS

1   Five to seven pairs of gill openings on each side of head (fig. 1), last two openings sometimes very close together and appearing as one **2**

   One external gill opening on each side of head (fig. 2) **43**

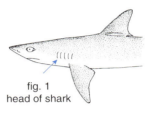
fig. 1
head of shark

2   Snout saw-like, flattened and armed with lateral teeth (figs 3, 5) **3**

   Snout not saw-like, no rostral teeth **4**

fig. 2
head of chimaera

3   Gill slits on undersurface of head (fig. 4); no barbels on snout (fig. 3) ....
   .... Pristidae (fig. 3; p. 360)
   (sawfishes)

   Gill slits on sides of head (fig. 5); rostral barbels present (fig. 5) ...... Pristiophoridae (fig. 5; p. 106)
   (sawsharks)

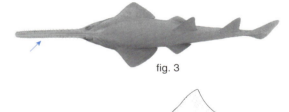
fig. 3

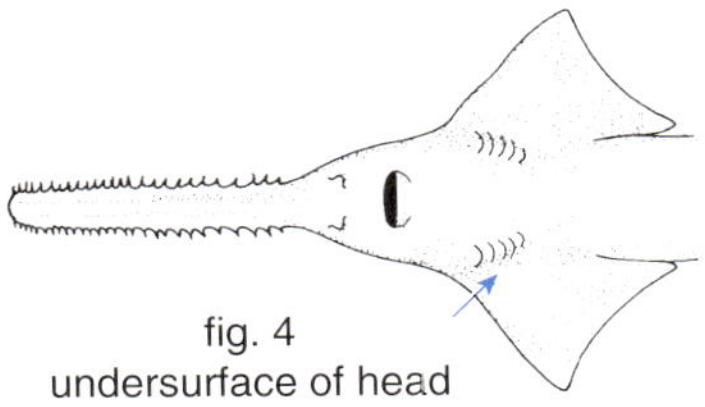

fig. 4
undersurface of head

4   Body flattened, ray-like (fig. 6); eyes on top of head, except in mobulids (fig. 29), myliobatidids (fig. 30) and rhinopterids (fig. 31) **5**

   Body more or less fusiform, shark-like (fig. 7); eyes on sides of head (fig. 7) **19**

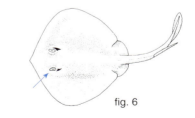
fig. 5

5   Gill openings partly on sides of head (fig. 8); pectoral fins clearly detached from head (front part of fin extending forward of fin origin) (fig. 8)
   .... Squatinidae (fig. 9; p. 277)
   (angel sharks)

fig. 6

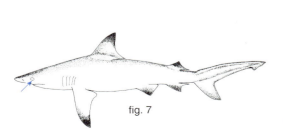
fig. 7

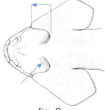
fig. 8
undersurface of head

fig. 9

28 Sharks and Rays of Australia

Gill openings entirely on undersurface of head (fig. 10); pectoral fins wholly or partly joined to head (fig. 10) **6**

6 Two prominent dorsal fins (fig. 11); first dorsal fin originating closer to insertion of pelvic fins than to tip of tail (fig. 11) **7**

Dorsal fins 0–2; origin of first dorsal fin closer to tail tip than to insertion of pelvic fins when two fins are present (fig. 10) **11**

7 Disc large relative to tail, its maximum width more than twice tail length behind pelvic-fin tips (figs 13, 14); dorsal fins close together (figs 13, 14) **8**

Disc smaller relative to tail, its maximum width about equal to or less than tail length behind pelvic-fin tips (fig. 12); dorsal fins widely separated (fig. 12) **9**

8 Caudal fin much larger than dorsal fins, about the same size as pelvic fins (fig. 13) .... Torpedinidae (fig. 13; p. 368)
(torpedo rays)

Caudal fin barely larger than dorsal fins, much shorter than pelvic fins (fig. 14) .... Hypnidae (fig. 14; p. 372)
(coffin rays)

9 Caudal fin with a well-developed ventral lobe (fig. 15); pectoral and pelvic fins not touching (fig. 15) .... Rhynchobatidae (fig. 15; p. 295)
(sharkfin guitarfishes)

Ventral lobe of caudal fin not well defined (fig. 16); pectoral and pelvic fins touching or overlapping (fig. 16) **10**

10 Snout wedge-shaped, forming an acute angle at tip (fig. 16) or snout broadly rounded; thorns or fine denticles present on body or tail (surface rough); no electric organs .... Rhinobatidae (fig. 16; p. 283)
(shovelnose rays)

Snout broadly rounded (fig. 17); body surface entirely smooth; electric organs present .... Narcinidae (fig. 17; p. 374)
(numbfishes)

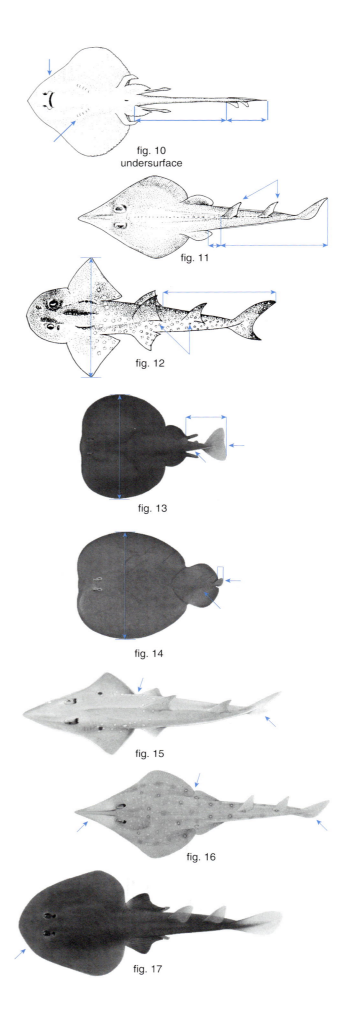

fig. 10 undersurface

fig. 11

fig. 12

fig. 13

fig. 14

fig. 15

fig. 16

fig. 17

11 Pelvic fin divided into two distinct lobes (fig. 18); no enlarged stinging spine on tail (fig. 18) **12**

Pelvic fin with one lobe (fig. 19); 1–2 enlarged, serrated stinging spines usually present on tail (deep scar visible when spine absent) (fig. 19) **13**

12 Thorns or fine denticles (rough to touch) present on dorsal surface (fig. 20); preorbital snout less than 8 times eye diameter; tail slender but not filamentous (fig. 20) .... Rajidae (fig. 20; p. 299)
(skates)

Entire dorsal surface smooth (except for alar thorns of male) (fig. 21); preorbital snout more than 8 times eye diameter; tail very short, thin and filamentous (fig. 21) ....
.... Anacanthobatidae (fig. 21; p. 357)
(leg skates)

13 Six gill slits (fig. 22) ....
.... Hexatrygonidae (fig. 23; p. 445)
(sixgill stingrays)

Five gill slits (fig. 19) **14**

14 Anterior part of head not extended beyond disc (fig. 24); eyes located dorsally and well inward from disc margin (fig. 24) **15**

Anterior part of head extended beyond disc (fig. 25); eyes located laterally on side of head (fig. 25) **17**

15 Disc very broad, width more than 1.5 times length (fig. 26); tail extremely short and filamentous (fig. 26) .... Gymnuridae (fig. 26; p. 443)
(butterfly rays)

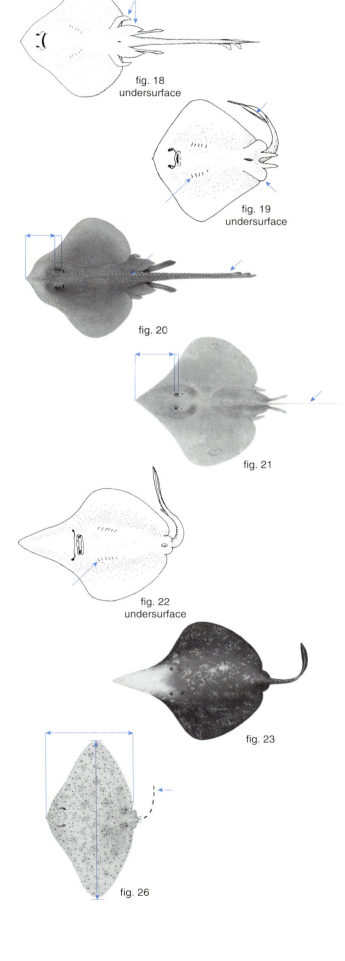

fig. 24
head

fig. 25
head

30 Sharks and Rays of Australia

Disc width less than 1.5 times length (fig. 27); tail moderately (fig. 28) to very (fig. 27) long **16**

16 Caudal fin absent (fig. 27); central disc and dorsal surface of tail normally with some thorns or tubercles (fig. 27) ..... Dasyatididae (fig. 27; p. 380)
(stingrays)

Caudal fin present (fig. 28); no thorns or tubercles on disc or tail (completely smooth) .... Urolophidae (fig. 28; p. 416)
(stingarees)

17 A pair of long, paddle-like flaps situated laterally on front of head (fig. 29); teeth minute, in many rows, more than 10 rows in each jaw .... Mobulidae (fig. 29; p. 468)
(devilrays)

No lateral, paddle-like processes on head, instead with a single, fleshy, subrostral lobe (fig. 30) or pair of broadly rounded lobes (fig. 31); teeth large, less than 10 rows in each jaw **18**

18 Margin of subrostral lobe rounded (fig. 30); floor of mouth with fleshy papillae .... Myliobatididae (fig. 30; p. 447)
(eagle rays)

Margin of subrostral lobe with a deep central notch (fig. 31); floor of mouth without papillae .... Rhinopteridae (fig. 31; p. 456)
(cownose rays)

19 A single dorsal fin (fig. 32); 6–7 gill openings (fig. 32) **20**

Two dorsal fins (fig. 33); 5 gill openings (fig. 33) **21**

20 Mouth terminal (fig. 34); first gill openings connected around throat (fig. 35); no sub-terminal notch on caudal fin (fig. 34) .... Chlamydoselachidae (fig. 34; p. 36)
(frilled sharks)

fig. 34

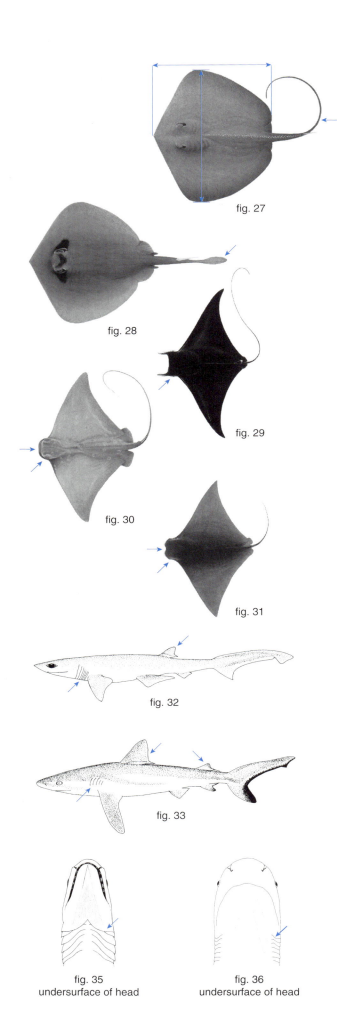

fig. 27

fig. 28

fig. 29

fig. 30

fig. 31

fig. 32

fig. 33

fig. 35
undersurface of head

fig. 36
undersurface of head

Mouth ventral (fig. 37); first gill openings not connected around throat (fig. 36); sub-terminal notch present on caudal fin (fig. 37) .... Hexanchidae (fig. 37; p. 38)
(sixgill & sevengill sharks)

21 Anal fin absent (figs 38, 39, 40) **22**

Anal fin present (fig. 41), sometimes small (fig. 42) **24**

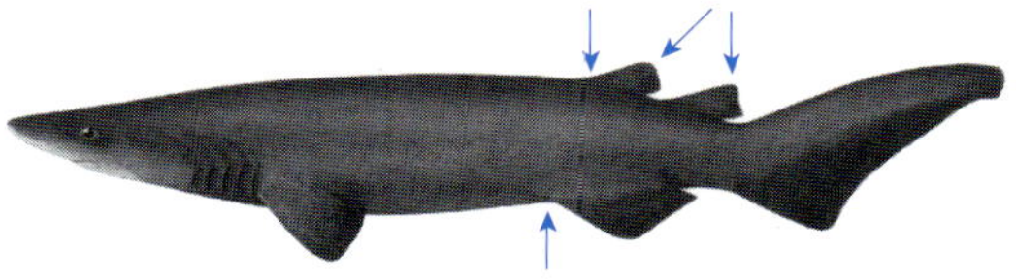

22 First dorsal fin originating behind pelvic-fin origins (fig. 38); dorsal fins located near caudal fin and almost touching each other (fig. 38); denticles extremely large .... Echinorhinidae (fig. 38; p. 44)
(bramble sharks)

First dorsal fin originating in advance of pelvic fins (figs 39, 40); dorsal fins well separated and located well forward of caudal fin; denticles not greatly enlarged (figs 39, 40) **23**

23 Trunk compressed, almost triangular in cross-section; fins tall, height of first dorsal fin more than or about equal to head length (fig. 39) .... Oxynotidae (fig. 39; p. 104)
(prickly dogfishes)

Trunk rounded or oval in cross-section; fins much lower, height of first dorsal fin much less than head length (fig. 40) ...... Squalidae (fig. 40; p. 47)
(dogfishes)

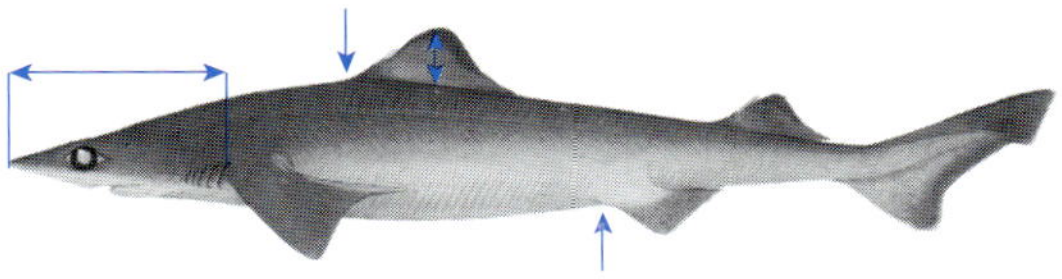

24 Head hammer-shaped (fig. 41); eyes located on outer margin of head (fig. 41) .... Sphyrnidae (fig. 41; p. 270)
(hammerhead sharks)

Head not hammer-shaped **25**

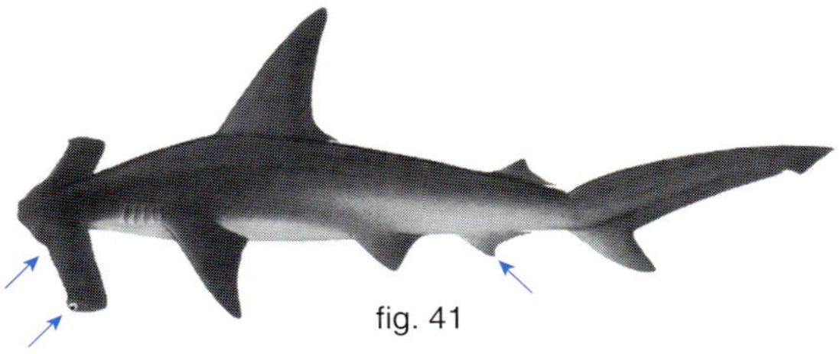

25 Length of caudal fin equal to or more than half total length (fig. 42); body not spotted or banded . ... Alopiidae (fig. 42; p. 154)
(thresher sharks)

Caudal fin much less than half total length (fig. 43) (caudal fin also long in *Stegastoma* but body spotted and/or banded, see fig. 52) **26**

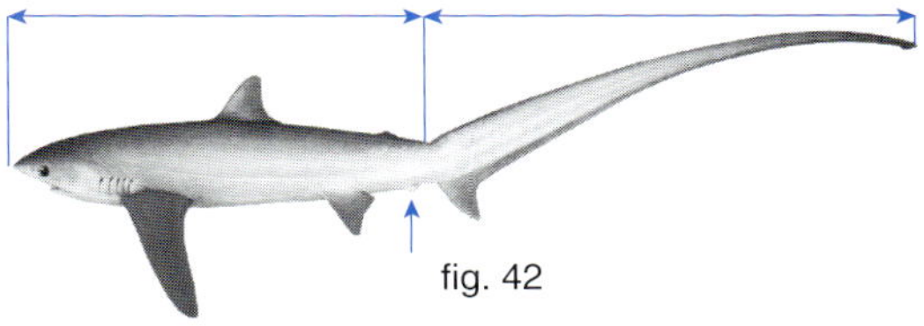

26 Dorsal-fin spines present (fig. 43) .... Heterodontidae (fig. 43; p. 112)
(horn sharks)

Dorsal-fin spines absent **27**

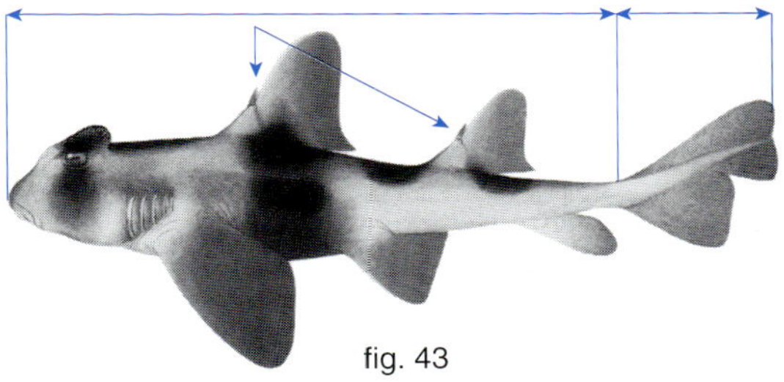

fig. 37

fig. 38

fig. 39

fig. 40

fig. 41

fig. 42

fig. 43

27 Snout extending above mouth as long, flattened, blade-like shelf (fig. 44); nostrils close to mouth (fig. 45) .... Mitsukurinidae (fig. 44; p. 148)
   (goblin sharks)

fig. 44

Snout not as above (extended slightly in some scyliorhinids, fig. 71), but nostrils well forward of mouth) 28

28 Whole mouth forward of front margin of eye (fig. 46) 29

Mouth partly beneath or behind front margin of eye (fig. 47) 35

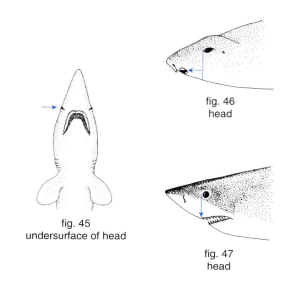
fig. 45
undersurface of head

fig. 46
head

fig. 47
head

29 Mouth very broad, terminal (fig. 48); caudal fin forked, upper and lower lobes tall (fig. 48); subterminal notch absent (fig. 48) .... Rhincodontidae (fig. 48; p. 142)
   (whale sharks)

Mouth smaller, subterminal (fig. 49); upper and lower lobes of caudal fin low (fig. 49); subterminal notch present (fig. 49) 30

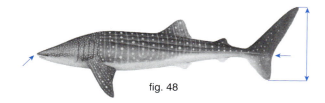
fig. 48

30 No fleshy lobe or groove on outer margin of nostril (fig. 50) 31

Fleshy lobe and groove present on outer margin of nostril (fig. 51) 32

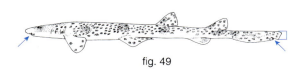
fig. 49

31 Caudal fin very long, almost as long as body (fig. 52); ridges present on side of body (fig. 52) .... Stegastomatidae (fig. 52; p. 138)
   (zebra sharks)

Caudal fin shorter, much less than half length of body (fig. 53); no ridges on side of body .... Ginglymostomatidae (fig. 53; p. 140)
   (nurse sharks)

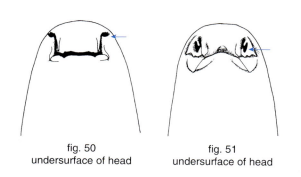
fig. 50
undersurface of head

fig. 51
undersurface of head

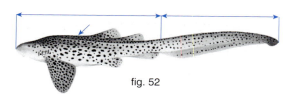
fig. 52

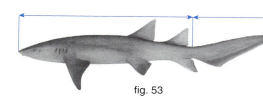
fig. 53

32 Origin of anal fin forward of origin of second dorsal fin (fig. 54); anal fin more than its base length from caudal fin (fig. 54) .... Parascylliidae (fig. 54; p. 117)
(collared carpet sharks)

Origin of anal fin well behind origin of second dorsal fin (fig. 55); anal fin adjacent to caudal fin (fig. 55) and sometimes barely distinguishable from it (fig. 59) **33**

33 Body strongly depressed anteriorly (fig. 56); skin flaps present along side of head behind nostrils (fig. 56); enlarged canine teeth at symphyses of both jaws .... Orectolobidae (fig. 58; p. 125)
(wobbegongs)

Body more or less cylindrical anteriorly (fig. 57); no skin flaps along side of head behind nostrils (fig. 57); teeth small, those at middle of jaws not distinctly larger than those adjacent **34**

34 Tail long, distance from vent to lower caudal-fin origin greater than distance from snout to vent (fig. 59); insertion of second dorsal fin well in front of anal-fin origin (fig. 59) .... Hemiscylliidae (fig. 59; p. 134)
(longtail carpet sharks)

Tail shorter, distance from vent to lower caudal-fin origin less than distance from snout to vent (fig. 60); insertion of second dorsal fin over or slightly behind origin of anal fin (fig. 60) .... Brachaeluridae (fig. 60; p. 122)
(blind sharks)

35 Caudal fin almost symmetrical, lunate, upper lobe less than 1.5 times longer than lower lobe (fig. 61) **36**

Caudal fin asymmetrical, upper lobe more than 1.5 times longer than lower lobe (fig. 62) **37**

36 Gill openings very long, extending on to both dorsal and ventral surfaces (fig. 63); first gill openings almost continuous on throat; more than 150 rows of small hook-like teeth in both jaws (fig. 65) .... Cetorhinidae (fig. 63; p. 159)
(basking sharks)

fig. 54

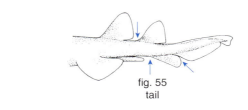
fig. 55
tail

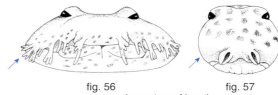
fig. 56   fig. 57
front view of head

fig. 58

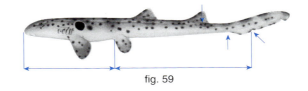
fig. 59

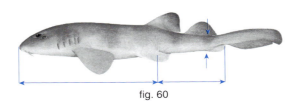
fig. 60

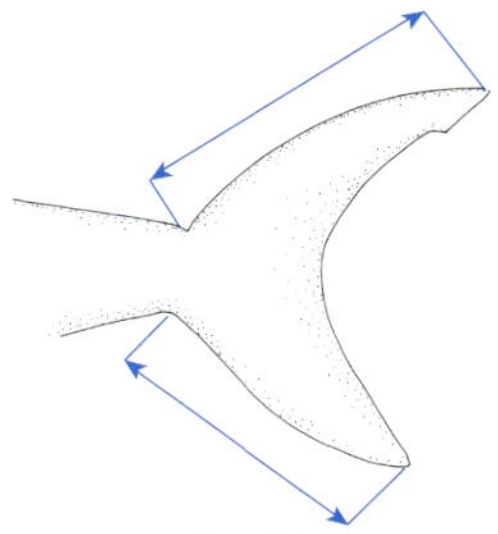

fig. 61
caudal fin (lunate)

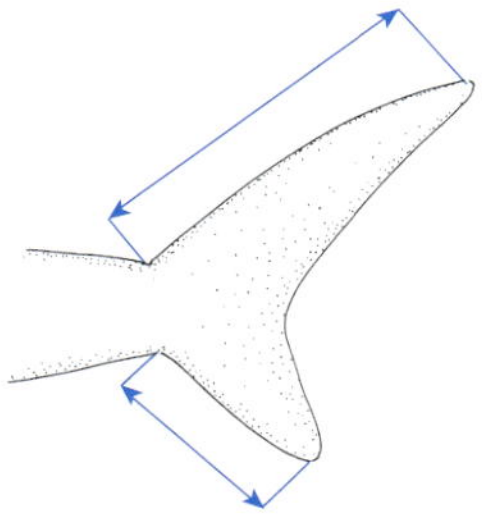

fig. 62
caudal fin (heterocercal)

fig. 63

Gill openings shorter, confined to sides (fig. 64); first gill openings widely separated on throat; less than 40 rows of sharp blade-like teeth in each jaw (fig. 66) .... Lamnidae (fig. 64; p. 161)
(mackerel sharks)

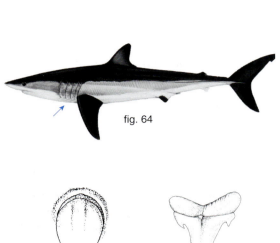
fig. 64

37 Mouth huge and terminal, lower jaw extending to snout tip (fig. 67); very large sharks .... Megachasmidae (fig. 67; p. 152)
(megamouth sharks)

Mouth located on undersurface of head, preoral distance distinctly longer than eye diameter (fig. 68) **38**

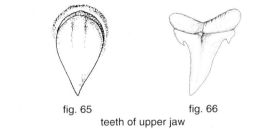
fig. 65   fig. 66
teeth of upper jaw

38 Eyes very large, more than half greatest height of snout (fig. 69); gill openings extending onto dorsal surface of head (fig. 69); caudal keels present (fig. 69) .... Pseudocarchariidae (fig. 69; p. 150)
(crocodile sharks)

fig. 67

Eyes smaller, less than half greatest height of snout (fig. 70); gill openings not extending onto dorsal surface of head (fig. 70); caudal keels absent in most species **39**

39 Eyelid fixed, not capable of closing over eye .... Odontaspididae (fig. 70; p. 144)
(grey nurse sharks)

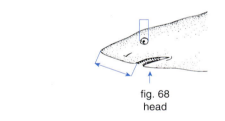
fig. 68
head

Eyelid capable of closing over eye (nictitating) **40**

40 First dorsal-fin origin well behind pelvic-fin origin (fig. 71) .... Scyliorhinidae (fig. 71; p. 167)
(catsharks)

fig. 69

First dorsal-fin origin well in front of pelvic-fin origin (fig. 72) **41**

41 Precaudal pits absent (fig. 73); leading edge of upper lobe of caudal fin smooth (fig. 73) .... Triakidae (fig. 72; p. 205)
(hound sharks)

fig. 70

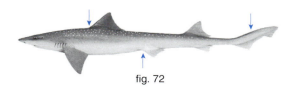
fig. 71   fig. 72

Precaudal pits present (fig. 74); leading edge of upper lobe of caudal fin usually rippled (fig. 74) **42**

42 Spiracles present (fig. 75); posterior margin of second dorsal fin deeply concave (fig. 75); intestine with a spiral valve (fig. 77) .... Hemigaleidae (fig. 75; p. 217) **(weasel sharks)**

Spiracles absent (except in *Galeocerdo* and sometimes in *Loxodon*, *Negaprion* and *Triaenodon*); posterior margin of second dorsal fin not deeply concave (slightly concave in *Negaprion* and *Triaenodon*) (fig. 76); intestine with a scroll valve (fig. 78) .... Carcharhinidae (fig. 76; p. 221) **(whaler sharks)**

43 Snout recurved with a hoe-shaped process at tip (fig. 79); caudal fin arched upward (fig. 79) .... Callorhinchidae (fig. 79; p. 465) **(elephant fishes)**

Snout straight, bluntly rounded or pointed (figs 80, 81); caudal-fin axis straight (figs 80, 81) **44**

44 Snout relatively short, tip bluntly rounded (fig. 80) .... Chimaeridae (fig. 80; p. 467) **(shortnose chimaeras)**

Snout very long, tip pointed (fig. 81) .... Rhinochimaeridae (fig. 81; p. 479) **(spookfishes)**

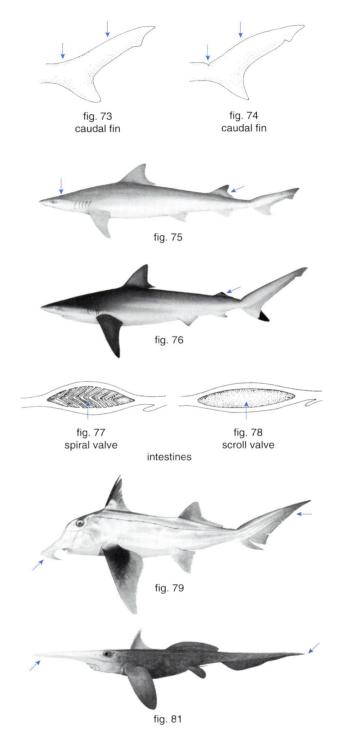

fig. 73 caudal fin

fig. 74 caudal fin

fig. 75

fig. 76

fig. 77 spiral valve

fig. 78 scroll valve

intestines

fig. 79

fig. 81

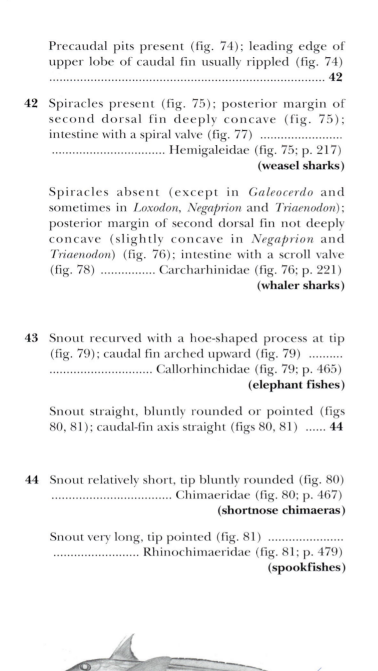

fig. 80

5 Family Chlamydoselachidae
FRILLED SHARKS

Frilled sharks have a distinctive eel-like body and an almost reptilian head with six pairs of gill slits, the first being continuous across the throat. They have a very short snout, small spiracles, long tricuspid teeth in both jaws, and the nostrils are not connected to a terminal mouth. The single, small, spineless dorsal fin is situated well back on the body just over a larger anal fin. The caudal fin has a weak ventral lobe and no subterminal notch.

The family shares the presence of a single dorsal fin and six pairs of gill slits with some members of the family Hexanchidae. The Chlamydoselachidae contains only one living species.

Frilled Shark — *Chlamydoselachus anguineus* Garman, 1884

5.1 *Chlamydoselachus anguineus* (Plate 1)   Fish Code: 00 006001

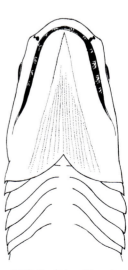

undersurface of head

- **Alternative names:** Frill shark, frill-gilled shark.
- **Field characters:** A distinctive shark with a greatly elongated body, six pairs of gill slits, a terminal mouth with widely spaced tricuspid teeth, one dorsal fin and a caudal fin without a subterminal lobe.
- **Distinctive features:** Body greatly elongate, almost eel-like. Snout short; mouth long and terminal. Six pairs of gill slits, first continuous across throat. Teeth in both jaws each have three slender, curved cusps and two intermediate cusplets. Dorsal fin small and lobe-like; origin varying from over pelvic-fin bases to behind anal-fin origin. Anal fin larger than dorsal fin. Pectoral fins small, paddle-shaped. Caudal fin with weak ventral lobe, without subterminal notch. Tooth count 24–27/21–26 [19–30/21–27]. Total vertebrae [146–171]*; precaudal [93–94]*; monospondylous [72–76]*.
- **Colour:** Dark chocolate brown in life, otherwise dorsal surfaces brownish grey or brownish black; ventral surfaces similar or somewhat paler.
- **Size:** Born at 40–60 cm and attains 196 cm. Males mature at about 117 cm and females at about 135 cm.
- **Distribution:** Wide ranging but patchy. Atlantic, southern Africa and the eastern, central and western Pacific including Australia (New South Wales and Tasmania). On or near the bottom of the outer continental and insular shelves and upper slopes

at depths of 120–1280 m. A Californian specimen was caught in a gillnet set at 20 m depth over water more than 1500 m deep.

- **Remarks:** Little is known of the biology of this species. Ovoviviparous with litters of 8–15. The few stomachs that have been examined have contained elasmobranchs; the highly distensible jaws and buccal cavity suggest that the frill shark is able to take relatively large prey.
- **References:** Nakaya and Bass (1978); Bass (1979).

third teeth from symphysis
(upper and lower jaw)

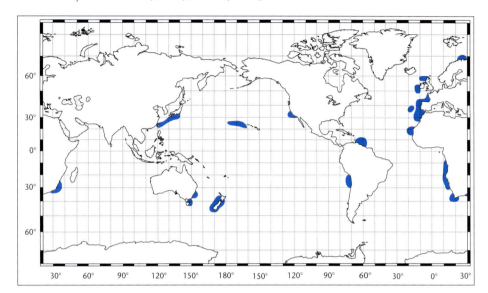

6

Family Hexanchidae

SIXGILL AND SEVENGILL SHARKS

Members of this small family, also known as cowsharks, can be easily distinguished by the presence of six or seven pairs of gill slits, none of which is continuous across the throat. They have a fusiform body, a ventrally situated mouth with differently shaped teeth in the upper and lower jaws (lowers comb-shaped) and a caudal fin with a distinct subterminal notch. They also have small spiracles, a single, spineless dorsal fin, and nostrils that are not connected to the mouth.

This family of medium to large (attaining 1.4–4.8 m) sharks has a worldwide distribution in cold temperate and tropical seas. They are usually found near the bottom in deepwater, although one species inhabits relatively shallow bays and estuaries. Reproduction is ovoviviparous. Their diet includes large mammalian prey and elasmobranchs. All four species in this family occur in the Australian region.

Key to hexanchids

1 Six pairs of gill slits 2
   Seven pairs of gill slits 3

2 Lower jaw with 6 rows of large comb-like teeth (fig. 1) on each side; space between dorsal-fin insertion and upper caudal-fin origin about equal to dorsal-fin base (fig. 2); large sharks up to 4.8 m .... *Hexanchus griseus* (fig. 2; p. 40)

   Lower jaw with 5 rows of large comb-like teeth on each side; space between dorsal-fin insertion and upper caudal-fin origin much greater than dorsal-fin base (fig. 3); smaller sharks up to 1.8 m .... *Hexanchus nakamurai* (fig. 3; p. 41)

3 Eyes large, head narrow and pointed (fig. 4); body colour plain; size to about 1.4 m .... *Heptranchias perlo* (fig. 6; p. 39)

   Eyes small, head broad and rounded (fig. 5); body colour usually with spots; size to about 3 m .... *Notorynchus cepedianus* (fig. 7; p. 42)

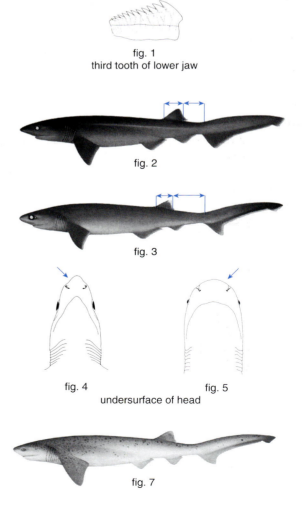

fig. 1
third tooth of lower jaw

fig. 2

fig. 3

fig. 4   fig. 5
undersurface of head

fig. 6   fig. 7

Sharpnose Sevengill Shark — *Heptranchias perlo* (Bonnaterre, 1788)

6.1 *Heptranchias perlo* (Plate 1) **Fish Code:** 00 005001

- **Alternative names:** One-finned shark, perlon shark, slender sevengill shark, sharpsnout sevengill shark.
- **Field characters:** A medium-sized shark with seven pairs of gill slits, large eyes, a narrowly pointed head, a single dorsal fin and a plainly coloured body.
- **Distinctive features:** Body fusiform and slender. Caudal peduncle long (distance from dorsal-fin insertion to upper caudal-fin origin more than twice length of dorsal-fin base). Head narrow, snout relatively pointed; preoral snout length more than 1.5 times distance between nostrils. Seven pairs of long gill slits; eyes large (greater than internasal distance); mouth narrowly pointed (in ventral view). Upper jaw teeth with a narrow, hooked cusp and small lateral cusplets (first two from jaw symphysis may lack cusplets). Lower jaw usually with five large teeth on each side; teeth wide, low and comb-shaped, comprising a main mesial cusp and several large distal cusplets. Dorsal fin small, origin over inner margins of pelvic fins. Anal fin small, height about 0.5–0.6 times dorsal-fin height. Tooth count (omitting very small lateral teeth) 18–24/11–13 [18–22/11]. Total vertebrae 143–161 [125–159]; precaudal 89–95 [54–90].
- **Colour:** Dorsal surfaces brownish grey, ventral surfaces paler. Upper caudal-fin tip and dorsal-fin tip and posterior margin dark in juveniles, becoming very faint in adults; pectoral, dorsal and upper caudal fins with white trailing margins in adults. Flanks sometimes with dark blotches in juveniles. The eyes are a fluorescent green in life.
- **Size:** Born at about 25 cm and attains 137 cm. Males mature at about 75 cm; females at about 100 cm.
- **Distribution:** Tropical and temperate areas of the Atlantic and Indian Oceans and the margins of the Pacific. Widely distributed off Australia where it is recorded from Cairns (Queensland) to Ashmore Reef (Western Australia), including Tasmania. On or near the bottom of the continental and insular shelves and upper slopes, usually in depths of 100–400 m, but has been recorded both close inshore and down to 1000 m.

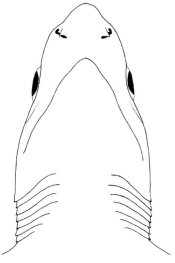

undersurface of head

third teeth from symphysis
(upper and lower jaw)

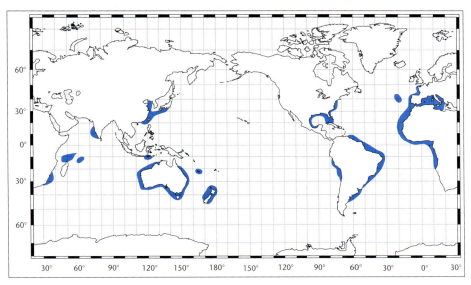

- **Remarks:** Ovoviviparous, with litters of 9–20 young. The diet consists mainly of bony fishes, cephalopods and some crustaceans.
- **Local synonymy:** *Heptranchias dakini* Whitley, 1931.
- **References:** Whitley (1931a); Garrick and Paul (1971a).

Bluntnose Sixgill Shark — *Hexanchus griseus* (Bonnaterre, 1788)

6.2 *Hexanchus griseus* (Plate 1)   **Fish Code:** 00 005005

- **Alternative names:** Sixgill shark, bull shark.
- **Field characters:** A large, heavy-bodied shark with six pairs of gill slits, small eyes, six rows of large comb-like teeth on each side of the lower jaw, a broadly, rounded snout, a short caudal peduncle and a single dorsal fin.
- **Distinctive features:** Body fusiform and relatively stout; caudal peduncle short (distance from dorsal-fin insertion to upper caudal-fin origin about equal to, or slightly longer than, dorsal-fin base). Head broad; snout broadly rounded. Six pairs of long gill slits; eyes small; mouth broadly rounded in ventral view. First two anterior upper-jaw teeth from symphysis each with a narrow, hooked cusp and no lateral cusplets; remaining large teeth becoming progressively wider and more comb-like (with more cusplets) laterally. Lower jaw usually with six large teeth on each side, teeth wide, low and comb-shaped, with a mesial cusp and several large cusplets. Dorsal fin relatively small, its origin over or behind pelvic-fin insertion. Anal fin somewhat smaller than dorsal fin. Tooth count (omitting very small lateral teeth) 20/13* [18–20/13]*. Total vertebrae [118–148]; precaudal [57]*; monospondylous [41–52].

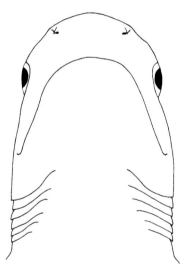

undersurface of head

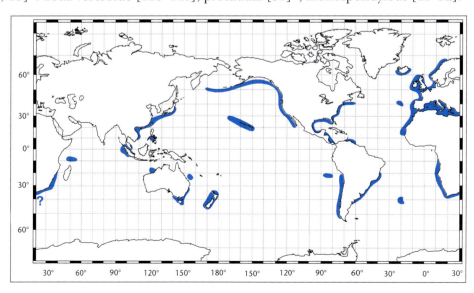

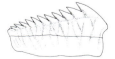

third teeth from symphysis
(upper and lower jaw)

- **Colour:** Dorsal surfaces pale or blackish grey to chocolate brown, usually with a paler streak along the lateral line. Ventral surfaces greyish-white. Fins with thin white posterior margins; neo-natals with whitish anterior fin margins. The eyes are a fluorescent green in life.
- **Size:** Born at 65–70 cm and attains 480 cm. Males mature at about 315 cm; females at around 420 cm.
- **Distribution:** Tropical and temperate areas of the Atlantic (including the Mediterranean), Indian and Pacific Oceans. In Australia, recorded from seamounts off Queensland and from a few specimens collected off New South Wales, Victoria, Tasmania and northern Western Australia. Continental and insular shelves and upper slopes from the surface to about 2000 m. Juveniles may occur close inshore while adults are normally taken on or near the bottom from deeper water.
- **Remarks:** Ovoviviparous, with large litters of 22–108 being reported. Stomachs have contained a wide variety of teleost and elasmobranch fishes, cephalopods, crustaceans, carrion and even seals. A relatively sluggish species that is a poor fighter when captured on hook and line. Utilised in some areas for its meat and liver oil.
- **Local synonymy:** *Hexanchus griseus australis* De Buen, 1960.
- **References:** De Buen (1960a); Bass (1979).

Bigeye Sixgill Shark — *Hexanchus nakamurai* Teng, 1962

6.3 *Hexanchus nakamurai* (Plate 1) **Fish Code:** 00 005004

- **Field characters:** A relatively slender shark with six pairs of gill slits, large eyes, five rows of large comb-like teeth on each side of the lower jaw, a moderately long and bluntly pointed snout, a long caudal peduncle and a single dorsal fin.
- **Distinctive features:** Body fusiform and relatively slender; caudal peduncle long (distance from dorsal-fin insertion to upper caudal-fin origin at least twice length of dorsal-fin base). Head dorsoventrally flattened; snout bluntly pointed. Six pairs of long gill slits; eyes large (about equal to internasal distance). First two anterior upper-jaw teeth from symphysis each with a narrow, hooked cusp and no lateral cusplets; remaining large teeth becoming progressively wider and more comb-like (with more cusplets) laterally. Lower jaw usually with five large teeth on each side; teeth wide, low and comb-shaped, comprising a main mesial cusp and several large cusplets. Dorsal fin relatively small; origin varying from over last half of pelvic-fin base to just behind pelvic-fin insertion. Anal fin somewhat smaller than dorsal fin. Tooth count (omitting very small lateral teeth) [18/11]*. Total vertebrae [155]*; precaudal [87]*.
- **Colour:** Dorsal surfaces brownish grey, ventral surfaces paler. Fins with white posterior margins. Juveniles with a black tip to the upper caudal fin. The eyes are a fluorescent green in life.

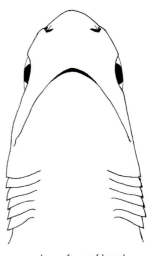

undersurface of head

- **Size:** Born at about 43 cm and attains 180 cm. Size at maturity unknown, but males and females are mature by 123 cm and 142 cm respectively.
- **Distribution:** Patchily distributed in tropical and warm temperate waters of the western Pacific, Atlantic and Indian Oceans. In Australia, recorded from Western Australia (the North West Shelf and Bunbury), Queensland and northern New South Wales. Continental and insular shelves and upper slopes, usually on or near the bottom at depths from 90 to 600 m.
- **Remarks:** Little is known of its biology. Ovoviviparous, with 13 embryos recorded from one litter. The few stomach contents examined have contained teleost fish (and in one case a crustacean). The teleosts included a tuna which suggests that this shark makes occasional excursions towards the surface.
- **Local synonymy:** *Hexanchus vitulus* Springer & Waller, 1969.
- **Reference:** Whitley (1931b).

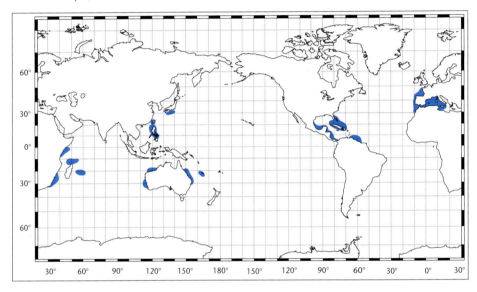

Broadnose Sevengill Shark — *Notorynchus cepedianus* (Péron, 1807)

6.4 Notorynchus cepedianus (Plate 1) **Fish Code:** 00 005002

- **Alternative names:** Seven-gilled shark, Tasmanian tiger shark, broad-snout, ground shark, cowshark.
- **Field characters:** A large shark with seven pairs of gill slits, small eyes, a single dorsal fin and a wide head with a short, blunt snout. The body is usually peppered with numerous small, dark and/or white spots.
- **Distinctive features:** Body fusiform. Caudal peduncle short (distance from dorsal-fin insertion to upper caudal-fin origin from 1–1.5 times length of dorsal-fin base).

Head wide, snout short and blunt; preoral snout length less than 1.5 times distance between nostrils. Seven pairs of long gill slits; eyes small (about 2 times in internasal distance); mouth broadly rounded in ventral view. First few upper-jaw teeth from symphysis with a narrow cusp and reduced or small cusplets; remaining large teeth progressively wider and more comb-like (with more cusplets) laterally. Large lower-jaw teeth relatively high, short and comb-shaped, comprising a main mesial cusp and several large cusplets. Dorsal fin relatively small; origin varying from over insertion to free rear tips of pelvic fins. Anal fin somewhat smaller than dorsal fin. Tooth count (omitting very small lateral teeth) [15–16/13]. Total vertebrae [123–157]; precaudal [71]*; monospondylous [48–57].

- **Colour:** Dorsal surfaces varying from silvery grey to brownish, ventral surfaces white. Dorsal body surfaces and fins speckled with small black and white spots. Juveniles with white posterior fin margins and dorsal-fin tip, and with a black stripe extending from the caudal peduncle to the caudal-fin tip.

- **Size:** Born at 40–45 cm and attains 300 cm. Males mature at about 150 cm and females at about 220 cm.

- **Distribution:** Temperate waters of the south Atlantic, Pacific and Indian Oceans. In Australasia, recorded from New Zealand and from Sydney (New South Wales) to Esperance (Western Australia), including Tasmania. From close inshore in bays and estuaries to at least 136 m on the continental shelves. Usually on or near the bottom but may come to the surface in shallow areas.

- **Remarks:** Ovoviviparous, with large litters of up to 82 young. Although relatively common in parts of southern Australia, nothing is known of its local biology. It is a powerful shark with a rather generalised diet that includes other sharks, bony fishes, seals and carrion. Solitary sharks use stealth rather than speed to catch seals but pack-hunting of seals has also been observed. A potentially dangerous species, although no verified attacks on people (other than attacks on divers in aquaria) have been recorded. The flesh is of good quality and there is a small-scale commercial fishery for this species in California. In other areas it is also taken for its hide and liver oil.

- **Local synonymy:** *Notorynchus macdonaldi* Whitley, 1931.

- **Reference:** Whitley (1931b).

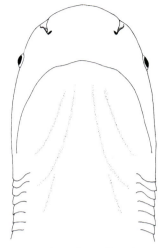
undersurface of head

third teeth from symphysis (upper and lower jaw)

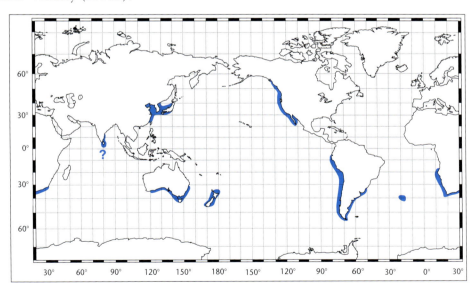

7

Family Echinorhinidae

BRAMBLE SHARKS

The two species in this family can be easily distinguished by their fusiform body-shape, moderately large to very large thorn-like denticles, the absence of an anal fin and subterminal notch on the caudal fin, and the presence of two, small, spineless dorsal fins situated close together and posteriorly on the body (origin of the first behind the pelvic-fin origin). They also have small spiracles, very short labial furrows and the teeth in both jaws have a central, oblique, blade-like cusp with up to three cusplets on each side (absent in juveniles).

These large, ovoviviparous sharks (attaining 2.6–4 m) are widely distributed in temperate and tropical seas, living in deep-water on or near the bottom. They are thought to use suction to catch their prey by rapidly expanding their mouth and pharynx. Bramble sharks are closely related to dogfishes.

Key to echinorhinids

1 Denticles on body very large, sparse, irregularly distributed and their bases with smooth margins (fig. 1); some denticles fused into compound plates .... ***Echinorhinus brucus* (fig. 3, p. 45)**

2 Denticles on body moderately large, numerous, regularly distributed and their bases with scalloped margins (fig. 2); denticles never fused into compound plates ....
.... ***Echinorhinus cookei* (fig. 4, p. 46)**

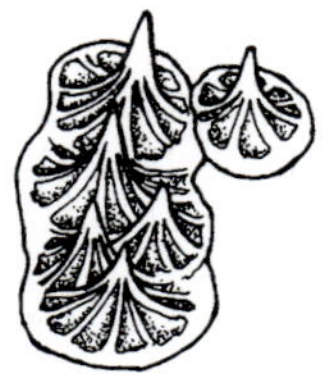

fig. 1
denticles

fig. 2
denticle

fig. 3

fig. 4

Bramble Shark — *Echinorhinus brucus* (Bonnaterre, 1788)

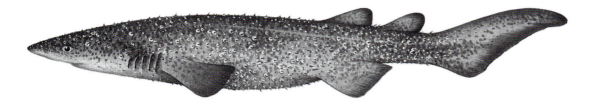

7.1 *Echinorhinus brucus* (Plate 11) **Fish Code:** 00 022001

- **Alternative name:** Spinous shark.
- **Field characters:** A large shark with no anal fin, the first dorsal fin originating behind the pelvic-fin origin and with very large, sparse, irregularly distributed, thorn-like denticles with regular bases that are sometimes fused into compound plates.
- **Distinctive features:** Body stout. Snout relatively short. Teeth in both jaws alike; primary cusp oblique, blade-like, with up to three cusplets on each side. Denticles very large (single denticles up to about 15 mm in basal diameter in adults), sparse, irregularly distributed and thorn-like with smooth basal margins; some bases fused into compound plates. Dorsal fins small, close together, situated posteriorly on body; the first dorsal fin originating behind pelvic-fin origin. Caudal fin without subterminal notch. Tooth count [20–26/21–26]*.
- **Colour:** Dorsal surfaces dark purplish grey to brown with whitish denticles, paler ventrally; dark spots may be present on the back and sides.
- **Size:** Attains at least 260 cm; birth size unknown, between 30 and 90 cm. Size at maturity unknown but adult males of 150 cm and adult females of 213 cm have been reported.
- **Distribution:** Virtually circumglobal, but patchy. Australian records are from Victoria and the Great Australian Bight. On or near the bottom of continental and insular shelves and slopes mainly at depths of 400–900 m, but occasionally taken shallower.
- **Remarks:** A sluggish, primarily deepwater species. Ovoviviparous with litters of 15–24. Stomach contents have contained fish (including small sharks) and crabs. Utilised for fishmeal in the eastern Atlantic and for liver oil in South Africa; not used in Australia where it is caught infrequently.
- **Local synonymy:** *Echinorhinus (Rubusqualus) mccoyi* Whitley, 1931.
- **References:** Whitley (1931a); Garrick (1960a).

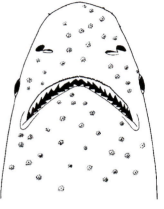

undersurface of head

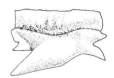

third teeth from symphysis (upper and lower jaw)

flank denticles

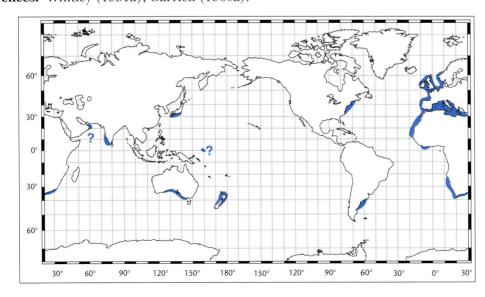

Prickly Shark — *Echinorhinus cookei* Pietschmann, 1928

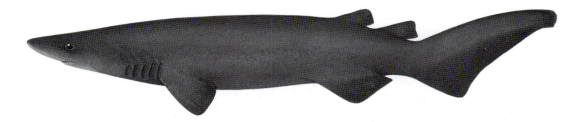

7.2 *Echinorhinus cookei* (Plate 11)   **Fish Code:** 00 022002

- **Alternative name:** Cooks bramble shark.
- **Field characters:** A large shark with no anal fin, the first dorsal fin originating behind the pelvic-fin origin and with moderately large, numerous, regularly distributed, thorn-like denticles with crenate bases that are not fused into compound plates.
- **Distinctive features:** Body stout. Snout relatively short. Teeth in both jaws alike; primary cusp oblique and blade-like with up to three cusplets on each side. Denticles small (single denticles up to about 4 mm in basal diameter in adults), numerous, regularly distributed and thorn-like with uneven basal margins; bases not fused into compound plates. Dorsal fins small, close together and situated posteriorly on body; first dorsal fin originating behind the pelvic-fin origin. Caudal fin without subterminal notch. Tooth count 21–23/20–22*. Total vertebrae 89*; precaudal 59*.
- **Colour:** Greyish brown with black distal fin margins; white around mouth and on ventral surface of snout.
- **Size:** Born at 40–45 cm and attains about 400 cm. Some males mature by 198 cm; females mature between 250 and 300 cm.
- **Distribution:** Tropical and temperate areas of the western Pacific including Australia (Victoria) and New Zealand. On or near the bottom of continental and insular shelves and slopes at depths from 11 to 424 m (mostly taken at 70 m or more).
- **Remarks:** A sluggish, bottom-living species which is captured infrequently. Presumably ovoviviparous. Stomach contents have contained fish (including elasmobranchs) and cephalopods.
- **References:** Garrick (1960a); Garrick and Moreland (1968).

undersurface of head

flank denticle

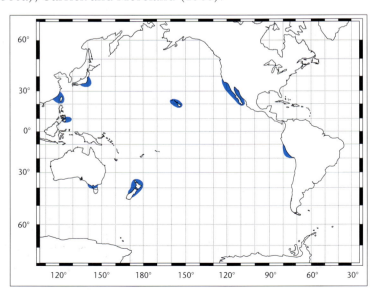

8 Family Squalidae
DOGFISHES

Dogfishes are highly varied in morphology and include both the smallest and largest of sharks (0.2–6 m long at maturity). They are distinguished from related families in being cylindrical in cross-section, lacking an anal fin and a saw-like snout, and having two dorsal fins (the first of which originates in advance of the pelvic fins), usually preceded by spines. They have single or multicuspid teeth of varied shapes (these vary within and between species), large spiracles, variable denticle sizes and shapes, and may have a subterminal notch in the caudal fin. The lengths of the labial furrows, which vary from short to almost connected, are often useful in distinguishing between species.

Squalids, as a group, have the widest depth and geographical distributions of all sharks. They are represented in all oceans and, although most species occur in deepwater beyond the continental shelf, they have been observed or caught from near the shore to depths exceeding 6 kilometres. They comprise the largest elasmobranch family in the Australian region, with 14 genera and at least 40 species. Some of these are of commercial value as food and others could be exploited for a high-quality oil, squalene, which is stored in the liver. All are ovoviviparous and are reported to have litters of 1–36 or more.

Key to squalids

1   Fin spines present at origin of both dorsal fins (sometimes broken or small but seldom impossible to detect) (fig. 1) 2

   Fin spines absent at origin of second dorsal fin; mostly absent at first dorsal-fin origin (except in *Squaliolus*) 35

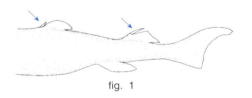

fig. 1

2   Teeth in upper jaw almost identical in shape and size to those of lower jaw (figs 2, 4) 3

   Teeth in upper jaw markedly different in shape and/or size from those of lower jaw (fig. 5) 13

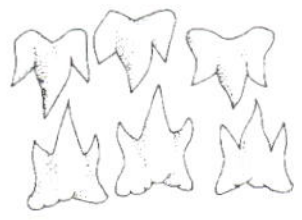

3   Teeth with more than one cusp, not blade-like, adjacent teeth not overlapping (fig. 2); skin ragged and peeling, mostly lacking denticles ....
   .... *Centroscyllium kamoharai* (fig. 3; p. 61)

   Teeth with one broad, blade-like cusp, adjacent teeth overlapping (fig. 4); skin covered in denticles, not deciduous 4

fig. 2
upper and lower teeth
(*Centroscyllium*)

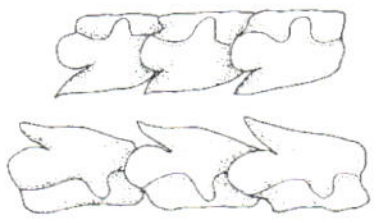

fig. 4
upper and lower teeth
(*Squalus*)

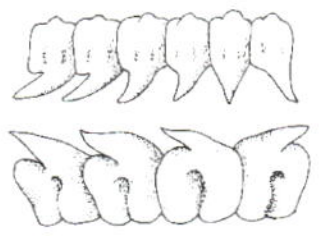

fig. 5
upper and lower teeth

fig. 3

4   Anterior nasal flaps with elongate barbels ....
    .... **Cirrhigaleus barbifer (fig. 6; p. 68)**

    Anterior nasal flaps lacking elongate barbels (fig. 9)
    .... **5**

5   First dorsal-fin spine origin distinctly behind free rear tips of pectoral fins (fig. 9); anterior nasal flap single-lobed (fig. 7); usually bluish grey with white spots .... **Squalus acanthias (fig. 9; p. 98)**

    First dorsal-fin spine origin over free rear tips of pectoral fins or further forward (fig. 17); anterior nasal flap with two lobes (fig. 8); greyish, lacking white spots **6**

6   Distance from snout tip to inner margin of nostril shorter than distance from inner edge of nostril to labial furrow (fig. 10) **7**

    Distance from snout tip to inner margin of nostril longer than distance from inner edge of nostril to labial furrow (fig. 11) **10**

7   First dorsal fin raked backwards slightly (fig. 12); denticles lanceolate (fig. 14); fewer than 87 precaudal vertebrae **8**

    First dorsal fin more upright (fig. 13); denticles with three posterior cusps (figs 15, 16); more than 87 precaudal vertebrae **9**

8   Dorsal-fin spines slender, narrow-based (height of undamaged second spine more than 4 times its base length), tapering gradually towards tip (fig. 17); precaudal vertebrae 78–82 ....
    .... **Squalus megalops (fig. 17; p. 99)**

    Dorsal-fin spines robust, broad-based (height of undamaged second spine less than 4 times its base length), tapering rapidly towards tip (fig. 18); precaudal vertebrae 83–86 ....
    .... **Squalus sp. D (fig. 18; p. 95)**

9   Dorsal-fin spines slender (fig. 19); prominent dark bar along base of lower caudal-fin lobe in small specimens (fig. 19); denticle crowns lacking lateral keels (fig. 16); precaudal vertebrae mostly 94–96
    .... **Squalus sp. A (fig. 19; p. 91)**

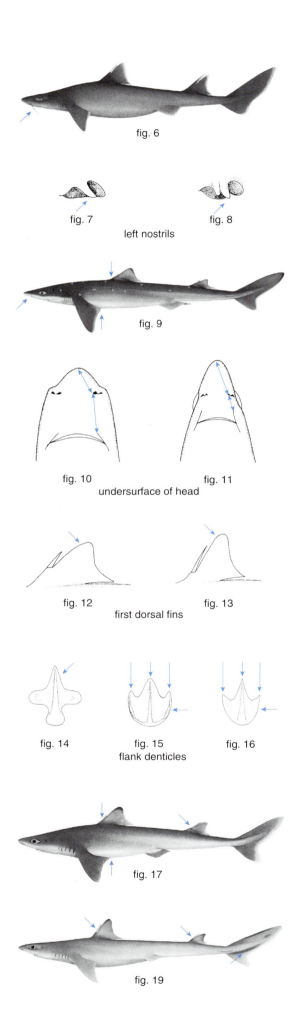

Dorsal-fin spines robust (fig. 20); no dark bar on base of lower caudal-fin lobe (fig. 20); denticle crowns with lateral keels (fig. 15); fewer than 94 precaudal vertebrae ....
.... ***Squalus* sp. B (fig. 20; p. 93)**

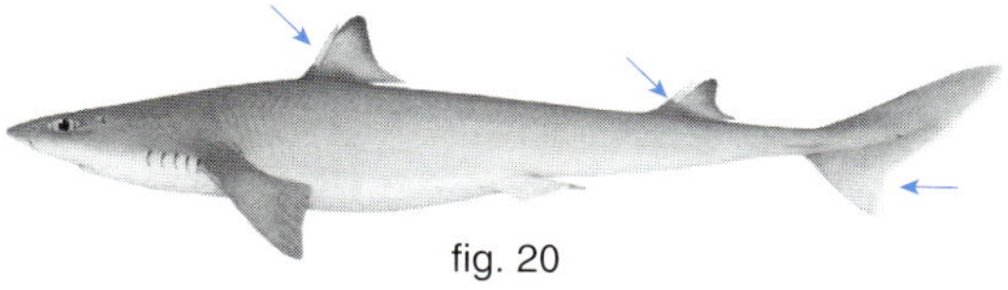

fig. 20

10 Dorsal-fin spines broad-based (fig. 21); no dark blotch on posterior notch of caudal fin (fig. 21); more than 88 precaudal vertebrae ....
.... ***Squalus* sp. C (fig. 21; p. 94)**

Dorsal-fin spines slender (fig. 24); black blotch on posterior notch of caudal fin (sometimes faint in adults) (fig. 25); less than 88 precaudal vertebrae
.... **11**

fig. 21

11 Head relatively broad, direct preorbital distance shorter than interorbital distance (fig. 22); mostly 43 or more monospondylous vertebrae ....
.... ***Squalus mitsukurii* (fig. 24; p. 101)**

Head relatively narrow, direct preorbital distance longer than interorbital distance (fig. 23); mostly 42 or fewer monospondylous vertebrae **12**

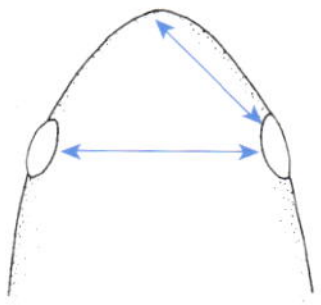

fig. 22 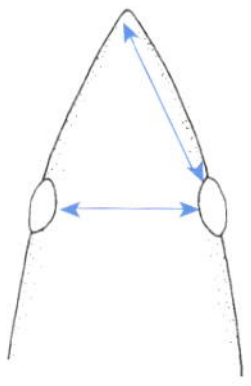   fig. 23
top of head

12 Snout very long, horizontal length more than 1.8 times eye diameter (fig. 25); first dorsal-fin spine much shorter than second dorsal-fin spine (fig. 25); monospondylous vertebrae 37–39 ....
.... ***Squalus* sp. E (fig. 25; p. 96)**

Snout shorter, horizontal length mostly less than 1.8 times eye diameter (fig. 26); first dorsal-fin spine slightly shorter than second dorsal-fin spine (fig. 26); monospondylous vertebrae 38–42 ....
.... ***Squalus* sp. F (fig. 26; p. 97)**

fig. 24

13 Upper teeth with more than one cusp (fig. 27) **14**
Upper teeth with a single cusp (fig. 28) **24**

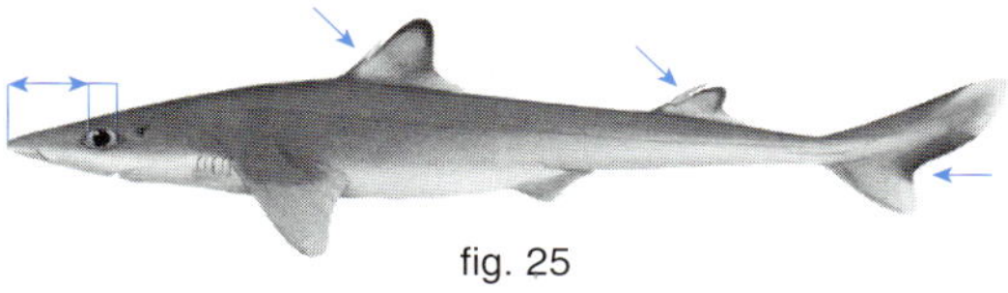

14 Denticles low, flat-topped, fixed and immovable (minute, never in distinct rows, barely distinguishable to the naked eye) (fig. 29) **15**

Denticles slender, upright, movable (small but distinguishable to the naked eye, sometimes in distinct rows) (fig. 30) **16**

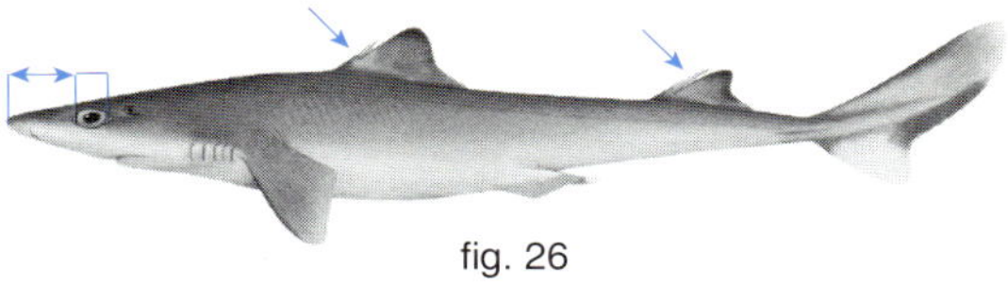

fig. 25

fig. 26

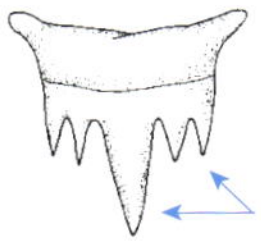

fig. 27
upper tooth

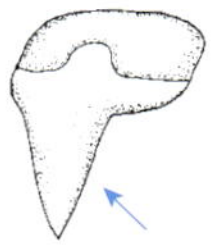

fig. 28
upper tooth

fig. 29
denticle

fig. 30
denticle

15 First dorsal fin less than an eye diameter behind free rear tip of pectoral fin (fig. 31); head length to first gill slit longer than distance from first gill slit to first dorsal-fin origin (fig. 31) .... *Etmopterus* **sp. A (fig. 31; p. 73)**

First dorsal fin more than an eye diameter behind free rear tip of pectoral fin (fig. 32); head length to first gill slit shorter than distance from first gill slit to first dorsal-fin origin (fig. 32) .... ***Etmopterus pusillus* (fig. 32; p. 84)**

16 Caudal peduncle short, distance from pelvic-fin insertion to lower lobe of caudal fin shorter than distance from snout tip to gill slits (fig. 35) **17**

Caudal peduncle long, distance from pelvic-fin insertion to lower lobe of caudal fin longer than distance from snout tip to gill slits (fig. 36) **19**

17 Upper surface pale, sharply demarcated from dark belly (fig. 33); pelvic flank and caudal markings distinct (fig. 33) ..... *Etmopterus* **sp. D (fig. 33; p. 76)**

Uniformly dark; pelvic flank and caudal markings absent or barely visible **18**

18 Denticles widely spaced; bases of dorsal fins with naked areas (fig. 34) .... ***Etmopterus granulosus* (fig. 34; p. 80)**

Denticles dense over entire surface (including bases of dorsal fins) .... *Etmopterus* **sp. B (fig. 35; p. 74)**

19 Posterior branch of pelvic flank marking short with broad, truncated tip (fig. 37); additional oval marking near base of upper lobe of caudal fin; first dorsal fin well forward of free rear tip of pectoral fin .... *Etmopterus* **sp. C (fig. 36; p. 75)**

Posterior branch of pelvic flank marking slender with pointed tip (fig. 38); no additional oval marking near base of upper lobe of caudal fin; first dorsal fin over or behind free rear tip of pectoral fin (fig. 39) **20**

20 Flank denticles not arranged in regular lines; dark ventral saddle on caudal peduncle and dark bands across middle and tip of upper caudal-fin lobe (fig. 39) **21**

fig. 31

fig. 32

fig. 33

fig. 34

fig. 35

fig. 36

fig. 37
pelvic flank marking

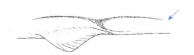
fig. 38
pelvic flank marking

Flank denticles mostly arranged in regular lines; no dark ventral saddle on caudal peduncle and no dark bands across middle and tip of upper caudal-fin lobe (fig. 41) **22**

21 Distance from pelvic-fin insertion to lower lobe of caudal fin subequal to distance from snout tip to pectoral-fin origin (fig. 39); flanks lacking rows of prominent, dark dashes ....
.... ***Etmopterus* sp. E (fig. 39; p. 77)**
Distance from pelvic-fin insertion to lower lobe of caudal fin longer than distance from snout tip to pectoral-fin insertion (fig. 40); flanks with prominent rows of dark dashes (fig. 40) ....
.... ***Etmopterus* sp. F (fig. 40; p. 78)**

22 Base of pelvic flank marking situated posteriorly, under exposed base of second dorsal-fin spine (fig. 41); anterior branch of flank marking usually longer than posterior branch ....
.... ***Etmopterus lucifer* (fig. 41; p. 81)**
Base of pelvic flank marking situated anteriorly, ahead of base of second dorsal-fin spine (fig. 42); anterior branch of flank marking shorter than posterior branch **23**

23 Precaudal marking longer than caudal marking ......
.... ***Etmopterus brachyurus* (fig. 42; p. 79)**
Precaudal marking shorter than caudal marking .....
.... ***Etmopterus molleri* (fig. 43; p. 83)**

24 Preoral snout much longer than distance from mouth to pectoral-fin origin (fig. 44); upper labial furrows more than nostril width apart (fig. 44); denticles on back with pitchfork-shaped crowns
.... **25**
Preoral snout shorter than distance from mouth to pectoral-fin origin (fig. 45) (if equal to or longer, then upper labial furrows less than nostril width apart, fig. 62); denticles on back unicuspidate or with plate-like crowns **26**

25 Dorsal fins dissimilar in size and shape (fig. 46); distance from exposed origin of first dorsal-fin spine to free tip of fin mostly exceeding distance from its free tip to exposed portion of second dorsal-fin spine (fig. 46); small specimens with dark posterior border to each dorsal fin (fig. 46) ....
.... ***Deania calcea* (fig. 46; p. 70)**
Dorsal fins similar in size and shape (fig. 47); distance from exposed origin of first dorsal-fin spine to free tip of fin less than distance from its free tip to exposed portion of second dorsal fin spine (fig. 47); small specimens with dark blotch near anterior margin of each dorsal fin ....
.... ***Deania quadrispinosa* (fig. 47; p. 71)**

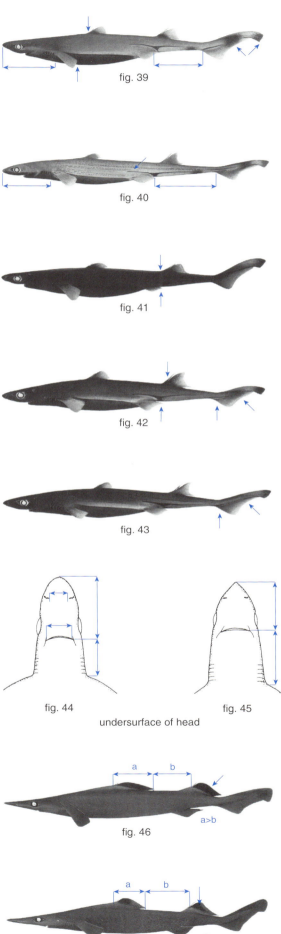

26 Upper teeth relatively broad, their bases overlapping to form an interlocking band (fig. 48); rear tips of pectoral fins angular (fig. 53) or elongated and pointed (fig. 54) **27**

Upper teeth relatively slender, their bases not overlapping (fig. 49); rear tips of pectoral fins rounded (fig. 60) **31**

27 Free rear tips of pectoral fins broadly angular (not extended into a lobe) (fig. 53); denticles with leaf-like crowns on narrow stalks (fig. 50) ....
.... ***Centrophorus squamosus*** **(fig. 53; p. 58)**

Free rear tips of pectoral fins extended into an acute lobe (fig. 54); denticles block-like, crowns not on stalks (figs 51, 52) **28**

28 Second dorsal fin relatively small, half height of first dorsal fin or less (fig. 54); second dorsal-fin spine origin behind tip of pelvic fin (fig. 54) ....
.... ***Centrophorus moluccensis*** **(fig. 54; p. 57)**

Second dorsal fin three-quarters of height of first dorsal fin or more (fig. 55); second dorsal-fin spine usually above inner margin of pelvic fin (fig. 56) **29**

29 Crowns of flank denticles with broadly rounded posterior margins that do not form cusps (fig. 51); preoral snout much shorter than width of head at mouth .... ***Centrophorus granulosus*** **(fig. 55; p. 55)**

Crowns of flank denticles angular, thorn-like with posterior margins forming a short cusp (sometimes visible only under magnification) (fig. 52); preoral snout equal to or longer than width of head at mouth **30**

30 Snout relatively long, slender, depth at front of mouth more than 1.9 in preoral distance; dorsal fins each with a dark anterior blotch (sometimes extending posteriorly as a submarginal bar and less obvious in adults) (fig. 56) ....
.... ***Centrophorus harrissoni*** **(fig. 56; p. 56)**

Snout relatively short, thick, depth at front of mouth less than 1.9 in preoral distance; dorsal fin apices and posterior margins dark (less obvious in adults (fig. 57) ....
.... ***Centrophorus uyato*** **(fig. 57; p. 60)**

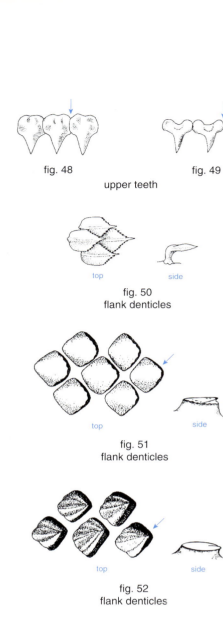

fig. 53

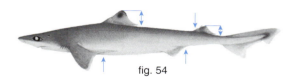

fig. 54

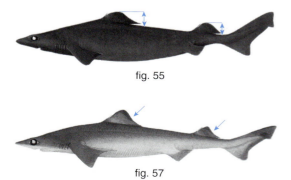

fig. 56 fig. 57

31 Cusps of teeth near middle of lower jaw rather upright (fig. 58) ....
.... ***Zameus squamulosus* (fig. 60; p. 102)**

Cusps of teeth near middle of lower jaw distinctly oblique (fig. 59) **32**

32 Upper labial furrows extremely long, separated by less than the distance between nostrils (fig. 62); preoral snout about equal in length to distance from mouth to pectoral-fin origins (fig. 61) ....
.... ***Centroscymnus crepidater* (fig. 61; p. 64)**

Upper labial furrows shorter, separated by more than the distance between nostrils (fig. 63); preoral snout much shorter than distance from mouth to pectoral-fin origins (fig. 64) **33**

33 Teeth near symphysis of upper jaw distinctly smaller than those in rows 4–8 either side (fig. 65); median ridge of flank denticles normally extending along entire length of crown (fig. 67); precaudal vertebrae more than 83 ....
.... ***Centroscymnus plunketi* (fig. 64; p. 66)**

Teeth near symphysis of upper jaw similar in size to those adjacent (fig. 66); median ridge of flank denticles extending for two-thirds or less of length of crown (fig. 68); precaudal vertebrae normally 82 or less **34**

34 Denticles above gill slits with tricuspidate and ridged crowns (fig. 70); mostly brownish black ....
.... ***Centroscymnus owstoni* (fig. 69; p. 65)**

Denticles above gill slits with smooth and leaf-shaped crowns (lacking ridges) (fig. 71); mostly golden brown ....
.... ***Centroscymnus coelolepis* (fig. 72; p. 62)**

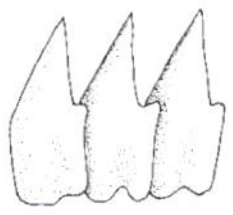

fig. 58
teeth near
mid-lower jaw

fig. 59
teeth near
mid-lower jaw

fig. 60

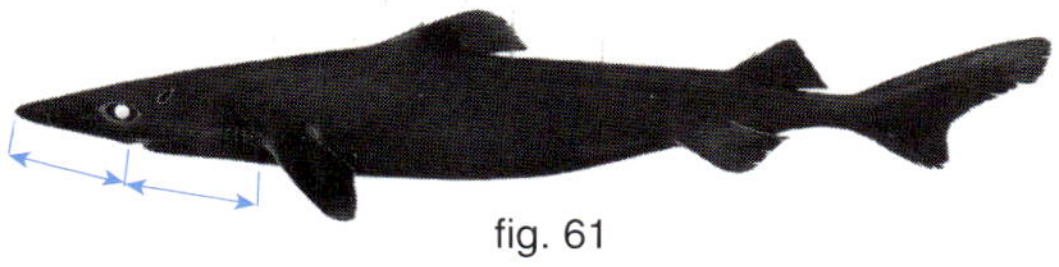

fig. 61

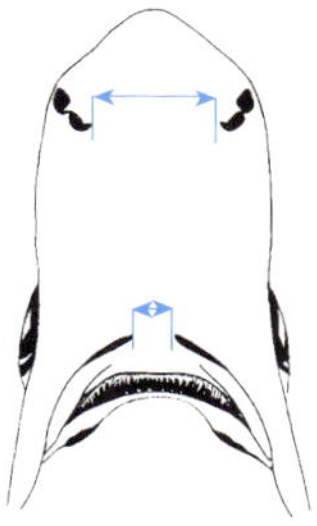

fig. 62 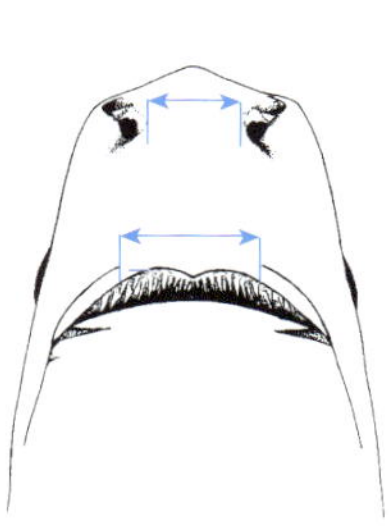   fig. 63
undersurface of head

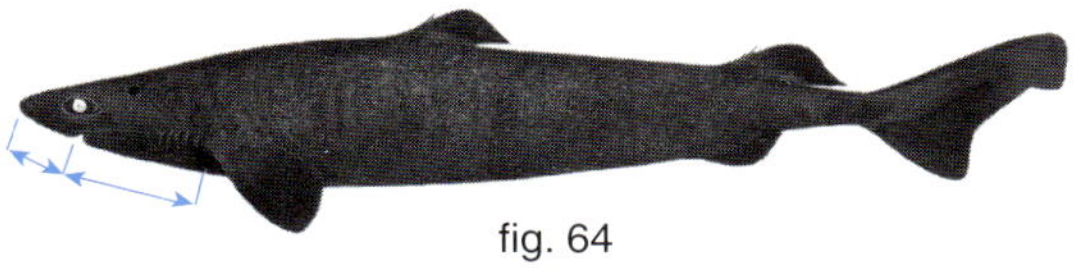

fig. 64

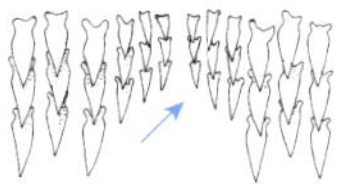

fig. 65
central teeth
of upper jaw

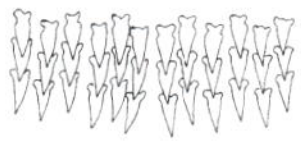

fig. 66
central teeth
of upper jaw

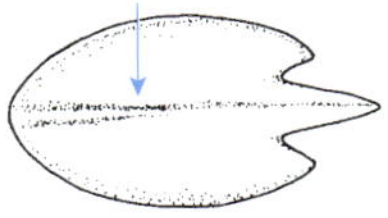

fig. 67
flank denticle

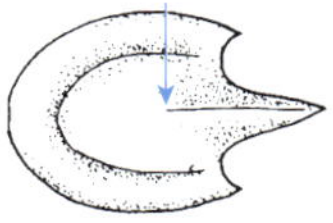

fig. 68
flank denticle

fig. 69

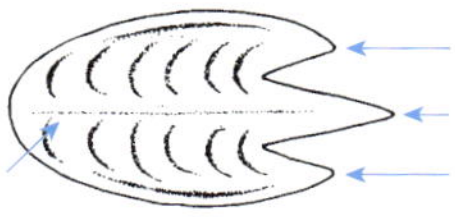

fig. 70
denticle from
above gill slit

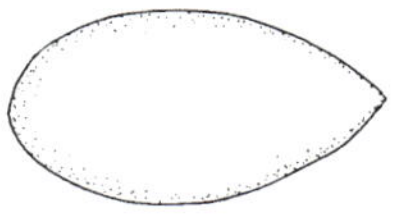

fig. 71
denticle from
above gill slit

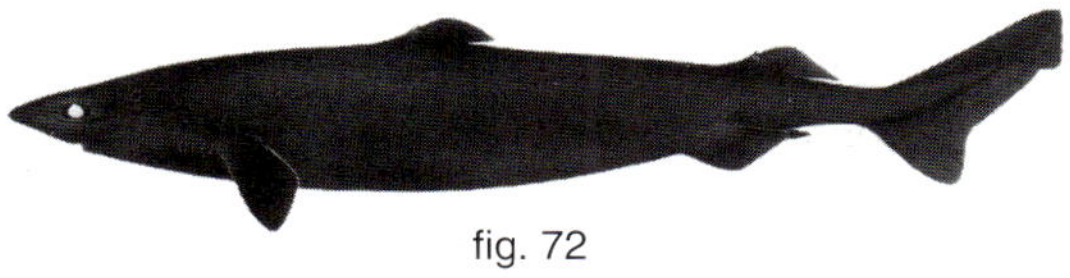

fig. 72

35 First dorsal fin over pelvic fin (fig. 73); distinct dark collar around head through gill region (fig. 73) .... *Isistius brasiliensis* (**fig. 73; p. 86**)

   First dorsal fin well in front of pelvic fin (fig. 74); no dark collar around head **36**

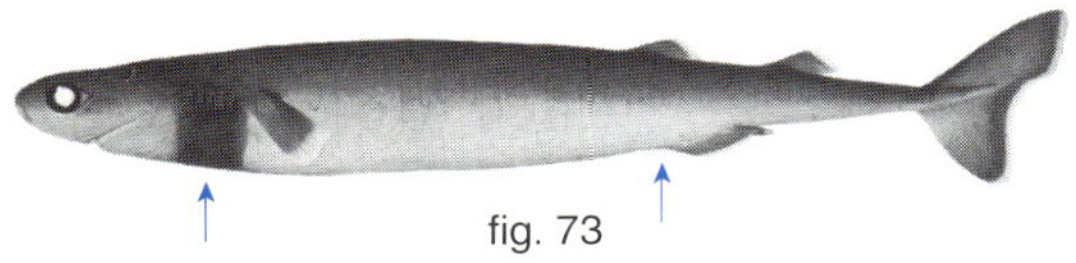

36 First dorsal-fin base less than half length of second dorsal-fin base (fig. 75); caudal-fin upper lobe short, barely longer than second dorsal-fin base (fig. 75) **37**

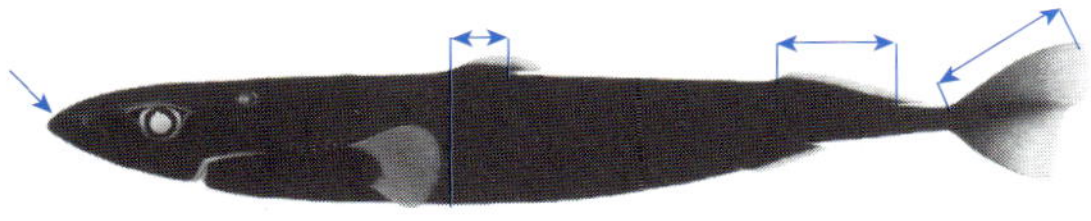

   First dorsal-fin base much greater than half length of second dorsal-fin base (fig. 76); caudal-fin upper lobe much longer than second dorsal-fin base (fig. 76) **38**

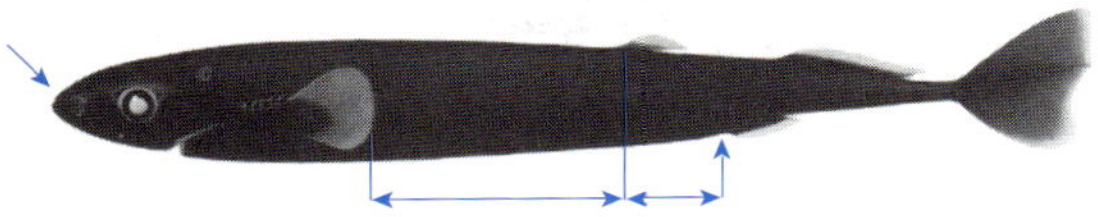

37 First dorsal-fin origin closer to pelvic-fin origin than to pectoral-fin free rear tip (fig. 74); snout bluntly rounded (fig. 74) .... *Euprotomicrus bispinatus* (**fig. 74; p. 85**)

   First dorsal-fin origin only slightly behind pectoral-fin free rear tip (fig. 75); snout elongate and pointed (fig. 75) .... *Squaliolus aliae* (**fig. 75; p. 90**)

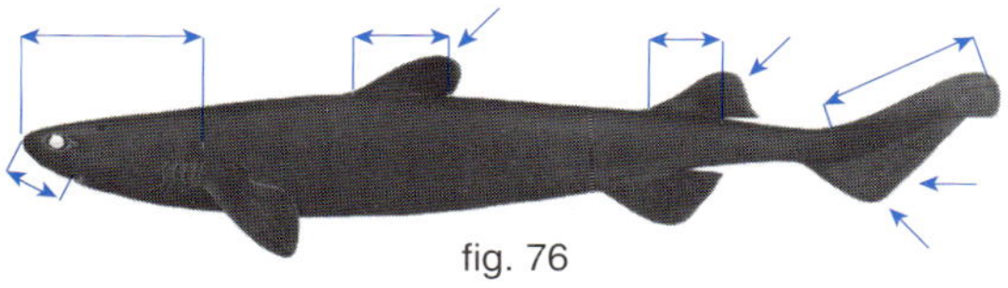

38 Preoral snout shorter than a third of head length (fig. 76); dorsal fins tall (fig. 76); lower lobe of caudal fin short, posterior margin of caudal fin straight (fig. 76) .... *Dalatias licha* (**fig. 76; p. 69**)

   Preoral snout longer than a third of head length (fig. 77); dorsal fins very low (fig. 77); lower lobe of caudal fin distinct, posterior margin of caudal fin concave (fig. 77) **39**

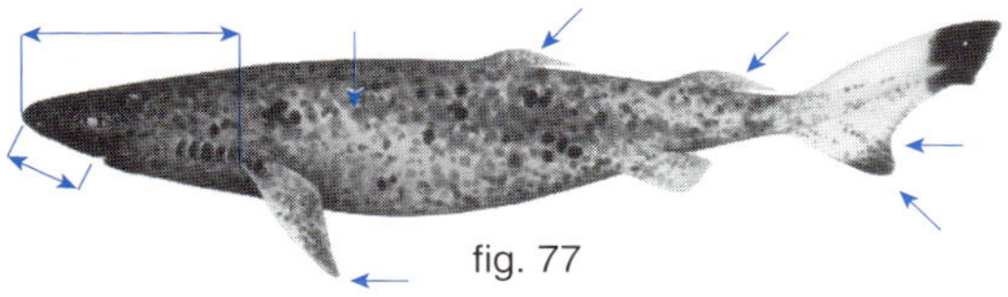

39 Pectoral fins slender, apex pointed (fig. 77); lower teeth with high, erect cusps; blotched black and white .... *Scymnodalatias albicauda* (**fig. 77; p. 88**)

   Pectoral fins broadly rounded (fig. 78); lower teeth with oblique cusps; uniform greyish pink .... *Somniosus pacificus* (**fig. 78; p. 89**)

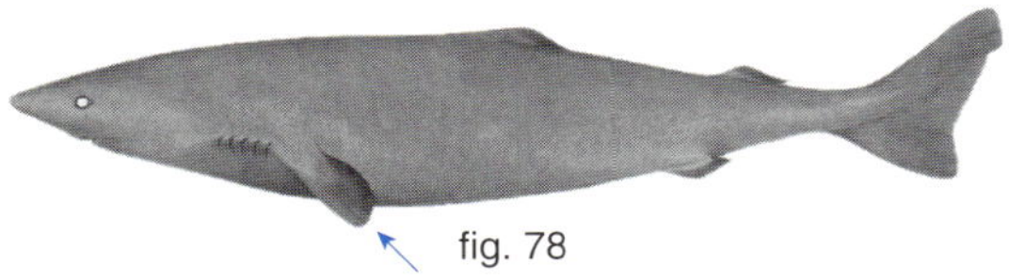

fig. 73

fig. 74

fig. 75

fig. 76

fig. 77

fig. 78

Gulper Shark — *Centrophorus granulosus* (Bloch & Schneider, 1801)

8.1 *Centrophorus granulosus* (Plate 4) Fish Code: 00 020023

- **Field characters:** A uniformly light greyish brown dogfish with a bulky snout, the second dorsal fin slightly smaller than the first dorsal fin, a rather deep ventral lobe to the caudal fin, and the free rear tips of the pectoral fins are extended into a distinct angular lobe. The preoral snout is shorter than the distance from the mouth to the pectoral-fin origins.

- **Distinctive features:** Body fusiform. Preoral snout rather short (less than distance from mouth to pectoral-fin origin) and fat (depth at front of mouth about 1.4 in distance from snout to mouth); labial furrows short, not extending far past mouth corners; eyes moderately large. Upper teeth erect to slightly oblique, triangular, relatively broad; bases occasionally overlapping. Lower teeth oblique, blade-like. Denticles of adults flat, almost circular, lacking cusps along their posterior margins, not overlapping. Fin spines short, robust. First dorsal-fin origin slightly behind pectoral-fin insertion. First dorsal fin relatively short and high (height about 2.7 in length measured from the exposed spine origin to the free rear tip); second dorsal fin about two-thirds to three-quarters height of first dorsal fin. Pectoral-fin free rear tip angular and moderately extended, reaching to about level of first dorsal-fin spine; lobe about half of second dorsal-fin height. Caudal-fin ventral lobe relatively deep. Tooth count [36–40/30–32].

- **Colour:** Dorsal surfaces light greyish brown, paler ventrally; eyes greenish. Juveniles with dark tips to the dorsal and caudal fins.

- **Size:** Born at about 35 cm and attains at least 160 cm.

- **Distribution:** Western North Atlantic (Gulf of Mexico), eastern Atlantic (Bay of Biscay to Zaire including the Mediterranean), Indian Ocean (Aldabra Islands) and the western Pacific from Japan, Papua New Guinea and tropical Australia (Queensland

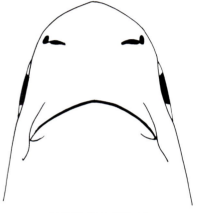

undersurface of head

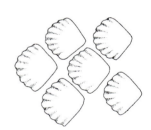

patch of flank denticles

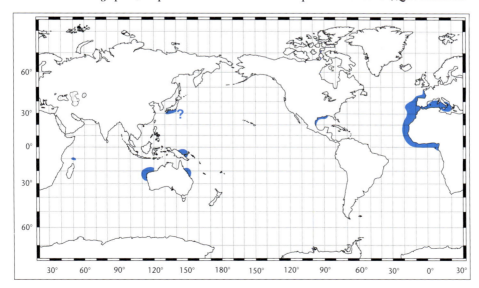

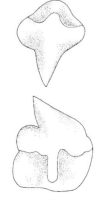

third teeth from symphysis
(upper and lower jaw)

and northern Western Australia). Demersal on continental shelves and slopes in depths from 100–1200 m.

- **Remarks:** Little is known of its biology other than that it is ovoviviparous and feeds mainly on bony fish. In the eastern Atlantic, it is smoked, dried and salted for human consumption and is used for fish meal and liver oil. The few known specimens from Australia appear to have a more posteriorly located dorsal fin than those from the Atlantic.

Harrisons Dogfish — *Centrophorus harrissoni* McCulloch, 1915

8.2 *Centrophorus harrissoni* (Plate 4)   **Fish Code:** 00 020010

- **Alternative names:** Dumb shark, dumb gulper shark, Harrisons deepsea dogfish.
- **Field characters:** A uniformly greyish dogfish with the free rear tips to the pectoral fins extended into a long angular lobe, a rather slender and depressed snout, the second dorsal fin slightly smaller than the first dorsal fin, and with a rather deep ventral lobe to the caudal fin. The preoral snout is mostly longer than the distance from the mouth to the pectoral-fin origins and about twice or more its depth at the front of the upper jaw.
- **Distinctive features:** Body fusiform. Preoral snout long (about equal to or greater than distance from mouth to pectoral-fin origin) and narrow (depth at front of mouth 2.0–2.3 in distance from snout to mouth); labial furrows short, not extending far past mouth corners; eyes moderately large. Upper teeth erect to slightly oblique, relatively broad and triangular; bases occasionally overlapping. Lower teeth oblique, blade-like. Denticles of adults flat, not overlapping. Fin spines short, robust. First dorsal-fin origin well behind pectoral-fin insertion. First dorsal fin relatively short and high (height 1.8–2.1 in length measured from the exposed spine origin to the free rear tip); second dorsal fin about two-thirds to three-quarters height of first dorsal fin. Pectoral-fin free rear tip angular and moderately lobe-like, reaching well beyond level of first dorsal-fin spine; lobe 0.8–1.2 times second dorsal-fin height. Caudal-fin lower lobe relatively deep. Total vertebrae 119–124*; precaudal 85–90*.
- **Colour:** Dorsal surfaces light greyish, paler ventrally. Each dorsal fin with a dark oblique blotch extending from the mid anterior margin to the insertion (most intense anteriorly and often faint in adults); a diffuse darker area on the terminal lobe of the caudal-fin; eye greenish.
- **Size:** Born at about 32 cm and attains 110 cm. Males mature at about 85 cm.
- **Distribution:** Possibly restricted to Australia where it has been recorded from Clarence River (New South Wales) to Maria Island (Tasmania) and from the Abrolhos Islands to Bunbury (Western Australia). Demersal on the continental slope in depths of 220–790 m.

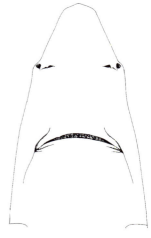

undersurface of head

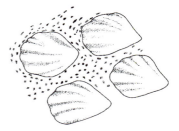

patch of flank denticles

- **Remarks:** Presumably ovoviviparous, but little is known of its biology. Taken by demersal trawlers but currently of little commercial interest. Eastern and western populations require closer comparison.
- **References:** McCulloch (1915); May and Maxwell (1986).

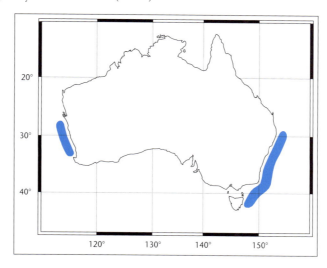

Endeavour Dogfish — *Centrophorus moluccensis* Bleeker, 1860

8.3 *Centrophorus moluccensis* (Plate 4) **Fish Code:** 00 020001

- **Alternative name:** Smallfin gulper shark.
- **Field characters:** A uniformly light greyish or brownish dogfish with very long and angular free rear tips to the pectoral fins, the second dorsal fin much smaller than the first dorsal fin, and with a rather deep ventral lobe to the caudal fin.
- **Distinctive features:** Body fusiform. Preoral snout relatively short (about equal to or shorter than distance from mouth to pectoral-fin origin; labial furrows short, not extending far past mouth corners; eyes relatively large. Upper teeth erect to slightly oblique, relatively broad and triangular; bases occasionally overlapping. Lower teeth oblique, blade-like. Denticles of adults flat, not overlapping. Fin spines short, robust. First dorsal-fin origin behind pectoral-fin insertion. First dorsal fin relatively short and high (height 1.6–2.1 in length measured from the exposed spine origin to the free rear tip); second dorsal fin about half height of first dorsal fin. Pectoral-fin free rear tips greatly elongated, reaching beyond level of first dorsal-fin spine; lobe 1.5–3.0 times second dorsal-fin height. Caudal fin ventral lobe moderately deep. Tooth count [36–45/31–35]. Total vertebrae [113–127]; precaudal 88* [84–90].

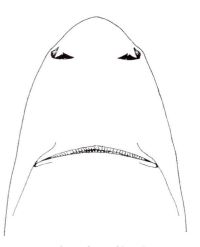

undersurface of head

- **Colour:** Dorsal surfaces light greyish brown, paler ventrally. Caudal fin (and sometimes pectoral and pelvic fins) with a narrow, pale posterior margin; free rear tips of pectoral fins and base of first dorsal-fin spine pale. Juveniles with a dark blotch near the apices of the first dorsal fin and the caudal-fin terminal lobe (markings paler in adults); leading edge to the second dorsal fin and area around the gill slits dark; eye greenish.

- **Size:** Born at about 35 cm and attains 100 cm. Males mature at about 70 cm.
- **Distribution:** Western Indian Ocean off southern Africa and India, and some areas of the Western Pacific including the Philippines, Indonesia, Japan and Australia (Queensland, New South Wales, Victoria and Western Australia). Reports from Tasmania and the Great Australian Bight require validation. Demersal on outer continental and insular shelves and upper slopes at depths from 125–820 m. Locally most abundant in 300–500 m.
- **Remarks:** Ovoviviparous, mostly with litters of 2 pups. It eats mainly bony fish and cephalopods, but elasmobranchs and crustaceans are also taken. Caught by demersal trawlers in Australia but currently of little commercial interest. Rather rare off southern Australia.
- **Local synonymy:** *Centrophorus scalpratus* McCulloch, 1915.
- **References:** McCulloch (1915); Bass *et al.* (1976).

patch of flank denticles

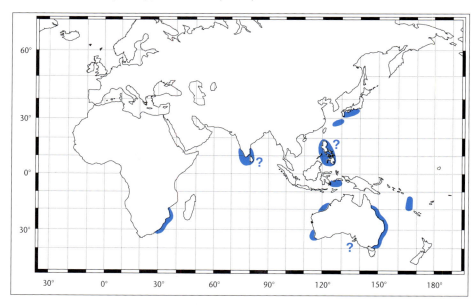

Leafscale Gulper Shark — *Centrophorus squamosus* (Bonnaterre, 1788)

8.4 *Centrophorus squamosus* (Plate 4) Fish Code: 00 020009

- **Alternative names:** Nilsons deepsea dogfish, deepwater spiny dogfish.
- **Field characters:** A uniformly greyish or brownish dogfish with imbricate, leaf-like denticles, short and angular free rear tips to the pectoral fins, the second dorsal fin

shorter and more triangular in shape than the first dorsal fin, and with a very deep ventral lobe to the caudal fin.

- **Distinctive features:** Body fusiform. Preoral snout short (about equal to distance from mouth to second gill slit); labial furrows short, not extending far past mouth corners; eyes large. Upper teeth erect to slightly oblique, triangular and relatively broad; bases occasionally overlapping. Lower teeth oblique, blade-like. Denticles of adults raised on pedicles, leaf-shaped, imbricate, ridged and with a serrated posterior margin; more bristle-like and tridentate in juveniles. Fin spines short, robust. First dorsal-fin origin varying from over inner margin to just behind free rear tip of pectoral fin; extending anteriorly as a low ridge. First dorsal fin relatively long and low (height 3.3–3.5 in length measured from the exposed spine origin to the free rear tip); second dorsal fin shorter-based, more triangular in shape; height 2.0–2.3 in length. Pectoral-fin free rear tip quadrate, occasionally very slightly extended. Caudal-fin ventral lobe deep. Tooth count 32–35/24–29* [30–38/27–32]. Total vertebrae 114* [106–107]*; precaudal 86* [82–83]*

- **Colour:** Uniformly light greyish brown to dark grey; eyes greenish.

- **Size:** Born at 35–40 cm and attains 160 cm. Males mature at about 100 cm and females at about 110 cm.

- **Distribution:** Eastern Atlantic (Iceland to southern Africa), Indian Ocean (South Africa and the Aldabra Group) and the western Pacific (Japan, Philippines, New Zealand and Australia). Locally from Tasmania, Victoria and New South Wales in 870–920 m. Elsewhere on or near the bottom of the continental and insular slopes in 230–2400 m. Also caught in the upper 1250 m of oceanic water in depths to 4000 m.

- **Remarks:** Probably more wide-spread in Australia than records suggest. Ovoviviparous with litters of 5–8 pups. Nothing is known about the diet of this shark, but it probably consists of fish and cephalopods. Currently of little commercial interest in Australia, but elsewhere the meat is dried and salted for human consumption and is used also for fishmeal.

- **Local synonymy:** *Centrophorus foliaceus* Günther, 1877; *Centrophorus nilsoni* Thompson, 1930.

- **References:** Garrick (1959a); Bass *et al.* (1976).

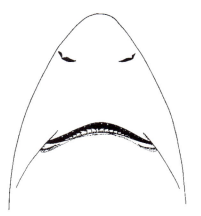

undersurface of head

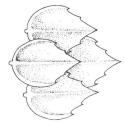

patch of flank denticles

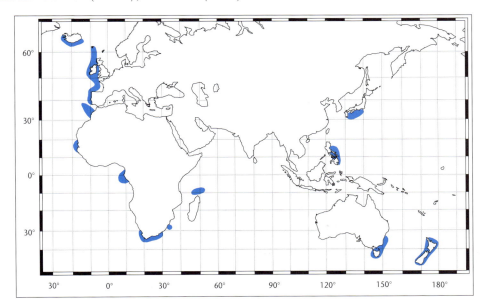

Southern Dogfish — *Centrophorus uyato* (Rafinesque, 1810)

8.5 *Centrophorus uyato* (Plate 4)　　　　　　　　　　**Fish Code:** 00 020011

- **Alternative name:** Little gulper shark.

- **Field characters:** A uniformly light greyish brown dogfish with a rather bulky snout, the second dorsal fin slightly smaller than the first dorsal fin, a moderately deep ventral lobe to the caudal fin, and the free rear tips of the pectoral fins extended into a distinct angular lobe. The preoral snout is mostly shorter than the distance from the mouth to the pectoral-fin origins and less than twice its depth at the front of the upper jaw.

- **Distinctive features:** Body fusiform. Preoral snout short (less than or equal to distance from mouth to pectoral-fin origin) and fat (depth at front of mouth 1.5–1.8 in distance from snout to mouth); labial furrows short, not extending far past mouth corners; eyes moderately large. Upper teeth erect to slightly oblique, relatively broad and triangular; bases occasionally overlapping. Lower teeth oblique, blade-like. Denticles of adults flat with very short cusps on their posterior edges, not overlapping. Fin spines short, robust. First dorsal-fin origin behind pectoral-fin insertion. First dorsal fin relatively short and high (height 1.7–2.3 in length measured from the exposed spine origin to the free rear tip); second dorsal fin about two-thirds to three-quarters height of first dorsal fin. Pectoral-fin free rear tip angular and moderately extended, reaching beyond level of first dorsal-fin spine; lobe 0.7–1.3 times second dorsal-fin height. Caudal-fin ventral lobe relatively deep. Tooth count [33–39/29–35]. Total vertebrae 116–126 [113–122]; precaudal 83–92 [82–90].

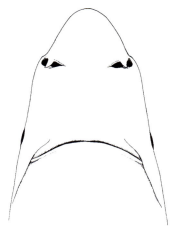

undersurface of head

- **Colour:** Dorsal surfaces light greyish brown (with a bluish sheen in life), paler ventrally. Juveniles with apices and most of posterior margins of dorsal fins dark; outer half of dorsal caudal-fin margin dark with a dark blotch above the subterminal notch; outer margin white with a dark submarginal border; eyes greenish. Dark markings fading in adults.

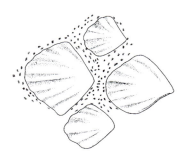

patch of flank denticles

- **Size:** Born at about 35 cm and attains 100 cm. Size at maturity unknown, although males are mature by 80 cm.

- **Distribution:** Western North Atlantic (Gulf of Mexico), eastern Atlantic (Portugal to Namibia including the Mediterranean), Indian Ocean (southern Mozambique and possibly India), and the western Pacific (Taiwan). Temperate Australia from Albany to Geraldton (Western Australia) and from Fowlers Bay (South Australia) to Port Stephens (New South Wales), including Tasmania. Demersal on continental shelves and slopes in depths of 50–1400 m; locally mainly from 400 to 650 m.

- **Remarks:** Ovoviviparous, usually producing one pup. The diet consists mainly of bony fish and cephalopods. In Australia, common throughout its range but currently of little commercial interest; dried and salted for human consumption in the eastern Atlantic. The taxonomy of this genus is unstable and some recent authors have suggested that the type specimen of *Centrophorus uyato* may belong to the genus *Squalus*.

- **References:** Rafinesque (1810a); Bass *et al.* (1976).

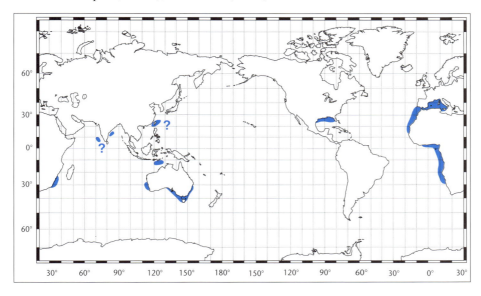

Bareskin Dogfish — *Centroscyllium kamoharai* Abe, 1966

8.6 *Centroscyllium kamoharai* (Plate 9)   **Fish Code:** 00 020024

- **Field characters:** A black dogfish with similar teeth in both jaws, each with a slender cusp (not blade-like) and prominent cusplets, and relatively long dorsal-fin spines. The skin is smooth and deciduous with few denticles (trawl-caught specimens appear ragged where skin is removed).

- **Distinctive features:** Body stout and soft. Preoral snout relatively long, about three-fifths of distance from mouth to pectoral-fin origins. Mouth very short and broadly arched; teeth small, similar in both jaws, cusps slender and with prominent cusplets. First dorsal-fin origin posterior to free rear tips of adpressed pectoral-fin. Second dorsal fin slightly larger than first dorsal fin; second dorsal-fin spine considerably longer than first dorsal-fin spine. Caudal peduncle short, distance from second dorsal-fin insertion to upper caudal-fin origin about equal to distance from eye to third gill slit. Flank denticles sparse; skin and denticles easily removed, mostly ragged.

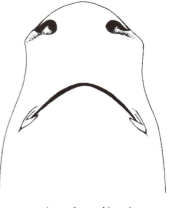

undersurface of head

- **Colour:** Uniformly black when undamaged; white where skin is chafed.

- **Size:** Attains at least 60 cm.

- **Distribution:** Indo–West Pacific from Japan, and eastern (Port Macquarie, New South Wales, to southern Tasmania) and Western Australia (North West Cape to Bunbury). Possibly throughout the Great Australian Bight. Demersal on the continental slope at depths of 730–1200 m but mostly deeper than 900 m.
- **Remarks:** This rather smooth, soft-bodied dogfish is easily damaged when caught in trawl nets and most specimens held by Australian museums are in poor condition. Little is known of its biology; presumably ovoviviparous.

third teeth from symphysis
(upper and lower jaw)

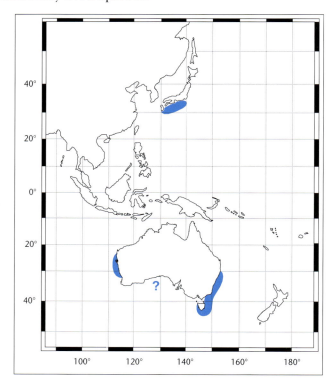

Portuguese Dogfish — *Centroscymnus coelolepis* (Bocage & Capello, 1864)

8.7 *Centroscymnus coelolepis* (Plate 10)   **Fish Code:** 00 020025

- **Field characters:** A uniformly golden to dark brown dogfish with dorsal fins of a similar size and shape, short dorsal-fin spines and very large, smooth denticles resembling a snake's skin. Denticles from above the gill slits are typically smooth, leaf-shaped and lack ridges.
- **Distinctive features:** Body stout. Preoral snout about equal to distance from mouth to first gill slit; upper labial furrows short, their length much less than distance between their anterior ends. Upper teeth dagger-like, erect towards the symphysis,

more oblique laterally; those in first 8 rows usually similar in size; lower teeth oblique and blade-like. Denticles from above gill slits leaf-shaped, flat but with a median circular depression, their posterior margins smooth or tridentate; 21–26 denticle rows between lateral line and pelvic-fin insertion in males; about 30–32 in females. Fin spines small and slender, only tips protruding through skin. First dorsal-fin spine origin slightly behind free rear tips of adpressed pectoral-fins. First dorsal fin low and lobe-like, height 2.3–2.8% of total length, origin distinct; second dorsal-fin shape similar to first dorsal fin but slightly taller (height 2.8–3.7% of total length). Pectoral fins broad, maximum width 1.6–2.1 in anterior margin. Tooth count [43–68/29–42]. Total vertebrae 105–110* [102–114]; precaudal 79–82* [68–84].

- **Colour:** Uniformly golden brown to dark brown.
- **Size:** Born at 30 cm and attains 120 cm. Males mature at about 85 cm and females at 100 cm.
- **Distribution:** Eastern Atlantic (Iceland to South Africa), western North Atlantic (Grand Banks to Delaware Bay), and the Pacific off Japan, New Zealand and Australia (from Cape Hawke (New South Wales) to Beachport (South Australia), including Tasmania). On or near the bottom of the continental slope and abyssal plain in depths from 270–3700 m; locally in 770–1400 m.
- **Remarks:** The Portuguese dogfish is found in water temperatures of 5–13°C. Little is known of its biology other than that it is ovoviviparous with 13–17 young per litter and its diet consists mainly of fish (including sharks) and cephalopods; some stomachs examined contained benthic invertebrates and cetacean pieces (presumably taken as carrion). Off Japan, females have been caught deeper than males. It is taken as bycatch in the Australian orange roughy fishery but not utilised (the flesh is high in mercury). Elsewhere it is used for its liver oil (which is high in squalene), for fishmeal, or dried and salted for human consumption.

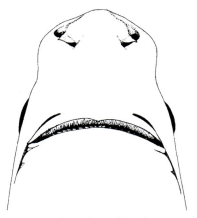
undersurface of head

denticle above gill slits

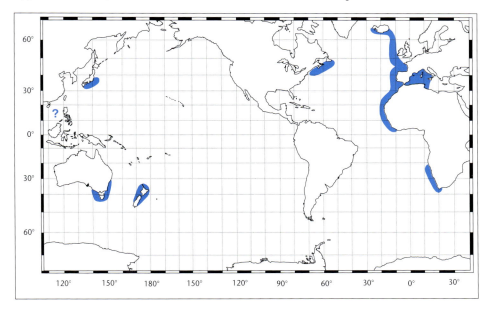

third teeth from symphysis (upper and lower jaw)

Golden Dogfish — *Centroscymnus crepidater* (Bocage & Capello, 1864)

8.8 *Centroscymnus crepidater* (Plate 10) Fish Code: 00 020012

- **Alternative names:** Deepwater dogfish, longnose velvet dogfish.
- **Field characters:** A uniformly dark brown to black dogfish with a rather long snout, small dorsal-fin spines, and distinctive upper labial furrows that are very long and almost meet at their anterior ends.
- **Distinctive features:** Body slender. Preoral snout relatively long, equal to or slightly greater than distance from mouth to pectoral-fin origin; upper labial furrows long, nearly meeting at anterior ends. Upper teeth slender and dagger-like, erect towards symphysis, more oblique laterally, first 8 rows usually similar in size; lower teeth blade-like, erect near symphysis, becoming oblique laterally. Denticles from above gill slits distinctly tridentate, with 3 ridges (median ridge shortest, confined to posterior half or two-thirds of crown behind an anterior depression); flank denticles similar but sometimes with 5–7 ridges in adult specimens. Fin spines small and slender, only tips protruding through skin. First dorsal-fin spine origin about over free rear tips of adpressed pectoral fins. Dorsal fins relatively large, about equal in size and height (first and second dorsal-fin heights 3.9–5.5% and 4.2–5.9% of total length, respectively). Pectoral fins paddle-shaped, maximum width 1.6–2.4 in anterior margin. Tooth count 36–38/32–33* [38–51/30–36]. Total vertebrae 105–119 [106]*; precaudal 73–85 [77]*.
- **Colour:** Uniformly dark brown to black.
- **Size:** Born at 30–35 cm and attains 105 cm. Males mature at about 60 cm and females at about 80 cm.
- **Distribution:** Eastern Atlantic (Iceland to southern Africa), Indian Ocean (Aldabra Islands and India), eastern Pacific (northern Chile) and the western Pacific, from New Zealand and southern Australia, on or near the bottom of continental and

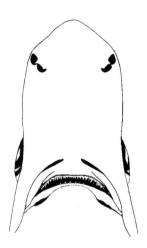

undersurface of head

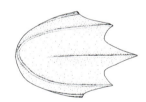

denticle from above gill slits

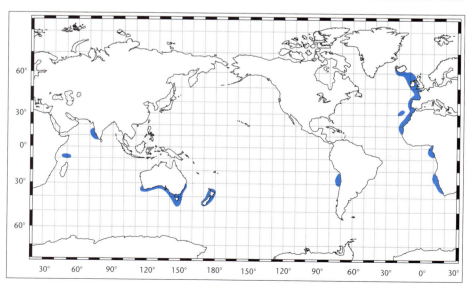

insular slopes in depths of 270–1300 m. Locally from Sydney (New South Wales) to Perth (Western Australia), including Tasmania and the southern seamounts, and most common in 780–1100 m.

- **Remarks:** A fairly common but little studied species. Ovoviviparous, with 4–8 pups in a litter; females appear to breed throughout the year. Diet consists mainly of fish and cephalopods. Taken as a bycatch in the Australian orange roughy fishery but not normally utilised; the flesh is high in mercury. Elsewhere used for fishmeal and for its liver oil (which is high in squalene).

- **References:** Garrick (1959b); Bass (1979); Yano and Tanaka (1984a); Davenport and Deprez (1989).

Owstons Dogfish — *Centroscymnus owstoni* (Garman, 1906)

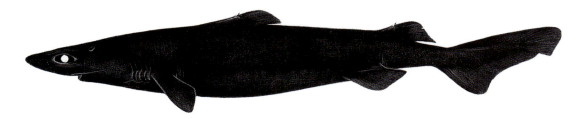

8.9 *Centroscymnus owstoni* (Plate 10) Fish Code: 00 020019

- **Alternative names:** Roughskin dogfish, deepwater dogfish, Owstons spiny dogfish.

- **Field characters:** A uniformly dark brownish or black dogfish with rather large, flat denticles giving it a smooth-skinned appearance (denticles from above the gill slits tridentate and ridged). It has short dorsal-fin spines, a low and lobe-like first dorsal fin, and the second dorsal fin is taller and more triangular in shape than the first dorsal fin.

- **Distinctive features:** Body stout. Preoral snout about equal to distance from mouth to second gill slit; upper labial furrows short, their length much less than distance between anterior ends. Upper teeth dagger-like, erect towards the symphysis, more oblique laterally, first 8 rows usually similar in size; lower teeth oblique and blade-like. Denticles from above gill slits tridentate and ridged (median ridge shortest, confined to posterior half or two-thirds of crown behind an anterior depression); flank denticles of juveniles similar, but flat, smooth and leaf-shaped in adults. About 25–36 denticle rows between lateral line and pelvic-fin insertion. Fin spines small and slender, only tips protruding through skin. First dorsal-fin spine origin well behind free rear tips of adpressed pectoral fins. Dorsal fins long, low anteriorly, bases indistinct. First dorsal fin low and lobe-like, height 2.6–3.2% of total length; second dorsal fin taller than first dorsal fin and more triangular in shape (height 3.8–4.8% of total length). Pectoral fins paddle-shaped, maximum width 2.0–2.5 in anterior margin. Tooth count 37–39/35–37* [36–37/32–40]. Total vertebrae 101–114 [96–108]; precaudal 74–82 [71–79].

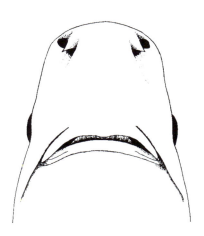

undersurface of head

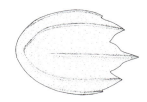

denticle from above gill slits

- **Colour:** Uniformly dark brown to black (sometimes light greyish brown).

- **Size:** Born at about 30 cm and attains 120 cm. Males mature at about 70 cm and females at 100 cm.

- **Distribution:** Western North Atlantic (northern Gulf of Mexico) and western Pacific from southern Japan, New Zealand and southern Australia: from Cape Hawke (New South Wales) to Shark Bay (Western Australia), including Tasmania and the southern seamounts. Demersal on the upper continental slope at depths of 500–1400 m but most common locally in 730–900 m.
- **Remarks:** Reproduction is ovoviviparous and females contain as many as 34 eggs, each about 6.5 cm in diameter and weighing up to 100 g (the ovary may be about a quarter of the body weight). The diet appears to consist of fish and cephalopods. Off Japan, females are caught at greater depths than males. Locally, Owstons dogfish is common in the bycatch of trawlers fishing for orange roughy (*Hoplostethus atlanticus*) or occasionally by Japanese tuna longliners from midwater. The flesh is high in mercury and is not utilised in Australia. Elsewhere small quantities are used for fishmeal and liver oil, which is high in squalene.
- **References:** Garrick (1959b); Bass (1979); Yano and Tanaka (1984a); Davenport and Deprez (1989).

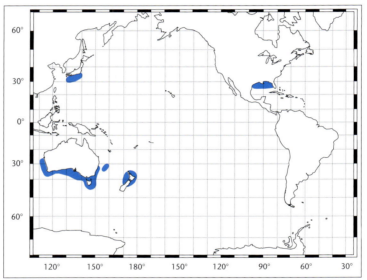

Plunkets Dogfish — *Centroscymnus plunketi* (Waite, 1910)

8.10 *Centroscymnus plunketi* (Plate 10)   Fish Code: 00 020013

- **Alternative names:** Lord Plunkets shark, Plunkets shark, Waites deepsea dogfish.
- **Field characters:** A uniformly greyish brown dogfish with the first dorsal fin slightly smaller than the second dorsal fin, very broad (almost round) pectoral fins and with

upper jaw teeth near the symphysis noticeably smaller than in the fourth to eighth rows on either side of the symphysis.

- **Distinctive features:** Body stout. Preoral snout short, about equal to or slightly less than distance from mouth to first gill slit; upper labial furrows short, their length much less than distance between anterior ends. Upper teeth dagger-like, erect towards the symphysis, more oblique laterally; teeth near the symphysis usually noticeably smaller than in fourth to eighth rows on each side of symphysis; lower teeth oblique and blade-like. Denticles from above gill slits and on flank similar, somewhat deciduous, tridentate and ridged (median ridge extending along entire length of crown). Fin spines small but robust, only tips protruding through skin. First dorsal-fin spine origin over or slightly behind free rear tips of appressed pectoral-fins. Dorsal fins long, low anteriorly; bases indistinct. First dorsal fin slightly lower than second dorsal fin (heights 3.0–3.6% and 3.9–4.7% of total length, respectively). Pectoral fins very broad, maximum width 1.7–1.8 in anterior margin. Tooth count 48/32–35*. Total vertebrae 114–115*; precaudal 84–85*.

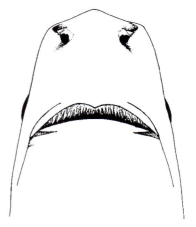
undersurface of head

denticle above gill slits

- **Colour:** Uniformly greyish brown, juveniles more greyish.
- **Size:** Born at 32–36 cm and attains 170 cm. Males mature at about 110 cm and females at about 130 cm.
- **Distribution:** Temperate western South Pacific including New Zealand and Australia. Recorded locally from Portland (Victoria) to Port Macquarie (New South Wales), including Tasmania and nearby seamounts. Records from Western Australia have yet to be validated. Demersal on the continental and insular slopes in depths of 240–1550 m.
- **Remarks:** This species, placed in the genus *Scymnodon* by some authors, is common but little studied. It may occur in large schools segregated by size and sex. Ovoviviparous, with large litters of up to 36 young. Diet consists mainly of fish and cephalopods. Not utilised in Australia; elsewhere used for fishmeal and liver oil.
- **Local synonymy:** *Centrophorus waitei* Thompson, 1930.
- **References:** Bigelow and Schroeder (1957); Garrick (1959c).

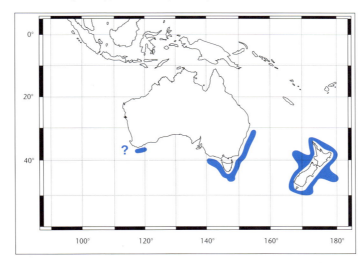

Mandarin Shark — *Cirrhigaleus barbifer* Tanaka, 1912

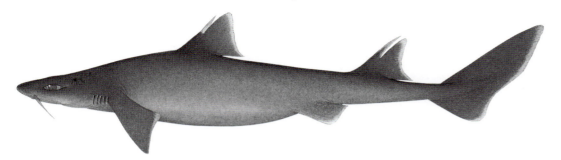

8.11 *Cirrhigaleus barbifer* (Plate 6) **Fish Code:** 00 020026

- **Field characters:** A highly distinctive, stout-bodied dogfish, resembling a spurdog in shape, that has long dorsal-fin spines and greatly elongated, barbel-like nasal lobes.
- **Distinctive features:** Body stout, lateral keels on caudal peduncle. Snout short and bluntly pointed; very long nasal barbels reaching to level of mouth; teeth similar in both jaws, oblique and blade-like. First dorsal-fin origin just behind free rear tips of pectoral fins. Dorsal fins large, about equal in size and with stout spines. Subterminal notch weak or absent. Tooth count 23–27/23–24* [26/22–26]*. Total vertebrae 114–115*; precaudal 85–87*.
- **Colour:** Dorsal surfaces greyish brown, pale ventrally. Posterior fin margins, and insertions of pectoral and pelvic fins, white.
- **Size:** Attains 126 cm; males mature at about 85 cm and females at about 110 cm.
- **Distribution:** Western Pacific from Japan, Torres Island, New Zealand and Australia (New South Wales). Demersal on the outer continental shelves and upper slopes at depths of 146–640 m.
- **Remarks:** Ovoviviparous with about 10 embryos. The long nasal barbels are presumably used to detect prey. Not utilised commercially, although the liver is relatively high in squalene oil.
- **Reference:** Garrick and Paul (1971b).

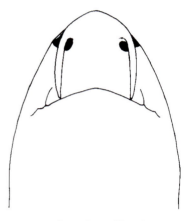

undersurface of head

third teeth from symphysis
(upper and lower jaw)

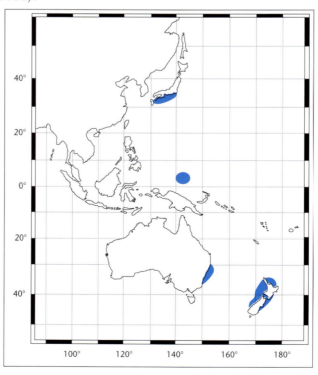

Black Shark — *Dalatias licha* (Bonnaterre, 1788)

8.12 *Dalatias licha* (Plate 10)　　　　　　　　　　　　　Fish Code: 00 020002

- **Alternative names:** Kitefin shark, seal shark.
- **Field characters:** A uniformly dark dogfish with no fin spines, a short snout, and with thick, fleshy lips. The teeth are dissimilar between jaws and those of the lower jaw are large, triangular and serrated.
- **Distinctive features:** Body fusiform. Preoral snout short and conical, its length considerably less than distance from mouth to first gill slit; labial furrows short, hidden when mouth is closed by thick, fleshy lips. Eye large, 2.2–4.3% of total length. Upper teeth erect to oblique, slender, awl-shaped and smooth-edged; lowers large, triangular and serrated. Flank denticles loosely spaced, ridged, with a thorn-like crown. First dorsal-fin origin just behind free rear tips of adpressed pectoral fins; second dorsal-fin origin forward of pelvic-fin insertion. First dorsal fin lobe-like; height 3.8–5.5% of total length. Second dorsal fin slightly larger and more triangular in shape than first dorsal fin, height 4.4–6.5% of total length. Pectoral fins paddle-shaped, maximum width about half their length. Tooth count 18–19/19 [16–21/17–20]. Total vertebrae 69–82 [75–85]; precaudal 48–50 [47–55].
- **Colour:** Mostly uniformly black, sometimes light grey or brown; lips pale; posterior margins of most fins translucent.
- **Size:** Born at about 30 cm and attains 160 cm. Males mature at about 100 cm; females at about 120 cm.
- **Distribution:** Eastern Atlantic (Scotland to Cameroon), western Atlantic (Georges Bank and northern Gulf of Mexico), western Indian Ocean (southern Africa), and western and central Pacific from Hawaii, Japan and New Zealand. In Australia, from Swain Reefs (Queensland) to Port Hedland (Western Australia) including Tasmania and adjacent seamounts. Mainly demersal (sometimes pelagic) on the outer continental and insular shelves and slopes from 40–1800 m but mainly 450–850 m locally.

undersurface of head

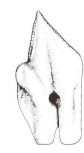

third teeth from symphysis
(upper and lower jaw)

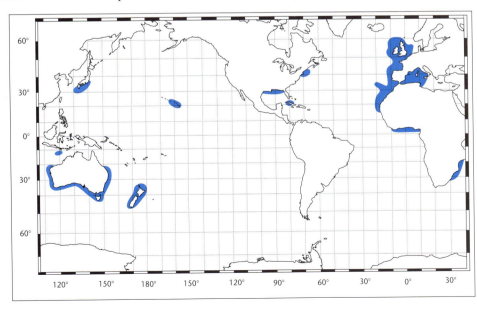

- **Remarks:** This dogfish, which is rarely caught shallower than 200 m, may be solitary or occur in small schools. Its oily liver makes it neutrally buoyant and, like several other dogfishes, it can move slowly above the bottom without using much energy. Ovoviviparous with litters of 10–16 young. The diet consists mainly of bony fish but also includes elasmobranchs, cephalopods and crustaceans. Often chunks of flesh taken from large fish (including fast-swimming species) are found in the stomachs, suggesting that it may also feed as a 'cookie-cutter' (see 8.27). Several large adult specimens caught recently in a trawl net off Western Australia had severely cannibalised each other and totally shredded a large ray through this feeding strategy. Not of commercial value in Australia; elsewhere used for its squalene liver oil, leather and meat, as well as for fishmeal.
- **Local synonymy:** *Scymnorhinus phillippsi* Whitley, 1931.
- **References:** Whitley (1931a); Garrick (1960b); Bass *et al.* (1976).

Brier Shark — *Deania calcea* (Lowe, 1839)

8.13 *Deania calcea* (Plate 3)   **Fish Code:** 00 020003

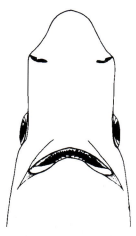

undersurface of head

- **Alternative names:** Dorian Grey, shovelnose spiny dogfish, Thompsons shark, Thompsons deepsea dogfish, birdbeak dogfish.
- **Field characters:** A dogfish with a very long, spatulate snout, deciduous skin, and an extremely long, low first dorsal fin that is different in size and shape to the second dorsal fin. Juveniles have a dark posterior border to each dorsal fin.
- **Distinctive features:** Body slender; trunk width of adults usually more than 1.9 in trunk depth at level of second dorsal-fin spine origin. Snout extremely long and flattened, preoral length 12.6–16.3% of total length. Eye large, its diameter 4.1–6.2% of total length. Denticles fairly small, pitchfork-shaped and mostly tridentate; cusps directed posteriorly (some individuals, particularly larger ones, have a vertically directed spine preceding the central cusp, or all cusps). Upper teeth with single, erect, awl-shaped cusps; lower teeth with single, smooth-edged, blade-like cusps that are oblique and laterally notched (males may have more erect cusps). Dorsal-fin spines grooved, second spine usually longer than first. First dorsal-fin origin over pectoral-fin base; distance from first dorsal-fin spine origin to free rear tip of fin usually greater than distance from its free rear tip to second dorsal-fin spine origin; second dorsal-fin insertion over lower caudal-fin origin. First dorsal fin long and low, height 3.5–5.8 in length (from junction with spine to free rear tip); second dorsal fin dissimilar in size and shape, shorter and taller than first dorsal fin. Tooth count 25–35/27–33 [28–31/27–28]*. Total vertebrae 121–127 [118–126]; precaudal 87–95 [85–92].
- **Colour:** Adults varying from uniform light or dark grey to dark brown. Juveniles have darker patches above the eye and gill regions, darker caudal-fin lobes and broad, uniformly darker posterior margins to the dorsal fins.

- **Size:** Born at about 30 cm and attains 113 cm. Both sexes mature at about 70 cm.
- **Distribution:** Eastern Atlantic (Iceland to southern Africa), eastern Pacific (Chile) and Western Pacific (Japan, temperate Australia and New Zealand). Locally between Coffs Harbour (New South Wales) and Green Head (southern Western Australia), including Tasmania. A deepwater demersal species of the continental and insular slopes and outer shelves, usually in depths between 400 and 900 m but recorded in 70–1450 m.
- **Remarks:** Ovoviviparous, with pregnant females rarely caught in Australian waters but elsewhere having litters of 6–12. Diet consists of fish and crustaceans. Caught overseas by line and trawl for its squalene-rich liver. Rather surprisingly, although catch rates of up to 500 kg/h are possible in some areas, this market is little exploited in Australia.
- **Local synonymy:** *Centrophorus kaikourae* Whitley, 1934.
- **References:** Whitley (1934a); Garrick (1960b); Bass *et al.* (1976).

third teeth from symphysis
(upper and lower jaw)

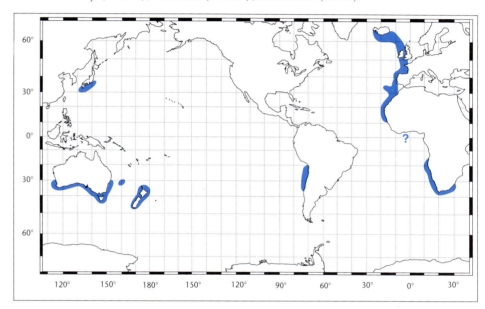

Longsnout Dogfish — *Deania quadrispinosa* (McCulloch, 1915)

8.14 *Deania quadrispinosa* (Plate 3) **Fish Code:** 00 020004

- **Alternative name:** Long-snouted dogfish.
- **Field characters:** A dogfish with a very long, spatulate snout, deciduous skin, and with a short, high first dorsal fin which is similar in size and shape to the second dorsal fin. Juveniles have a dark blotch near the anterior border of each dorsal fin.

- **Distinctive features:** Body slender; trunk width of adults usually less than 1.9 in trunk depth at level of second dorsal-fin spine origin. Snout extremely long and flattened, preoral length 12.5–15.6% of total length. Eye large, diameter 3.6–5.0% of total length. Denticles fairly large, pitchfork-shaped and usually quadriradiate with 3 posteriorly directed cusps and one short, anteriorly directed cusp. Upper teeth with single, erect to oblique, awl-shaped cusps; lower teeth with single, smooth-edged, blade-like cusps that are oblique and laterally notched (cusps sometimes more erect in adult males). Dorsal-fin spines grooved, second spine usually longer than first. First dorsal-fin origin over pectoral-fin inner margins; distance from first dorsal-fin spine origin to free rear tip of fin usually less than distance from its free rear tip to second dorsal-fin spine origin; second dorsal-fin insertion almost above lower caudal-fin origin. First dorsal fin short and tall, height 3.1–4.0 in length (from junction with spine to free rear tip); second dorsal-fin similar in size and shape and only slightly taller than first dorsal fin. Tooth count [28–33/29–31]*. Total vertebrae 118–127* [125–128]*; precaudal 86–89* [91–92]*.

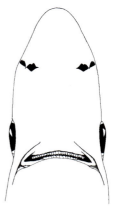

undersurface of head

- **Colour:** Dorsal surfaces brownish, greyish or black, paler ventrally. Fins sometimes white-edged. Juveniles with darker caudal-fin lobes and a dark blotch on anterior margin of each dorsal fin.
- **Size:** Attains 115 cm; a male of 87 cm was mature.
- **Distribution:** South Africa, New Zealand and Australia. Found locally off southern Australia from Moreton Island (Queensland) to Perth (Western Australia), including Tasmania; also off Port Hedland (Western Australia). A demersal species of the outer continental shelf and slope occurring in 150–820 m depths but mainly deeper than 400 m.
- **Remarks:** Little is known of the biology of this species other than it is ovoviviparous and feeds on fish. Not caught in sufficient quantities to be of interest to fisheries.
- **Local synonymy:** *Acanthidium quadrispinosum* McCulloch, 1915.
- **Reference:** Bass *et al.* (1976).

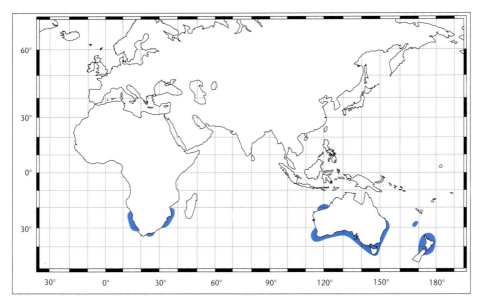

Smooth Lantern Shark — *Etmopterus* sp. A

8.15 *Etmopterus* sp. **A** (Plate 8) Fish Code: 00 020027

- **Field characters:** A medium-sized, greyish lantern shark with a firm body, a relatively short caudal peduncle, a long head, obscure flank markings and with the dorsal fin located near the pectoral-fin tip. Its low, flat-topped and widely spaced denticles (not thorn or bristle-like) are not arranged in regular rows.

- **Distinctive features:** Body fusiform, slender; caudal peduncle short (distance from pelvic-fin insertion to lower lobe of caudal fin shorter than distance from snout tip to gill slits); head relatively long, distance from snout tip to first gill slit greater than distance from first gill slit to dorsal-fin origin. Snout subtriangular, distinctly longer than eye diameter; eyes large. Upper teeth relatively large, erect, multicuspid, widely spaced; lowers interlocking, blade-like with oblique cusps. Denticles blunt, low, widely spaced. First dorsal-fin origin above (or less than an eye diameter behind) pectoral-fin tip. Second dorsal-fin origin behind pelvic-fin insertion. Second dorsal fin about twice height of first dorsal fin, subequal in height to second dorsal spine; spines about equal in length. Pectoral fin relatively large, broad. Pelvic flank marking mostly indistinct, expanded and truncated posteriorly; precaudal marking faint, caudal marking shorter than half eye diameter.

- **Colour:** Pale greyish (juveniles darker); belly usually darker than upper surface; broad black tip on upper caudal lobe; remaining fin tips pale.

- **Size:** Attains 67 cm. Males mature at 46–48 cm.

- **Distribution:** Probably widely distributed in the Southern Hemisphere. So far known from the Abrohlos Islands to Bunbury (Western Australia) and from Sydney (New South Wales) to Maria Island (Tasmania) in 435–760 m.

- **Remarks:** Lantern sharks are small, poorly described fishes of the deep sea. They have small, erect, multicuspid teeth in the upper jaw and blade-like slicing teeth in the lower jaw. Most of the species have light organs, visible mainly as dark markings on the abdomen, flank, caudal fin and peduncle. This species, which is closely related to the slender lantern shark (8.25), appears to live mainly in midwater. Little is known of its biology or distribution.

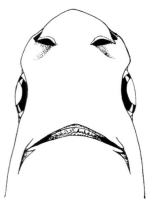

undersurface of head

third teeth from symphysis (upper and lower jaw)

patch of flank denticles

flank denticle

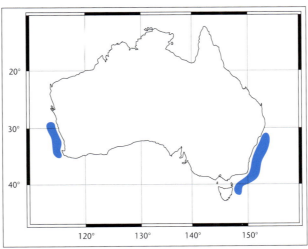

Bristled Lantern Shark — *Etmopterus* sp. B

8.16 *Etmopterus* sp. **B** (Plate 8) **Fish Code:** 00 020022

- **Field characters:** A large, robust lantern shark with a short caudal peduncle, dense bristle-like flank denticles (irregularly distributed), denticles over the dorsal-fin bases, and faint pelvic, flank and precaudal markings.

- **Distinctive features:** Body fusiform, stout (females more robust); caudal peduncle short (distance from pelvic-fin insertion to lower lobe of caudal fin shorter than distance from snout tip to gill slits). Snout short, broadly rounded; eyes large. Upper teeth erect, multicuspid; lowers inter-locking, blade-like, cusps oblique. Denticles of adults not arranged in regular rows (in rows on tail of juveniles). First dorsal-fin origin well behind pectoral-fin tip. Second dorsal-fin origin mostly over posterior half of pelvic fin. First dorsal fin long, low; fin spine very short, broad, about half height of fin. Second dorsal fin about twice height of first dorsal fin; fin spine strong, extending over about three-quarters of anterior margin of fin. Pelvic flank marking indistinct; caudal marking distinct, generally shorter than eye diameter; precaudal marking faint. Total vertebrae 82–89; precaudal 57–64; monospondylous 43–46.

- **Colour:** Uniformly dark brown to brownish black; denticles not deciduous.

- **Size:** Born at about 17 cm and attains 69 cm; males smaller (to about 63 cm) and mature by 53 cm.

- **Distribution:** Off southern Australia from Perth (Western Australia) to Taree (New South Wales), including Tasmania and the seamounts to the south (Cascade Plateau and South Tasman Rise) in 750–1380 m.

- **Remarks:** This large lantern shark is similar to a North Pacific relative, *Etmopterus unicolor*, and may turn out to be the same species. It is common on the mid-continental slope where it is frequently caught, together with the southern lantern shark (8.22), by vessels trawling for orange roughy (*Hoplostethus atlanticus*). It may live partly in midwater as specimens have been caught within 120 m of the surface by Japanese longliners in the open ocean.

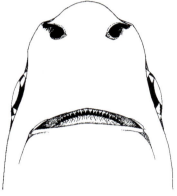

undersurface of head

patch of flank denticles

flank denticle

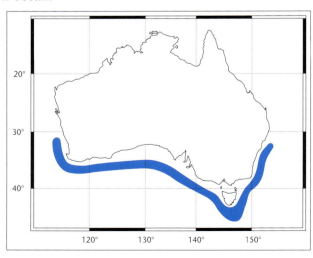

Pygmy Lantern Shark — *Etmopterus* sp. C

8.17 *Etmopterus* sp. C (Plate 8)   **Fish Code:** 00 020028

- **Field characters:** A small, firm-bodied lantern shark with an elongate caudal peduncle, flank denticles arranged in regular rows, blunt-ended posterior pelvic flank and caudal peduncle markings, and an additional oval marking near the base of the upper lobe of the caudal fin.

- **Distinctive features:** Body fusiform, subcylindrical; caudal peduncle moderately long (distance from pelvic-fin insertion to lower lobe of caudal fin about equal to distance from snout tip to pectoral-fin origin). Snout short, subconical, shorter than eye diameter; eyes large. Upper teeth minute, erect, multicuspid; lowers interlocking, blade-like and oblique. Denticles on flank single-cusped, erect, arranged in rows. First dorsal-fin origin well forward of free rear tip of pectoral fin. Second dorsal-fin origin well behind insertion of pelvic fin. First dorsal fin small, low; spine short, less than half height of fin. Second dorsal fin more than twice height of first dorsal fin; subequal in height to long slender spine. Pelvic flank marking distinct, with short anterior branch (reaching only slightly in advance of pelvic-fin origin); posterior branch longer, truncate posteriorly. Caudal marking short, located near posterior margin of fin; peduncle marking broad, truncate anteriorly; additional oval marking on central base of upper caudal fin.

- **Colour:** Dark greyish or black; flank and caudal markings faint but distinct; fins with dark margins, paler centrally; caudal fin with broad black tip on upper lobe.

- **Size:** Known from two mature males of 25 and 26 cm.

- **Distribution:** Only known from the continental slope off Broome (Western Australia) in 410 m.

- **Remarks:** This small species, which resembles some of the pelagic dogfishes in having a firm cylindrical body, is distinct among lantern sharks. More specimens are required.

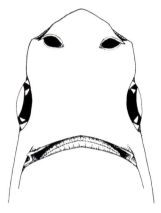

undersurface of head

patch of flank denticles

flank denticle

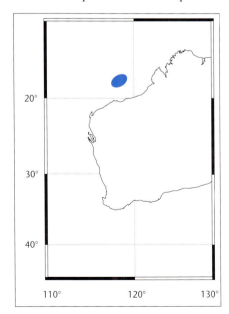

Pink Lantern Shark — *Etmopterus* sp. **D**

8.18 *Etmopterus* sp. **D** (Plate 8) Fish Code: 00 020029

- **Field characters:** A rather stout lantern shark with a distinctive colour pattern, moderately short caudal peduncle, fine denticles that are not arranged in regular rows, and a pelvic flank marking with a truncated posterior branch and an anterior branch extending to the pelvic-fin origin.

- **Distinctive features:** Body fusiform, moderately stout; caudal peduncle short (distance from pelvic-fin insertion to lower lobe of caudal fin equal to distance from snout tip to about first gill slit). Snout short, subconical; eyes large. Upper teeth erect, multicuspid; lowers interlocking, blade-like, cusps oblique. Denticles bristle-like. First dorsal-fin origin well behind free rear tip of pectoral fin. Second dorsal-fin origin over or slightly in front of pelvic-fin insertion. First dorsal fin small, low; spine relatively short, reaching about half of distance to fin tip. Second dorsal fin less than twice size of first dorsal fin; spine long, nearly reaching level of fin tip. Pelvic flank marking with anterior branch mostly reaching level of pelvic-fin origin; posterior branch short, extending to just beyond second dorsal-fin insertion. Caudal marking 0.8–1.4 times length of precaudal marking. Total vertebrae 80–85; precaudal 56–60; monospondylous 40–43.

- **Colour:** Light pinkish to brownish grey dorsally, dusky to black ventrally. Distinct black markings behind the pelvic fin, on the caudal peduncle and upper caudal fin. Upper caudal-fin tip dark.

- **Size:** Attains at least 41 cm. Males mature at about 37 cm.

- **Distribution:** Known only from off Cairns, northern Queensland, near the bottom on the upper continental slope in 800 to 880 m.

- **Remarks:** Little is known of the distribution or biology of this lantern shark. Specimens are required.

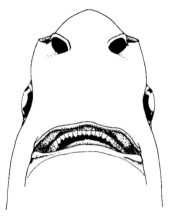

undersurface of head

patch of flank denticles

flank denticle

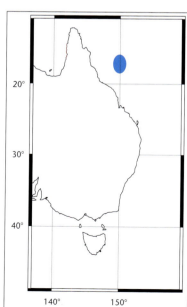

Blackmouth Lantern Shark — *Etmopterus* sp. E

8.19 *Etmopterus* sp. E (Plate 7) **Fish Code:** 00 020030

- **Field characters:** A small, pale lantern shark with a moderately elongate caudal peduncle with a dark ventral blotch, the posterior denticles arranged in regular rows, and dark bands through the middle and tip of the upper caudal-fin lobe.

- **Distinctive features:** Body fusiform, slender, soft-bodied; caudal peduncle moderately long (distance from pelvic-fin insertion to lower lobe of caudal fin about equal to distance from snout tip to pectoral fin). Snout short, subconical, eyes large. Upper teeth erect, multicuspid; lowers interlocking, blade-like, cusps oblique. Denticles arranged in distinct rows along dorsal midline and on caudal peduncle. First dorsal-fin origin almost over free rear tip of pectoral fin. Second dorsal-fin origin from slightly in front to slightly behind pelvic-fin insertion. First dorsal fin small, low; spine short, slender, reaching about three-quarters of distance to fin tip. Second dorsal fin about twice size of first dorsal fin; spine long, slender, tip curved posteriorly and reaching level of fin tip. Pelvic flank marking with anterior branch about equal to or shorter than posterior branch; its base inserted under or slightly anterior to second dorsal-fin spine. Precaudal marking very elongate, 4–5 times length of caudal-fin marking. Total vertebrae 79*; precaudal 55–56*; monospondylous 37*.

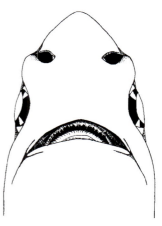

undersurface of head

patch of flank denticles

flank denticle

- **Colour:** Light brown dorsally, darker ventrally. Dark borders around the mouth, above the eyes and sometimes around the gill slits. Distinct black markings behind the pelvic fin, and on the caudal peduncle and upper caudal fin. A dark blotch anterior to the lower caudal-fin origin, a dark band across the caudal fin and a dark upper caudal-fin tip.

- **Size:** Attains at least 29 cm.

- **Distribution:** Known only from a few specimens collected off the Rowley Shoals and Ashmore Reef, Western Australia, in depths of 430–440 m.

- **Remarks:** Little is known of the distribution or biology of this lantern shark. Additional specimens are required.

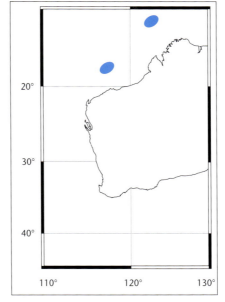

Lined Lantern Shark — *Etmopterus* sp. F

8.20 *Etmopterus* sp. F (Plate 7) **Fish Code:** 00 020031

- **Field characters:** An elongate lantern shark with a very long caudal peduncle, flank denticles not arranged in regular rows and with a pattern of dark broken lines (resembling dashes) along the upper flanks. The base of the pelvic flank marking is well in front of the origin of the second dorsal-fin spine, the anterior branch of the pelvic flank marking is shorter than the posterior branch, and the caudal marking is shorter than the caudal peduncle marking.

- **Distinctive features:** Body very elongate; caudal peduncle extended (distance from pelvic-fin insertion to lower lobe of caudal fin about equal to distance from snout tip to pectoral-fin insertion). Snout short, subconical; eyes large. Upper teeth erect, multicuspid; lowers interlocking, blade-like, cusps oblique. Denticles bristle-like. First dorsal-fin origin slightly behind free rear tip of pectoral fin. Second dorsal-fin origin about over or slightly behind free rear tips of pelvic fin. First dorsal fin small, low; spine short, reaching about three-quarters of distance to fin tip. Second dorsal fin about twice size of first dorsal fin, its height 1.4–1.8 in base; fin spine long, relatively straight and reaching about three-quarters of distance to fin tip. Pelvic flank marking with anterior branch shorter than posterior branch; its base inserted well in advance of second dorsal-fin spine origin. Caudal marking 0.6–0.9 times length of precaudal marking. Total vertebrae 89–95; precaudal 65–70; monospondylous 42–45.

- **Colour:** Dorsal surfaces and flanks light silvery brown with a striking longitudinal pattern of small black dots and dashes. Darker ventrally with distinct black markings near the pelvic fin and on the caudal peduncle and upper caudal fin. Dusky patches on the upper caudal-fin tip, on the mid caudal fin and on the ventral surface of the caudal peduncle.

- **Size:** Attains at least 45 cm. Males mature at about 34 cm.

- **Distribution:** Currently known only from off northern Queensland between Cairns and Rockhampton. On or near the bottom of the upper continental slope at depths of 590–700 m.

- **Remarks:** This rather attractive lantern shark is known only from a handful of specimens collected during the initial survey of continental slope fishes of northeastern Australia. Little is known of its biology.

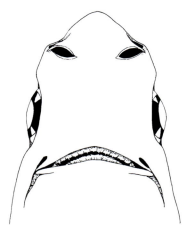
undersurface of head

patch of flank denticles

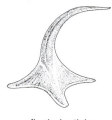
flank denticle

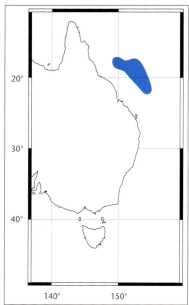

Short-tail Lantern Shark — *Etmopterus brachyurus* Smith & Radcliffe, 1912

8.21 *Etmopterus brachyurus* (Plate 7) Fish Code: 00 020032

- **Field characters:** A small, slender lantern shark with a moderately elongate caudal peduncle and flank denticles mostly arranged in regular rows. The insertion of the base of the pelvic flank marking is in advance of the second dorsal-fin spine, the anterior branch is shorter than the posterior branch, and the precaudal marking is about twice as long as the caudal marking.

- **Distinctive features:** Body fusiform, slender; caudal peduncle moderately long (distance from pelvic-fin insertion to lower lobe of caudal fin about equal to distance from snout tip to pectoral-fin origin). Snout short, subconical, eyes large. Upper teeth erect, multicuspid; lowers interlocking, blade-like, cusps oblique. Denticles mostly in rows. First dorsal-fin origin slightly behind free rear tip of pectoral fin. Second dorsal-fin origin slightly behind pelvic-fin insertions. First dorsal fin small, low; spine short, reaching about three-quarters of distance to fin tip. Second dorsal fin about twice size of first dorsal fin, its height about 1.5 in base; spine long, relatively straight and reaching about three-quarters of distance to fin tip. Pelvic flank marking with anterior branch shorter than posterior branch; its base inserted in advance of second dorsal-fin spine origin. Caudal marking about half length of precaudal marking. Total vertebrae 86*; precaudal 61* [57–68]; monospondylous 40*.

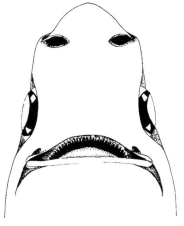

undersurface of head

patch of flank denticles

- **Colour:** Light brown dorsally with a purplish sheen in life; darker brown on flanks merging to blackish ventrally with a greenish sheen in life. Distinct dark markings behind pelvic fin, on caudal peduncle and upper caudal fin. Pale stripe mostly extending between each pectoral and pelvic fin and surrounding the dark pelvic flank markings. Mid-dorsal line from the inter-orbital area to the upper caudal-fin origin pale. Eye green, iris yellow.

- **Size:** Born at about 15 cm and attains 50 cm.

- **Distribution:** Indo–West Pacific. Recorded from Japan, the Philippines and southern Africa, although its actual distribution is uncertain. In Australia, on or near the bottom of the upper continental slope off Shark Bay, central Western Australia, in depths of 400–610 m.

flank denticle

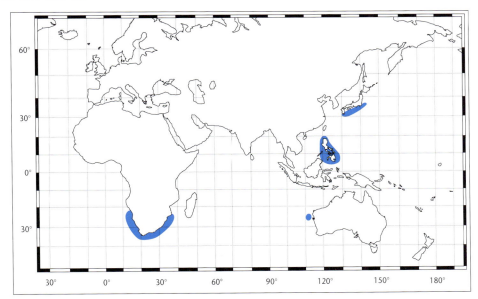

- **Remarks:** This species is closely related to the black-belly (8.23) and Mollers (8.24) lantern sharks with which it has been confused in the past. Uncommon locally and of no commercial value elsewhere.
- **Reference:** Yamakawa *et al.* (1986).

Southern Lantern Shark — *Etmopterus granulosus* (Günther, 1880)

8.22 *Etmopterus granulosus* (Plate 9)   Fish Code: 00 020021

- **Alternative name:** New Zealand lantern shark.
- **Field characters:** A large, robust lantern shark with a short caudal peduncle, slender widely spaced flank denticles (irregularly distributed in adults), naked skin patches at the base of each dorsal fin, and indistinct pelvic flank and precaudal markings.
- **Distinctive features:** Body fusiform, stout (females more robust); caudal peduncle short (distance from pelvic-fin insertion to lower lobe of caudal fin shorter than distance from snout tip to gill slits). Snout short, broadly rounded; eyes large. Upper teeth erect, multicuspid; lowers interlocking, blade-like, cusps oblique. Denticles of adults not arranged in regular rows (in rows on tail of juveniles). First dorsal-fin origin mostly behind pectoral-fin free rear tip. Second dorsal-fin origin over or slightly in advance of pelvic-fin insertion. First dorsal fin long, low; spine short, broad, equal in height to fin. Second dorsal fin of similar height to first dorsal fin; fin spine robust, long, mostly curving over fin tip. Pelvic flank marking usually indistinct; caudal marking about equivalent in length to eye diameter; precaudal marking indistinct or absent.
- **Colour:** Uniform brownish black to black; pale where denticles and skin removed; flank markings indistinct (more obvious in young).
- **Size:** Born at about 18 cm; females attain 75 cm and males 62 cm. Males mature at about 46 cm.
- **Distribution:** Circumglobal in the temperate Southern Hemisphere: South America, southern Africa (including the Walters Shoals), New Zealand and the Falkland Islands in 220–1430 m. Records from Sierra-Leone require validation. In Australia, off southern New South Wales, Victoria and Tasmania (including seamounts to the south) in 830–1200 m.
- **Remarks:** Possibly the largest lantern shark. It is among the most abundant bottom-dwelling sharks on the mid-continental slope of south-eastern Australia. Ovoviviparous, with litters of 10–13 pups. High in squalene oil but currently of no commercial value.

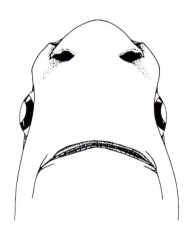

undersurface of head

patch of flank denticles

flank denticle

- **Local synonymy:** *Etmopterus baxteri* Garrick, 1957.
- **References:** Gunther (1880a); Garrick (1957); Tachikawa *et al.* (1989).

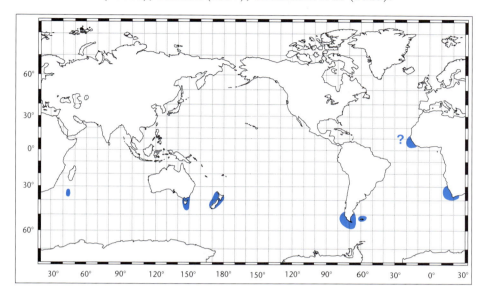

Blackbelly Lantern Shark — *Etmopterus lucifer* Jordan & Snyder, 1902

8.23 *Etmopterus lucifer* (Plate 7) Fish Code: 00 020005

- **Alternative name:** Lucifer shark.
- **Field characters:** A small, slender lantern shark with a moderately elongate caudal peduncle and with flank denticles mostly arranged in regular rows. The base of the pelvic flank marking is under the second dorsal-fin spine, the anterior branch is usually longer than the posterior branch, and the caudal marking is considerably longer than the precaudal marking.
- **Distinctive features:** Body fusiform, slender; caudal peduncle moderately long (distance from pelvic-fin insertion to lower lobe of caudal fin about equal to distance from snout tip to gill slits). Snout short, subconical; eyes large. Upper teeth erect, multicuspid; lowers interlocking, blade-like, cusps oblique. Denticles mostly in rows. First dorsal-fin origin about over or slightly behind free rear tip of pectoral fin. Second dorsal-fin origin over insertion of pelvic fin in females, somewhat behind in males. First dorsal fin small, low; spine short, reaching about three-quarters of

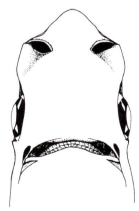

undersurface of head

distance to fin tip. Second dorsal fin about twice size of first dorsal fin; its height 1.4–2.1 in its base; fin spine long, tip curved posteriorly and nearly reaching fin tip. Pelvic flank marking usually with anterior branch longer than posterior branch; its base under second dorsal-fin spine. Caudal marking 1.5–2.4 times length of caudal peduncle marking. Tooth count 21–26/29–39 [21–24/33–40]. Total vertebrae 80–91 [83–91]; precaudal 55–64 [55–67]; monospondylous 42–47.

patch of flank denticles

- **Colour:** Light brown dorsally, merging from darker brown on the flanks to blackish ventrally. Distinct black markings behind the pelvic fin, on the caudal peduncle and upper caudal fin. Pale stripe sometimes present along the mid-dorsal line extending from between the eyes to the upper caudal-fin origin. Upper caudal-fin tip dark.

- **Size:** Born at about 15 cm and attaining 47 cm. Males mature at about 30 cm; females at about 34 cm.

- **Distribution:** Unclear. Currently recorded from the South Atlantic (Uruguay, Argentina and possibly Namibia), western Indian Ocean (southern Africa) and the western Pacific (from Japan to New Zealand). Widespread around southern Australia from Cairns (Queensland) to Perth (Western Australia), including Tasmania. On or near the bottom of outer continental and insular shelves and upper slopes in 183–1000 m; locally mainly between 400 and 800 m.

- **Remarks:** Represented by at least two morphs in Australian waters that may form part of a complex of very similar species. Mollers (8.24) and the short-tail (8.21) lantern sharks have also been confused with these forms in the past. Reproduction is presumably ovoviviparous and the diet consists of squid, bony fish (mainly lanternfish) and crustaceans. Not utilised commercially.

- **Local synonymy:** *Etmopterus abernethyi* Garrick, 1957.

- **Reference:** Yamakawa *et al.* (1986).

flank denticle

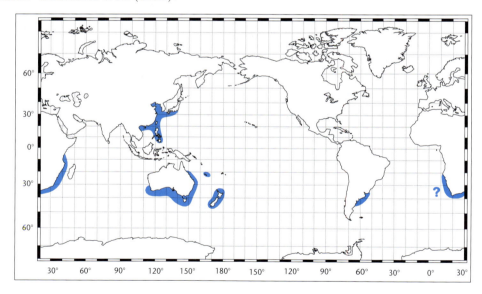

Mollers Lantern Shark — *Etmopterus molleri* Whitley, 1939

8.24 *Etmopterus molleri* (Plate 7)   Fish Code: 00 020033

- **Alternative name:** Mollers deepsea shark.
- **Field characters:** A small, slender lantern shark with a moderately elongate caudal peduncle and with flank denticles mostly arranged in regular rows. The base of the pelvic flank marking is inserted in advance of the origin of the second dorsal-fin spine, the anterior branch is shorter than the posterior branch, and the caudal marking is longer than the precaudal marking.
- **Distinctive features:** Body fusiform, moderately slender; caudal peduncle moderately long (distance from pelvic-fin insertion to lower lobe of caudal fin about equal to distance from snout tip to pectoral-fin origin). Snout short, subconical; eyes large. Upper teeth erect, multicuspid; lowers interlocking, blade-like, cusps oblique. Denticles mostly in rows. First dorsal-fin origin almost over free rear tip of pectoral fin. Second dorsal-fin origin slightly behind pelvic-fin insertion in females, well behind in males. First dorsal fin small, low; spine short, reaching about three-quarters of distance to fin tip. Second dorsal fin about twice size of first dorsal fin, its height 1.2–1.5 in its base; spine long, tip curved posteriorly and reaching level of fin tip. Pelvic flank marking with anterior branch shorter than posterior branch; its base inserted in advance of second dorsal-fin spine origin. Caudal marking 1.3–2.0 times length of precaudal marking. Total vertebrae 90–93; precaudal 63–66 [59–70]; monospondylous 41–43.
- **Colour:** Light brown dorsally, merging from darker brown on flanks to blackish ventrally; usually with a pale stripe extending from the pectoral-fin insertions to the pelvic-fin origins and surrounding the dark pelvic flank marking; mid-dorsal line pale

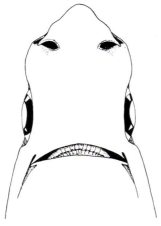

undersurface of head

patch of flank denticles

flank denticle

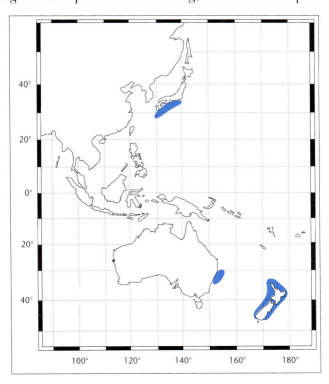

from the interorbital area to the upper caudal-fin origin. Black markings clearly demarcated near the pelvic fin, and on the caudal peduncle and upper caudal fin. Upper caudal-fin tip dark.

- **Size:** Born at about 15 cm and attains 46 cm. Males are mature by 33 cm.
- **Distribution:** Uncertain, although records from Japan, New Zealand and Australia appear to be valid. Locally, known from the type locality (off New South Wales) where it is demersal on the upper continental slope in depths of 250–480 m.
- **Remarks:** This lantern shark, which was described from specimens collected off Sydney, has only recently been recognised as distinct from the blackbelly lantern shark (8.23). Little is known of its biology and distribution.
- **References:** Whitley (1939a); Yamakawa *et al.* (1986).

Slender Lantern Shark — *Etmopterus pusillus* (Lowe, 1839)

8.25 *Etmopterus pusillus* (Plate 8)   **Fish Code:** 00 020015

- **Field characters:** A small, brownish lantern shark with a rather short caudal peduncle, firm body, obscure flank markings and with the first dorsal fin located well behind the pectoral-fin tip. Its low, flat-topped and widely spaced denticles (not thorn or bristle-like) are not arranged in regular rows.
- **Distinctive features:** Body fusiform, slender; caudal peduncle short (distance from pelvic-fin insertion to lower lobe of caudal fin shorter than distance from snout tip to gill slits); head relatively short, distance from snout tip to first gill slit equal to or less than distance from first gill slit to first dorsal-fin origin. Snout bluntly pointed, equal to or slightly longer than eye diameter; eyes large. Upper teeth relatively small, slender, erect, multicuspid; lowers interlocking, blade-like with oblique cusps. Denticles blunt, low, widely spaced. First dorsal-fin origin behind pectoral-fin tip by an eye diameter or more. Second dorsal-fin origin behind pelvic-fin insertion. Second dorsal fin almost twice height of first dorsal fin, subequal in height to second dorsal spine. Pectoral fin relatively short, slender. Pelvic flank marking mostly indistinct, expanded and truncated posteriorly; precaudal marking indistinct, caudal marking shorter than eye diameter.
- **Colour:** Pale brownish or chocolate brown; abdomen usually slightly darker than upper surface; fins mostly pale.
- **Size:** Attains at least 48 cm. Males are mature by 45 cm.
- **Distribution:** Widely distributed in the North and South Atlantic, western Indian Ocean (off South Africa) and the Indo–West Pacific (from Japan and Australia). Locally from off Tasmania and central New South Wales. On or near the bottom of continental and insular slopes at depths from 275–1000 m (possibly to 2000 m); oceanic in the South Atlantic from the surface to 710 m.

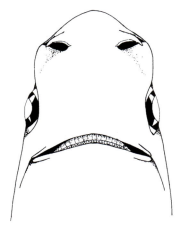

undersurface of head

flank denticle

patch of denticles

- **Remarks:** This species and the closely related smooth lantern shark (8.15) have been confused in the past so their recorded distributions may not be reliable. The smooth lantern shark seems to prefer warmer water than its smaller relative; both occur in midwater.
- **Reference:** Bass *et al.* (1976).

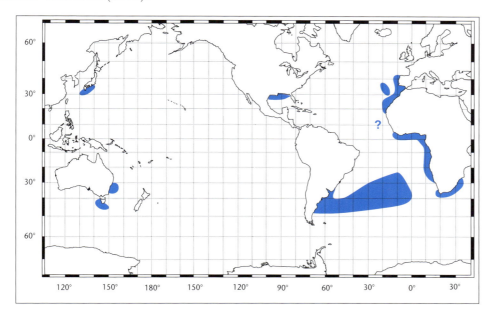

Pygmy Shark — *Euprotomicrus bispinatus* (Quoy & Gaimard, 1824)

8.26 *Euprotomicrus bispinatus* (Plate 9)   Fish Code: 00 020034

- **Field characters:** A very small dogfish with a bulbous head, a very small spineless first dorsal fin located well behind the pectoral fins, a low, spineless second dorsal fin with a very long base, a caudal fin with nearly symmetrical upper and lower lobes, and low keels on the caudal peduncle.
- **Distinctive features:** Body cigar-shaped, tapering posteriorly; caudal peduncle with low keels. Head conical and bulbous; gill slits small. Upper teeth with narrow, erect cusps; lowers oblique and blade-like. First dorsal fin very small and spineless, closer to pelvic fins than to pectoral fins. Second dorsal fin spineless, low, long (base about four times longer than base of first dorsal fin). Caudal fin with nearly symmetrical lobes. Tooth count [21/19–23]. Total vertebrae [60–70]; precaudal [46–52].
- **Colour:** Dorsal surfaces black, slightly paler below; fins pale or with pale posterior margins.
- **Size:** Born at about 8 cm and attains 27 cm. Males mature at about 17 cm and females at about 23 cm.

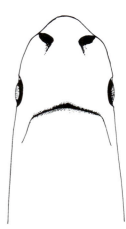

undersurface of head

- **Distribution:** Widespread, circumglobal in the open ocean (rare outside tropical and warm temperate regions). Migrates between the surface at night to below 400 m (possibly as deep as 1800 m) in the day. In Australia, recorded from Western Australia between Perth and the Rowley Shoals.
- **Remarks:** Ovoviviparous, with litters of eight pups. The diet includes bony fish, cephalopods and crustaceans. The luminous organs, which are best developed on the ventral surface, may help camouflage the shark from predators below, when it is in water shallow enough to receive filtered light from the surface. The organs may also assist in social recognition and feeding.
- **Reference:** Bass *et al.* (1976).

third teeth from symphysis
(upper and lower jaw)

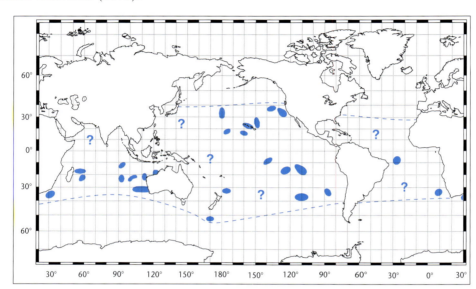

Cookie-cutter Shark — *Isistius brasiliensis* (Quoy & Gaimard, 1824)

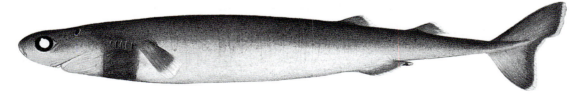

8.27 *Isistius brasiliensis* (Plate 3)   **Fish Code:** 00 020014

- **Alternative name:** Luminous shark.
- **Field characters:** A small dogfish with a cigar-shaped body, a distinct dark collar marking around the gill region, a conical snout, a caudal fin with nearly symmetrical upper and lower lobes, fleshy lips, enlarged triangular lower teeth and two very small, spineless dorsal fins.
- **Distinctive features:** Body cigar-shaped, tapering posteriorly; caudal peduncle with low keels. Snout conical and bulbous; gill slits small; lips fleshy and suctorial. Upper teeth small and erect, lowers large and triangular. First dorsal fin situated far back

on body, its insertion over pelvic-fin bases; both dorsal fins small and spineless (second slightly larger than first); dorsal-fin bases similar in length. Caudal fin with nearly symmetrical lobes. Tooth count [31–37/25–31]. Total vertebrae [81–89]; precaudal [60–66].

- **Colour:** Dorsal surface dark brown, paler ventrally; a prominent dark brown collar-like marking around the gill region (more distinct around the ventral surface). Fin tips with translucent posterior margins. Caudal-fin lobes dark with translucent posterior margins.
- **Size:** Attains about 50 cm, males mature at about 38 cm and females at about 40 cm.
- **Distribution:** Widespread oceanic in temperate and tropical regions. Off Australia, recorded from isolated localities off Queensland, New South Wales, Tasmania, Western Australia and off Lord Howe Island. Makes diurnal vertical migrations probably from below 1000 m in the day to near the surface at night.
- **Remarks:** This shark is often ectoparasitic on large fish and marine mammals, to which it attaches itself with its suctorial lips and modified pharynx. It then spins, boring out a plug of flesh with its large lower teeth and leaving a crater-shaped wound on its victim. This behavior pattern has given rise to its common name. The cookie cutter has a large, oily liver that makes it neutrally buoyant and able to hang motionless in the water; victims may be lured by the shark's strong luminescence. Cookie cutters have optimistically attacked nuclear submarines, leaving crater-marks on the rubber sonar domes. The diet also includes whole prey, particularly squid, some of which are nearly as large as their captors. It apparently swallows its own lower teeth during their replacement, possibly to maintain its calcium levels. Ovoviviparous, but litter sizes are unknown.
- **Local synonymy:** *Leius ferox* Kner, 1865.
- **References:** Bass *et al.* (1976); Jahn and Haedrich (1987).

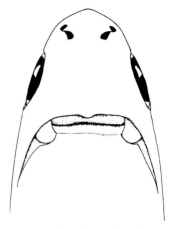

undersurface of head

third teeth from symphysis
(upper and lower jaw)

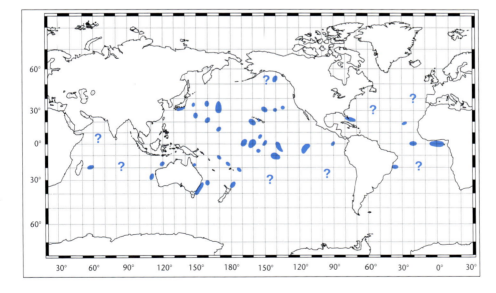

Whitetail Dogfish — *Scymnodalatias albicauda* Taniuchi & Garrick, 1986

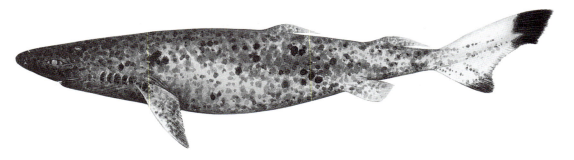

8.28 *Scymnodalatias albicauda* (Plate 9) **Fish Code:** 00 020035

- **Field characters:** A dark greyish and white mottled dogfish with two small, spineless dorsal fins (the first originating about midway between the pectoral and pelvic fins), angular pectoral fins, and an asymmetric caudal fin with a dark-tipped upper lobe.

- **Distinctive features:** Body stout without keels on the caudal peduncle. Snout slightly depressed, short and conical (preoral length shorter than mouth width); upper teeth erect with a long, slender, somewhat hooked cusp; lowers large, semi-erect and blade-like. Flank denticles with prominent longitudinal ridges and smaller transverse ridges on their outer surfaces. Origin of first dorsal fin about midway between pectoral-fin free rear tips and pelvic-fin origin. Dorsal fins low (much longer than their height), spineless; second dorsal fin slightly larger than first, its posterior tip almost reaching origin of caudal fin. Pectoral-fin tips pointed. Caudal fin asymmetric with a subterminal notch. Tooth count 57–62/35*. Total vertebrae 82–84*; precaudal 57–61*.

- **Colour:** Uniformly mottled light grey and white, with large, dark brown spots and blackish markings on the head. Basal half of caudal-fin upper lobe and most of lower lobe white, with a few greyish spots; posterior half of upper lobe and tip of lower lobe black. Dorsal fins and both surfaces of pectoral and pelvic fins mottled grey and white.

- **Size:** Attains at least 107 cm. A female of 74 cm was mature.

undersurface of head

third teeth from symphysis
(upper and lower jaw)

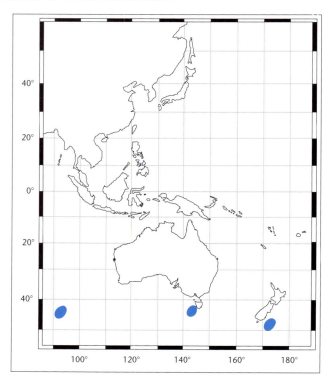

- **Distribution:** New Zealand and Australia (Tasmania and the Southern Ocean off Western Australia). Oceanic in depths of 150–510 m.
- **Remarks:** This rather rare shark is known from only three specimens, caught by Japanese tuna longliners in 150–200 m off Australasia. A single New Zealand specimen was trawled in about 500 m. It is thought to be ovoviviparous. A closely related species, *S. sherwoodi*, occurs off New Zealand and may appear off Australia. It differs from the whitetail dogfish in lacking the pale coloration and in having the free rear tip of the second dorsal fin falling well short of the caudal-fin origin. Further comparisons need to be made between these two poorly known species.
- **Reference:** Taniuchi and Garrick (1986).

Pacific Sleeper Shark — *Somniosus pacificus* Bigelow & Schroeder, 1944

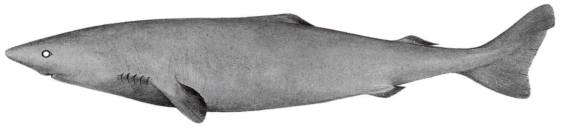

8.29 *Somniosus pacificus* (Plate 3) Fish Code: 00 020036

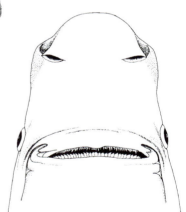

undersurface of head

- **Field characters:** A gigantic dogfish with two small, spineless dorsal fins (the first originating behind the free rear tip of the pectoral fin) and an asymmetrical caudal fin with a well-developed lower lobe.
- **Distinctive features:** Body stout, almost cylindrical. Snout rounded, short and conical; preoral length subequal to mouth width. Upper teeth long, slender; lower teeth short, low, with very oblique cusps. Flank denticles large, widely spaced, arising obliquely from the skin. Insertion of first dorsal fin closer to pelvic fins than to pectoral fins; predorsal distance 45–48% of total length. Dorsal fins small, spineless; second subequal in size to first. Pectoral-fin free rear tips broadly rounded. Caudal fin asymmetrical, with a subterminal notch; keels present on caudal peduncle. Tooth count 42/55–59*[35–45/53–58].
- **Colour:** Uniformly greyish pink with bluish black fins; usually covered with dark brown mucous; underside of snout, upper lip and mouth lining dark. Possibly with small white spots on the dorsal surface in life.
- **Size:** Attains at least 430 cm but reputed to attain 600 cm or more. Its size at maturity is unknown. Smallest known specimen (42 cm) had an unhealed umbilical scar.
- **Distribution:** Japan to California with sporadic records from the South Pacific. In Australasian waters, known from the seamounts south of Tasmania, the Challenger Plateau (off eastern New Zealand), and possibly from Macquarie Island in 400–1100 m.
- **Remarks:** This dogfish is the Pacific equivalent of the Greenland shark (*Somniosus microcephalus*). Neither the number of sleeper shark species nor their distribution in the Southern Hemisphere is known; only recently have specimens been collected

third teeth from symphysis (upper and lower jaw)

from this part of the world. A second species, the longnose sleeper shark (*S. rostratus*), occurs off New Zealand. It differs from the Pacific sleeper shark in having small, closely spaced denticles, more erect tooth cusps in the lower jaw, and a shorter predorsal distance (less than 45% of total length). Considered to be harmless to humans.

- **Local synonymy:** *Somniosus antarcticus* Whitley, 1939
- **References:** Whitley (1939b); Francis *et al.* (1988).

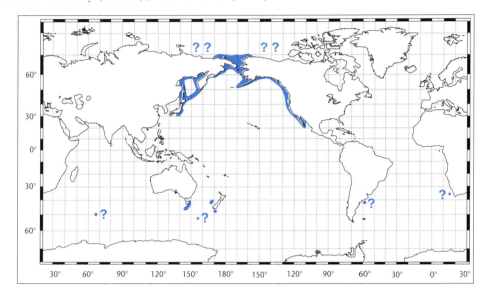

Smalleye Pygmy Shark — *Squaliolus aliae* Teng, 1959

8.30 *Squaliolus aliae* (Plate 9)   Fish Code: 00 020017

- **Field characters:** A very small dogfish with the first dorsal fin closer to the pectoral fin than the pelvic fin, a first dorsal-fin spine but no second dorsal-fin spine, a pointed snout, nearly symmetrical upper and lower caudal-fin lobes, and a long-based second dorsal fin that is about twice the length of the first dorsal-fin base.

- **Distinctive features:** Body cigar-shaped, tapering posteriorly; caudal peduncle slender and with low keels. Snout bulbous, conical; gill slits very small; eye small for this genus (diameter 46–70% of interorbital width); orbit upper margin a slightly inverted v-shape. Upper teeth small, narrow and erect; lower teeth larger, blade-like and semi-erect. First dorsal-fin origin almost over free rear tips of pectoral fin. First dorsal fin and spine small (spine sometimes concealed by skin); second dorsal fin spineless, long and low (base about twice as long as base of first dorsal fin). Caudal fin with nearly symmetrical fin lobes, subterminal notch present.

- **Colour:** Dark brown to black, fin margins pale.

- **Size:** Attains at least 22 cm; males mature at about 15 cm.

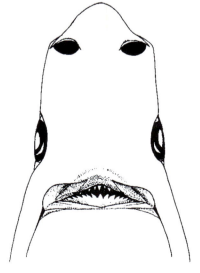

undersurface of head

- **Distribution:** Indo–West Pacific from Japan to Australia. Locally from off northwestern Australia and New South Wales. Epipelagic or mesopelagic near continental and island land masses. Thought to undertake diurnal vertical migrations from within 200 m of the surface at night down to about 2000 m during the day.
- **Remarks:** Possibly the smallest living shark. A close relative, *S. laticaudus*, with a larger eye (diameter 73–86% of interorbital width) may also occur in this area. Unlike the pygmy shark (8.26), which is primarily oceanic, these sharks occur in association with continental slopes. The luminous organs, which are best developed on the ventral surface, may be used to camouflage the shark from predators. Ovoviviparous, but litter sizes are unknown. The diet consists mainly of cephalopods and small midwater bony fish, including lanternfish.
- **Reference:** Sasaki and Uyeno (1987).

third teeth from symphysis
(upper and lower jaw)

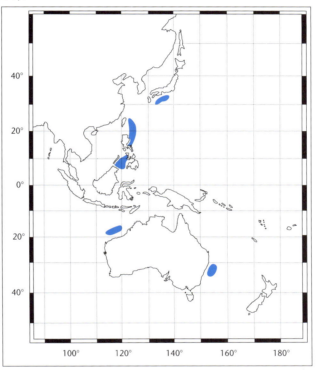

Bartail Spurdog — *Squalus* sp. **A**

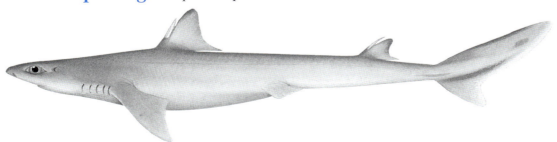

8.31 *Squalus* sp. **A** (Plate 6)   Fish Code: 00 020037

- **Field characters:** A medium-sized dogfish with prominent dorsal-fin spines (the first originating just in advance of the free rear tips of the pectoral fins), a tall, upright first dorsal fin, a rather long, deeply concave pectoral fin, similar blade-like teeth in both jaws, a relatively short snout, and no white spots on the body. The caudal fin is relatively long and has a distinctive dark bar through the ventral lobe and a white posterior margin.

- **Distinctive features:** Body moderately elongate, fusiform; head rather robust. Snout short, rounded; distance from snout to inner nostril less than distance from nostril to upper labial furrow; upper labial furrows relatively long; anterior nasal flap well developed, bifurcated. Denticles tricuspidate, without lateral keels. First dorsal fin upright, tall, its height about 72–77*% of its length from spine origin to free rear tip; dorsal-fin spines well developed, relatively slender, without lateral grooves; first dorsal-fin spine shorter than second dorsal-fin spine and much lower than associated fin; second dorsal-fin spine long and slender (base 6.3–6.5* in height), extending over second dorsal-fin. Pectoral-fin apex narrow, lobe-like; posterior margin deeply concave; inner margin 1.7–2.2* in anterior margin. Caudal fin relatively elongate, length of upper lobe 4.1–4.4* in total length; upper caudal lobe without subterminal notch; caudal keels (sometimes weak in juveniles) and precaudal pits present. Precaudal vertebrae 94–96*.

- **Colour:** Dorsal surface pale greyish brown, white ventrally; dorsal-fin apices (and often posterior margin) dark, free rear tips white. Posterior margin of pectoral fin white. Caudal fin with a dark anterior margin (often extending from a central dark blotch on the upper lobe) and a dark bar extending through base of lower lobe to posterior notch; remainder of posterior margin pale. Dark markings more prominent in juveniles.

- **Size:** Largest known specimen is a 62 cm immature male.

- **Distribution:** Known only from a few specimens collected off Queensland between Cairns and Rockhampton in 220–450 m.

- **Remarks:** Spurdogs are small to medium sized dogfishes with prominent fin spines without grooves in each dorsal fin, similar blade-like teeth with broad oblique cusps in both jaws, an upper precaudal pit and lateral keels on the caudal peduncle. Little is known of this species. More specimens are required.

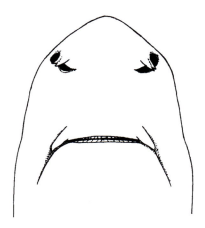

undersurface of head

third teeth from symphysis
(upper and lower jaw)

flank denticle

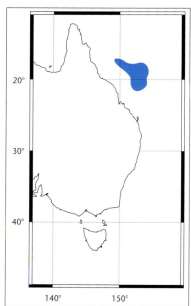

Eastern Highfin Spurdog — *Squalus* sp. **B**

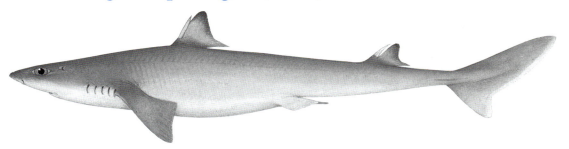

8.32 *Squalus* sp. **B** (Plate 6)　　　　　　　　　　　　　**Fish Code:** 00 020038

- **Field characters:** A medium-sized dogfish with prominent dorsal-fin spines (the first originating just in advance of the free rear tips of the pectoral fins), an upright first dorsal fin, a rather long, slightly concave pectoral fin, similar blade-like teeth in both jaws, a relatively short snout and no white spots on the body. The caudal fin is relatively short with a white posterior margin but lacks a distinctive dark bar through its ventral lobe.

- **Distinctive features:** Body moderately elongate, fusiform; head relatively robust. Snout short, rounded; distance from snout to inner nostril less than distance from nostril to upper labial furrow; upper labial furrows relatively long; anterior nasal flap well developed, bifurcated. Denticles tricuspidate, with lateral keels. First dorsal fin upright, relatively tall, its height about 61–64*% of its length from spine origin to free rear tip. Dorsal-fin spines well developed, moderately robust, without lateral grooves, subequal in height; base of first spine usually slightly more robust than base of second (base width of first about three quarters of spiracle length, base width of second 5.0–5.3* in its height); first dorsal-fin spine lower than associated fin; second dorsal-fin spine slender, tapering gradually, subequal in height to fin. Pectoral-fin apex broad, lobe-like; inner margin about 1.8–1.9* in anterior margin. Caudal fin relatively broad, short, length of upper lobe 4.6–5.0* in total length; upper caudal lobe without subterminal notch; caudal keels and precaudal pits present. Precaudal vertebrae 90–93*.

- **Colour:** Dorsal surface pale greyish; white ventrally. Posterior margins of dorsal fins and anterior margin of caudal fin dark-edged; posterior margin of caudal fin pale, lacking a dark bar through its lower base.

- **Size:** Attains at least 65 cm; smallest mature male examined was 62 cm.

- **Distribution:** Upper continental slope off eastern Australia, from the Queensland Plateau (off Cairns) to Byron Bay (New South Wales), in 240–450 m.

- **Remarks:** This species occurs together with the bartail spurdog (8.31) from which it differs in having relatively shorter first dorsal and caudal fins, a more robust second dorsal-fin spine, no dark caudal bar and fewer vertebrae. More specimens are required.

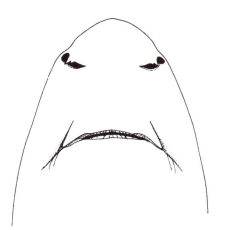

undersurface of head

flank denticle

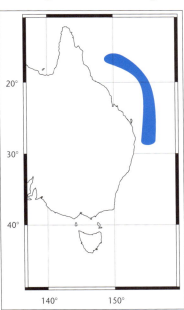

Western Highfin Spurdog — *Squalus* sp. C

8.33 *Squalus* sp. C (Plate 6)　　　　　　　　　　　Fish Code: 00 020018

- **Field characters:** A medium-sized dogfish with large dorsal-fin spines (the first originating near the free rear tips of the pectoral fins), a short, upright first dorsal fin, the second dorsal-fin spine more slender based than the first, a rather long, deeply concave pectoral fin, similar blade-like teeth in both jaws, a medium length snout and no white spots on the body. The caudal fin is relatively short, has a whitish posterior margin and sometimes has a faint dark bar through its lower lobe.

- **Distinctive features:** Body elongate, fusiform. Snout moderate, rather pointed; distance from snout to inner nostril more than distance from nostril to upper labial furrow; direct snout length equal to or slightly less than interorbital distance; upper labial furrows relatively long; anterior nasal flap well developed, bifurcated. Denticles tricuspidate, with lateral keels. First dorsal fin upright, of moderate height (about 59–63% of its length from spine origin to free rear tip). Dorsal-fin spines well developed, robust, without lateral grooves, subequal in height; concealed portion low (about spiracle width from fin base); first dorsal-fin spine base much larger than second dorsal-fin spine base (base of first about three quarters of spiracle diameter, base of second 6.0–6.5 in its height); first dorsal-fin spine lower than associated fin; second dorsal fin slightly taller than spine. Pectoral-fin inner margin 1.7–2.2 in anterior margin. Caudal fin moderate (length of upper lobe 4.8–5.2 in total length), lower lobe narrow; upper caudal lobe without subterminal notch; caudal keels and precaudal pits present. Precaudal vertebrae 79–83.

- **Colour:** Dorsal surfaces greyish, lacking white spots or distinct white markings; greyish under head and sometimes on belly and undersurface of tail. Fins greyish with pale tips; posterior margin of caudal fin with narrow pale border; lower caudal lobe often with a faint dark bar (less obvious in adults).

- **Size:** Reaches at least 78 cm; males mature at about 56 cm.

- **Distribution:** Continental slope off Western Australia from Rottnest Island to North West Cape in 220–510 m.

- **Remarks:** This species is rather similar to the eastern highfin spurdog (8.32) from which it differs in dorsal fin coloration, and in the relative sizes of the head, second dorsal-fin spine and caudal fin. Often aggregates by sex.

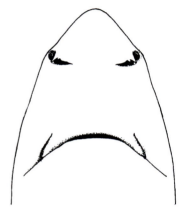

undersurface of head

flank denticle

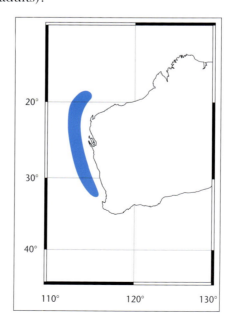

Fatspine Spurdog — *Squalus* sp. **D**

8.34 *Squalus* sp. **D** (Plate 6) Fish Code: 00 020039

- **Field characters:** A medium-sized dogfish with very broad based dorsal-fin spines (the first originating over the pectoral fins), the first dorsal fin slightly raked, a rather long, concave pectoral fin, similar blade-like teeth in both jaws, a relatively short snout and without white spots. The caudal fin is moderately elongate with a pale posterior margin and lacks a dark bar through the ventral lobe.

- **Distinctive features:** Body moderately elongate, fusiform. Snout short, not especially rounded; distance from snout to inner nostril less than distance from nostril to upper labial furrow; upper labial furrows relatively long; anterior nasal flap well developed, bifurcated. Denticles lanceolate, unicuspidate, with lateral keels. First dorsal fin raked backward slightly; dorsal-fin spines broad-based and tapering rapidly, robust (base of second 2.9–3.6* in height, almost equal to spiracle width), without lateral grooves, subequal in size; first dorsal-fin spine lower than associated fin; second dorsal fin and spine about equal in height. Pectoral-fin inner margin 1.8–2* in anterior margin. Caudal fin relatively broad, moderately elongate, length of upper lobe 4.5–4.7* in total length; upper lobe without subterminal notch; caudal keels and precaudal pits present. Precaudal vertebrae 83–86*.

- **Colour:** Dorsal surfaces and flanks pale greyish; fins paler, extreme tips of dorsal fins dusky. Caudal-fin anterior margin dusky, upper tip white, posterior margin pale; no bar on tail. Pale underneath.

- **Size:** Attains about 56 cm. Males mature at about 44 cm.

- **Distribution:** Known from the upper continental slope between North West Cape and Port Hedland, northwestern Australia, in 180–210 m.

- **Remarks:** This rather stocky spurdog has very distinctive dorsal fin spines that are broad-based and which taper rapidly. As with most other members of the family, little is known of its biology.

undersurface of head

flank denticle

Western Longnose Spurdog — *Squalus* sp. E

8.35 *Squalus* sp. E (Plate 5)

Fish Code: 00 020040

- **Field characters:** A small, slender dogfish with prominent dorsal-fin spines (the first originating near the pectoral-fin rear tips), similar blade-like teeth in both jaws, a long and slender snout (horizontal length mostly more than 1.9 times eye diameter, direct length exceeding interorbital width) and lacking white spots. The caudal fin is short (mostly more than 5 times in total length) with a dark marginal bar above the posterior notch.

- **Distinctive features:** Body moderately elongate, fusiform; head very slender. Snout relatively very elongate and pointed; direct preorbital length more than interorbital width; distance from snout to inner nostril exceeding distance from nostril to upper labial furrow; upper labial furrows relatively long; anterior nasal flap well developed, bifurcated. Denticles tricuspidate. Dorsal-fin spines slender, without lateral grooves; first spine much shorter than second spine, lower than associated fin; second dorsal fin and spine about equal in height. Caudal fin lanceolate, relatively short, length of upper lobe 5–5.5 in total length (about 5–5.2 in specimens shorter than 40 cm); upper caudal lobe without subterminal notch; caudal keels and precaudal pits present. Precaudal vertebrae 79–83; monospondylous 37–39.

- **Colour:** Dorsal surfaces and flanks pale greyish; whitish ventrally. Upper surface of pectoral fins greyish, with paler posterior margins; apices of dorsal fins dark. Caudal fin with dark marginal band above posterior notch and dark central blotch on upper lobe (less obvious in adults); tips of upper and lower lobes white.

- **Size:** Attains at least 55 cm; males mature at about 50 cm.

- **Distribution:** Continental slope off Western Australia from the Scott Reef to Perth, in depths of 300–510 m.

- **Remarks:** Unlike most spurdogs, which occur in schools, this species appears to be solitary. Little else is known of its biology.

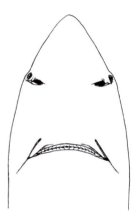

undersurface of head

flank denticle

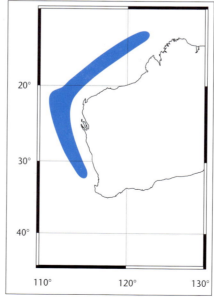

Eastern Longnose Spurdog — *Squalus* sp. F

8.36 *Squalus* sp. F (Plate 5)   **Fish Code:** 00 020041

- **Field characters:** A small, slender dogfish with prominent dorsal spines (the first originating near the pectoral-fin rear tips), similar blade-like teeth in both jaws, a long and pointed snout (horizontal length less than 1.9 times eye diameter, direct length exceeding interorbital width) and lacking white spots. The caudal fin is relatively short (mostly less than 5 times in total length) with a dark marginal bar above the posterior notch.

- **Distinctive features:** Body moderately elongate, fusiform; head relatively slender. Snout elongate, rather pointed; direct preorbital length more than interorbital width; distance from snout to inner nostril exceeding distance from nostril to upper labial furrow; upper labial furrows relatively long; anterior nasal flap well developed, bifurcated. Denticles tricuspidate. Dorsal-fin spines slender, without lateral grooves; first spine slightly shorter than second spine, lower than associated fin; second dorsal fin and spine about equal in height. Caudal fin rather short, length of upper lobe 4.5–5.3 in total length (less than 4.6 in specimens shorter than 40 cm); upper caudal lobe without subterminal notch; caudal keels and precaudal pits present. Precaudal vertebrae 81–86; monospondylous 38–42.

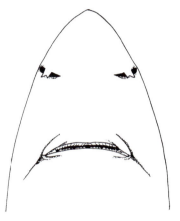

undersurface of head

- **Colour:** Dorsal surfaces and flanks pale greyish; paler ventrally. Posterior margins of pectoral fins white; unpaired fins pale grey, fin apices dark, free tips white; dark stripe between second dorsal fin and caudal fin. Caudal fin with dark marginal bar on posterior notch (about equal to snout length); tip of upper and lower lobes white; anterior margin and centre of upper lobe often dark. Darker markings better defined in small specimens.

flank denticle

- **Size:** To about 64 cm (females appear to be larger than males); males mature by 52 cm; free swimming by 22 cm.

- **Distribution:** Continental slope off Queensland between Cape York and Rockhampton in 220–500 m.

- **Remarks:** This species differs from the closely related western longnose spurdog (8.35) in having a relatively shorter snout, longer first dorsal spine, larger caudal fin and higher average vertebral count. More specimens are required from outside its known distribution.

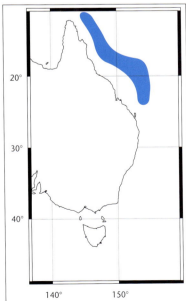

White-Spotted Spurdog — *Squalus acanthias* Linnaeus, 1758

8.37 *Squalus acanthias* (Plate 5)   **Fish Code:** 00 020008

- **Alternative names:** Piked dogfish, spiny dogfish, spotted spiny dogfish, spurdog, white-spotted dogfish, Victorian spotted dogfish.

- **Field characters:** A slender dogfish with prominent dorsal-fin spines (the first originating just behind the pectoral-fin free rear tips), the anterior nasal flap with a single lobe, similar blade-like teeth in both jaws, a relatively long and pointed snout, and with a scattering of white spots on the upper surface and flanks.

- **Distinctive features:** Body slender, fusiform. Snout moderately elongate (eye diameter about half of snout length), rather sharply pointed; direct preorbital length equal to or slightly less than interorbital width; distance from snout to inner nostril more than distance from nostril to upper labial furrow; upper labial furrows relatively long; anterior nasal flap well developed, single-lobed. Denticles tricuspidate (simple in juveniles). First dorsal fin raked slightly backward; dorsal-fin spines slender, without lateral grooves; first dorsal-fin spine much shorter than second dorsal-fin spine, less than half height of associated fin; second dorsal fin and spine about equal in height. Caudal fin relatively broad, short, length of upper lobe about 5.2* in total length; upper caudal lobe without subterminal notch; caudal keels and precaudal pits present. Total vertebrae [96–117]; precaudal vertebrae 75–79 [68–85].

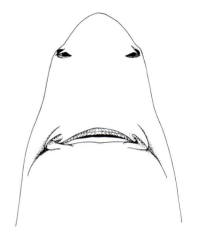

undersurface of head

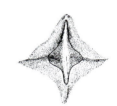

flank denticle

- **Colour:** Dorsal surfaces and flanks bluish to grey with an irregular array of poorly defined white spots (about size of pupil). Pectoral, pelvic and caudal fins with broad, pale posterior margins; undersurface of pectoral fins dusky; apex of first dorsal fin dark, free rear tip pale; apex and free rear tip of second dorsal fin pale. Undersurface of body pale.

- **Size:** Varies considerably on a regional basis; reported to reach 160 cm in the eastern North Pacific; much smaller in other regions. Locally attaining at least 100 cm; males maturing at about 59 cm; born at about 22 cm.

- **Distribution:** Antitropical. Widely distributed in the North Atlantic and Pacific, and around the southern tips of South America, Africa and New Zealand. Reports from off New Guinea are doubtful. In Australian waters, common around Tasmania and Victoria (also recorded from the Great Australian Bight) inshore in bays and estuaries.

- **Remarks:** Considered to be one of the most abundant living sharks, this species penetrates well into brackish water and is occasionally intertidal in southern Tasmania. It breeds inshore in large bays and estuaries and may produce litters of up to 20 young. The reported gestation period of 18–24 months is among the longest of all chondrichthyans. It is also very long-lived, first maturing at 10–25 years and with age estimates up to 70 years. Of little value locally as a food fish because the flesh is considered to be rather coarse. Elsewhere, it is an important commercial species with the annual European catch reaching up to 34 000 tonnes. It is consumed fresh, smoked, boiled, marinated, dried salted or as fish cakes. It is also used for its oil and in the manufacture of leather, pet food, fishmeal and fertiliser. Main prey consists of small fish and crustacea but they will eat molluscs including small scallops.

- **Local synonymy:** *Acanthias vulgaris* Risso, 1826; *Squalus fernandinus*: McCulloch, 1929; *Squalus kirki* Phillipps, 1931; *Squalus whitleyi* Phillipps, 1931.
- **References:** McCulloch (1930); Garrick (1960c).

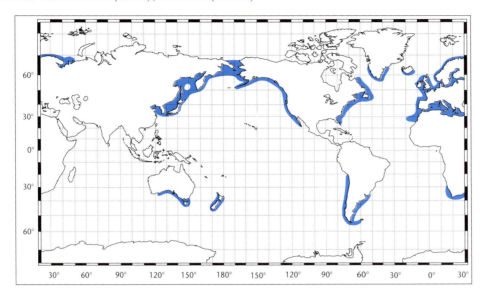

Piked Spurdog — *Squalus megalops* (Macleay, 1881)

8.38 *Squalus megalops* (Plate 5) Fish Code: 00 020006

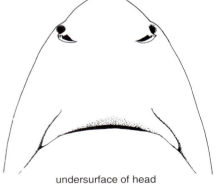
undersurface of head

flank denticle

- **Alternative names:** Dogshark, piked dogfish, skittle dog, shortnose spurdog, shortnose spiny dogfish, spiked dogfish, spurdog, Tasmanian dogfish.
- **Field characters:** A small dogfish with slender dorsal-fin spines (the first originating over the pectoral fins), a raked first dorsal fin, a rather short, deeply concave pectoral fin, similar blade-like teeth in both jaws, a relatively short snout and no white spots on the body. The caudal fin is relatively short and has a white posterior margin (occasionally interrupted by a bar through the base of the lower caudal lobe).
- **Distinctive features:** Body moderately elongate, fusiform. Snout short, bluntly pointed to rounded; distance from snout to inner nostril less than distance from nostril to upper labial furrow; upper labial furrows relatively long; anterior nasal flap well developed, bifurcated. Denticles lanceolate, unicuspidate, without lateral keels. First dorsal fin raked slightly backward; dorsal-fin spines without lateral grooves; first dorsal-fin spine feeble, lower than associated fin, much shorter than second dorsal-fin spine; second dorsal-fin spine slender and narrow-based, equal in height or taller than second dorsal fin. Pectoral-fin posterior margin deeply concave, apex broadly rounded, free rear tip angular; inner margin 1.2–1.5 in anterior margin. Caudal fin

relatively short, length of upper lobe 4.7–5.1 in total length; upper caudal lobe without subterminal notch; caudal keels and precaudal pits present. Precaudal vertebrae 78–82.

- **Colour:** Dorsal surfaces and fins light greyish brown to brownish, paler ventrally. Anterior margin and tip of dorsal fins dark, particularly in juveniles. Caudal-fin anterior margin not dark; posterior margin with a white edge (broadest at apex of lower lobe); sometimes with a dusky bar through the lower lobe which may interrupt the pale posterior margin at the notch. Pectoral fins with broad pale posterior margins. Dorsal midline posterior to first dorsal fin dark (black posterior to second dorsal fin) in new-born young.
- **Size:** Females reach 62 cm, males much smaller (to 47 cm). Males mature at about 34 and 41 cm in tropical and southern waters respectively. Born at about 20–24 cm.
- **Distribution:** Unconfirmed reports off southern Africa, Indo China, New Caledonia and New Hebrides. Locally off southern Australia from Carnarvon (Western Australia) to Townsville (Queensland), including Tasmania. Confined to the continental shelf in southern localities but deeper on the upper slope (to 510 m) in warm temperate and tropical areas.
- **Remarks:** Dogfishes presently referred to in the current literature as *Squalus megalops* appear to belong to a complex of similar species. *S. megalops* appears to be an Australian endemic although specimens from the different regions are still to be compared carefully. Nevertheless, the apparent widespread Australian distribution of the piked spurdog and its absence from New Zealand are surprising. Small quantities are sold at major Australian markets. The flesh is most appealing dried, salted or smoked. Aggregations of a single sex are encountered regularly by trawl fishermen.
- **Local synonymy:** *Squalus tasmaniensis* Rivero, 1936.
- **Reference:** Macleay (1881a).

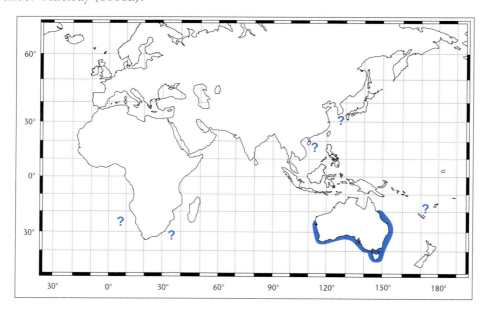

Greeneye Spurdog — *Squalus mitsukurii* Jordan & Snyder, 1903

8.39 *Squalus mitsukurii* (Plate 5) **Fish Code:** 00 020007

- **Alternative names:** Green-eyed dogfish, grey spiny dogfish, shortspine spurdog, spiny dogfish.
- **Field characters:** A medium-sized dogfish with prominent dorsal-fin spines (the first originating near the pectoral-fin rear tips), similar blade-like teeth in both jaws, a broadly pointed snout (direct length less than interorbital width) and lacking white spots. The caudal fin is relatively short (less than 5 times in total length) with a dark marginal bar above the posterior notch.
- **Distinctive features:** Body moderately elongate, fusiform; head relatively robust. Snout moderately elongate, rather pointed; direct preorbital length shorter than interorbital width; distance from snout to inner nostril about equal to distance from nostril to upper labial furrow; upper labial furrows relatively long; anterior nasal flap well developed, bifurcated. Denticles tricuspidate. Dorsal-fin spines slender, without lateral grooves; first spine shorter than second spine, less than three-quarters height of associated fin; second dorsal fin slightly taller than spine. Caudal fin variable, length of upper lobe 4–5 in total length; upper caudal lobe without subterminal notch; caudal keels and precaudal pits present. Precaudal vertebrae 84–87*; monospondylous 43–46*.
- **Colour:** Dorsal surfaces and flanks greyish, paler ventrally. Posterior notch of caudal fin and centre of upper lobe usually dark (sometimes barely evident in adults but pronounced in juveniles); upper half of dorsal fins mostly dark.
- **Size:** Attains at least 76 cm (reported elsewhere to 110 cm); males mature at about 61 cm; about 22 cm at birth.
- **Distribution:** Inadequately defined. Considered to be widely distributed in temperate and subtropical parts of most oceans but possibly consisting of a species complex. The Australian form occurs around southern and eastern Australia from off Townsville (Queensland) to Shark Bay (Western Australia) on the continental slope in 180–600 m.

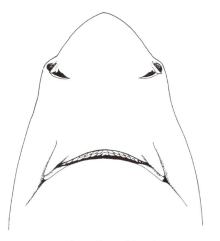

undersurface of head

flank denticle

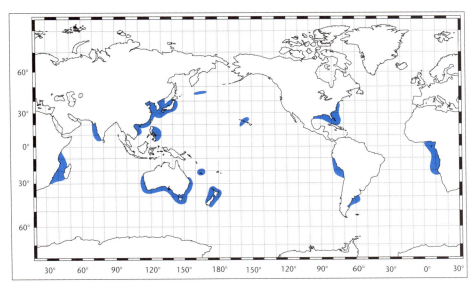

- **Remarks:** Little is known of the biology of this species in Australian waters. A related form off southern Africa is reported to give birth to 4–9 young in autumn after a gestation period of 1–2 years. It is carnivorous on small fish, cephalopods and crustaceans. Small quantities are sold in Australia under the marketing name of 'greeneye dogfish'.
- **Local synonymy:** *Squalus blainvillei*: Last *et al.*, 1983.
- **References:** Garrick (1960c); Chen *et al.* (1979).

Velvet Dogfish — *Zameus squamulosus* (Günther, 1877)

8.40 *Zameus squamulosus* (Plate 3) **Fish Code:** 00 020042

- **Field characters:** A dark brownish to black dogfish with very small dorsal-fin spines, rather rounded pectoral fins, an asymmetric caudal fin that is uniformly dark and has a deep subterminal notch, and with large, semi-erect, blade-like lower teeth.
- **Distinctive features:** Body fusiform, no keels on caudal peduncle. Snout relatively long (preoral length greater than mouth width); upper teeth small, narrow and erect; lower teeth larger, blade-like and semi-erect. Flank denticles with prominent longitudinal ridges and smaller transverse ridges on their outer surfaces. Origin of first dorsal fin closer to free rear tips of pectoral-fin than to pelvic-fin origin; second dorsal-fin origin over pelvic-fin base. Dorsal fin spines small; second dorsal fin larger and more triangular than first dorsal fin. Pectoral fin lobe-like, apex rounded. Caudal fin with asymmetric fin lobes and a strong subterminal notch. Tooth count [47–60/32–38]. Total vertebrae 102* [93–105]; precaudal 73* [66–76].

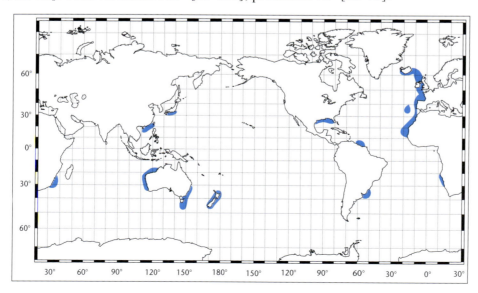

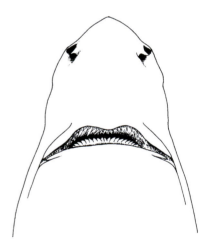

undersurface of head

third teeth from symphysis (upper and lower jaw)

- **Colour:** Uniformly chocolate brown or blackish; no pale areas on the caudal fin.
- **Size:** Attains at least 84 cm. Males mature at about 47 cm.
- **Distribution:** Eastern Atlantic (Iceland to Senegal), western Atlantic (Gulf of Mexico, Surinam and southern Brazil), western Indian Ocean (southern Africa) and the western Pacific from Japan, South China Sea, New Zealand and Australia (from New South Wales to Tasmania and off Western Australia). Demersal or pelagic near continental slopes and seamounts in depths of 550–2000 m.
- **Remarks:** Caught off southern Australia by Japanese longliners in midwater but rarely taken by demersal trawlers. Off northern Western Australia, however, often taken near the bottom. In the eastern Atlantic, used dried and salted for human consumption and for fish meal.
- **Local synonymy:** *Scymnodon squamulosus* (Günther, 1877).
- **References:** Yano and Tanaka (1984b); Taniuchi and Garrick (1986).

9

Family Oxynotidae

PRICKLY DOGFISHES

Members of this family, also known as rough sharks, are unmistakable in having a hump-backed body that is nearly triangular in cross-section, prominent ridges between the pectoral and pelvic fins, rough skin and no anal fin. The dorsal fins are large and tall and have spines. They also have a slot-like mouth with very long labial furrows (almost encircling the mouth), large spiracles, lanceolate teeth in the upper jaw and blade-like teeth in the lower jaw.

Prickly dogfishes share a number of important features with the family Squalidae, to which they are very closely related. The group is represented worldwide by four species of which one occurs in Australian waters.

Prickly Dogfish — *Oxynotus bruniensis* (Ogilby, 1893)

9.1 *Oxynotus bruniensis* (Plate 11) **Fish Code:** 00 021001

- **Field characters:** A bizarre hump-backed dogfish with rough skin, prominent abdominal ridges and sail-like dorsal fins with short spines.

- **Distinctive features:** Body humped behind head, almost triangular in cross-section; abdominal ridge present between each pectoral and pelvic fin. Snout short and blunt, mouth slot-like; nostrils broad. Upper teeth lanceolate; lowers blade-like. Skin very rough; denticles with an elongate median blade-like cusp. First dorsal fin consisting of a large, fleshy portion covered in denticles anteriorly with an upright, sail-like portion behind the dorsal spine; fin origin over pectoral-fin base. Second dorsal fin upright, triangular; smaller than first dorsal fin. First dorsal-fin spine inclined forwards (partly concealed beneath skin but can be felt); second dorsal-fin spine inclined backwards; spines barely protruding. Caudal fin with subterminal notch. Tooth count 14–18/11–13*. Total vertebrae 87–94*; precaudal 58–63*.

- **Colour:** Uniformly brownish or greyish; tips of dorsal fins and posterior margins of pectoral and pelvic fins white or translucent.
- **Size:** Born at about 24 cm and attains at least 72 cm; males mature at about 60 cm.
- **Distribution:** Confined to temperate waters off New Zealand and southern Australia from Newcastle (New South Wales) to the western Great Australian Bight, including Tasmania. Lives near the bottom on continental and insular shelves and slopes at depths of 45–650 m; most frequently taken in 350–650 m.
- **Remarks:** Little is known of the biology of this species. Ovoviviparous; one female contained seven embryos. Not utilised commercially.
- **Reference:** Garrick (1960b).

undersurface of head

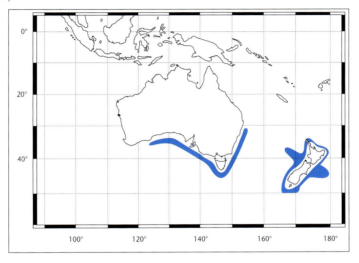

third teeth from symphysis (upper and lower jaw)

10 Family Pristiophoridae
SAWSHARKS

Sawsharks have elongate, subcylindrical to slightly depressed bodies and a blade-like snout armed with sharp, lateral, tooth-like denticles (rostral teeth) and a pair of barbels on the ventral surface. Their eyes are situated dorsolaterally and they have five or six pairs of gill slits, two dorsal fins without spines and they lack an anal fin. The caudal fin axis is almost straight, and both upper and lower lobes are slender. The spiracles, located behind and above the eyes, are large. The caudal peduncle has lateral ridges and the pectoral fins are well developed, but are not ray-like. Unlike sawfishes (family Pristidae), which also have a blade-shaped snout with rostral teeth, the gills are situated on the side of the head rather than on the undersurface.

This group consists of two genera and seven or more species. All but one of these (a Caribbean species) occur in the Indo–Pacific. Four members of the genus *Pristiophorus* are found only on the continental shelf and upper slopes of temperate and tropical Australia. Sawsharks may grow to 1.4 m. They are considered to be harmless, although the rostral teeth, which consist of enlarged embryological denticles interspersed with smaller teeth that form after birth, are sharp so specimens should be handled carefully. All species are ovoviviparous.

Key to pristiophorids

1 Nostrils situated about halfway from barbels to corner of mouth (fig. 1); width at nostrils less than 4.5 in preoral snout in adults; nostrils distinctly oval .... **Pristiophorus nudipinnis (fig. 3; p. 111)**

Nostrils situated about two-thirds way from barbels to corner of mouth (fig. 2); width at nostrils more than 4.5 in preoral snout in adults; nostrils almost circular 2

2 Body with a pattern of dark blotches and spots (occasionally faint) (fig. 5); preoral snout length more than 2.3 times distance from barbels to snout tip (fig. 4) .... **Pristiophorus cirratus (fig. 5; p. 109)**

Body uniform in colour; preoral snout length less than 2.3 times distance from barbels to snout tip (fig. 6) 3

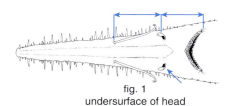
fig. 1
undersurface of head

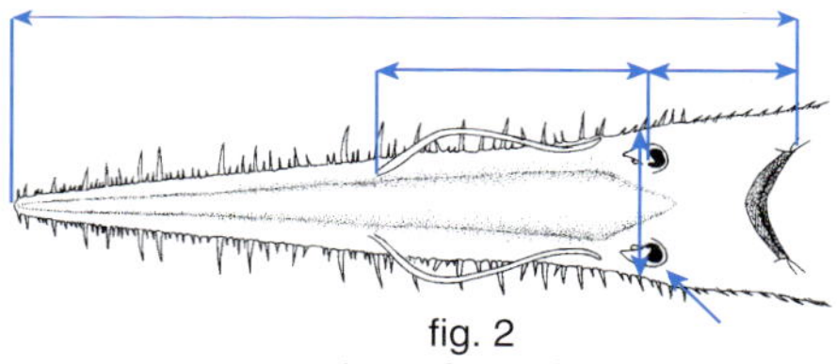

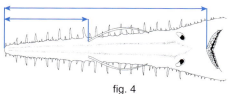
fig. 2
undersurface of head

fig. 4
undersurface of head

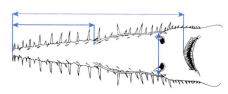
fig. 6
undersurface of head

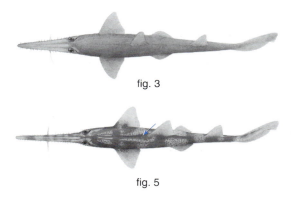
fig. 3

fig. 5

3   Greyish; width of snout at nostrils less than 5.1 in preoral snout length (fig. 6); New South Wales .... ***Pristiophorus* sp. A (fig. 7; p. 107)**

Yellowish; width of snout at nostrils more than 5.1 in preoral snout length (fig. 2); Queensland .... ***Pristiophorus* sp. B (fig. 8; p. 108)**

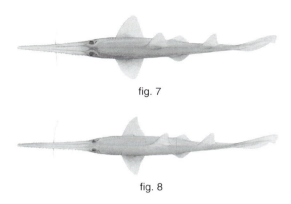

fig. 7

fig. 8

Eastern Sawshark — *Pristiophorus* sp. A

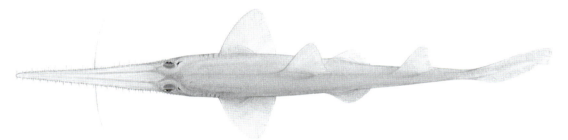

10.1 *Pristiophorus* sp. A (Plate 44)   **Fish Code:** 00 023003

- **Field characters:** A medium-sized, uniformly grey shark with a long, tapering, saw-shaped snout and gills on the side of the head. The barbels are located slightly closer to the snout tip than to the mouth.
- **Distinctive features:** Body slender (trunk often robust), head depressed. Snout very long (preoral length 25–30*% of total length), moderately broad based, tapering evenly; width at nostrils 4.7–5.1* in preoral length. Barbels extending to less than an eye diameter short of nostrils in adults (reaching nostrils in juveniles); distance from snout tip to barbels about 44–47%* of preoral snout length. Rostral teeth extending for about an eye diameter behind eye; about 20–25* enlarged teeth on each side of head (about 11–15* in front of barbels); juveniles rarely have more than one smaller tooth between large teeth. Five pairs of gill slits. Nostrils large, oval, with a prominent anterior lobe; distance from mouth to nostrils 1.3* or more times internarial space. Spiracle large, about three-quarters eye diameter. Mouth large, extending forward to near hind margin of eye; labial furrows short, directed posteriorly. Oral teeth short, closely spaced, with short single cusps; 34–40* rows in upper jaw. Skin velvety; denticles minute, with single, plate-like cusps; outer margins of dorsal and pectoral fins covered with denticles. Pectoral fins large (half of preoral length or slightly less). First dorsal-fin origin located behind free rear tips of pectoral fins by eye diameter or less, its insertion in advance of pelvic-fin origin; second dorsal fin slightly smaller than first dorsal fin; distance from pelvic-fin insertion to second dorsal-fin origin equal to or less than length of first dorsal-fin base.
- **Colour:** Uniformly greyish brown above, white below. Darker brownish stripes extending along middle of saw; edge of saw darker. Juveniles with narrow dark anterior margin to pectoral and dorsal fins; no other fin markings.
- **Size:** Females to at least 107 cm. No males examined.

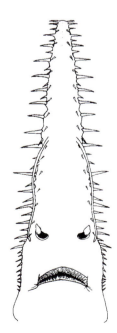

undersurface of head

- **Distribution:** Warm temperate eastern Australia from Lakes Entrance (Victoria) to Coffs Harbour (New South Wales) in 100–630 m depth.

- **Remarks:** This species appears to be confined to the continental shelf and upper slope off New South Wales. It is similar to another undescribed species that occurs off northern Queensland, but attains a larger size, is greyish in colour (rather than yellowish) and has a relatively shorter and broader snout. Little is known of its biology.

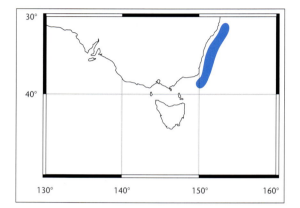

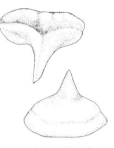

anterior oral teeth
(upper and lower)

Tropical Sawshark — *Pristiophorus* sp. **B**

10.2 *Pristiophorus* sp. **B** (Plate 44) **Fish Code:** 00 023004

- **Field characters:** A small, uniformly yellowish brown sawshark with barbels located almost half way along the prenasal length, and gills on the side of the head. The snout is very long, slender and tapers only slightly.

- **Distinctive features:** Body very slender, head depressed. Snout extremely long (preoral length 29–31% of total length), very slender, tapering slightly; width at nostrils 5.2–6 in preoral length. Barbels extending to less than an eye diameter short of nostrils in adults (reaching nostrils in juveniles); distance from snout tip to barbels about 45–51% of preoral snout length. Rostral teeth extending for about an eye diameter behind eye; usually more than 15* enlarged teeth on each side of head (about 12–15* in front of barbels); juveniles often with 2–3 smaller teeth between large teeth. Five pairs of gill slits. Nostrils large, oval, with a prominent anterior lobe; distance from mouth to nostrils more than 1.3 times internarial space. Spiracle moderately large, less than half eye diameter. Mouth large, extending forward to hind margin of eye; labial furrows short, directed posteriorly. Oral teeth short, closely spaced; 38–40* rows in upper jaw. Skin velvety; denticles minute, with single plate-like cusps; outer margins of dorsal and pectoral fins covered with denticles. Pectoral fins rather large (much less than half preoral length). First dorsal-fin origin located only slightly behind free rear tips of pectoral fins (by less than an eye diameter), its insertion in advance of pelvic-fin origin; second dorsal fin slightly smaller than first dorsal fin; distance from pelvic-fin insertion to second dorsal-fin origin exceeding length of first dorsal-fin base.

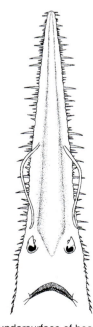

undersurface of head

- **Colour:** Uniformly pale yellowish brown above, white below. Rostral teeth not dark-edged; fins uniform, without dark markings.
- **Size:** To at least 84 cm. Males mature at about 62 cm.
- **Distribution:** Known to occur on the continental slope off tropical northeastern Australia between Rockhampton and Cairns in 300–400 m depth.
- **Remarks:** Several specimens of this undescribed species were collected during a recent exploratory fishing survey to northern Queensland. Nothing is known of its distribution outside the survey area.

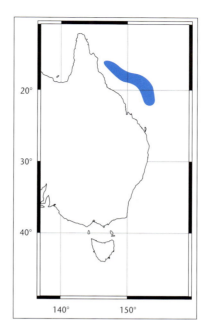

Common Sawshark — *Pristiophorus cirratus* (Latham, 1794)

10.3 *Pristiophorus cirratus* (Plate 44) **Fish Code:** 00 023002

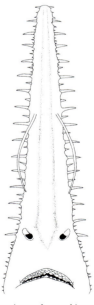

undersurface of head

- **Alternative name:** Longnose sawshark.
- **Field characters:** A medium-sized shark with a saw-shaped snout, a pattern (sometimes dull) of brownish spots and blotches on a sandy background, and gills on the side of the head. The barbels are located about half way along the prenasal length.
- **Distinctive features:** Body slender (more robust in larger adults), head depressed. Snout very long (preoral length 26–30% of total length), relatively slender, tapering slightly; width at nostrils 5.1–5.6 in preoral length. Barbel tip more than an eye diameter short of nostrils in adults (barbels longer in juveniles and extending to or beyond mouth in new born); distance from snout tip to barbels about 40–43% of preoral snout length. Rostral teeth extending for about 1–1.5 eye diameters behind eye; about 19–21 enlarged teeth on each side of head when undamaged (9–11 in front of barbels); juveniles with 2–3 smaller teeth between large teeth. Five pairs of gill slits. Nostrils large, oval, with a prominent anterior lobe; distance from mouth to nostrils more than 1.3 times internarial space. Spiracle large, about three-quarters eye diameter. Mouth large, extending forward to hind margin of eye; labial furrows short, directed posteriorly. Oral teeth short, closely spaced; 33–39* rows in upper jaw. Skin velvety; denticles minute, with single, plate-like cusps; outer margins of dorsal and pectoral fins covered with denticles. Pectoral fins large (mostly about half

preoral length but slightly less than half in smaller individuals). First dorsal-fin origin located behind free rear tips of pectoral fins (sometimes by an eye diameter or less), its insertion in advance of pelvic-fin origin; second dorsal fin equal to or slightly smaller than first dorsal fin; distance from pelvic-fin insertion to second dorsal-fin origin exceeding length of first dorsal-fin base.

- **Colour:** Upper surface pale yellowish to greyish brown; mostly with darker bands between pectoral-fin bases, over gill slits, between spiracles and below dorsal fins; paler areas mostly with irregular peppering of large dark brown spots and blotches; similar fainter, markings on dorsal and caudal fins; other fins uniformly greyish brown. Saw pinkish with two parallel, dark median stripes and blackish lateral margins; rostral teeth with dark margins. Ventral surface uniformly white, abruptly demarcated from darker upper surface.
- **Size:** To at least 134 cm. Males mature at about 97 cm. Born at about 38 cm.
- **Distribution:** A widely distributed temperate and subtropical Australian endemic with a poorly defined distribution. Possibly occurring along the southern coast from Eden (New South Wales) to Jurien Bay (Western Australia), including Tasmania. Demersal on the continental shelf in 40–310 m depth.
- **Remarks:** The largest and most widely distributed of the Australian sawsharks. It overlaps in distribution with the greyish coloured, southern sawshark (10.4) but lives mostly in deeper water. Small quantities reach fish markets from the bycatch of trawlers and gillnetters. The flesh is of good quality and the saw is sometimes used for ornamental purposes.
- **Local synonymy:** *Squalus anisodon* Lacépède, 1802; *Squalus tentaculatus* Shaw, 1804.

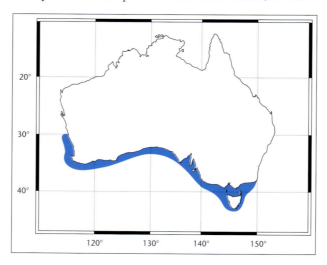

Southern Sawshark — *Pristiophorus nudipinnis* Günther, 1870

10.4 *Pristiophorus nudipinnis* (Plate 44) **Fish Code:** 00 023001

- **Field characters:** A medium-sized, uniformly grey shark with a comparatively short, broad-based, saw-shaped snout, with barbels distinctly closer to the mouth than to the snout tip, and gills on the side of the head.

- **Distinctive features:** Body slender, head depressed. Snout long (preoral length 22–24% of total length), relatively broad-based, abruptly tapering; width at nostrils 3.7–4.3 in preoral length. Barbels extending posteriorly to at least level of nostrils in adults (extending beyond mouth in new born); distance from snout tip to barbels about 58–60% of preoral snout length. Rostral teeth extending for slightly more than an eye diameter behind eye; about 17–19* enlarged teeth on each side of head (about 12–13* in front of barbels); juveniles with mostly one smaller tooth between each pair of large teeth. Five pairs of gill slits. Nostrils large, obliquely ovoid, with a prominent anterior lobe; distance from mouth to nostrils 1–1.2 times internarial space. Spiracle large, about three-quarters eye diameter. Mouth large, extending forward to beneath posterior half of eye; labial furrows short, directed posteriorly. Oral teeth short, closely spaced; 32–37* rows in upper jaw. Skin velvety; denticles minute, with single plate-like cusps; outer margins of dorsal and pectoral fins frequently smooth in adults. Pectoral fins large (about half preoral length). First dorsal-fin origin located behind free rear tips of pectoral fins by eye diameter or more, its insertion in advance of pelvic-fin origin; second dorsal fin about equal in size to first dorsal fin; distance from pelvic-fin insertion to second dorsal-fin origin almost twice length of first dorsal-fin base.

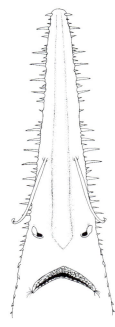

undersurface of head

- **Colour:** Uniformly slate grey, fins lacking spots or markings. Snout with indistinct median stripes; lateral margins of both saw and rostral teeth sometimes dusky. Ventral surface pale.

- **Size:** To at least 99 cm; males are mature by 90 cm; born at about 25 cm.

- **Distribution:** An Australian endemic occurring off Tasmania and Victoria through to the western sector of the Great Australian Bight. Inner continental shelf to 70 m depth.

- **Remarks:** Caught commercially in small quantities. Occasionally ventures into shallow bays in Tasmania, where it is occasionally snared by gillnets. Sawsharks use their sensory barbels to locate food in the sand and then uproot their prey with vigorous movements of the snout.

- **Local synonymy:** *Pristiophorus owenii* Günther, 1870.

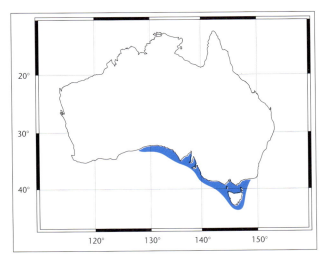

11

Family Heterodontidae

HORN SHARKS

Distinctive sharks with enlarged, rather blunt heads that have a prominent crest above each eye, spiny dorsal fins, an anal fin, and a small, almost terminal mouth. Their nostrils are connected to the mouth by a groove and they have small spiracles, molar-like posterior teeth and rough skin.

These medium-sized sharks (up to 1.6 m) generally have restricted distributions in tropical and warm temperate seas. They are oviparous and live on or near the bottom, usually in fairly shallow water. Three of the eight recognised species occur in Australian waters.

Key to heterodontids

1 Supra-orbital crests very high and ending abruptly behind the eyes (fig. 1); an indistinct colour pattern of broad, dark bars and saddles (fig. 1) .... ***Heterodontus galeatus* (fig. 1; p. 113)**

Supra-orbital crests relatively low and not ending abruptly behind the eyes (fig. 2); a distinctive colour pattern of narrow bands or stripes (figs 2, 3) **2**

2 A colour pattern of numerous, narrow, dark bands on a pale background (fig. 2) .... ***Heterodontus zebra* (fig. 2; p. 115)**

A colour pattern of dark, oblique stripes beyond the first dorsal fin (fig. 3) .... ***Heterodontus portusjacksoni* (fig. 3; p. 114)**

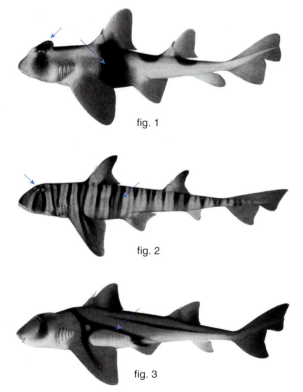

fig. 1

fig. 2

fig. 3

Crested Horn Shark — *Heterodontus galeatus* (Günther, 1870)

11.1 *Heterodontus galeatus* (Plate 2)

Fish Code: 00 007003

- **Alternative names:** Crested bullhead shark, crested Port Jackson shark.
- **Field characters:** A horn shark with a large, blunt head, high supra-orbital crests terminating abruptly behind the eyes, dorsal-fin spines, an anal fin, and with broad, dark bars, and saddles that do not form a harness-like pattern.
- **Distinctive features:** Body stocky. Head large and blunt; high supra-orbital crests with nearly vertical posterior edges; nostrils connected to mouth by grooves; gill slits moderately large; mouth small, almost terminal. Teeth similar in both jaws; anteriors small, pointed, multicuspid in juveniles; posteriors molar-like with medial ridges. Skin rough, denticles enlarged. First dorsal-fin origin over pectoral-fin bases. Both dorsal fins preceded by spines. Dorsal fins very high and with rounded tips in juveniles, relatively lower in adults. Pectoral fins very broad, tips broadly rounded. Anal fin present; caudal fin with distinct ventral and subterminal lobe.
- **Colour:** Mostly yellowish brown; interorbital region and back anterior to first dorsal fin dark; broad dark bars on cheek and below first dorsal fin; dark saddles between dorsal fins and behind insertion of second dorsal fin; lacking a harness-like pattern.
- **Size:** Young hatch at about 22 cm and reported to attain 130 cm. Males mature at about 60 cm and females at about 70 cm.
- **Distribution:** Cape Moreton (southern Queensland) to Batemans Bay (New South Wales), possibly also Cape York. Continental shelf from close inshore to about 90 m depth.
- **Remarks:** The egg-cases, which are are spirally flanged and have long tendrils at their apices, are laid in 20–30 m amongst seaweed or sponges during July or August. The eggs hatch after about 8 months. One individual kept in an aquarium grew at about 5 cm per year, and laid eggs at 11 years of age when it was 70 cm long. The diet consists of invertebrates, including echinoids, crustaceans and molluscs, as well as small fish. Not utilised commercially in Australia.
- **Local synonymy:** *Molochophrys galeatus* (Günther, 1870).
- **Reference:** McLaughlin and O'Gower (1971).

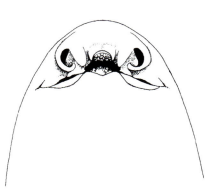

undersurface of head

anterior tooth of upper jaw

posterior tooth of upper jaw

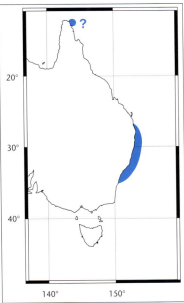

Port Jackson Shark — *Heterodontus portusjacksoni* (Meyer, 1793)

11.2 *Heterodontus portusjacksoni* (Plate 2)

Fish Code: 00 007001

- **Alternative names:** Bullhead, oyster crusher, tabbigaw.
- **Field characters:** A distinctive horn shark with a large, blunt head, supra-orbital crests that slope away gradually behind the eyes, dorsal-fin spines, an anal fin, and dark markings on the body in the shape of a harness.
- **Distinctive features:** Body stocky. Head large and blunt; supra-orbital crests with gradually sloping posterior edges; nostrils connected to mouth by grooves; gill slits large; mouth small, almost terminal. Teeth similar in both jaws; anteriors small, pointed, multicuspid in juveniles, unicuspid in adults; posteriors large, flat, molar-like and without medial ridges. Skin rough, denticles enlarged. First dorsal-fin origin over pectoral-fin bases. Both dorsal fins preceded by spines. Dorsal fins high with narrowly pointed tips in juveniles, relatively lower and with rounded tips in adults. Pectoral fins very broad, tips rather pointed. Anal fin present; caudal fin with distinct ventral and subterminal lobe.
- **Colour:** Mostly greyish (sometimes brownish) with a dark bar between the eyes extending down the cheeks; dark harness-like bars below the first dorsal fin, and with a few oblique stripes posteriorly (markings often pale in large adults).
- **Size:** Hatches at about 23 cm and reported to attain 165 cm (normally much smaller). Males mature at about 75 cm and females between 80 and 95 cm.
- **Distribution:** Southern Australia from Byron Bay (New South Wales) to the Houtman Abrolhos (Western Australia), including Tasmania. Questionable records from York Sound (northern Western Australia) and Moreton Bay (Queensland). A single record from New Zealand. Continental shelf from close inshore down to 275 m.

undersurface of head

anterior tooth of upper jaw of adult

posterior tooth of upper jaw

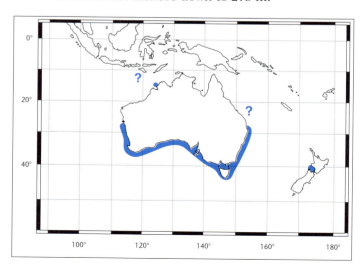

- **Remarks:** Research has shown that this species segregates by sex and maturity stage. Sharks found on New South Wales sublittoral reefs consist mainly of adult females. They forage at night, returning to particular caves and rocky gullies during the day; these resting sites may be reused in the future. On the central New South Wales coast, breeding occurs in late winter and spring, when females lay 10–16 eggs in rock fissures on inshore reefs (usually less than 5 m depth). The eggs are about 15 cm long, 8 cm wide and have spiral flanges (which help retain them in the rock fissures) but lack tendrils. The young hatch after about 12 months and grow to maturity in 8–10 years (males) and 11–14 years (females). Adult females migrate south in summer, some as far as 800 km, returning north in winter. They feed mainly on benthic invertebrates, with echinoderms the most important prey. Caught as bycatch in the southern shark gillnet fishery but not utilised for food.
- **Local synonymy:** *Squalus jacksoni* Suckow, 1799; *Squalus philippi* Bloch & Schneider, 1801; *Squalus philippinus* Shaw, 1804; *Cestracion heterodontus* Sherrard, 1896; *Heterodontus bonaespei* Ogilby, 1908.
- **References:** Ogilby (1908a); McLaughlin and O'Gower (1970, 1971); O'Gower and Nash (1978).

Zebra Horn Shark — *Heterodontus zebra* (Gray, 1831)

11.3 *Heterodontus zebra* (Plate 2)　　　　　　　　　　　　　　　**Fish Code:** 00 007002

- **Alternative names:** Zebra bullhead shark, zebra Port Jackson shark, bullhead shark.
- **Field characters:** A horn shark with a large, blunt head, low supra-orbital crests sloping away gradually behind the eyes, dorsal fin spines, an anal fin, and a striking zebra-pattern of dark, narrow bands on a pale background.
- **Distinctive features:** Body stocky. Head large and blunt; low supra-orbital crests with gradually sloping posterior edges; nostrils connected to mouth by grooves; moderate-sized gill slits; mouth small, almost terminal. Teeth similar in both jaws; anteriors small, pointed, multicuspid in juveniles; posteriors molar-like with medial ridges. Skin rough, denticles enlarged. First dorsal-fin origin over pectoral-fin bases. Both dorsal fins preceded by spines. Dorsal fins very high and with rounded tips in juveniles, relatively lower in adults. Pectoral fins very broad, tips narrowly rounded. Anal fin present; caudal fin with distinct ventral and subterminal lobe.
- **Colour:** Pale brownish to white with a dense pattern of narrow, dark vertical bands; bands frequently extending onto fins.
- **Size:** Attains about 120 cm. Males mature at about 84 cm.

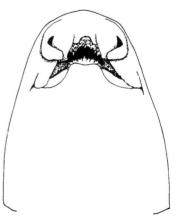

undersurface of head

- **Distribution:** Western Pacific from Japan, Korea, China, Vietnam and Indonesia. Recorded recently from the continental shelf of northern Western Australia in 150–200 m (elsewhere mostly found in depths shallower than 50 m).
- **Remarks:** Oviparous, but little is known of its biology. Probably feeds on bottom invertebrates and small fishes. Not utilised commercially.
- **Reference:** Sainsbury *et al.* (1985).

anterior tooth of upper jaw

posterior tooth of upper jaw

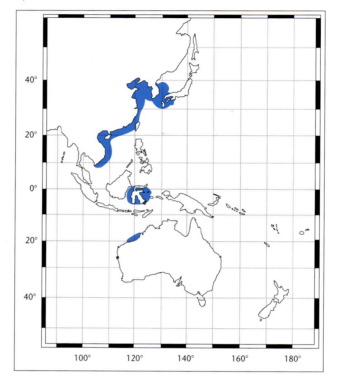

12 Family Parascylliidae
COLLARED CARPET SHARKS

The members of this family of small (shorter than 1 m), slender sharks have a tiny, subterminal mouth situated well in front of the eyes, a first dorsal-fin origin behind the pelvic-fin bases, short nasal barbels (but no throat barbels) and very small spiracles. The mouth is connected to the nostrils by a shallow groove located beneath the nasal flap (nasoral grooves), and smaller fleshy folds and grooves skirt the nostrils laterally. Their anal fin originates well in front of the second dorsal-fin origin but its distance from the lower caudal-fin lobe is at least equal to the length of the anal-fin base. The caudal fin, which lacks a ventral lobe but has a subterminal notch, is much shorter than the rest of the body. These sharks also have five gill slits, spineless dorsal fins, and lack dermal lobes around the head, ridges on the body and caudal keels.

The family is comprised of two genera but is represented in Australian waters by a single genus. All four local species are endemic, occurring on or near the bottom on the continental shelf and upper slope. They resemble some of the catsharks of the family Scyliorhinidae but differ from them in having the mouth well in front of the eyes, and in having nasoral grooves.

Key to parascylliids

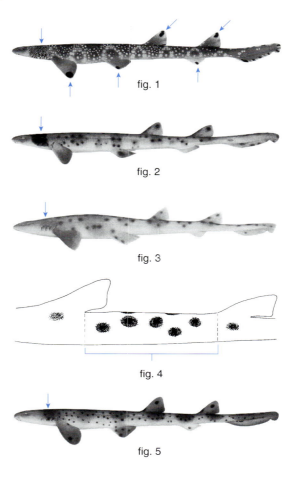

fig. 1

fig. 2

fig. 3

fig. 4

fig. 5

1   Dark collar around gill region peppered with dense white spots (fig. 1); white spots over body; bold black blotches near fin margins (fig. 1) .... ***Parascyllium variolatum* (fig. 1; p. 121)**

   With or without a dark collar around gill region but without dense white spots (figs 2, 3); only dark spots over body; no bold black blotches near fin margins **2**

2   A conspicuous dark brownish collar-like marking around the gill region (fig. 2); collar sharply demarcated from anterior part of head .... ***Parascyllium collare* (fig. 2; p. 119)**

   Collar-like marking around the gill region absent (fig. 3), pale or indistinct (fig. 5); collar not sharply demarcated from anterior part of head **3**

3   Less than six dark spots on each flank between the dorsal fins (fig. 4); few (if any) discrete spots along dorsal midline; restricted to deepwater (>200 m) .... ***Parascyllium* sp. A (fig. 3; p. 118)**

   More than six spots on each flank between the dorsal fins; at least some discrete spots along dorsal midline; restricted to shallow water (< 200 m) .... ***Parascyllium ferrugineum* (fig. 5; p. 120)**

Ginger Carpet Shark — *Parascyllium* sp. A

12.1 *Parascyllium* sp. A (Plate 16) **Fish Code:** 00 013018

- **Field characters:** An elongate, light brown carpet shark with an indistinct half 'collar' around the gills, indistinct dusky saddles, and large, dark, diffuse spots and blotches on the body and fins (fewer than 6 on the flanks between the dorsal fins).
- **Distinctive features:** Body elongate and tubular. Head and snout slightly depressed; nasal barbels short; nasoral and circumnarial grooves present. Eyes small, oval and entirely posterior to a small mouth. Teeth slender and triangular in both jaws. First dorsal-fin origin behind pelvic-fin bases; second dorsal-fin origin over, or slightly in front of anal-fin insertion. Dorsal fins about equal in size; pectoral fins rounded; caudal fin with distinct subterminal lobe and undeveloped ventral lobe.
- **Colour:** Pale brownish or greyish above, paler ventrally. An indistinct dark saddle through the gill region and about six faint saddles on body. Only a few darker spots and enlarged blotches on body and fins, those on the dorsal midline usually coalescing to form elongated blotches.
- **Size:** Attains at least 79 cm.
- **Distribution:** Currently known from only three specimens taken on the continental slope off Western Australia between Lancelin and Bunbury in depths of 245–435 m.
- **Remarks:** Unlike other Australian members of the family, this species appears to occur mainly on the continental slope rather than on the shelf. It is presumably oviparous. More specimens are required.

undersurface of head

third teeth from symphysis (upper and lower jaw)

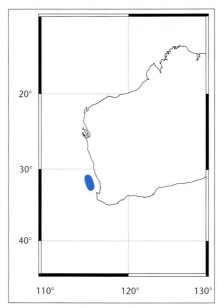

Collared Carpet Shark — *Parascyllium collare* Ramsay & Ogilby, 1888

12.2 *Parascyllium collare* (Plate 16) **Fish Code:** 00 013002

- **Alternative name:** Collared catshark.
- **Field characters:** An elongate, light brown carpet shark with a broad, dark, unspotted 'collar' through the gills, dusky saddles, and large, dark, diffuse spots on the body and fins.
- **Distinctive features:** Body elongate and tubular. Head and snout slightly depressed; nasal barbels short; nasoral and circumnarial grooves present. Eyes small, oval and entirely posterior to small mouth. Teeth slender and triangular in both jaws. First dorsal-fin origin behind pelvic-fin bases; second dorsal-fin origin over to slightly forward of anal-fin insertion. Dorsal fins about equal in size; pectoral fins rounded; caudal fin with distinct subterminal lobe and undeveloped ventral lobe.
- **Colour:** Yellowish brown above (light brown in preservative), paler below. A dark, unspotted 'collar' around gill region, 4–5 dusky saddles on body and large, dark spots on body and fins.
- **Size:** Attains about 86 cm.
- **Distribution:** From Mooloolaba (southern Queensland) to Gabo Island (Victoria). No positive records from Tasmania. Demersal on the continental shelf in depths from 20–160 m.
- **Remarks:** This species appears to have been confused with other dark spotted members of the genus and is less widely distributed than first thought. It is oviparous. Not utilised commercially in Australia.

undersurface of head

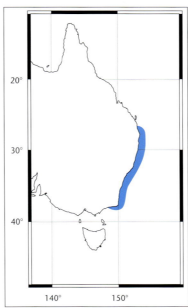

Rusty Carpet Shark — *Parascyllium ferrugineum* McCulloch, 1911

12.3 *Parascyllium ferrugineum* (Plate 16)   **Fish Code:** 00 013005

- **Alternative name:** Rusty catshark.
- **Field characters:** An elongate, greyish or brownish carpet shark with an indistinct dusky 'collar' around the gills, 5 or 6 dusky saddles on the dorsal surface of the body and tail, and small, sharply defined brown spots on the body and fins (more than 6 on the flanks between the dorsal fins).
- **Distinctive features:** Body elongate and tubular. Head and snout slightly depressed; nasal barbels short; nasoral and circumnarial grooves present. Eyes small, oval and entirely posterior to a small mouth. Teeth slender and triangular in both jaws. First dorsal-fin origin behind pelvic-fin bases; second dorsal-fin origin about over anal-fin insertion. Dorsal fins about equal in size; pectoral fins rounded; caudal fin with distinct subterminal lobe and undeveloped ventral lobe.
- **Colour:** Greyish to brown above, paler below. An indistinct dusky 'collar' around the gill region and 5 or 6 dusky saddles on dorsal surface of body and tail interspersed with fainter blotches on flanks. Brown spots on the body and fins; spot density and size variable between individuals.
- **Size:** Attains about 80 cm. Males mature by 60 cm.
- **Distribution:** From Gabo Island (Victoria) to Albany (southern Western Australia), including Tasmania. Continental shelf in depths from 5–150 m.
- **Remarks:** Oviparous, but the biology of this species is poorly known. Tasmanian specimens have a variable but greater average density of spots which has led to their recognition as a separate species. Not utilised commercially in Australia.
- **Local synonymy:** *Parascyllium multimaculatum* Scott, 1935.
- **References:** Scott (1935); Edgar *et al.* (1982).

undersurface of head

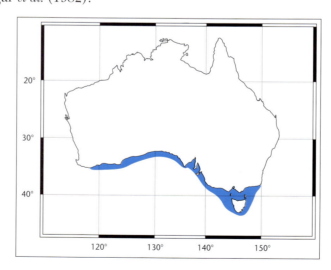

Varied Carpet Shark — *Parascyllium variolatum* (Duméril, 1853)

12.4 *Parascyllium variolatum* (Plate 16)　　　　　　　　　**Fish Code:** 00 013004

- **Alternative names:** Southern catshark, necklace carpet shark, varied catshark.
- **Field characters:** An elongate carpet shark with a striking colour pattern consisting of a broad, dark brownish or black 'collar' (peppered with white spots) through the gill area, white spots on the body, and dark brown blotches and white spots near the fin margins.
- **Distinctive features:** Body elongate and tubular. Head and snout slightly depressed; nasal barbels short; nasoral and circumnarial grooves present. Eyes small, oval and entirely posterior to small mouth. Teeth slender and triangular in both jaws. First dorsal-fin origin behind pelvic-fin bases; second dorsal-fin origin over, or slightly in front of anal-fin insertion. Dorsal fins about equal in size; pectoral fins rounded; caudal fin with distinct subterminal lobe and undeveloped ventral lobe.
- **Colour:** Strongly decorated with a broad, dark brown 'collar' with white spots around the gill area, dark brown blotches and white spots on the fins, and white spots on a greyish or brownish body.
- **Size:** Attains 90 cm; a 47 cm male was immature.
- **Distribution:** Lakes Entrance (Victoria) to Dongara (southern Western Australia), including the Bass Strait coast of Tasmania. Continental shelf in depths down to 180 m.
- **Remarks:** Little is known of the biology of this species other than it is oviparous. Another white-spotted form occurring off southern Western Australia may be an additional undescribed species. Specimens are required from this area.
- **Local synonymy:** *Parascyllium nuchalis* McCoy, 1874.

undersurface of head

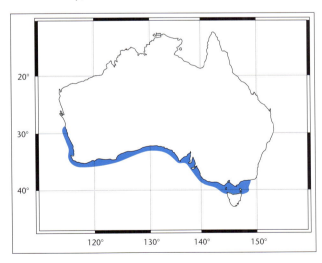

13 Family Brachaeluridae
BLIND SHARKS

The members of this family of stout, small to medium-sized sharks (attaining 0.6–1.2 m) have a small, subterminal mouth situated well in front of the eyes, long nasal barbels (but no throat barbels), very large spiracles, and the first dorsal fin originates over or just behind the pelvic-fin bases. The mouth is connected to the nostrils by a shallow groove located beneath the nasal flap (nasoral grooves) and smaller fleshy folds and grooves skirt the nostrils laterally. The anal fin originates behind the second dorsal-fin origin and is separated from the lower caudal-fin lobe by less than the length of the anal-fin base. The caudal fin, which lacks a ventral lobe but has a subterminal notch, is much shorter than the body. These sharks also have five gill slits, no dermal lobes around the head, spineless dorsal fins, and lack both ridges on the body and caudal keels.

Blind sharks differ from the related long-tailed carpetsharks (Hemiscylliidae) in having longer nasal barbels and the precloacal length longer than the distance from the cloaca to the lower caudal-fin lobe. The two Australian species, both endemics, live on or near the bottom on the continental shelf. Compagno (1984) separated the family into two monospecific genera, *Brachaelurus* and *Heteroscyllium*, based mainly on the presence or absence of a median symphysial groove on the chin. This groove, however, is present in both species so only *Brachaelurus* is recognised herein.

Key to brachaelurids

1   Nostrils inferior (fig. 1); nasal barbel bifurcated; anal-fin origin under or in front of second dorsal-fin insertion (fig. 3); no white spots on body .... ***Brachaelurus colcloughi* (fig. 3; p. 123)**

   Nostrils almost terminal (fig. 2); nasal barbel single lobed; anal-fin origin just behind second dorsal-fin insertion (fig. 4); mostly with white spots on body .... ***Brachaelurus waddi* (fig. 4; p. 124)**

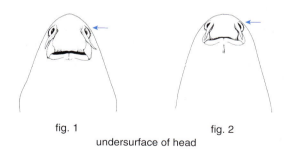

fig. 1   fig. 2
undersurface of head

fig. 3   fig. 4

Colcloughs Shark — *Brachaelurus colcloughi* Ogilby, 1908

13.1 *Brachaelurus colcloughi* (Plate 17) **Fish Code:** 00 013013

- **Alternative name:** Bluegrey carpet shark.
- **Field characters:** A stout shark with long nasal barbels (weakly bilobed), nostrils situated well behind the snout tip, large spiracles, no dermal lobes on the side of the head, dorsal fins close together, a short caudal peduncle, and small, dorsolaterally situated eyes. The second dorsal-fin origin is in advance of the anal-fin origin.
- **Distinctive features:** Body stout. Head slightly depressed, small supra-orbital ridges. Nasal barbels long; nasoral and circumnarial grooves present. Nostrils inferior (rather than terminal); spiracles very large, close to eyes; eyes small, situated dorsolaterally, entirely behind mouth; mouth narrow. Teeth similar in both jaws, erect and tricuspidate; central awl-shaped cusp very long compared with lateral cusps. Skin smooth, denticles very small, flat and not overlapping. Dorsal fins about equal in size; first dorsal-fin origin about over midbase of pelvic fins; interdorsal distance less than length of first dorsal-fin base. Anal-fin origin under or in front of second dorsal-fin insertion. Distance between vent and lower caudal-fin origin less than distance from snout to vent. Caudal fin with a subterminal lobe; ventral lobe not developed.
- **Colour:** Greyish brown dorsally; dark saddles on body fading with size and may be absent on adults. Body and tail with faint dark bands in juveniles, about 4 between eye and first dorsal fin. Pale ventrally.
- **Size:** Attains about 60 cm.
- **Distribution:** Continental shelf off southern Queensland between Gladstone and Coolangatta, and off York Peninsula and the Great Barrier Reef.
- **Remarks:** A poorly known species of no commercial value. Presumably ovoviviparous.
- **Local synonymy:** *Heteroscyllium colcloughi* (Ogilby, 1908).
- **Reference:** Ogilby (1908a).

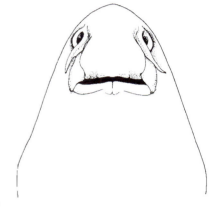

undersurface of head

third teeth from symphysis
(upper and lower jaw)

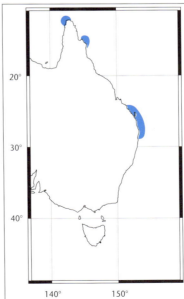

Blind Shark — *Brachaelurus waddi* (Bloch & Schneider, 1801)

13.2 *Brachaelurus waddi* (Plate 17)   **Fish Code:** 00 013007

- **Alternative name:** Brown catshark.
- **Field characters:** A stout shark with long, single-lobed nasal barbels, nostrils situated near the snout tip, large spiracles, no dermal lobes on the side of the head, dorsal fins close together, a very short caudal peduncle, and small, dorsally situated eyes. The second dorsal-fin insertion is in advance of the anal-fin origin.
- **Distinctive features:** Body stout. Head depressed, supra-orbital ridges absent. Nasal barbels long; nasoral and circumnarial grooves present. Nostrils terminal (rather than inferior); spiracles large, close to eyes; eyes small, situated dorsally, entirely behind mouth; mouth short. Teeth similar in both jaws, erect and tricuspidate; central awl-shaped cusp very long compared with lateral cusps. Skin smooth, denticles very small, flat and not overlapping; juveniles with rows of enlarged, paired denticles on the dorsolateral surface. Dorsal fins about equal in size; first dorsal-fin origin over to in front of pelvic-fin insertion; interdorsal distance less than length of first dorsal-fin base. Anal-fin origin just behind second dorsal-fin insertion. Distance between vent and lower caudal-fin origin less than distance from snout to vent. Caudal fin with a subterminal lobe; ventral lobe not developed. Total vertebrae 142*.
- **Colour:** Light brown to black dorsally; usually with pale spots and about 11 faint dark saddles, which fade with size and may be absent on adults. Body and tail with dark bands in juveniles (about 5 between eye and first dorsal fin); pale ventrally.
- **Size:** Born at about 17 cm and reported to attain almost 120 cm. Normally much smaller; a male of 60 cm and a female of 66 cm were both sexually mature.
- **Distribution:** Eastern Australia from Moreton Bay (Queensland) to Jervis Bay (New South Wales). Reports from Western Australia and the Northern Territory require confirmation. Continental shelf from the intertidal zone to about 140 m.
- **Remarks:** Often caught by rock fishermen, it is reported to live for a long time out of water. Its habit of closing its eyes when caught has led to its common name of blind shark. It has often been confused with the grey carpetshark (15.1). Ovoviviparous with litters of 7–8 that are born around November. The diet consists of reef invertebrates and small fish. Not utilised commercially in Australia.
- **Local synonymy:** *Chiloscyllium modestum* Günther, 1871; *Chiloscyllium furvum* Macleay, 1881; *Chiloscyllium fuscum* Parker & Haswell, 1897.
- **References:** Macleay (1881a); Grant (1978).

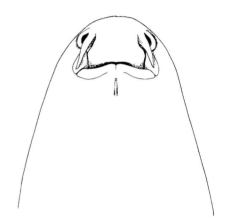

undersurface of head

upper tooth from near symphysis

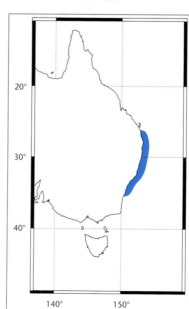

14 Family Orectolobidae
WOBBEGONGS

Wobbegongs are small to large, bottom-dwelling sharks (attaining 0.6–3 m) with a flattened head and body, large, nearly terminal mouths situated in front of the eyes, nasoral and circumnarial grooves, nasal barbels, and narrow flaps of skin (dermal lobes) around the mouth and sides of the head. Their anal fin originates behind the second dorsal-fin origin and is situated less than the length of the anal-fin base from the lower lobe of the caudal fin. The caudal fin, which lacks a ventral lobe but has a subterminal notch, is much shorter than the rest of the body. Wobbegongs also have large spiracles, very sharp, dagger-like teeth in both jaws, two spineless dorsal fins (about equal in size and situated behind the pelvic-fin origin), five gill slits, and no caudal keels or ridges on the body. Most species have highly ornamented colour patterns. Angel sharks (Squatinidae), which also have depressed bodies and possess many of these characteristics, differ from wobbegongs in having much larger pectoral fins and no anal fin.

Wobbegongs occur on the continental shelves of the western Pacific in warm temperate and tropical waters. Six of the seven known species are found in the Australian region. They are ovoviviparous and females have large litters with 20 or more young.

Key to orectolobids

1 Dermal lobes mostly extensively branched, present on sides of head and on chin (fig. 3) .... *Eucrossorhinus dasypogon* (fig. 1; p. 127)

Dermal lobes mostly simple, present on sides of head but not on chin (fig. 4) **2**

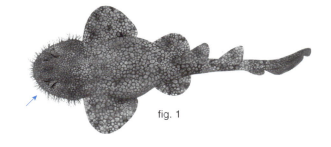
fig. 1

2 Head and body covered in large bony tubercles resembling warts (fig. 2); dorsal fins long and low, height about two-thirds or less of base length (fig. 5) .... *Sutorectus tentaculatus* (fig. 2; p. 133)

Head and body smooth (minute tubercles may be present in juveniles); dorsal fins taller, height much more than two-thirds of base length (fig. 6) **3**

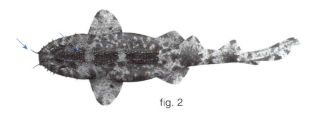
fig. 2

fig. 3
undersurface of head

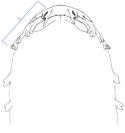
fig. 4
undersurface of head

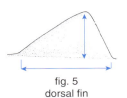
fig. 5   fig. 6
dorsal fin   dorsal fin

3 Nasal barbels not branched (fig. 11); 2–3 dermal lobes on each side in front of eye (fig. 11); colour pattern relatively simple with 3 large, dark, pale-edged blotches on dorsal midline, anterior to first dorsal fin (fig. 7) .... **Orectolobus wardi (fig. 7; p. 132)**

Nasal barbels branched (fig. 12); more than 5 dermal lobes on each side in front of eye; colour pattern complex (figs 8, 9, 10) **4**

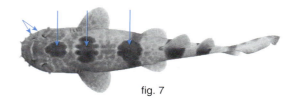

fig. 7

fig. 8

4 Dermal lobes behind spiracle unbranched and slender (fig. 13) .... **Orectolobus sp. A (fig. 8; p. 128)**

Dermal lobes behind spiracle branched, or if unbranched, very broad (fig. 12) **5**

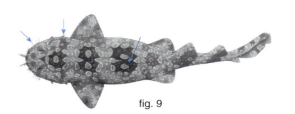

fig. 9

5 Body with dark saddles overlain with white rings and blotches (fig. 9); usually 6–10 dermal flaps on each side in front of eye (fig. 12)...
.... **Orectolobus maculatus (fig. 9; p. 129)**

Body with dark saddles overlaid with white rings and blotches, but saddles, rings and blotches edged in black (fig. 10); usually 5–6 dermal flaps on each side in front of eye (fig. 4) ....
.... **Orectolobus ornatus (fig. 10; p. 131)**

fig. 10

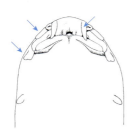

fig. 11

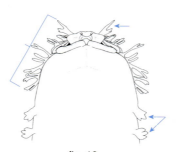

fig. 12
undersurface of head

fig. 13

Tasselled Wobbegong — *Eucrossorhinus dasypogon* (Bleeker, 1867)

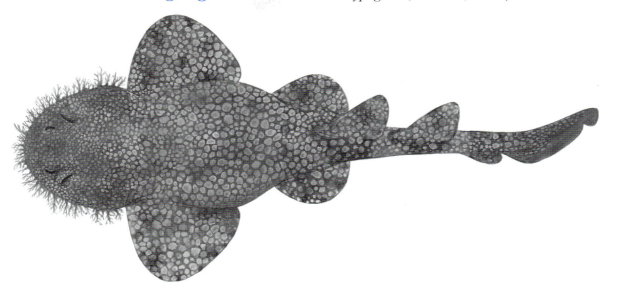

14.1 *Eucrossorhinus dasypogon* (Plate 25) **Fish Code:** 00 013011

- **Alternative name:** Ogilbys wobbegong.
- **Field characters:** A very wide-bodied wobbegong with broad pectoral fins, a dense tassel of dermal lobes around the head margin, and a striking reticulated colour pattern.
- **Distinctive features:** Body moderately depressed, very wide, tapering abruptly behind pelvic fins; no dermal ridges and tubercles on back. Head broad, width greater than length (from snout tip to fifth gill opening) in adults; enlarged tubercles above eyes (may be indistinguishable in small juveniles); dense beard-like tassel of branched dermal lobes fringing head from snout tip to pectoral-fin origins. Nasal barbels branched, with multiple lobes; nasoral and circumnarial grooves present. Teeth similar in both jaws, long (longest near symphysis) and slender. First dorsal-fin origin over posterior quarter of pelvic-fin bases; interdorsal distance longer than inner margin of first dorsal fin. Dorsal fins high and short, first slightly larger than second. Pectoral and pelvic fins large and broad; pectoral-fin width about equal to head length. Total vertebrae 162*; precaudal 100*.
- **Colour:** Greyish to yellowish brown (sometimes with darker bands) with a striking pattern of fine reticulations and blotches which form a complex mosaic; pattern usually present on upper surface and fins, also on undersurface of tail and outer half of undersurface of pectoral and pelvic fins; remainder of undersurface white.

undersurface of head

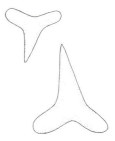

third teeth from symphysis
(upper and lower jaw)

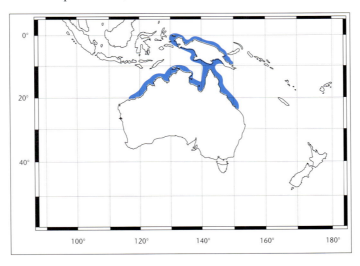

- **Size:** Born at about 20 cm and attains about 125 cm. A record of a 366 cm specimen is probably erroneous.
- **Distribution:** Indonesia, Papua New Guinea and northern Australia from Barrow Island (Western Australia) to Bundaberg (Queensland). Shallow areas of the continental shelf and offshore reefs.
- **Remarks:** The most common tropical wobbegong that is frequently observed by divers among coral-heads on northern parts of the Great Barrier Reef. Little is known of the biology of this species but it is presumably ovoviviparous and feeds on bottom invertebrates and fish. Currently of little commercial value, although the skin is occasionally used for leather.
- **Local synonymy:** *Orectolobus ogilbyi* Regan, 1909.
- **Reference:** Stead (1963).

Western Wobbegong — *Orectolobus* sp. A

14.2 *Orectolobus* sp. A (Plate 26)   Fish Code: 00 013016

- **Field characters:** A relatively slender wobbegong with only a few unbranched dermal lobes on the side of the head, moderately tall dorsal fins, and no warty tubercles along the back of adults (small ones present in juveniles). The colour pattern is yellowish to brownish dorsally with darker corrugated saddles that lack white spots and blotches.
- **Distinctive features:** Body depressed, relatively slender, tapering behind pelvic fins; two rows of very small warty tubercles extending from head along body in juveniles (absent in adults). Dermal lobes sparse, unbranched, two behind spiracle; no dermal lobes on chin. Nasal barbels with one small branch; nasoral and circumnarial grooves present. Teeth similar in both jaws, long and slender (longest near symphysis). First dorsal-fin origin over posterior third of pelvic-fin base; interdorsal distance shorter than inner margin of first dorsal fin. Dorsal fins rather tall, about equal in size; height of first about equal to or slightly shorter than its base. Pectoral and pelvic fins broad.
- **Colour:** Chocolate brown to yellowish brown dorsally, ornamented with darker saddles that have indistinct, corrugated margins; saddles contain paler ocelli or blotches; no whitish spots or blotches; pale ventrally.
- **Size:** Born at about 22 cm and attains 200 cm. An 85 cm male was mature.

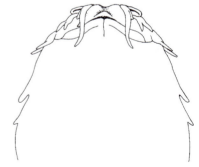

undersurface of head

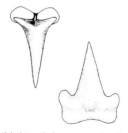

third teeth from symphysis
(upper and lower jaw)

- **Distribution:** Western Australia from Cape Leeuwin to Coral Bay. Inshore on the continental shelf.
- **Remarks:** Occurs in coastal areas on reefs and amongst seagrass, but little is known of its biology. It is ovoviviparous and presumably feeds on invertebrates and small fish. Small quantities are taken as a bycatch of the Western Australian shark fishery and are mostly used for human consumption.
- **Local synonymy:** *Orectolobus* sp.: Hutchins & Swainston, 1986.
- **Reference:** Hutchins and Swainston (1986).

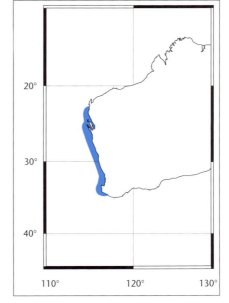

Spotted Wobbegong — *Orectolobus maculatus* (Bonnaterre, 1788)

14.3 *Orectolobus maculatus* (Plate 26) Fish Code: 00 013003

- **Alternative name:** Wobbegong.
- **Field characters:** A robust wobbegong with a few branched dermal lobes along the head margin, moderately tall dorsal fins, and no warty tubercles along the back. The ornate colour pattern is pale yellowish or greenish brown with darker saddles superimposed and bordered with white rings and blotches.
- **Distinctive features:** Body moderately depressed, elongate; no dermal ridges and tubercles on back. Head with 6–10 long, coarsely branched dermal lobes on each side in front of eye; dermal lobes absent from chin. Nasal barbels simple with 1–2 basal branches; nasoral and circumnarial grooves present. Teeth similar in both

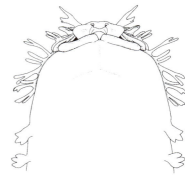

undersurface of head

jaws, long and slender (longest near symphysis). First dorsal-fin origin over posterior third of pelvic-fin base; interdorsal distance longer than first dorsal-fin inner margin; first dorsal-fin height about equal to its base. Pectoral and pelvic fins broad. Total vertebrae 156*; precaudal 106*.

- **Colour:** Upper surface pale yellowish to greenish brown, distinctly patterned with darker saddles and numerous white rings formed from a chain of small white spots and flecks. Predorsal surface with four dark brown saddles, each with a distinctive white margin; white-edged rings present within saddles. Other parts of head, body and fins with some dark blotches, including a distinctive brown triangle between eyes.
- **Size:** Born at about 20 cm and attains at least 300 cm. Males may mature at about 60 cm.
- **Distribution:** Southern coast of Australia from Fremantle (Western Australia) to Moreton Island (southern Queensland). Records from Japan and the South China Sea need to be confirmed. Tasmanian records are probably invalid. Most common inshore but trawled to 110 m.
- **Remarks:** This large, well camouflaged wobbegong has been seen by divers resting during the day in shallow water on reefs or sand, in caves or under piers. Ovoviviparous, with litter sizes up to 37. It is a nocturnal feeder, preying on such large bottom-dwelling animals as crabs, rock lobsters, octopuses and reef fishes. It can be aggressive to humans if molested or when in the presence of speared fish. The flesh is highly regarded but the species is of limited commercial value. Small quantities, however, are taken by dropline off New South Wales. The attractive skin makes an excellent decorative leather.
- **Local synonymy:** *Squalus barbatus* Gmelin, 1789; *Squalus lobatus* Bloch & Schneider, 1801; *Squalus appendiculatus* Shaw & Nodder, 1806.
- **Reference:** Coleman (1980).

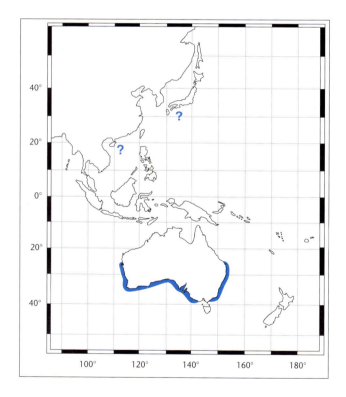

Banded Wobbegong — *Orectolobus ornatus* (de Vis, 1883)

14.4 *Orectolobus ornatus* (Plate 26)   **Fish Code:** 00 013001

- **Alternative names:** Ornate wobbegong, carpet shark, gulf wobbegong.
- **Field characters:** A robust wobbegong with several branched dermal lobes along the head margin, moderately tall dorsal fins, and no warty tubercles along the back. The colour pattern is strongly ornamented with the margins of all corrugated saddles highlighted with black borders.
- **Distinctive features:** Body moderately depressed, tapering behind the pelvic fins; no dermal ridges or tubercles on back. Head with 5–6 dermal lobes on each side in front of eye; no dermal lobes on chin. Nasal barbels simple, with two small basal branches; nasoral and circumnarial grooves present. Teeth similar in both jaws, long and slender (longest near symphysis). First dorsal-fin origin over posterior third of pelvic-fin bases; interdorsal distance longer than inner margin of first dorsal fin; dorsal fins tall, height of first about equal to its base length. Pectoral and pelvic fins broad. Tooth count 26–28/22–26*.

undersurface of head

- **Colour:** Extremely ornate and variegated; yellowish brown to greyish brown with darker, corrugated saddles; each saddle, and the many paler bluish to whitish patches, bordered with lines of small, black spots; a prominent white spot behind each spiracle; pale ventrally.
- **Size:** Born at about 20 cm and attains 290 cm. Normally maturing at about 175 cm but a Queensland male was mature at 63 cm.
- **Distribution:** Indonesia, Papua New Guinea and Australia; records for Japan are doubtful. Recorded locally from tropical eastern Australia to the Houtman Abrolhos (Western Australia), south to Flinders Island (Bass Strait). Inshore on the continental shelf to at least 100 m.
- **Remarks:** Although relatively common, the biology of this species is poorly known. The small size at maturity of some male specimens is unusual. A nocturnal feeder on bottom invertebrates and fish, it rests on the bottom during the day. Large individuals have attacked divers. Currently of little commercial value, although small quantities are taken as bycatch in the Western Australian shark fishery and by dropline off New South Wales. The skin is occasionally used for leather. Ovoviviparous.
- **Local synonymy:** *Orectolobus devisi* Ogilby, 1916; *Orectolobus ornatus halei* Whitley, 1940.
- **References:** Stead (1963); Coleman (1980).

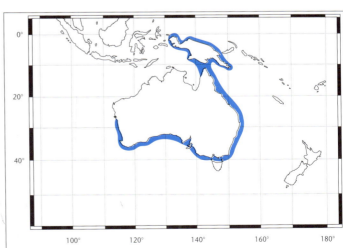

Northern Wobbegong — *Orectolobus wardi* Whitley, 1939

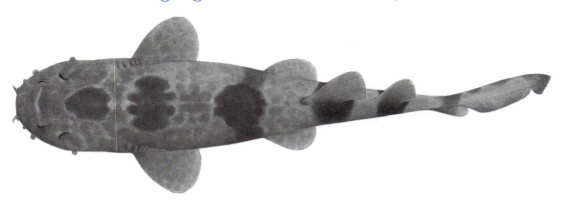

14.5 *Orectolobus wardi* (Plate 25)   Fish Code: 00 013017

- **Alternative name:** Wards wobbegong.
- **Field characters:** A small, comparatively drab, wobbegong with very few dermal lobes along the head margin (mostly simple and with two before eye), tall dorsal fins, and no warty tubercles along the back. There are three large, dark blotches (each edged with white) on the back before the first dorsal fin.
- **Distinctive features:** Body moderately depressed, tapering behind pelvic fins; no dermal ridges or tubercles on back. Head with two dermal lobes in front of eye and one broad dermal lobe behind spiracle; no dermal lobes on chin. Nasal barbels simple, not branched or with one or two indistinct branches; nasoral and circumnarial grooves present. Teeth similar in both jaws, long and slender (longest near symphysis). First dorsal-fin origin over or just anterior to pelvic-fin insertion; interdorsal distance longer than inner margin of first dorsal fin; dorsal fins tall, about equal in size; height of first about equal to its base length. Pectoral and pelvic fins broad.
- **Colour:** Brownish dorsally with three large, dark brown blotches on the back before the first dorsal fin; blotches with pale margins; additional dark bands encircling tail under each dorsal fin and above anal fin. Fainter spots and blotches on upper surface and fins. Pale ventrally.
- **Size:** Attains at least 63 cm; a 45 cm male was mature.
- **Distribution:** Inshore on the continental shelf of northern Australia from Fraser Island (Queensland) to Onslow (Western Australia).
- **Remarks:** Little is known of the biology of this species. Presumably ovoviviparous and feeds on bottom invertebrates and fish. Not utilised in Australia.
- **Local synonymy:** *Sutorectus wardi* (Whitley, 1939).
- **Reference:** Whitley (1939a).

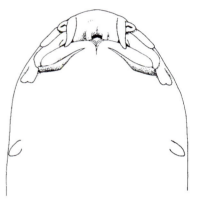

undersurface of head

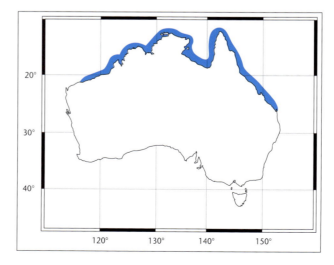

Cobbler Wobbegong — *Sutorectus tentaculatus* (Peters, 1864)

14.6 *Sutorectus tentaculatus* (Plate 25) **Fish Code:** 00 013012

- **Alternative name:** Cobbler carpet shark.
- **Field characters:** A small, relatively slender wobbegong with rows of warty tubercles along the back, low and long-based dorsal fins, only a few, unbranched dermal lobes on the side of the head, and a variegated colour pattern overlaid with dark, corrugated saddles and small black spots.
- **Distinctive features:** Body depressed, relatively slender, tapering behind pelvic fins; rows of warty tubercles on upper surface. Head relatively narrow, maximum width slightly less than head length (from snout tip to first gill slit). Dermal lobes sparse and unbranched; no dermal lobes on chin. Nasal barbels simple, without branches; nasoral and circumnarial grooves present. Teeth similar in both jaws, long and slender (longest near symphysis). First dorsal-fin origin over posterior half of pelvic-fin base; interdorsal distance shorter than inner margin of first dorsal fin. Dorsal fins long, low and about equal in size; height of first is half to two-thirds its base length. Pectoral and pelvic fins broad.
- **Colour:** Strongly ornamented dorsally; pale brownish with darker brown saddles and blotches, and fine, black spots on upper surface and fins; no white spots. Pale ventrally.
- **Size:** Born at about 22 cm and attains 92 cm. Males mature at about 65 cm.
- **Distribution:** Inshore on the continental shelf of southern Australia from Adelaide (South Australia) to the Houtman Abrolhos (Western Australia).
- **Remarks:** Occurs in coastal areas on reefs and amongst weed, but little is known of its biology. Ovoviviparous. Not utilised commercially in Australia.

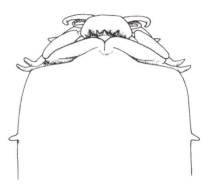

undersurface of head

third teeth from symphysis
(upper and lower jaw)

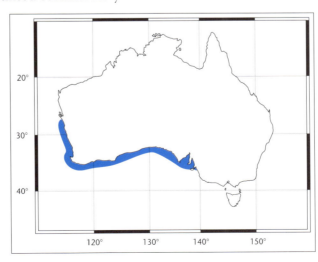

15

Family Hemiscylliidae

LONGTAIL CARPET SHARKS

These small, slender sharks (up to 1 m long), have a small, subterminal mouth situated well in front of the eyes, nasoral and circumnarial grooves, short nasal barbels (but no throat barbels) and very large spiracles. The first dorsal-fin origin is over or behind the pelvic-fin bases, and the anal fin originates well behind the second dorsal-fin origin and is separated from the lower caudal-fin origin by less than the anal-fin base length. The caudal fin, which lacks a ventral lobe but has a subterminal notch, is much shorter than the body. They have five gill slits (the fourth and fifth situated close together), no dermal lobes around the head, no caudal keels or ridges on the body, and spineless dorsal fins. These carpet sharks are similar to blind sharks (family Brachaeluridae), but have shorter nasal barbels and, as their common name implies, are relatively long-tailed; the distance from the vent to the lower caudal-fin origin is longer than the precloacal length (rather than shorter as in blind sharks).

The three Australian hemiscylliids, represented by two genera, are all bottom-living in shallow, tropical waters.

Key to hemiscylliids

1   Nostrils subterminal on snout (fig. 1); no black ocellus behind fifth gill slit (fig. 3) ....
    .... ***Chiloscyllium punctatum*** (fig. 3; p. 135)

    Nostrils nearly terminal on snout (fig. 2); a large black ocellus behind fifth gill slit (figs 4, 5) **2**

2   Body with rather large, widely spaced, dark spots, not forming a mosaic pattern (fig. 4); dark ocellus behind gills not bordered posteriorly with a cluster of fine black spots (fig. 4) ....
    .... ***Hemiscyllium ocellatum*** (fig. 4; p. 136)

    Body peppered with numerous fine, dark spots that form a mosaic pattern (fig. 5); dark ocellus behind gills bordered posteriorly with a cluster of fine spots that form dark blotches (fig. 5) ....
    .... ***Hemiscyllium trispeculare*** (fig. 5; p. 137)

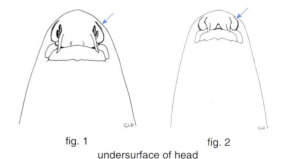

fig. 1   fig. 2
undersurface of head

fig. 3

fig. 4

fig. 5

Grey Carpet Shark — *Chiloscyllium punctatum* Müller & Henle, 1838

15.1 *Chiloscyllium punctatum* (Plate 17) **Fish Code:** 00 013008

- **Alternative names:** Brown-banded catshark, brown-spotted catshark, spotted catshark, brown-banded bamboo shark.
- **Field characters:** A slender shark with relatively long barbels, two equal-sized dorsal fins with bases almost equal to the interdorsal space, the first dorsal fin partly over the pelvic-fin base, and the anal-fin origin under or slightly behind the free rear tip of the second dorsal fin.
- **Distinctive features:** Body slender. Snout with rather long nasal barbels; nasoral grooves present; nostrils subterminal; large spiracles. Teeth similar in both jaws, small, cusps broadly triangular. Dorsal fins about equal in size; first dorsal-fin origin over anterior base of pelvic fin; interdorsal distance about equal to or slightly longer than length of first dorsal-fin base. Anal-fin origin about under or just behind free rear tip of second dorsal fin. Tooth count 31–33/30–33. Total vertebrae 126–135, precaudal 62–73.
- **Colour:** Large adults uniformly brownish or greyish above (often shiny); pale ventrally and on gill slit margins. Juveniles (to about 29 cm) pale with about 10 dark, wavy bands; sometimes peppered with small dark spots.
- **Size:** Hatches at about 17 cm and attains about 105 cm. Males mature at about 70 cm.
- **Distribution:** Indo–West Pacific, from India and Japan through the Philippines, Indonesia and New Guinea to northern Australia. Locally inshore on the continental shelf from Moreton Bay (Queensland) to Shark Bay (Western Australia) in depths to at least 85 m.
- **Remarks:** Oviparous, laying rounded eggs about 11 cm long by 5 cm wide. Commonly found in tide pools on coral reefs, where it can survive considerable periods out of water. Probably feeds on bottom invertebrates and small fish. Its flesh is reported to be good eating; used for food in India and Thailand but seldom in Australia.
- **References:** Müller and Henle (1838a); Dingerkus and DeFino (1983).

undersurface of head

anterior teeth
(upper and lower jaw)

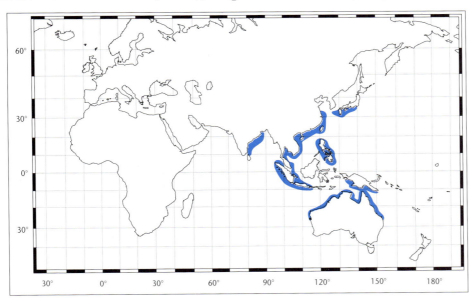

Epaulette Shark — *Hemiscyllium ocellatum* (Bonnaterre, 1788)

15.2 *Hemiscyllium ocellatum* (Plate 16) **Fish Code:** 00 013014

- **Alternative name:** Blind shark.
- **Field characters:** A small, slender shark with a large black ocellus above each pectoral fin, widely spaced, dark spots on the body, short barbels, two equal-sized dorsal fins set posteriorly, and an anal fin originating well behind the free rear tip of the second dorsal fin.
- **Distinctive features:** Body long and slender. Snout with short nasal barbels; nasoral grooves present; nostrils nearly terminal; large spiracles. Teeth similar in both jaws, small, broad-based, cusps narrowly triangular. Dorsal fins about equal in size; first dorsal-fin origin about over or just behind pelvic-fin insertion; interdorsal distance longer than first dorsal-fin base. Anal-fin origin well posterior to second dorsal-fin free rear tip. Tooth count 38/30*. Total vertebrae 190–191*.
- **Colour:** Upper body yellowish or brownish (mostly shiny), with many rather large, widely spaced, dark brown spots; a very large black ocellus (with a white margin) located on each flank behind the gill slits. Juveniles have several darker brown saddles on the back and tail.
- **Size:** Hatches at about 15 cm and attains 107 cm. Males mature at about 60 cm.
- **Distribution:** New Guinea and northern Australia; records from Malaysia and Sumatra require verification. Locally, mainly from shallow, inshore waters between Port MacQuarie (New South Wales) and Shark Bay (Western Australia). Museum records suggest that it occasionally ventures as far south as Sydney (New South Wales).
- **Remarks:** Common in shallow water on coral reefs. Oviparous, feeding mainly on benthic invertebrates. This species survives well in aquaria, where females have been observed to lay about 50 eggs annually. The eggs are ellipsoid in shape (about 10 cm long and 4 cm wide) and take about 130 days to hatch. In a study at the Taronga Zoo aquarium, 37% of the hatchlings survived for six months. Juveniles grow slowly; about 3 cm per year. Other than the aquarium trade, not used commercially.
- **Local synonymy:** *Squalus oculatus* Gray, 1827.
- **References:** Grant (1978); Dingerkus and DeFino (1983); West and Carter (1990).

undersurface of head

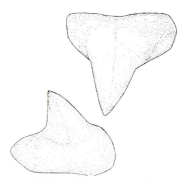

third teeth from symphysis
(upper and lower jaw)

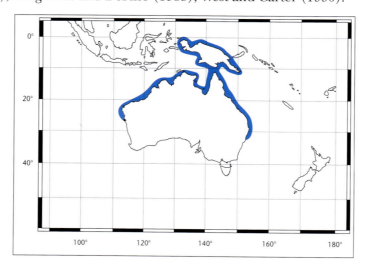

Speckled Carpet Shark — *Hemiscyllium trispeculare* Richardson, 1843

15.3 *Hemiscyllium trispeculare* (Plate 16)　　　　Fish Code: 00 013015

- **Alternative names:** Speckled catshark, marbled catshark.
- **Field characters:** A small, slender shark with a large, black ocellus above each pectoral fin (with dark blotches behind), a dense mosaic of numerous fine spots on the body, short barbels, two equal-sized dorsal fins set posteriorly, and an anal fin originating well behind the free rear tip of the second dorsal fin.
- **Distinctive features:** Body rather slender. Snout with short nasal barbels; nasoral grooves present; nostrils nearly terminal; large spiracles. Teeth similar in both jaws, small, broad-based, cusps narrowly triangular. Dorsal fins about equal in size; first dorsal-fin origin over to slightly behind pelvic-fin insertion; interdorsal distance much longer than first dorsal-fin base. Anal-fin origin well behind second dorsal-fin free rear tip.
- **Colour:** Dorsal surface and fins yellowish, covered with a dense mosaic of very fine, brownish spots; a very large, black ocellus (bordered in white) and flanked posteriorly by 2–3 dark blotches is located behind the gill slits on each side; brownish bands are evident on the body and tail. Pale ventrally
- **Size:** Attains at least 65 cm. Males mature at about 53 cm.
- **Distribution:** Indonesia (possibly) and northern Australia, where it is recorded from inshore waters of the continental shelf from (Rockhampton) Queensland to Ningaloo (Western Australia).
- **Remarks:** Particularly common around Darwin where it is frequently found in shallow water. Specimens from the Northern Territory appear to have much smaller and denser spots than those from Western Australia. Oviparous, probably feeding mainly on benthic invertebrates. Not used commercially.
- **Reference:** Dingerkus and DeFino (1983).

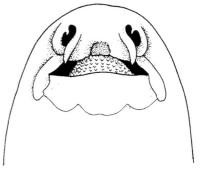

undersurface of head

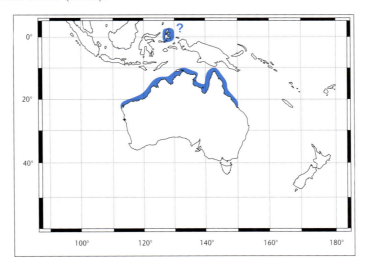

16

Family Stegastomatidae
ZEBRA SHARKS

This monospecific family is easily recognised by the following combination of characters: a very long caudal fin (which is almost as long as the rest of the body), nasoral grooves, and firm ridges running along the upper body. The zebra shark also has a subterminal mouth situated well in front of the eyes, a caudal fin without a ventral lobe but with a subterminal notch, short nasal barbels, large spiracles, 4 main gill openings (gill slits 4 and 5 overlap), and two closely spaced, spineless dorsal fins.

Sometimes placed with the wobbegongs in the family Orectolobidae.

Zebra Shark — *Stegastoma fasciatum* (Hermann, 1783)

16.1 *Stegastoma fasciatum* (Plate 15) **Fish Code:** 00 013006

- **Alternative name:** Leopard shark.
- **Field characters:** An unmistakable shark with a very long, blade-like caudal fin, ridges along the upper surface and flanks, and a yellowish brown coloration peppered with numerous, dark brown spots (juveniles less than 70 cm long are dark with white bars and spots).
- **Distinctive features:** Body moderately stout with prominent ridges on dorsal surface and flanks. Head broad; snout bluntly rounded; fourth and fifth gill slits overlapping; nasoral grooves present; barbels short; mouth transverse. Teeth similar in both jaws, small and tricuspid. First dorsal fin larger than second dorsal fin; free rear tip of first dorsal fin close to second dorsal-fin origin; first dorsal-fin insertion over or slightly anterior to pelvic-fin insertion. Anal-fin insertion about opposite upper caudal-fin origin. Pectoral fins large and broadly rounded. Caudal fin about as long as rest of body, with a terminal lobe but no ventral lobe. Tooth count 28/28* [28–33/22–34]. Total vertebrae [217–234]; precaudal [81–101].
- **Colour:** Adults are yellowish to dark brown on the back, sides and fins; covered with numerous, rather large, dark brown spots; spots sharply-defined and almost round. Juveniles smaller than about 70 cm, markedly different; dark with white bars and spots. Pale ventrally.
- **Size:** Hatches at about 20 cm, matures at about 170 cm and attains at least 235 cm; reports to 354 cm need to be validated.

- **Distribution:** Indo–West Pacific, from South Africa and the Red Sea to India, Thailand, Japan, Indonesia and northern Australia. Inshore waters of the continental and insular shelves. Locally, northern Australia from Sydney (New South Wales) to Port Gregory (Western Australia).
- **Remarks:** Lays large egg cases (13–17 cm long and about 8 cm wide) that are dark brown or purplish black with longitudinal striations. Egg cases are anchored to the substrate by lateral bunches of hair-like fibres. Probably more active at night, resting on the bottom during the day. The diet consists mainly of gastropods and bivalves, but also crustaceans and fish. Not of commercial value in Australia; elsewhere caught for its meat and fins.
- **Local synonymy:** *Stegastoma tigrinum naucum* Whitley, 1939.
- **References:** Whitley (1939b); Grant (1978).

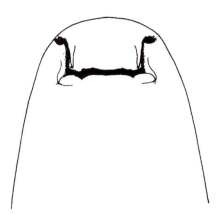

undersurface of head

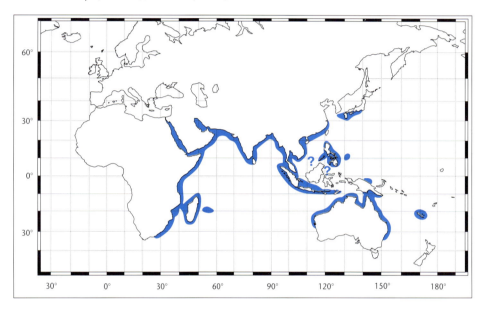

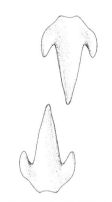

teeth from near symphysis
(upper and lower jaw)

17   Family Ginglymostomatidae
NURSE SHARKS

Nurse sharks have a relatively small, subterminal mouth situated well in front of the eyes, nasoral but no circumnarial grooves, nasal barbels and small spiracles. The first dorsal fin originates slightly in front of or over the pelvic-fin bases, the anal fin originates slightly behind the second dorsal-fin origin and is separated from the lower caudal-fin origin by less than the anal-fin base length The caudal fin, which has a subterminal notch and a weakly developed ventral lobe, is much shorter than the rest of the body. These sharks also have five gill slits, no dermal lobes around the head, no caudal keels or ridges on the body, and spineless dorsal fins.

This small family, containing only two genera, is circumglobal in tropical and temperate seas. The single Australian representative can be identified by its nasoral grooves, compressed, multicuspid teeth, mouth situated in front of the eyes, and its two similar-sized, posteriorly placed, angular dorsal fins.

Tawny Shark — *Nebrius ferrugineus* (Lesson, 1830)

17.1 *Nebrius ferrugineus* (Plate 17)   **Fish Code:** 00 013010

- **Alternative names:** Tawny nurse shark, spitting shark, sleepy shark, madame X.
- **Field characters:** A large, sandy to greyish brown shark with nasal barbels and nasoral grooves, two angular, similar-sized dorsal fins set well back on the body, and small multicuspid teeth.
- **Distinctive features:** Body moderately fusiform. Head with relatively short nasal barbels; nasoral but no circumnarial grooves; eyes small; fourth and fifth gill slits close together. Small, compressed, multicuspid teeth in both jaws. Dorsal fins large and angular; first slightly larger than second; first dorsal-fin base over pelvic-fin base. Anal-fin origin posterior to second dorsal-fin origin; anal fin similar in size and shape to second dorsal fin. Pectoral fins falcate. Caudal fin relatively long; subterminal lobe distinct, ventral lobe weak. Tooth count [28–30/25–26]*. Total vertebrae [188–189]*; precaudal [95–101]*.
- **Colour:** Sandy brown to greyish brown dorsally (possibly changeable depending on habitat), paler ventrally. The fins, particularly the tips, may be darker.
- **Size:** Born at about 40 cm and attains 320 cm, although most are smaller. Males mature at about 225 cm and females at about 230 cm.

- **Distribution:** Indo–West and Central Pacific. From south eastern Africa (including Madagascar and Aldabra) and the Red Sea, through to India, Indo–China, Japan, Indonesia, New Guinea, New Caledonia, Samoa, Palau, Marshall Islands and Tahiti. In tropical Australia, known from Rockhampton (Queensland) to Ningaloo (Western Australia). Inshore waters of the continental and insular shelves down to at least 70 m.

- **Remarks:** More active at night, often resting in groups in caves and rocky crevices during the day. Feeds over sandy areas near reefs preying on cephalopods (particularly octopuses), other invertebrates and reef fish. It uses its pharynx as a suction pump to suck up prey that is out of reach or hiding in rocky crevices. When caught on a line it is often sluggish but can reverse the sucking action to blast streams of water at its captor. Ovoviviparous, retaining the egg cases until the young hatch, with litters of about eight. No commercial value in Australia; elsewhere used for its meat, liver oil, fins and hide.

- **Local synonymy:** *Nebrius concolor* Rüppell, 1837; *Nebrodes concolor ogilbyi* Whitley, 1934.

- **Reference:** Whitley (1934a).

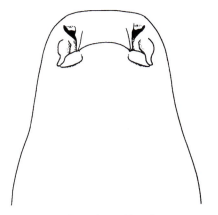

undersurface of head

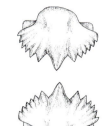

teeth from near symphysis
(upper and lower jaw)

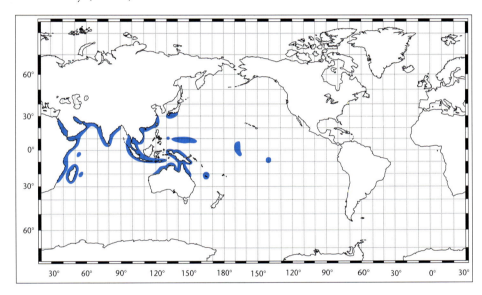

18 Family Rhincodontidae
WHALE SHARKS

This family contains a single, huge species (attaining 12 m) that has an enormous, transverse mouth situated well in front of the eyes (near the snout tip), minute teeth, large gill slits with internal filter screens, longitudinal body ridges, caudal keels, large first and small second dorsal fins, a semi-lunate caudal fin (except in small juveniles) without a subterminal notch, and a checkerboard colour pattern of light spots and stripes on a dark background.

These sharks, which are very large but harmless to humans, have also been erroneously referred to as rhiniodontids. They are planktonic feeders, occurring worldwide in tropical and warm temperate seas, near the coast and in the open ocean.

Whale Shark — *Rhincodon typus* (Smith, 1828)

18.1 *Rhincodon typus* (Plate 15) **Fish Code:** 00 014001

- **Field characters:** A huge filter-feeding shark with a broad, flattened head, a very large, nearly terminal mouth, a semi-lunate tail, a checkerboard pattern of light spots and stripes on a dark background, minute teeth, filter screens on its gill slits, and prominent ridges on its flanks.

- **Distinctive features:** Body fusiform, moderately stout with prominent longitudinal ridges on its upper flanks. Head depressed, broad and flattened. Mouth transverse, nearly terminal; nostrils with a rudimentary barbel. Gill slits very large, modified internally into filtering screens. Teeth minute, about 300 rows in each jaw, each comprising a single, hooked cusp. Caudal peduncle with lateral keels and a distinct upper precaudal pit. First dorsal fin much larger than second dorsal fin; set posteriorly on body, its insertion over the pelvic-fin bases. Anal-fin origin under front of second dorsal-fin base; these fins about equal in size. Pectoral fins falcate; caudal fin semi-lunate (except in small juveniles where upper lobe is considerably longer than lower lobe) with an indistinct terminal lobe. Total vertebrae [at least 153]*; precaudal [81]*.

teeth from near symphysis (upper and lower jaw)

- **Colour:** Greyish, bluish or brownish above, white ventrally; upper surface pattern of creamy white spots between pale, vertical and horizontal stripes resembles a checkerboard.

- **Size:** Free-swimming at 40–50 cm and attaining 1200 cm or more.
- **Distribution:** Cosmopolitan in tropical and warm temperate seas. In Australia, occurs mainly off the Northern Territory, Queensland and northern Western Australia. Isolated records from New South Wales, Victoria, South Australia and the western fringe of the Great Australian Bight. Epipelagic, oceanic and coastal.
- **Remarks:** The whale shark is the largest living fish. It feeds on a wide variety of planktonic and nektonic prey, including small crustaceans, small schooling fishes, and occasionally on tuna and squid. It does not rely on forward motion for filtration, but can hang vertically in the water and suction feed by opening its mouth and allowing water to rush in. Sea temperatures in the 21–25°C range, in the vicinity of cold water upwellings, are preferred because these conditions are probably optimal for its prey. Adults occur either singly or in aggregations of up to hundreds of individuals. They are highly migratory and small aggregations occur near the coast of central Western Australia each winter. Their movements are thought to be related to local productivity and are often associated with schools of pelagic fish. Some uncertainty exists over the whale shark's mode of reproduction, but it is thought to be ovoviviparous, retaining its egg cases until hatching. An egg case found in the Gulf of Mexico, 30 cm long by 14 cm wide and containing a near-term 36 cm embryo, may have been aborted. A young shark has survived for several years in an oceanarium in Japan. Small harpoon fisheries exist elsewhere but not exploited locally. Harmless to humans.
- **Local synonymy:** *Rhiniodon typus* Smith, 1828.
- **References:** Stead (1963); Grant (1978); Wolfson (1986).

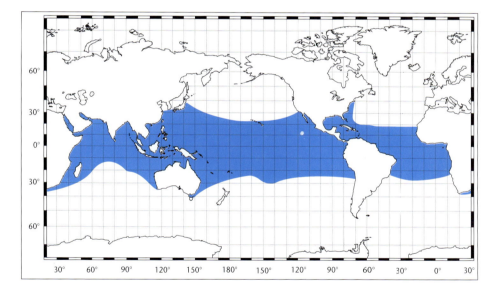

19

Family Odontaspididae

GREY NURSE SHARKS

Grey nurse sharks have large fusiform bodies (attaining 3.6 m), a conical snout, small eyes without nictitating membranes, two large, spineless dorsal fins, an anal fin, an upper precaudal pit, and five, rather long gill slits that do not extend onto the dorsal surface. They lack a lower precaudal pit and caudal keels. The caudal fin is heterocercal with a short ventral lobe and a subterminal notch. The mouth, which is armed with long, slender, lanceolate teeth with lateral cusplets, extends behind the front of the eyes and lacks nasoral grooves.

The family is represented worldwide by 2 genera and 4 species. Two widespread species occur in Australian waters. Despite their somewhat fearsome appearance these relatively sluggish sharks are generally harmless, although they should never be provoked.

Key to odontaspidids

Dorsal fins about equal in size, origin of first well behind free rear tips of pectoral fins (fig. 4); three rows of large teeth on each side of the upper jaw symphysis (fig. 3); snout short (fig. 1) ....
.... *Carcharias taurus* (**fig. 4; p. 145**)

First dorsal fin noticeably larger than second dorsal fin, origin of first about over free rear tips of pectoral fins (fig. 6); one row of small, then two rows of large, teeth on each side of the upper jaw symphysis (fig. 5); snout long (fig. 2) ....
.... *Odontaspis ferox* (**fig. 6; p. 146**)

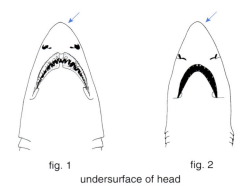

fig. 1   fig. 2

undersurface of head

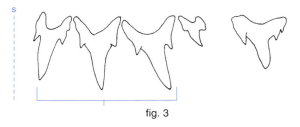

fig. 3

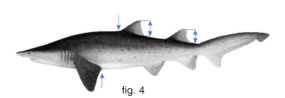

fig. 4

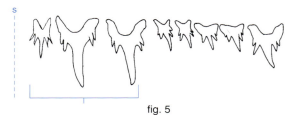

fig. 5

teeth next to upper jaw symphysis (s)

fig. 6

Grey Nurse Shark — *Carcharias taurus* Rafinesque, 1810

19.1 *Carcharias taurus* (Plate 15)   **Fish Code:** 00 008001

- **Alternative names:** Sand tiger shark, spotted ragged-tooth.
- **Field characters:** A large, rather stout shark with both dorsal fins and the anal fin of similar size, the first dorsal fin closer to the pelvic fins than the pectoral fins, long awl-like teeth with single lateral cusplets, and an asymmetrical caudal fin.
- **Distinctive features:** Body fusiform, moderately stout; caudal peduncle without keels. Snout relatively short, almost conical. Eyes without nictitating membrane. Teeth similar in both jaws; long, awl-shaped and with a relatively short lateral cusplet on either side of the main cusp; posterior teeth very small and lacking cusps. First dorsal fin closer to pelvic fins than to pectoral fins (its free rear tip almost over pelvic-fin origin). Similar-sized dorsal and anal fins. Caudal fin asymmetric. Tooth count [36–54/32–46]. Total vertebrae 165* [156–186]; precaudal 81* [80–97].
- **Colour:** Dorsal surfaces bronzy; paler ventrally (junction between surfaces may be abrupt). Juveniles with reddish or brownish spots on the caudal fin and posterior half of the body; spots fading with size, but sometimes still evident on adults.
- **Size:** Born at about 100 cm and attains 318 cm. Both sexes mature at about 220 cm.
- **Distribution:** Tropical and temperate parts of the North and South Atlantic, Indian, and western Pacific Oceans. Recorded in Australia from all States except Tasmania; rare in the Northern Territory. Continental shelf from the surf zone down to at least 190 m; usually lives near the bottom.
- **Remarks:** In the Atlantic and off South Africa, grey nurse sharks make seasonal migrations associated with reproduction, but nothing is known of movements of this species in Australia. Reproduction is oviphagous, with cannibalism inside the egg case and in the uterus, resulting in two young in a litter (one in each uterus). After eating other ova inside their egg case, surviving embryos (at about 55 mm) hatch

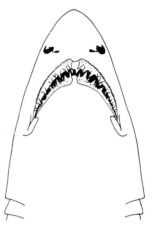

undersurface of head

second teeth from symphysis (upper and lower jaw)

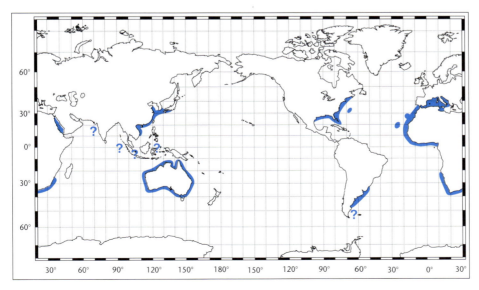

into the uterus. They then develop teeth (at about 10 cm) and begin to hunt and consume other embryos in the uterus. The single remaining embryo in each uterus then feeds on unfertilised ova. Gestation takes 9–12 months. Adults feed mainly on a variety of teleost and elasmobranch fishes. They can achieve near neutral buoyancy by swallowing air at the surface and holding it in their stomachs, enabling the shark to hang almost motionless above the bottom. Until recently, the grey nurse had an undeserved reputation in Australia as a 'maneater' which led to indiscriminate killing of the species by divers and line fishermen. It is now protected in New South Wales making it the first elasmobranch anywhere in the world to receive this status. Not currently exploited in Australia; elsewhere used for its meat (particularly in Japan), liver oil and fins.

- **Local synonymy:** *Odontaspis cinerea* Ramsay, 1880; *Carcharias arenarius* Ogilby, 1911; *Carcharius tricuspidatus*: McCulloch, 1929 (misidentification); *Eugomphodus taurus*: Compagno, 1984.

- **Reference:** Rafinesque (1810a).

Sand Tiger Shark — *Odontaspis ferox* (Risso, 1810)

19.2 *Odontaspis ferox* (Plate 15)
Fish Code: 00 008003

undersurface of head

- **Alternative names:** Herbsts nurse shark, smalltooth sand tiger, bumpytail raggedtooth.
- **Field characters:** A large, rather stout shark with the first dorsal fin noticeably larger than the second dorsal and anal fins, the first dorsal fin closer to the pectoral fins than the pelvic fins, long awl-like teeth with 2–3 pairs of relatively long, lateral cusplets, and an asymmetrical caudal fin.
- **Distinctive features:** Body fusiform, moderately stout; caudal peduncle without keels. Snout relatively long and bulbous. Eyes without nictitating membrane. Teeth similar in both jaws; long, awl-shaped and with relatively long lateral cusplets (usually 2 or 3 on each side of the main cusp); posterior teeth with very small cusps. First dorsal fin closer to pectoral fins than to pelvic fins (its origin almost over free rear tip of pectoral fin). First dorsal fin noticeably larger than second dorsal and anal fins. Caudal fin asymmetrical. Tooth count 48–51/40–42* [46–54/36–48]. Total vertebrae [177–183]*; precaudal [95–98]*.
- **Colour:** Greyish brown above, paler below; no sharp demarcation between the dorsal and ventral surfaces. Occasionally with dark reddish spots scattered on the body. Tips of some fins dark in juveniles.
- **Size:** Born at about 100 cm and attains at least 360 cm. Size at maturity unknown.

second teeth from symphysis (upper and lower jaw)

- **Distribution:** Known from a few tropical and warm temperate locations in the eastern Atlantic, Mediterranean, southwestern Indian and Pacific Oceans. In Australasia, recorded from off New South Wales, northwestern Australia and New Zealand. Lives on or near the bottom of the outer continental and insular shelves and upper slopes down to 420 m. Occasionally found in shallow water.
- **Remarks:** Little is known of the biology of this shark. Reproduction is presumably similar to that of the grey nurse shark (19.1). Stomachs examined have contained small bony fish, cephalopods and crustaceans. The large oily liver probably has a hydrostatic function. Used for its oil, and meat, which is considered by the Japanese to be inferior to that of the grey nurse.
- **Local synonymy:** *Odontaspis herbsti* Whitley, 1950.
- **References:** Whitley (1950a); Garrick (1974).

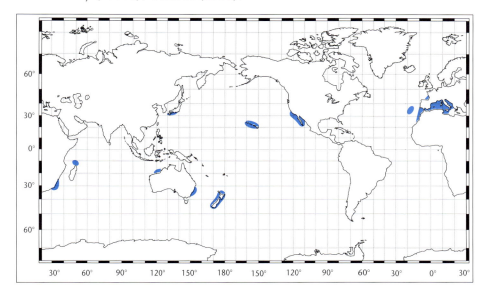

20 Family Mitsukurinidae
GOBLIN SHARKS

This family contains a single living species of bizarre appearance. It has a large, flabby body (attaining 3.8 m) with an elongate, depressed, blade-like snout, small eyes without nictitating membranes, and a caudal fin with a weakly developed ventral lobe and a small subterminal notch. The mouth, which is protrusile and extends posteriorly behind the front of the eyes, is armed with long, slender, lanceolate teeth. This shark also has exposed gill filaments, small lobe-like dorsal fins of a similar size, an anal fin, and small, broad pectoral fins. Nasoral grooves, precaudal pits and caudal keels are absent.

Poorly known but possibly widely distributed, the goblin shark is rarely caught. It lives near the bottom in deep water on the outer continental shelf and upper slope.

Goblin Shark — *Mitsukurina owstoni* Jordan, 1898

20.1 *Mitsukurina owstoni* (Plate 13)   **Fish Code:** 00 009002

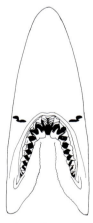

undersurface of head

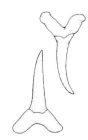

second teeth from symphysis
(upper and lower jaw)

- **Field characters:** A large shark with a bizarre head, very elongated snout forming a flat, blade-like rostrum, an elongate, asymmetrical caudal fin without a ventral lobe, partly exposed gill filaments, short rounded fins, and highly protrusile jaws with long, slender, awl-like teeth.
- **Distinctive features:** Body moderately slender and flabby. Snout greatly elongated into a flattened rostrum. Eyes very small; gill filaments partly exposed; jaws greatly protrusile. Teeth similar in both jaws; first 12–15 long, slender and awl-shaped; smaller and flatter towards side of mouth. Dorsal and pectoral fins rather small and rounded; pelvic and anal fins larger and more broadly rounded. Caudal fin long; no ventral lobe. Tooth count 51/55* [48–52/42–50]. Total vertebrae [122–124]*; precaudal [53–56]*.
- **Colour:** Uniformly greyish to reddish brown.
- **Size:** Attains 385 cm; birth size unknown (smallest recorded specimen 107 cm); a male of 264 cm was mature.
- **Distribution:** Recorded from scattered localities in the Atlantic, Indian and Pacific Oceans. Locally from a few specimens collected off New South Wales, eastern Bass Strait and possibly South Australia. Demersal on the outer continental shelves and upper slopes to a depth of about 1200 m. Very occasionally recorded from shallow inshore waters.

- **Remarks:** The biology of this shark is poorly known. Its soft, flabby body, small paired fins and weak caudal fin, support suggestions that it is almost neutrally buoyant and slow moving. The small eyes, raptorial jaws, grasping teeth, and the overhanging snout covered with sensory cells, suggests that it hunts its prey by the detection of electric fields. Also, its teeth have been found imbedded in submarine cables. Its reproductive method is unknown but it is presumed to be oviphagous. The majority of captures have been reported from Sagami Bay, Japan. It is harmless to humans.
- **Local synonymy:** *Scapanorhynchus owstoni* (Jordan, 1898).
- **Reference:** Stevens and Paxton (1985).

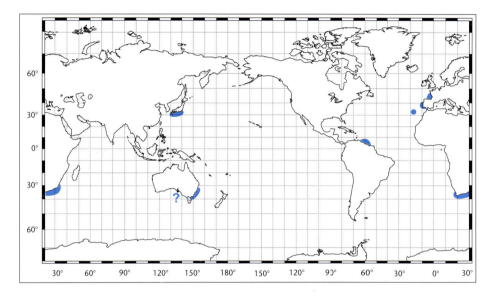

21

Family Pseudocarchariidae
CROCODILE SHARKS

The family Pseudocarchariidae contains a single species of medium-sized sharks (attaining 1.1 m) with a fusiform body, a small anal fin, a caudal fin with a strongly developed ventral lobe and a subterminal notch, a conical snout, very large eyes without a nictitating membrane, lanceolate teeth, five long gill slits, two rather low, angular dorsal fins (the first much larger than the second), and caudal keels. They also have small, broad pectoral fins and precaudal pits.

Crocodile sharks are somewhat similar in appearance to grey nurse sharks and were formerly placed in the family Odontaspididae. They are widespread in equatorial seas in epipelagic or mesopelagic habitats.

Crocodile Shark — *Pseudocarcharias kamoharai* (Matsubara, 1936)

21.1 *Pseudocarcharias kamoharai* (Plate 14) **Fish Code:** 00 009003

- **Field characters:** A medium-sized, pelagic shark with very large eyes lacking a nictitating membrane, long gill slits, lanceolate teeth, weak keels on the caudal peduncle, and an asymmetrical caudal fin.
- **Distinctive features:** Body fusiform; caudal peduncle with weak keels. Snout moderately long and conical, its tip rounded. Eyes very large with no nictitating membrane. Gill slits long, extending onto dorsal surface of head. Teeth similar in both jaws; long, slender and smooth-edged, becoming more oblique laterally. First dorsal fin low, its origin well posterior to pectoral-fin rear tips. Second dorsal fin less than half the size of first dorsal fin but larger than anal fin. Pectoral fins short and broadly rounded. Caudal fin asymmetrical with a distinct terminal lobe. Tooth count [26–29/19–24]*. Total vertebrae [152–153]*; precaudal [84–87]*.
- **Colour:** Dorsal surfaces dark brown; paler ventrally (sometimes with a few dark blotches on flanks and ventral surface); fins with narrow translucent or white margins.
- **Size:** Born at about 40 cm and attains 110 cm. Mature males of 74 cm and mature females of 89 cm have been reported.
- **Distribution:** Tropical and subtropical waters of all oceans. In Australia, recorded from off the Queensland coast. Oceanic, found between the surface and at least 590 m; occasionally occurring inshore.

- **Remarks:** The liver of the crocodile shark is high in low-density squalene oil, which suggests it is used to achieve near-neutral buoyancy. The large eyes suggest a nocturnal activity pattern or deepwater existence. Reproduction is oviphagous with a litter size of 4. Little is known of its diet but its grasping dentition suggests that it feeds on small mesopelagic teleosts, cephalopods and crustaceans. Although most frequently caught in oceanic longline fisheries, it is usually discarded because of its small size relative to other species in the catch.
- **Reference:** Matsubara (1936a).

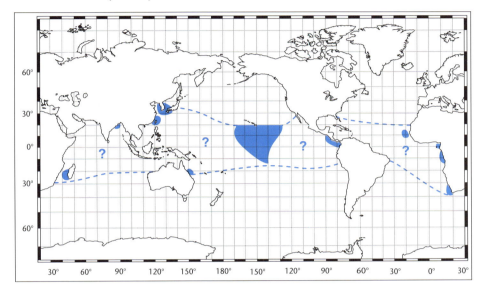

undersurface of head

second teeth from symphysis
(upper and lower jaw)

22 Family Megachasmidae
MEGAMOUTH SHARKS

This monospecific family contains the large megamouth shark (reaching more than 5 m). It has a large blubbery head with a short, bulbous snout, no nictitating eye membranes, a huge, terminal mouth (exceeding 1 m wide in adults) that extends behind the eyes when fully open, and jaws which are highly protrusile with small, hooked teeth. It has a small anal fin, a non-lunate caudal fin with a short but pronounced ventral lobe, an upper but no lower precaudal pit, two relatively low, angular dorsal fins (the first considerably larger than the second), long, narrow pectoral fins, and no caudal keels.

When the first megamouth specimen was caught in 1976, a new shark family had to be defined. Another four have been caught since from other parts of the Indo–Pacific. Pelagic in deepwater, it is one of only three planktivorous sharks.

Megamouth Shark — *Megachasma pelagios* Taylor, Compagno & Struhsaker, 1983

22.1 *Megachasma pelagios* (Plate 13)   **Fish Code:** 00 009001

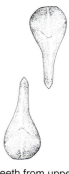

teeth from upper and lower jaw

- **Field characters:** A very large filter-feeding shark with a huge, blubbery head and terminal mouth, a strongly heterocercal caudal fin, moderately long gill slits, and lacking a pattern of light spots and/or stripes.
- **Distinctive features:** Body stout, tapering posteriorly. Head bulbous, wide and long; snout very short and broadly rounded. Gill slits moderately long but not reaching dorsal surface of head. Mouth broad, terminal, corner extending behind eyes. Numerous small, hooked teeth in jaws. Caudal peduncle without keels or ridges. First dorsal-fin origin closer to pectoral-fin bases than to pelvic-fin bases; dorsal fins relatively low; second dorsal fin less than half size of first dorsal fin. Pectoral fins shorter than head length in adults. Caudal fin asymmetrical, with pronounced ventral lobe. Tooth count [108/128]*.
- **Colour:** Dark grey to bluish black dorsally, lighter on the flanks; mostly pale grey below; posterior margins of most fins, and the pectoral- and pelvic-fin apices white; mouth lining silvery, black in preservative.
- **Size:** Attains at least 515 cm; males mature by 400 cm.

- **Distribution:** At present known only from Hawaii, southern California, Western Australia (Mandurah) and Japan, but probably wide-ranging. Oceanic, possibly occurring in depths between 150 and 1000 m.

- **Remarks:** The soft, flabby body and relatively small gill openings suggest that the megamouth is probably much less active than either the basking (24.1) or whale sharks (18.1). The combination of a large liver, high in low density squalene, poor calcification, and a flabby body, probably helps it to achieve neutral buoyancy. Nothing is known of reproduction in this species. The stomach contents have contained planktivorous prey such as euphausid shrimps, copepods and jellyfish. The only Australian specimen was beach-stranded.

- **Reference:** Berra and Hutchins (1990).

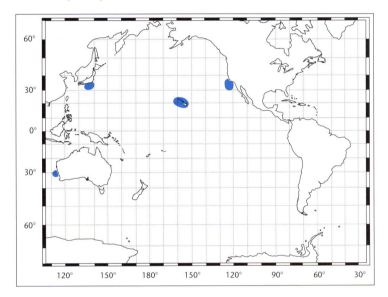

23

Family Alopiidae

THRESHER SHARKS

The threshers are large, distinctively shaped sharks (attaining 3.3–5.5 m) that are readily identified by the combination of an extremely elongated, scythe-like, upper caudal-fin lobe (equal in length to the rest of the body), and the absence of nasoral grooves and nasal barbels. They also have relatively stout but fusiform bodies, a conical snout, a relatively small mouth that extends backward of eye level, large eyes with no nictitating membranes, five relatively short gill slits, two spineless dorsal fins, long and narrow pectoral fins, upper and lower precaudal pits, no caudal keels, and very small second dorsal and anal fins.

These strong-swimming, highly active sharks have worldwide distributions in tropical and temperate seas. They are mainly pelagic, from the coast to the open ocean, where they feed on small schooling fishes and cephalopods. Prey are rounded up near the surface and stunned by the shark's thrashing tail. The group is comprised of a single genus and three species. All occur in Australian waters.

Key to alopiids

1   A deep groove extending on each side of the head from behind the eyes to above the gill slits (fig. 1); very large eyes extending onto the dorsal head surface (fig. 1); teeth relatively large, less than 25 tooth rows in each jaw; mid dorsal-fin base closer to pelvic-fin origin than pectoral-fin free rear tips (fig. 1) .... ***Alopias superciliosus*** **(fig. 1; p. 156)**

   No deep groove extending on each side of the head from behind the eyes to above the gill slits (fig. 2); smaller eyes not extending onto the dorsal head surface (fig. 2); teeth small, at least 30 tooth rows in each jaw; mid dorsal-fin base closer to pectoral-fin free rear tips than to pelvic-fin origin (fig. 2) **2**

2   Labial furrows absent (fig. 5); lateral teeth with distinct cusplets (fig. 3); pectoral fins nearly straight and broad-tipped (fig. 2); terminal lobe of caudal fin about equal in length to second dorsal-fin base (fig. 2); sides above pectoral-fin bases dark (fig. 2) .... ***Alopias pelagicus*** **(fig. 2; p. 155)**

   Labial furrows present (fig. 6); lateral teeth without distinct cusplets (fig. 4); pectoral fins falcate and narrow-tipped (fig. 7); terminal lobe of caudal fin over twice the length of second dorsal-fin base (fig. 7); sides above pectoral-fin bases white (fig. 7) .... ***Alopias vulpinus*** **(fig. 7; p. 157)**

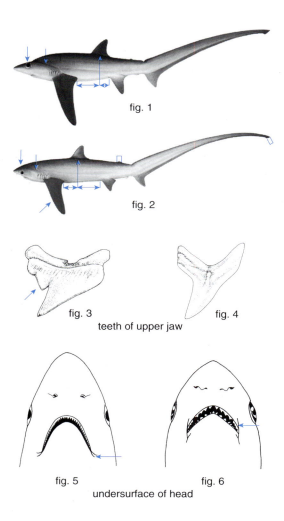

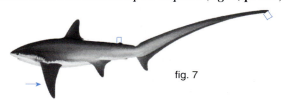

Pelagic Thresher — *Alopias pelagicus* (Nakamura, 1935)

23.1 *Alopias pelagicus* (Plate 12) **Fish Code:** 00 012003

- **Field characters:** A thresher shark with an extremely long upper caudal-fin lobe, relatively large eyes (not extending onto the dorsal head surface), straight pectoral fins, and the white ventral coloration not extending above the pectoral and pelvic-fin bases. No labial furrows or deep grooves behind the eyes.

- **Distinctive features:** Body fusiform, moderately stout; snout relatively short and conical. Eyes moderately large, but not extending onto dorsal head surface. Teeth similar in both jaws; small, oblique, blade-like, smooth-edged and all but first four to five rows with well-developed lateral cusplets. Free rear tip of first dorsal fin well in advance of pelvic-fin origin. Second dorsal and anal fins minute. Pelvic fins large. Pectoral fins straight; apices broadly rounded. Caudal-fin upper lobe enormously long; length (from caudal fork to upper caudal-fin tip) about as long as or longer than remaining body; terminal lobe about equal to length of second dorsal-fin base. Tooth count [43/42–48]*. Total vertebrae [472]*; precaudal [126]*.

- **Colour:** Pale grey dorsally and white ventrally; area above the gills and flank region may have a metallic silvery hue.

- **Size:** Born at about 100 cm and attains at least 330 cm. Smallest mature female recorded was 264 cm.

- **Distribution:** Tropical and subtropical Indo–Pacific. Recorded from the North West Shelf off Western Australia. Mainly oceanic from the surface to at least 150 m deep.

- **Remarks:** Relatively few reliable records of this species exist, partly because of confusion with *A. vulpinus*. Little is known of its biology. Reproduction is oviphagous and it usually has two pups. Presumably feeds on small fishes and cephalopods, which it stuns with its tail before capture. Caught incidentally in oceanic longline fisheries and used for its meat and fins.

- **Reference:** Sainsbury *et al.* (1985).

undersurface of head

ninth teeth from symphysis (upper and lower jaw)

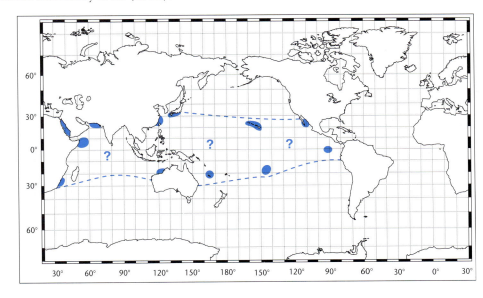

Bigeye Thresher — *Alopias superciliosus* (Lowe, 1839)

23.2 *Alopias superciliosus* (Plate 12) **Fish Code:** 00 012002

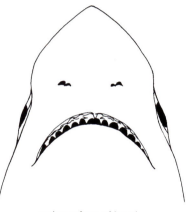

undersurface of head

- **Field characters:** A thresher shark with an extremely long upper caudal-fin lobe, huge eyes (extending onto the dorsal head surface), and pronounced lateral grooves on the top of the head.
- **Distinctive features:** Body fusiform, moderately stout; snout relatively long and bulbous. Lateral grooves originating in midline of head behind orbits and terminating above gill region. Eyes huge, extending onto dorsal surface of head. Teeth relatively large, similar in both jaws; cusps long, slender, smooth-edged and without lateral cusplets, first two or three near centre of jaw almost erect, remainder oblique. Free rear tip of first dorsal fin about over pelvic-fin origin. Second dorsal and anal fins minute. Pelvic fins large. Pectoral fins weakly falcate, apices relatively broad. Caudal-fin upper lobe very long; length (from caudal fork to upper caudal-fin tip) mostly shorter than preanal length; terminal lobe much longer than second dorsal-fin base. Tooth count 22/19* [19–24/20–24]. Total vertebrae [278–308]; precaudal [98–106].
- **Colour:** Purple to violet-grey dorsally; creamy white ventrally; flank region may have a metallic sheen.
- **Size:** Born at less than 100 cm (possibly as small as 65 cm) and attains 460 cm. Males mature at about 270 cm; females at about 300 cm.
- **Distribution:** Occurs in all tropical and warm temperate seas. Locally recorded from Western Australia (North West Shelf), Queensland (Middleton Reef), New South Wales and South Australia. Oceanic and coastal from the surface to at least 500 m deep.

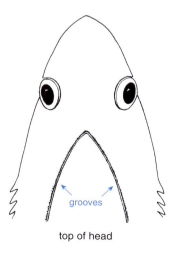

top of head

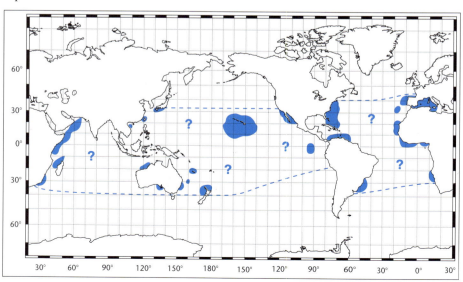

fifth tooth from symphysis (upper jaw)

- **Remarks:** Like the common thresher shark (23.3), the bigeye thresher may be able to maintain body temperatures above that of the surrounding sea water. Reproduction is oviphagous with litter sizes of 2–4. Feeds mainly on benthic and pelagic teleosts, and cephalopods, which it stuns with its tail. Caught incidentally in oceanic longline fisheries and used for its meat and fins.

Thresher Shark — *Alopias vulpinus* (Bonnaterre, 1788)

23.3 *Alopias vulpinus* (Plate 12)   **Fish Code:** 00 012001

undersurface of head

fifth tooth from symphysis (upper jaw)

- **Alternative names:** Thintail thresher, fox shark.
- **Field characters:** A thresher shark with an extremely long upper caudal-fin lobe, relatively large eyes (not extending onto the dorsal head surface), labial furrows, falcate pectoral fins, and the white ventral coloration extending above the pectoral and pelvic-fin bases. No deep grooves behind eyes.
- **Distinctive features:** Body fusiform, moderately stout; snout relatively short and conical. Eyes moderately large but not extending onto dorsal surface of head. Teeth relatively small, similar in both jaws; smooth-edged and narrowly triangular, usually without lateral cusplets; first three near centre of jaw erect, remainder oblique. Free rear tip of first dorsal fin well in advance of pelvic-fin origin. Second dorsal and anal fins minute. Pelvic fins large. Pectoral fins falcate, apices pointed. Caudal-fin upper lobe enormously long; length (from caudal fork to upper caudal-fin tip) about as long as or longer than remaining body; terminal lobe over twice length of second dorsal-fin base. Tooth count 38–40/35–41 [38–52/40–51]. Total vertebrae 343–356* [322–364]; precaudal 112–116* [112–121].
- **Colour:** Blue-grey above with a metallic lustre when alive. Ventral surface white; pale areas extending above pelvic and pectoral fin bases.
- **Size:** Born between 115 and 150 cm; attains about 550 cm. In Australia, males mature at about 340 cm although they may mature as small as 260 cm in other areas. Females mature between 350 and 400 cm.
- **Distribution:** Cosmopolitan in temperate and tropical seas. Locally more temperate, recorded from Brisbane (Queensland) to central Western Australia, including Tasmania. Coastal and oceanic from the surface to 370 m.
- **Remarks:** A powerful swimmer that, like the mackerel sharks, has a heat-exchanging circulatory system enabling it to maintain body temperatures higher than that of the surrounding water. Reproduction is oviphagous, with a litter size of 2–4. The diet consists mainly of small schooling fishes, which it herds and stuns with its tail (thresher sharks are often tail-hooked on longlines). The meat of this shark is

of high quality, and a target gillnet fishery for them has recently been developed off southern California. Also taken as a bycatch of the Japanese tuna longline fishery in southern Australia and retained for its meat and fins.

- **Local synonymy:** *Alopias caudatus* Phillipps, 1932; *Alopias greyi* Whitley, 1937.
- **Reference:** Whitley (1937a).

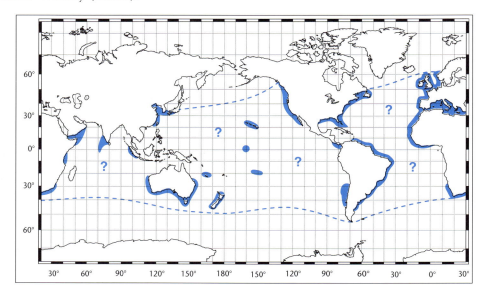

24 BASKING SHARKS

Family Cetorhinidae

The single living species in this family is a huge, plankton-feeder (attaining 10 m) with a relatively stout and fusiform body, a ventrally located mouth (extending well past the eyes and bearing numerous, minute teeth in the jaws), five very long gill slits (containing internal filter screens) that nearly encircle the head, caudal keels, and a lunate caudal fin. They also have a relatively long and conical snout, small eyes, two spineless dorsal fins (the first much bigger than the second), and precaudal pits.

Basking sharks are coastal pelagic planktivores of temperate seas. They bear a superficial resemblance to adult white sharks (25.1), but can be distinguished from them by their minute teeth and very long gill slits.

Basking Shark — *Cetorhinus maximus* (Gunnerus, 1765)

24.1 *Cetorhinus maximus* (Plate 13)   Fish Code: 00 011001

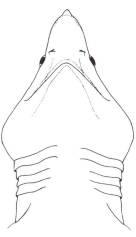

undersurface of head

- **Field characters:** A monstrous shark with extremely long gill slits that almost encircle the head, minute teeth, modified gill rakers, and a crescent-shaped caudal fin.
- **Distinctive features:** Body fusiform, moderately stout; caudal peduncle with distinct keel. Snout short, conical in specimens above about 400 cm (longer, forming an almost cylindrical proboscis in juveniles). Teeth minute, with a single, conical, hooked cusp; more than 100 on each side of jaw. First dorsal-fin origin behind rear tips of pectoral fins. Second dorsal-fin origin ahead of anal-fin origin. Second dorsal and anal fins moderately large, less than half the size of first dorsal fin. Caudal fin crescent-shaped. Total vertebrae [107–110]*; precaudal [50–53]*.
- **Colour:** Dorsal surface greyish brown (sometimes with lighter patches); ventral surface pale.
- **Size:** Size at birth unknown (but probably between 150 and 200 cm); attains 1000 cm. Males mature at 400–500 cm; the size at which females mature is unknown.

- **Distribution:** Anti-tropical, widespread in temperate coastal regions. Recorded infrequently from southern localities between Port Stephens (New South Wales) and Busselton (Western Australia), including Tasmania. Frequently coming close to shore.

- **Remarks:** Rarely encountered in Australian waters. Elsewhere, often seen swimming slowly at the surface, usually in groups of three or four but sometimes up to 100 have been reported together. Their seasonal occurrence in North Atlantic waters has been related to either a north–south migration or, more likely, to an inshore–offshore movement. Little is known of reproduction and no reliable records of pregnant females exist, although they are probably oviphagous. Their food consists mainly of plankton, which they ingest by swimming open-mouthed at slow speed with their gill rakers forming a filtering sieve. Their large livers (up to 25% of body weight) are high in squalene, a low-density hydrocarbon, which endows them with near-neutral buoyancy. It has been suggested that the gill rakers are lost in winter during low plankton densities and that the shark rests on the bottom in deep water. Alternatively, they may rely on food reserves in their large livers to support a reduced activity level. Periodic harpoon fisheries for basking sharks are located in Norway, the United Kingdom, California, South America, China and Japan. Their most valuable component is squalene oil, which is used as a lubricant and in the cosmetic industry, although the meat, fins and hide are also used.

- **Local synonymy:** *Tetroras maccoyi* Barrett, 1933; *Halsydrus maccoyi*: Whitley, 1940.

- **Reference:** Scott (1976).

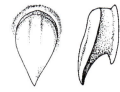
upper jaw tooth (dorsal and lateral views)

lower jaw tooth

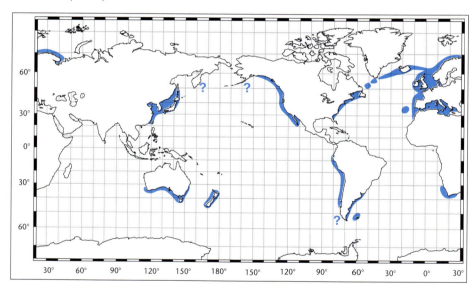

25

Family Lamnidae

MACKEREL SHARKS

Mackerel sharks are large, active and powerful sharks (attaining 3–6 m in length) with a stout, fusiform body and a conical snout. They have two spineless dorsal fins, minute second dorsal and anal fins, a lunate caudal fin, precaudal pits, and a laterally expanded caudal peduncle with well-developed caudal keels. The mouth, which is ventrally situated, extends well past the eyes and contains moderate to large-sized teeth. They also have relatively large eyes that lack nictitating membranes, and five long gill slits that extend onto the dorsal head surface but which do not almost encircle the head.

Mackerel sharks are distributed worldwide in tropical and temperate seas and live in both coastal and oceanic habitats. They have a heat-exchanging circulatory system that enables them to maintain body temperatures above that of the surrounding water. This family contains the white shark, usually considered to be the fish species most dangerous to humans. Three of the five species occur in the Australian region.

Key to lamnids

1 Teeth serrated, uppers broad, flat and triangular (narrower in small juveniles) (fig. 1) .... *Carcharodon carcharias* (**fig. 4; p. 162**)

   Teeth smooth-edged, uppers not flattened or broadly triangular (fig. 2) **2**

2 Dorsal coloration blue; no secondary keel below primary keel on caudal fin (fig. 6); teeth without lateral cusplets (fig. 2) .... *Isurus oxyrinchus* (**fig. 5; p. 163**)

   Dorsal coloration grey; a secondary keel below primary keel on caudal fin (fig. 7); teeth with lateral cusplets (fig. 3) .... *Lamna nasus* (**fig. 8; p. 165**)

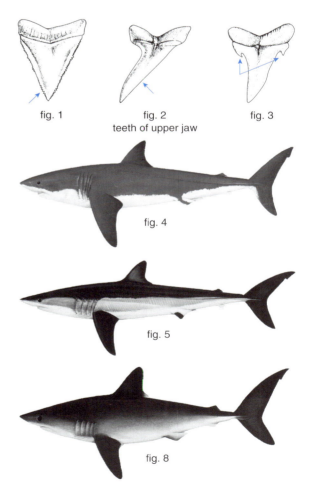

fig. 1   fig. 2   fig. 3
teeth of upper jaw

fig. 4

fig. 5

fig. 8

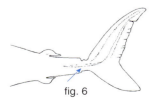

fig. 6

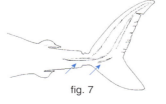

fig. 7

White Shark — *Carcharodon carcharias* (Linnaeus, 1758)

25.1 *Carcharodon carcharias* (Plate 14) Fish Code: 00 010003

- **Alternative names:** White pointer, great white shark, white death.
- **Field characters:** A mackerel shark with large, serrated triangular teeth, a crescent-shaped caudal fin, and minute second dorsal and anal fins.
- **Distinctive features:** Body fusiform, moderately stout; caudal peduncle with distinct keel. Snout relatively short and bluntly conical. Teeth large, erect, triangular and serrated; relatively more slender in lower jaw; juveniles less than 200 cm with more slender teeth (sometimes with lateral cusplets and lacking serrations on some cutting edges). First dorsal-fin origin over pectoral-fin inner margins. Second dorsal-fin origin in advance of anal-fin origin; second dorsal and anal fins minute. Caudal fin crescent-shaped, without a secondary keel below extension of caudal peduncle keel. Tooth count [23–28/21–25]. Total vertebrae [172–187]; pre-caudal [100–108].
- **Colour:** Dorsal surface blue-grey to grey-brown, often bronzy; white ventrally; boundary between these tones is mostly abrupt. Ventral tips of pectoral fins dusky; a dark spot may be present at the pectoral-fin axil.
- **Size:** Born at about 130 cm and attains 600 cm. Males mature at about 350 cm and females at about 400 cm.
- **Distribution:** Cosmopolitan but mostly anti-tropical in temperate seas. Probably throughout Australian waters, but more common in the south, from southern Queensland to North West Cape (Western Australia). Normally found over the continental shelf and often close inshore; recorded from the surface down to 1280 m.

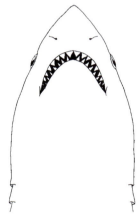

undersurface of head

second teeth from symphysis (upper and lower jaw)

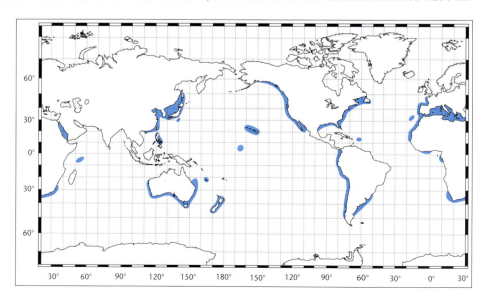

- **Remarks:** Despite its high-profile media image, little is known of the biology or behaviour of this awesome shark. It appears to be relatively scarce compared to most other widely distributed species, being most frequently reported from South Africa, the Great Australian Bight, northern California and the northeastern United States. Like other mackerel sharks, it has a heat-exchanging circulatory system enabling it to maintain body temperatures higher than that of the surrounding seawater. Aspects of reproduction remain an enigma as only two pregnant females have been recorded in contemporary times. Presumably oviphagous, a pregnant female from Japan had 7 pups. The diet of specimens smaller than about 300 cm consists mainly of a variety of teleost and elasmobranch fishes. Marine mammals comprise an important part of the diet of larger sharks. More attacks on humans, many of them fatal, have been attributed to the white shark than to any other aquatic animal. Occasionally used commercially for meat, fins, hide and oil.
- **Local synonymy:** *Carcharodon albimors* Whitley, 1939.
- **References:** Whitley (1939b); Sibley (1985).

Shortfin Mako — *Isurus oxyrinchus* (Rafinesque, 1810)

25.2 *Isurus oxyrinchus* (Plate 14) Fish Code: 00 010001

undersurface of head

- **Alternative names:** Mako shark, blue pointer, mackerel shark, snapper shark.
- **Field characters:** A mackerel shark with long, slender, pointed teeth that protrude noticeably from the mouth, short pectoral fins, minute second dorsal and anal fins, a crescent-shaped caudal fin, and indigo-blue dorsal surfaces and white undersides.
- **Distinctive features:** Body fusiform, moderately slender; caudal peduncle with a distinct keel. Snout relatively long, acutely conical. Teeth smooth-edged, similar in both jaws; anteriors long, slender and pointed with reflexed tips; lateral teeth progressively shorter and more triangular. First dorsal-fin origin over or behind rear tips of pectoral fins; first dorsal fin high, with a pointed apex in adults; relatively short with a rounded apex in juveniles. Second dorsal-fin origin well in front of anal-fin origin; second dorsal and anal fins minute. Pectoral fins considerably shorter than head length. Caudal fin crescent-shaped; no secondary keel below extension of caudal peduncle keel. Tooth count 24–26/22–28 [24–26/24–29]. Total vertebrae 186–193 [183–196]; precaudal 109–111 [107–112].

- **Colour:** Dorsal surfaces indigo-blue, merging abruptly from lighter blue on the flanks to white ventrally; undersurface of snout white. Dorsal surfaces becoming dark grey after preservation.
- **Size:** Born at about 70 cm and attains 394 cm. Males mature at about 195 cm and females at 280 cm.
- **Distribution:** Cosmopolitan in tropical and temperate seas. Widespread in Australian waters (with the exception of the Arafura Sea, Gulf of Carpentaria and Torres Strait). Oceanic and pelagic from the surface to at least 150 m. Occasionally found close inshore.
- **Remarks:** Seldom found in waters below 16°C. The shortfin mako, like other lamnids, has a heat-exchanging circulatory system. This enables it to maintain body temperatures higher than that of the surrounding seawater permitting a higher level of activity. It is probably the fastest of all sharks, capable of spectacular jumps clear of the water when hooked. Reproduction is oviphagous (embryos feed on eggs continuously ovulated by the female). Litters of 4–16 pups are born off New South Wales around November. The diet consists mainly of teleost fish and cephalopods. Individuals over 3 m may take larger prey such as billfish and small cetaceans. It is potentially dangerous, sometimes attacking boats. Of minor commercial importance in Australia, small sharks are sometimes retained in the southern and western shark fisheries. Larger quantities are taken by Japanese longliners. Elsewhere taken for its good-quality flesh and for its oil, fins, hide and teeth (for curios). Large numbers are caught by sport fishermen in Australia. The more tropical, longfin mako (*I. paucus*), which has longer pectoral fins (equal to its head length), still remains unrecorded locally but almost certainly occurs in oceanic waters off northern Australia.
- **Local synonymy:** *Isuropsis mako* Whitley, 1929.
- **References:** Rafinesque (1810a); Whitley (1929a); Stevens (1983, 1984).

second teeth from symphysis (upper and lower jaw)

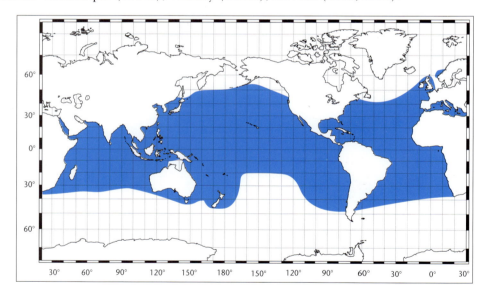

Porbeagle — *Lamna nasus* (Bonnaterre, 1788)

25.3 *Lamna nasus* (Plate 14) **Fish Code:** 00 010004

- **Alternative name:** Mackerel shark.
- **Field characters:** A mackerel shark with a crescent-shaped caudal fin that has a secondary keel below the extension of the caudal peduncle keel, minute second dorsal and anal fins, and moderately long, slender teeth with lateral cusplets.
- **Distinctive features:** Body fusiform and stout. Snout relatively long and conical. Teeth smooth-edged, similar in both jaws; moderately long, slender and pointed with a small basal cusplet on either side of main cusp. First dorsal-fin origin over or in front of inner margins of pectoral fins. Second dorsal-fin origin over or slightly ahead of anal-fin origin. Second dorsal and anal fins minute. Caudal fin crescent-shaped, with a secondary keel below extension of caudal peduncle keel. Tooth count 30–31/27–29* [28–29/26–27]. Total vertebrae 162* [150–162]; precaudal [84–91].
- **Colour:** Dorsal surfaces bluish grey; white ventrally, without dusky blotches; first dorsal fin with a pale, free rear tip. Juveniles have dusky patches beneath the pectoral fins and around the undersurface of the gill slits.
- **Size:** Born at 70–80 cm and attains at least 300 cm. Females apparently mature as small as 152 cm.
- **Distribution:** Anti-tropical in the North and South Atlantic, South Pacific and southern Indian Oceans. Few Australian records but probably not uncommon; southern Australia from southern New South Wales to southern Western Australia. Mainly on the continental shelf but also oceanic; occurs from the surface down to 370 m.

undersurface of head

second teeth from symphysis
(upper and lower jaw)

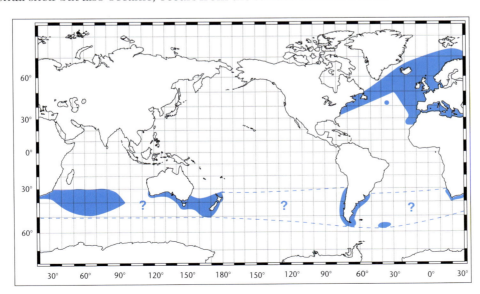

- **Remarks:** Probably confused in this region with the shortfin mako (25.2). Usually found in water cooler than 18°C and has been caught at high latitudes in 2°C. Like other mackerel sharks, it maintains a body temperature higher than that of the surrounding seawater. Reproduction is oviphagous with 1–5 pups in each litter. A pregnant female caught off Tasmania in June was 203 cm total length, weighed 140 kg and contained 4 full-term pups which had a combined weight of 17 kg. The diet consists mainly of teleost fishes. This shark supported a large fishery in the North Atlantic which peaked at 11 000 tonnes in 1964. Subsequently, catches have dropped alarmingly. It is sought for its good quality flesh and for its oil and fins. Locally, most common off Tasmania where it is taken as a bycatch by Japanese longliners fishing for southern bluefin tuna (*Thunnus maccoyii*).
- **Local synonymy:** *Lamna whitleyi* Phillipps, 1935.
- **Reference:** Stevens *et al.* (1983).

26 Family Scyliorhinidae
CATSHARKS

Catsharks have fusiform to slightly depressed bodies, slit-like eyes with nictitating eyelids, five pairs of gill slits, two dorsal fins without spines, small multicuspid teeth, and an anal fin. The first dorsal fin originates over or behind the pelvic-fin origin, the caudal fin is strongly heterocercal with a feeble lower lobe, and the caudal peduncle lacks pits and keels. The mouth is partly located beneath the eyes and the intestine has spiral valves.

This group, the largest of the shark families with up to 17 genera and about 100 species, is broadly distributed from tropical to arctic regions. Eight genera and 32 species are found off Australia. Many of these have narrow geographic ranges and several are endemic to the region. They mostly occur near the bottom in shallow water although a few genera contain species that occur along the continental slopes to depths exceeding 2000 m. Catsharks feed mainly on small fish and invertebrates and are harmless to humans. They are ovoviviparous or oviparous and the egg cases often have tendrils for attachment to the substrate.

Key to scyliorhinids

1 Upper labial furrows very long, more than 3 times spiracle diameter, mostly extending forward to front of eye (figs 1, 3, 4) **2**

   Upper labial furrows much shorter, less than 3 times spiracle diameter, never extending forward to front of eye (fig. 2) **12**

2 Anterior nasal flaps greatly expanded, reaching mouth (fig. 3); nasoral grooves present **3**

   Anterior nasal flaps much shorter, not reaching mouth (fig. 4); nasoral grooves absent (fig. 4) **4**

3 Ventral length of caudal fin shorter than head length (fig. 5); predorsal space lightly spotted and with 3–4 broad, dark saddles (fig. 5); posterior margins of dorsal fins straight (fig. 5) .... *Atelomycterus* **sp. A (fig. 5; p. 189)**

   Ventral length of caudal fin longer than head length (fig. 6); predorsal space heavily spotted and with about 7 broad, dark saddles (sometimes indistinct) (fig. 6); posterior margins of dorsal fins deeply concave (fig. 6) .... *Atelomycterus macleayi* **(fig. 6; p. 190)**

4 Anterior nasal flaps larger than nostrils (fig. 4); snout rather short and stout (figs 4, 7); anal fin

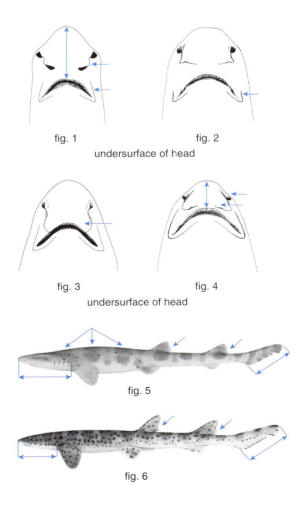

short–based, smaller in area than second dorsal fin (fig. 7) .... ***Aulohalaelurus labiosus* (fig. 7; p. 191)**

Anterior nasal flaps smaller than nostrils (fig. 1); snout rather long and flat (figs 1, 10); anal fin long-based, its area much greater than area of second dorsal fin (fig. 10) **5**

5 Snout very slender, length more than 4 times eye diameter and longer than wide (fig. 8) **6**

Snout not especially slender, length less than 4 times eye diameter and less than or equal to its width (fig. 9) **7**

6 Brownish; broad naked skin patches present between pectoral and pelvic fins (fig. 10); teeth enlarged, less than 45 rows in lower jaw .... ***Apristurus longicephalus* (fig. 10; p. 180)**

Greyish; no bare skin patches between pectoral and pelvic fins; teeth small, more than 45 rows in lower jaw .... ***Apristurus* sp. G (fig. 11; p. 179)**

7 Anal-fin margin distinctly rounded in shape (fig. 12); denticles mostly with single cusps (fig. 15), very widely spaced with naked areas visible between .... ***Apristurus* sp. D (fig. 19; p. 176)**

Anal-fin margin more or less triangular in shape (figs 13, 14); denticles tricuspidate (fig. 16), relatively closely spaced **8**

8 Anal-fin deep and short, its depth about equal to its base length (fig. 17); pectoral fin shorter than head width near mouth (fig. 21) **9**

Anal-fin low and long, its depth much less than its base length (fig. 18); pectoral fin longer than head width near mouth (fig. 22) **10**

9 Head moderately robust, preoral snout length less than 1.7 in maximum head width .... ***Apristurus* sp. E (fig. 20 p. 177)**

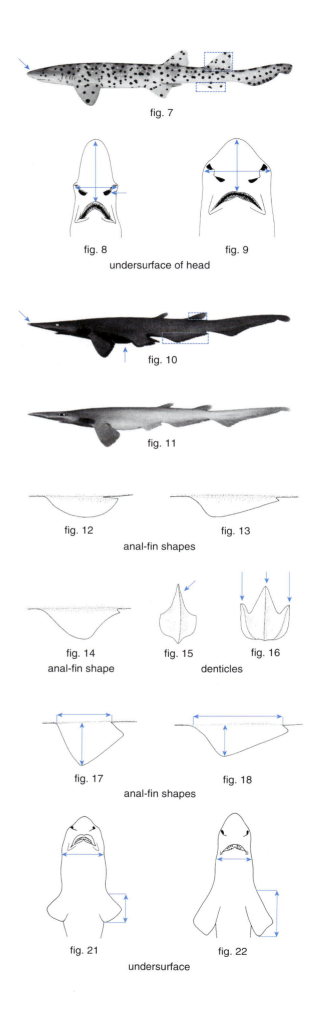

Head extremely robust, preoral snout length more than 1.7 in maximum head width ....
.... ***Apristurus* sp. F (fig. 23; p. 178)**

10 Pectoral fins large, trunk short (distance from adpressed pectoral-fin tips to pelvic-fin origin less than half length of anal-fin base) (fig. 24); monospondylous vertebrae less than 39 ....
.... ***Apristurus* sp. B (fig. 24; p. 174)**

Pectoral fins relatively short, trunk long (distance from pectoral-fin tips to pelvic-fin origin more than half length of anal-fin base) (fig. 25); monospondylous vertebrae more than 39 .... **11**

11 First dorsal fin distinctly smaller than second dorsal fin (fig. 25); teeth minute, central cusps barely larger than those laterally (difference in length not visible without magnification) (fig. 27) ....
.... ***Apristurus* sp. A (fig. 25; p. 173)**

Dorsal fins about equal in size (fig. 26); teeth relatively enlarged, central cusps distinctly larger than those laterally (difference in length visible without magnification) (fig. 28) ....
.... ***Apristurus* sp. C (fig. 26; p. 175)**

12 Upper margin of caudal fin with a crest of enlarged denticles (figs 29, 30) **13**

No crest of enlarged denticles on upper edge of caudal fin (fig. 31) **16**

13 Body soft; colour plain, without bands; pectoral fin width subequal to mouth width (fig. 34) ....
.... ***Parmaturus* sp. A (fig. 32; p. 204)**

Body firm; back patterned with light and dark bands (figs 33, 36, 37); pectoral fin broader than mouth width (fig. 35) **14**

14 Four small dark saddles on posterior half of body, none predorsally (fig. 35); no enlarged denticles along the ventral surface of caudal peduncle (fig. 29) .... ***Galeus* sp. A (fig. 33; p. 199)**

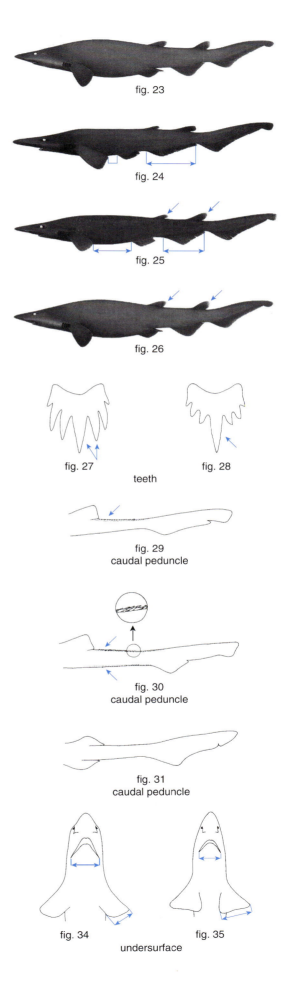

Several dark saddles on back before dorsal fin (figs 36, 37); ridge of enlarged denticles along the ventral surface of caudal peduncle (fig. 30) **15**

15 Predorsal space with 3 broad, dark saddles about equal in width to eye diameter (smaller bands also present) (fig. 36) .... *Galeus boardmani* **(fig. 36; p. 201)**

Saddles and bands in predorsal space narrower than eye diameter (10–16 small bands present) (fig. 37) .... *Galeus* **sp. B (fig. 37; p. 200)**

16 Second dorsal fin much smaller than first dorsal fin (fig. 38); origin of second dorsal fin only slightly behind origin of anal fin (fig. 38) **17**

Second dorsal fin about equal in size to first dorsal fin (fig. 39); origin of second dorsal fin behind midbase of anal fin (fig. 39) **23**

17 Pale with darker pattern of narrow transverse lines (figs 40, 41) **18**

Not as above **19**

18 Very narrow, transverse predorsal bars forming a pattern of rings, open saddle-like markings and blotches (fig. 40); northwestern Australia .... *Cephaloscyllium fasciatum* **(fig. 40; p. 197)**

Dorsal surface with 17–18 narrow, transverse predorsal bars, not connected to form rings or open saddles (fig. 41); northeastern Australia .... *Cephaloscyllium* **sp. D (fig. 41; p. 195)**

19 Body coloration relatively simple with only a few broad dark saddles on back and/or sides (figs 44, 45) **20**

Body with a strong colour pattern, mostly mottled with blotches and spots (figs 48, 51, 52) **21**

20 Head broad, more than 4.2 times preoral snout length (fig. 42); dorsal-fin origin located above the front half of the pelvic-fin bases (fig. 44); southern Australia .... *Cephaloscyllium* **sp. A (fig. 44; p. 192)**

Head broad, less than 4.2 times preoral snout length (fig. 43); dorsal-fin origin located above the back half of the pelvic fin bases (fig. 45); northeastern Australia .... *Cephaloscyllium* **sp. B (fig. 45; p. 193)**

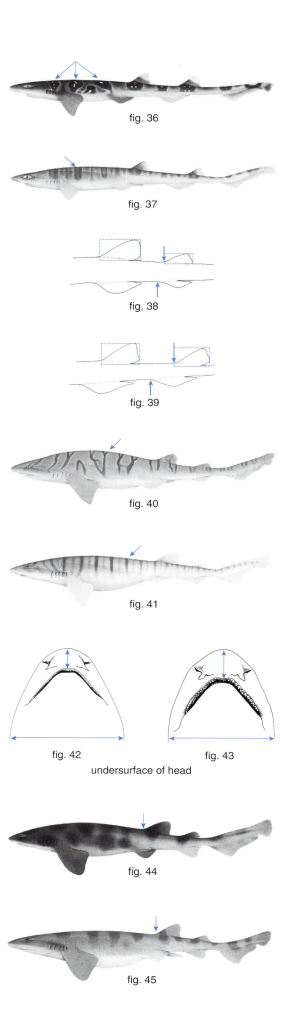

21 Anterior margin of pectoral fin longer than distance between insertion of pectoral fin and origin of pelvic fin (sometimes slightly shorter in large specimens) (fig. 46); prominent, broad dark band behind eyes (fig. 48); dark stripe usually present along midline of belly; egg case with ridges .... *Cephaloscyllium laticeps* (**fig. 48; p. 198**)

Anterior margin of pectoral fin shorter than distance between insertion of pectoral fin and origin of pelvic fin (fig. 47); no broad dark band behind eyes (or if present not distinctly broader than those behind) (fig. 51); no dark stripe along midline of belly; egg case lacking ridges **22**

22 Pectoral fin distinctly broader than mouth width (fig. 49); upper surface of body with dark saddles and light covering of pale flecks (fig. 51); temperate eastern Australia .... *Cephaloscyllium* sp. C (**fig. 51; p. 194**)

Pectoral fin width about equal to mouth width (fig. 50); upper surface of body heavily mottled, lacking distinct saddles (fig. 52); tropical Australia ....*Cephaloscyllium* sp. E (**fig. 52; p. 196**)

23 Body soft and flabby; normally hump-backed; uniformly dark .... *Halaelurus* sp. A (**fig. 53; p. 202**)

Body firm; not hump-backed; not uniformly dark (instead with spots and blotches) **24**

24 Anterior labial furrows very short, much shorter than spiracle diameter (fig. 56); gill slits elevated well above level of pectoral-fin origin (fig. 54) .... *Halaelurus boesemani* (**fig. 54; p. 203**)

Anterior labial furrows longer, 1–2 times spiracle diameter (fig. 57); anterior gill slits located partly below level of pectoral-fin origin (fig. 55) **25**

25 Body with large blotches but lacking spots (fig. 55); southwestern Australia .... *Asymbolus* sp. B (**fig. 55; p. 182**)

Body with spots and blotches (fig. 58) **26**

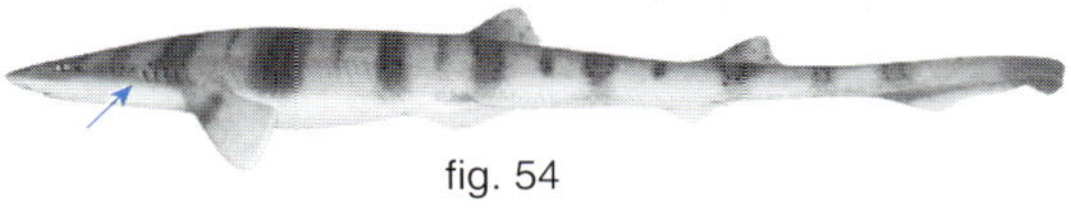

fig. 54

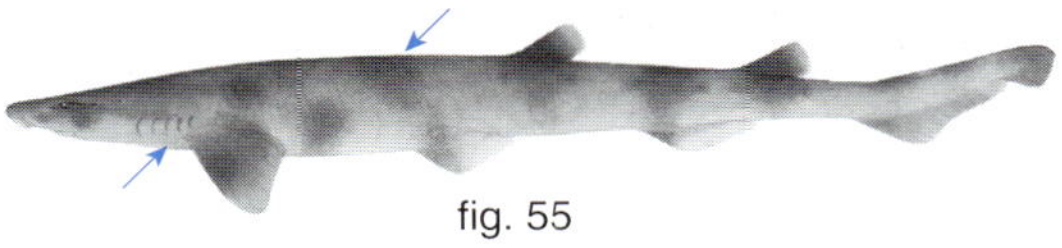

fig. 55

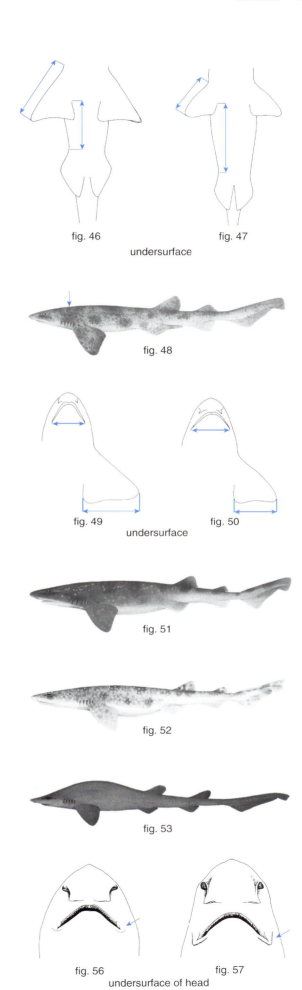

fig. 46   fig. 47
undersurface

fig. 48

fig. 49   fig. 50
undersurface

fig. 51

fig. 52

fig. 53

fig. 56   fig. 57
undersurface of head

26 Interdorsal distance less than 1.5 times total length of first dorsal fin (fig. 58); dorsal-fin tips broadly rounded (fig. 58); teeth large (length of longest exceeding half spiracle diameter); southern Australia .... ***Asymbolus* sp. C (fig. 58; p. 183)**

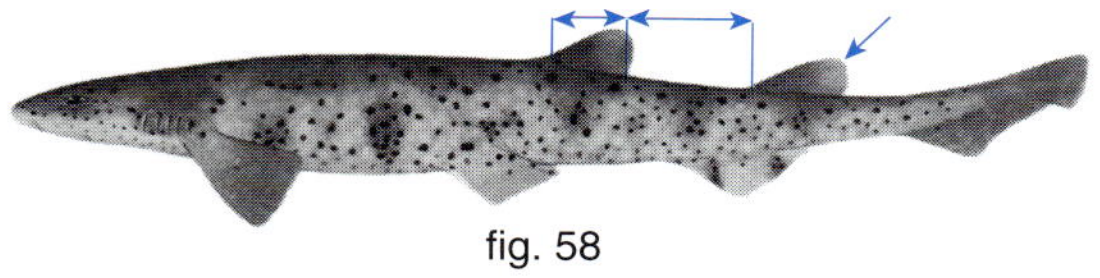

fig. 58

Interdorsal distance more than 1.5 times total length of first dorsal fin (fig. 59); dorsal-fin tips angular or only slightly rounded (fig. 59); longest teeth smaller than half spiracle diameter **27**

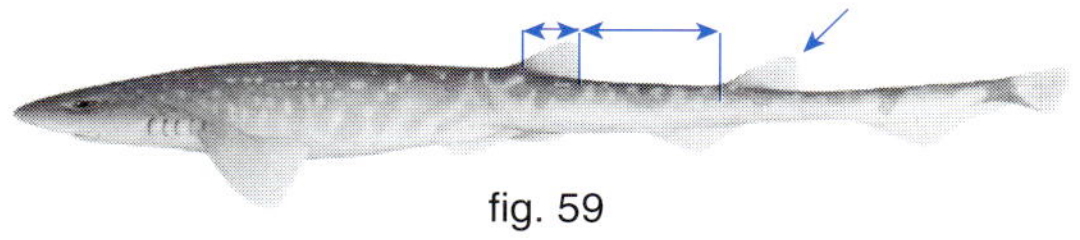

fig. 59

27 Body with more pale spots than dark spots **28**

Body with mainly dark spots **29**

28 Body pale brown with numerous white spots, lacking saddles (fig. 59); teeth mostly with 5 cusps; northwestern Australia ....***Asymbolus* sp. A (fig. 59; p. 181)**

fig. 60

Body mottled, greyish brown with faint white spots and darker saddles on back (fig. 60); teeth mostly with 3 cusps; southern Australia ....***Asymbolus vincenti* (fig. 60; p. 188)**

29 First dorsal fin height less than 4 in head length (fig. 61); spots distinctly orange-brown, many of those on flank distinctly larger with less well-defined borders than those on predorsal space (fig. 61); southeastern Australia ....***Asymbolus* sp. D (fig. 61; p. 184)**

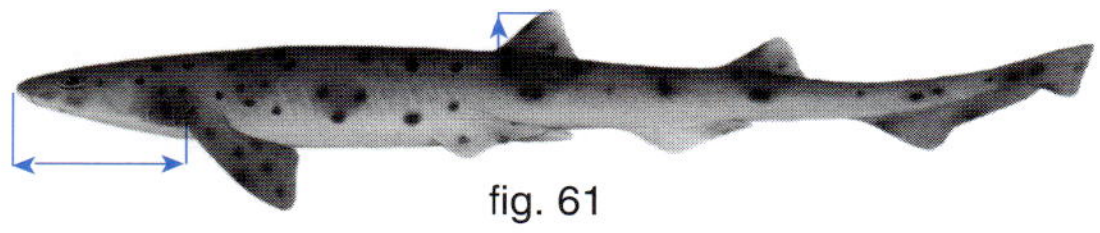

fig. 61

First dorsal fin height more than 4 in head length (fig. 62); spots dark brown or black; spots on flank and back similar in size and equally well defined **30**

30 Body dark greyish with poorly-defined, dark brown spots; white spots mostly also evident; eastern Australia .... ***Asymbolus analis* (fig. 62; p. 187)**

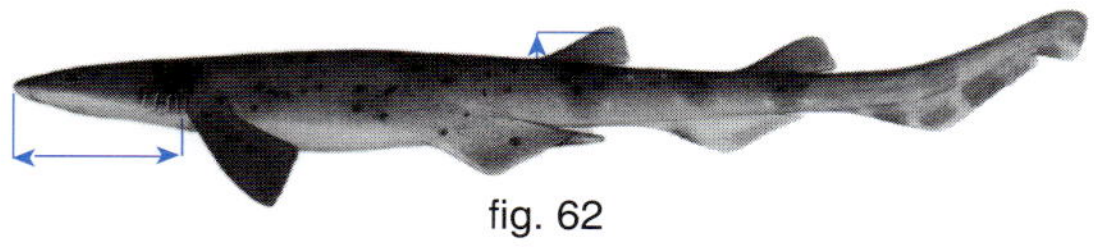

fig. 62

Body pale with well-defined, brownish black spots; no white spots **31**

31 No saddles or bands on dorsal surface (fig. 63); no dark spot beneath eye; mostly with a pair of spots on the dorsal midline preceding each dorsal fin and a single spot at the centre of each dorsal-fin base (fig. 63); northeastern Australia ....***Asymbolus* sp. E (fig. 63; p. 185)**

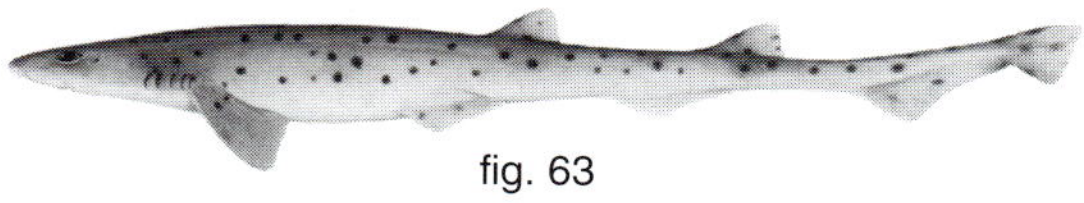

fig. 63

Distinct saddles present on dorsal surface (more prominent in juveniles) (fig. 64); usually a large spot beneath each eye; mostly with a single spot on the dorsal midline preceding each dorsal fin (fig. 64); Great Australian Bight and southwestern Australia .... ***Asymbolus* sp. F (fig. 64; p. 186)**

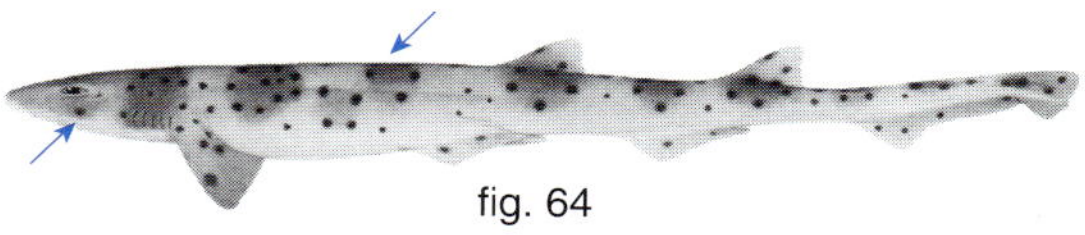

fig. 64

Freckled Catshark — *Apristurus* sp. A

26.1 *Apristurus* sp. A (Plate 19) **Fish Code:** 00 015014

- **Field characters:** A slender, brownish, deep-water catshark with a moderately elongate and flattened snout, very small, bristle-like teeth, and a long-based, subtriangular anal fin. The pectoral fins fall well short of the pelvic-fin bases, and most individuals have a scattering of pale flecks on the body.

- **Distinctive features:** Body slender, head not enlarged; tapering gradually forward of gills. Snout moderately long, broad, very flattened, slightly bell-shaped ahead of nostrils. Gill openings smaller than eye diameter. Eyes of moderate size, partly directed upwards, about 3–3.5 times in snout length. Nostrils large. Mouth moderately large, mouth and labial furrows extending to just in front of eyes. Teeth very small, bristle-like, closely spaced, with distinct ridges; mostly with 3–4 cusps, central cusps only slightly larger than those adjacent. Skin velvety; denticles on midflank minute, tricuspidate, broad-based, overlapping; throat with pleats; body lacking extensive naked areas (gill membranes scaled and small naked areas at insertions of pelvic, pectoral and dorsal fins); outer margins of fins naked. First dorsal-fin origin over or slightly behind pelvic-fin insertions; second dorsal fin much larger than first, its insertion in advance of anal-fin insertion; interdorsal space about twice first dorsal-fin base. Pectoral fin medium-sized; pectoral-pelvic space relatively long, much longer than preorbital length and mostly longer than pectoral-fin length. Pelvic fins relatively long and narrow, angular; anal fin long, deep, subtriangular. Caudal fin long, lower lobe relatively deep. Monospondylous vertebrae 40–42*.

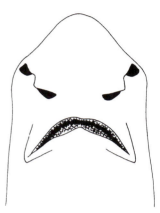

undersurface of head

- **Colour:** Dark brownish in adults, slightly more greyish in juveniles; mostly with pale, loosely scattered flecks; teeth greyish. Fin margins translucent or black.

- **Size:** To at least 74 cm. Males mature at about 64 cm off Tasmania (at about 51 cm in more northerly forms).

- **Distribution:** Known from southeastern Australia, from Newcastle (New South Wales) to Beachport (South Australia), and off Western Australia (North West Cape to Busselton and off Ashmore Reef) in 940–1290 m. Possibly more widespread; a New Zealand form may be the same species.

teeth from near symphysis (upper and lower jaw)

- **Remarks:** This species belongs to a genus of deepwater catsharks that contains a number of poorly known species. They have a drab, uniform coloration, a relatively long and expanded snout, enlarged nostrils, long labial furrows, small and posteriorly located dorsal fins, and a long anal fin that is almost connected to the caudal fin. The freckled catshark appears to be similar to *A. sinensis* from the western North Pacific. Several forms of this catshark occur in the Australian region and more than one species may be present. Further research is required to resolve species problems in this genus.

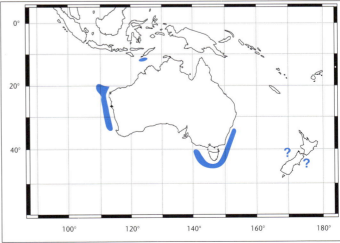

Bigfin Catshark — *Apristurus* sp. **B**

26.2 *Apristurus* sp. **B** (Plate 19)   **Fish Code:** 00 015015

- **Field characters:** A slender, rather compressed, greyish brown deepwater catshark with a moderately elongate and flattened snout, very small, bristle-like teeth, and a very long-based, subtriangular anal fin. The pectoral fins reach almost to the pelvic fins, and the body lacks pale flecks.

- **Distinctive features:** Body rather deep, very compressed behind head; head not enlarged, tapering gradually forward of gills. Snout moderately long, broad, very flattened; slightly bell-shaped ahead of nostrils. Gill openings small, much smaller than eye diameter. Eyes of moderate size, partly directed upwards, about 3–3.5 times in snout length. Nostrils large. Mouth moderately large, along with labial furrows extending to near front margin of eyes. Teeth very small, bristle-like, closely spaced, with distinct ridges; mostly with 4–5 cusps, central cusps only slightly larger than those adjacent. Skin velvety; denticles on midflank minute, tricuspidate, overlapping, bases of cusps forming a broad plate; throat mostly with pleats; body lacking extensive naked areas (gill membranes mostly naked and small to medium-sized naked areas at insertions of pelvic, pectoral and dorsal fins); naked areas on fin margins very narrow but mostly distinct. First dorsal-fin origin from slightly behind to well behind pelvic-fin insertions; second dorsal fin much larger than first, its insertion in advance of anal-fin insertion; interdorsal space about 2–3 times first dorsal-fin base. Pectoral fin large; pectoral–pelvic space short, about equal to or shorter than preorbital length and much less than length of pectoral fin. Pelvic fins relatively short and narrow, angular; anal fin very long, low, fin apex closer to its origin than to its insertion. Caudal fin long, lower lobe relatively deep. Monospondylous vertebrae 35–38.

undersurface of head

- **Colour:** Uniform medium greyish to dark brownish, teeth greyish; gill membranes and naked areas on fins black.

- **Size:** To at least 67 cm; males maturing between 53–57 cm.

- **Distribution:** Continental slopes of warm temperate and tropical Australia in 730–1000 m; eastern Australia from Ingham (Queensland) to Sydney (New South Wales), and off Geraldton (Western Australia).

- **Remarks:** A distinctive species (or group of species) with very large pectoral fins that is taken occasionally as a trawl bycatch. It is similar to *Apristurus acanutus* from the South China Sea and *A. platyrhynchus* from Japan. A closely related form has also been collected from New Caledonia.

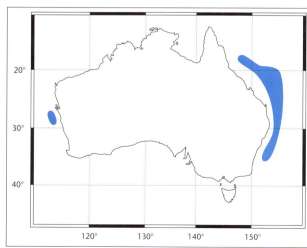

Fleshynose Catshark — *Apristurus* sp. C

26.3 *Apristurus* sp. C (Plate 20)　　　　　　　　　　Fish Code: 00 015016

- **Field characters:** A slender, dark brown, deepwater catshark with a moderately elongate and fleshy snout, relatively large teeth with central cusps 3–4 times longer than those adjacent, and a deep, long-based, triangular anal fin. The pectoral fins fall well short of the pelvic-fin bases, and most individuals have an irregular scattering of pale flecks on the body.

- **Distinctive features:** Body slender; head not enlarged, tapering rapidly from gills. Snout moderately long, slender, rather fleshy; tapering evenly or only slightly bell-shaped ahead of nostrils. Gill openings slightly smaller than eye diameter. Eyes of moderate size, directed sideways, about 3–4 times in snout length. Nostrils large. Mouth moderately large, mouth and labial furrows extending to just in front of eyes. Teeth of medium size, appearing widely spaced, lacking distinct ridges; mostly with 3–5 cusps, central cusps much larger than those adjacent. Skin velvety; denticles on midflank minute, tricuspidate, plate-like at apex in adults, overlapping; throat lacking pleats; body lacking extensive naked areas (gill membranes scaled but with small naked areas at insertions of pelvic, pectoral and dorsal fins and broad naked areas on fin margins). First dorsal-fin origin mostly slightly in front of pelvic-fin insertions; second dorsal fin only slightly larger than first, its insertion over or slightly posterior to anal-fin insertion; interdorsal space much less than twice first dorsal-fin base. Pectoral fins small, pectoral–pelvic interspace long, more than 1.25 times preorbital length. Pelvic fins relatively small, angular; anal fin long, deep, triangular. Caudal fin long, lower lobe moderately well developed. Monospondylous vertebrae 40–42.

- **Colour:** Uniform dark brown; teeth pale; naked tips of fins black.
- **Size:** To at least 71 cm. Smallest mature male examined 67 cm.
- **Distribution:** New Zealand and southern Australia from Eucla (Western Australia) to Broken Bay (New South Wales) in 900–1150 m. A similar form has been taken in the Indian Ocean well offshore from Perth.
- **Remarks:** This catshark belongs to a complex of species thought to be related to *Apristurus brunneus* from the western North Pacific.

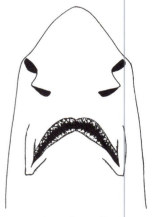

undersurface of head

upper tooth from near symphysis

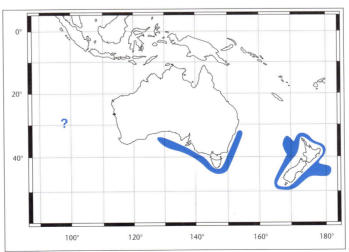

Roughskin Catshark — *Apristurus* sp. D

26.4 *Apristurus* sp. D (Plate 20) Fish Code: 00 015017

- **Field characters:** A large, dark deepwater catshark with a bulky head, unusually widely spaced denticles, very large teeth, and a relatively long-based and rounded anal fin.

- **Distinctive features:** Body relatively stout, head enlarged, tapering gradually forward of gills. Snout long, broad, slightly flattened, bell-shaped ahead of nostrils. Gill openings variable in size, mostly larger than eye length. Eyes small, directed sideways. Nostrils large. Mouth very large and broadly arched; mouth and labial furrows extending well in front of eyes. Teeth large, widely spaced, mostly tricuspidate (but often with 5 and sometimes with 7 cusps near back of jaws). Skin rough on midflank. Denticles with slender, single, erect cusp with narrow lateral keels (sometimes feebly tricuspidate), short, widely spaced, not overlapping; throat with pleats; extensive naked areas around insertion of pectoral fins, on gill membranes, behind and on outer margins of fins, and between the pelvic and anal fins. First dorsal-fin origin over posterior half of pelvic-fin bases; second dorsal fin only slightly larger than first; interdorsal space about twice first dorsal-fin base. Pectoral–pelvic space short, subequal to prespiracular length. Pelvic fins narrow, angular; anal fin relatively short, rounded. Caudal fin long and narrow, with a supracaudal crest; lower lobe low. Monospondylous vertebrae 33–36*.

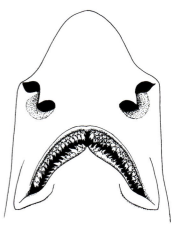

undersurface of head

- **Colour:** Dark brownish to black, mostly with irregular, scattered paler flecks and squiggles; outer tip of caudal fin whitish.
- **Size:** To at least 86 cm; males mature at about 67 cm.
- **Distribution:** Continental slope off New Zealand and temperate Australia in 840–1380 m. Locally recorded off Tasmania, the South Tasman Rise, and off Busselton (Western Australia).
- **Remarks:** This catshark is the largest member of the genus found in Australian seas and appears to extend further down the continental slope than most other *Apristurus* species.

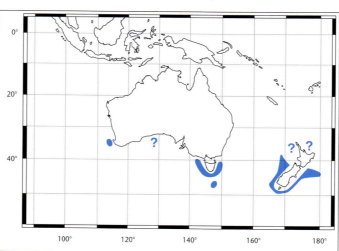

Bulldog Catshark — *Apristurus* sp. E

26.5 *Apristurus* sp. E (Plate 20)   Fish Code: 00 015018

- **Field characters:** A stout, brownish deepwater catshark with a stocky head and relatively short snout, long teeth, and a distinctive, deep, rather short-based, subtriangular anal fin. The preoral snout length is about two-thirds or more of the head width.

- **Distinctive features:** Body relatively robust, head enlarged, tapering rapidly from gills. Snout relatively short, broad, slightly flattened, bell-shaped ahead of nostrils. Gill openings mostly large, longer than eye length. Eyes small, directed sideways. Nostrils relatively small, length more than 1.5 in internarial space. Mouth large and broadly arched, extending well forward of eyes; labial furrows extending to anterior margin of eyes. Teeth widely spaced, mostly tricuspidate in males, with mostly 5–7 cusps in females; central cusps long, needle-like, much longer than adjacent cusps. Skin velvety on midflank; denticles with slender, single, erect cusp with narrow lateral keels (sometimes feebly tricuspidate); throat sometimes with pleats; lacking extensive naked areas (gill membranes and fin tips scaled but small naked areas behind pectoral and pelvic fins). First dorsal-fin origin over central part of pelvic-fin bases; second dorsal fin tall, distinctly larger than first; interdorsal space less than twice first dorsal-fin base. Pectoral fins short, relatively broad and rounded; pectoral–pelvic interspace rather long, almost twice preorbital length. Pelvic fins relatively broad, rounded; anal fin very deep, subtriangular, with posterior outer margin often slightly concave. Caudal fin short, lower lobe well developed. Total vertebrae 102–107*; monospondylous 34–35*.

undersurface of head

- **Colour:** Upper surface and fins medium brown with paler (often indistinct) speckles; undersurface darker brown; head often with irregular, pale blotches; teeth white.

- **Size:** To 63 cm. Males mature at about 50 cm.

- **Distribution:** Off southeastern Australia between Beachport (South Australia) and Broken Bay (New South Wales) from 1020 to at least 1500 m.

- **Remarks:** This species is most common off the continental slope of Victoria and Tasmania. Like other members of the genus, as well as several other deepwater sharks, the liver is rich in oil. It is, however, presently of no commercial value.

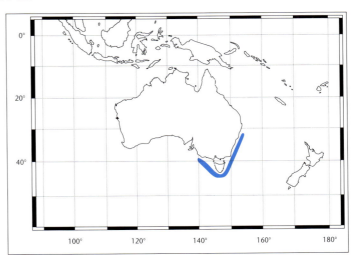

Bighead Catshark — *Apristurus* sp. F

26.6 *Apristurus* sp. F (Plate 20)　　　　　　　　　　　　　　Fish Code: 00 015019

- **Field characters:** A very stout, greyish brown, deepwater catshark with an extremely broad head and short snout, long teeth, and a deep, rather short-based, subtriangular anal fin. The preoral snout is about half of the maximum head width.

- **Distinctive features:** Body very robust, head enlarged, tapering gradually from gills. Snout relatively short, very broad, slightly flattened, bell-shaped ahead of nostrils. Gill openings large, about equal in length to eye diameter. Eyes small, directed sideways. Nostrils large. Mouth large and broadly arched, reaching well forward of eyes; labial furrows extending to anterior margin of eyes; teeth widely spaced, mostly tricuspidate in males, with mostly 5 cusps in females; central cusps long, needle-like, at least 3 times longer than cusps adjacent. Skin velvety on midflank; denticles tricuspidate, rather deciduous, dense and upright; throat with pleats; gill membranes scaled; no extensive naked areas (only small naked patches behind pectoral and pelvic fins and on margins of anal and caudal fins). First dorsal-fin origin over central part of pelvic-fin bases; second dorsal fin tall, mostly slightly larger than first; interdorsal space less than twice first dorsal-fin base. Pectoral fins short, relatively broad and rounded; pectoral–pelvic interspace rather long, exceeding preorbital length. Pelvic fins relatively broad, apices rounded; anal fin very deep, subtriangular, with posterior outer margin often slightly concave. Caudal fin short, lower lobe relatively well developed. Total vertebrae 117*; monospondylous 35*.

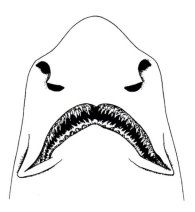

undersurface of head

- **Colour:** Uniform brownish to greyish brown, without flecks or reticulated markings; hind margins of caudal and anal fins black; teeth white.

- **Size:** To at least 73 cm.

- **Distribution:** Known from three specimens taken off Perth (Western Australia) in 1030–1050 m.

- **Remarks:** This catshark is similar to the bulldog catshark (26.5) but is more robust with a greatly enlarged head, reaches a larger size, and has a higher precaudal vertebral count than its eastern relative.

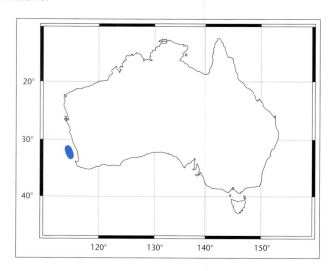

Pinocchio Catshark — *Apristurus* sp. G

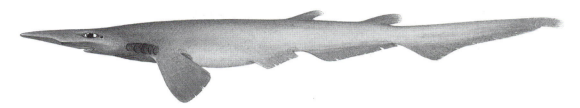

26.7 *Apristurus* sp. G (Plate 19) Fish Code: 00 015020

- **Field characters:** A pale greyish to brown, deepwater catshark with a very long and narrow snout, relatively small teeth, no naked skin patches between the pelvic and pectoral fins, and a low, very long-based anal fin.

- **Distinctive features:** Body slender, head long and narrow. Snout elongate, narrow, flattened, slightly bell-shaped ahead of nostrils. Gill openings small, narrower than eye length. Eyes small, directed somewhat upward, about 4–5 times in snout length. Nostrils large. Mouth moderate-sized, mouth and labial furrows extending to front margin of eyes; teeth fine, with fine basal ridges, closely set, mostly with 5 cusps anteriorly and 7 cusps in lateral teeth. Skin velvety; denticles short, close set but not overlapping, tricuspidate with narrow base, central cusp longest; throat lacking pleats; no extensive naked areas on body (gill membranes scaled but small naked patches at pectoral-fin insertions). First dorsal-fin origin near or slightly behind pelvic-fin insertions; second dorsal fin much larger than first; interdorsal space about twice first dorsal-fin base. Pectoral–pelvic space short, less than three-quarters of preorbital length. Pelvic fins narrow, angular; anal fin very long, low, angular. Caudal fin long and narrow; lower lobe prominent. Monospondylous vertebrae 31–34.

- **Colour:** Pale greyish to medium brown, throat darker; borders of anal, pelvic and pectoral fins, hind margin of caudal fin, and outer tips of dorsal fins white; northeastern populations have broader pale margins, particularly on the pectoral fins.

- **Size:** Attains 61 cm; males maturing at 51 cm.

- **Distribution:** Widely distributed along the Australian continental slope in 590–1000 m. From Shark Bay (Western Australia) to Cairns (Queensland), including seamounts to the south of Tasmania but not yet recorded from much of the Great Australian Bight. A similar, if not conspecific, species is known to occur off New Zealand.

- **Remarks:** This catshark is represented in Australia by several populations which are distinct from each other and may be separate species. The southern Australian form is usually more brownish in colour and has a relatively smaller abdomen and larger pectoral fins than those further north. Their relationship to the similar western North Pacific catshark, *Apristurus herklotsi*, has yet to be determined.

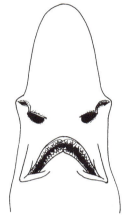

undersurface of head

teeth from near symphysis (upper and lower jaw)

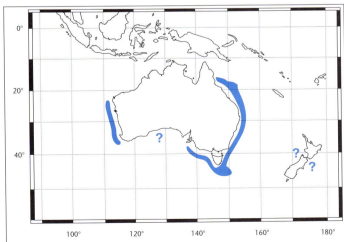

Smoothbelly Catshark — *Apristurus longicephalus* Nakaya, 1975

26.8 *Apristurus longicephalus* (Plate 19) Fish Code: 00 015021

- **Field characters:** A dark, deepwater catshark with a very long and narrow snout, small and widely spaced teeth, extensive naked skin patches between the pelvic and pectoral fins, and a low, very long-based anal fin.
- **Distinctive features:** Body slender, head long and narrow. Snout elongate, narrow, flattened, distinctly bell-shaped ahead of nostrils. Gill openings moderately large, about equal to eye length. Eyes small, directed somewhat upward, about 5 times in snout length. Nostrils large. Mouth moderate-sized; mouth and labial furrows extending well forward of eyes. Teeth fine, rather widely spaced, with indistinct basal ridges; 3 cusps, central cusp several times longer than those adjacent. Skin velvety; denticles short, overlapping, tricuspidate with broad bases; throat lacking pleats. Extensive naked areas on body; broad area extending diagonally above pectoral fin between gill slits and pelvic fin; interpectoral space, throat and around gills, and between pelvic and anal fins naked; broad naked areas along margins of fins. First dorsal-fin origin over insertions of pelvic fins; second dorsal fin much larger than first; interdorsal space more than twice first dorsal-fin base. Pectoral–pelvic space short, about two-thirds of preorbital length. Pelvic fins narrow, angular; anal fin long, angular. Caudal fin long and narrow; lower lobe prominent. Monospondylous vertebrae 31*.
- **Colour:** Dark brownish; naked areas black.
- **Size:** Attains at least 50 cm; male specimens up to 37 cm are sexually immature.
- **Distribution:** East China Sea, Japan, Seychelles, the Philippines and Australia. Locally from only three specimens trawled off Townsville (northern Queensland) in 880 m, and off Ashmore Reef and the North West Cape (northwestern Australia) in 680–900 m.
- **Remarks:** A distinctive deep-water catshark. Little is known of its biology.
- **References:** Nakaya (1988a, 1988b).

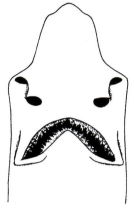

undersurface of head

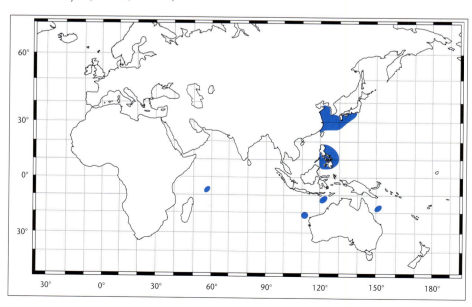

Dwarf Catshark — *Asymbolus* sp. A

26.9 *Asymbolus* sp. A (Plate 22)   **Fish Code:** 00 015022

- **Field characters:** A very small, pale brown catshark with numerous white spots (lacking dark markings), anterior nasal flaps enlarged but not reaching the mouth, relatively short labial furrows, and minute teeth.

- **Distinctive features:** Body slender, slightly compressed; head flattened; snout short, broadly rounded. Anterior nasal flaps expanded into a single pointed lobe; not reaching mouth. Labial furrows short, equal to or only slightly longer than spiracle diameter. Teeth minute, mostly with 5 cusps. Stomach not inflatable. Skin velvety; denticles minute with a plate-like crown and pronounced posterior cusp; no ridge of enlarged denticles along dorsal caudal-fin margin. First dorsal-fin origin over or slightly behind pelvic-fin insertion; second dorsal-fin origin just in advance of anal-fin insertion. First dorsal-fin base about half anal-fin base length; dorsal fins about equal in size. Pectoral fins short, broad with broadly rounded margins. Pelvic fins connected by a membrane, forming a partial apron over claspers of adult males; mature claspers extending about half their length past pelvic-fin tips.

- **Colour:** Upper surface pale brown with a dense array of white spots and flecks; no greyish or black markings. Two brownish bands mostly present on caudal fin and a single band at origin of each dorsal fin. Remainder of fins and ventral surface pale.

- **Size:** Reaches at least 33 cm and males mature at 28 cm.

- **Distribution:** Northwestern Australia from Dampier to the Buccaneer Archipelago in 59–252 m.

- **Remarks:** A small tropical species, little known and poorly represented in collections. Most abundant on the outer continental shelf.

undersurface of head

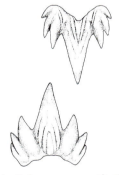
teeth from near symphysis (upper and lower jaw)

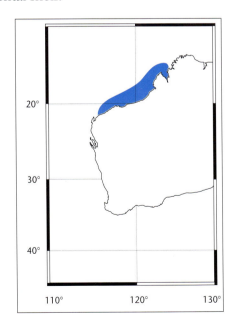

Blotched Catshark — *Asymbolus* sp. B

26.10 *Asymbolus* sp. B (Plate 22)   Fish Code: 00 015023

- **Field characters:** A small, medium brown catshark with a distinctive pattern of dark brown blotches and saddles (but lacking small spots), anterior nasal flaps enlarged but not reaching the mouth, relatively small labial furrows, and with greatly enlarged central tooth cusps.

- **Distinctive features:** Body slender, rather compressed; head moderately flattened; snout short, rounded. Anterior nasal flaps expanded into a truncate lobe, not reaching mouth; posterior nasal flaps relatively enlarged. Labial furrows short, about twice spiracle diameter. Teeth medium-sized (rather enlarged relative to other members of the genus), with 3 cusps, central cusp several times larger than minute lateral cusps (resembling a single cusp to naked eye). Stomach not inflatable. Skin velvety; denticles fine, tricuspidate with enlarged central cusp; no ridge of enlarged denticles along dorsal caudal-fin margin. First dorsal-fin origin over pelvic-fin insertion; second dorsal-fin origin over posterior third of anal-fin base. First dorsal-fin base about half length of anal-fin base; dorsal fins about equal in size. Pectoral fins short, broad with broadly rounded posterior margins. Pelvic fins connected by a membrane, forming a partial apron.

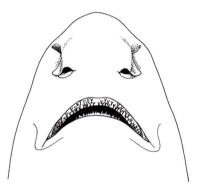
undersurface of head

- **Colour:** Medium brown above with large, dark brown blotches and saddles (markings spaced irregularly); 3 predorsal saddles, additional bars beneath each dorsal fin and one interdorsally; ventral surface only slightly paler.

- **Size:** Attains at least 44 cm.

- **Distribution:** Single record from off the Recherche Archipelago, Western Australia, in 195 m.

- **Remarks:** This species is known only from a single female specimen. It is easily distinguished from other *Asymbolus* species from the same geographic area in being deeper-bodied and lacking small spots. Specimens are required.

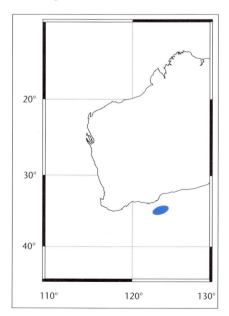

Variegated Catshark — *Asymbolus* sp. C

26.11 *Asymbolus* sp. C (Plate 21)   **Fish Code:** 00 015010

- **Alternative name:** Saddled catshark.
- **Field characters:** A small, greyish brown catshark with numerous, small black spots and darker, rusty-brown and bluish grey saddle-like blotches. The anterior nasal flaps are enlarged but do not reach the mouth, the labial furrows are short, the teeth are relatively large, and the interdorsal space is narrow (about 1.5 times first dorsal-fin base).
- **Distinctive features:** Body elongate, rather compressed; head flattened but deep; snout short, broadly rounded. Anterior nasal flaps expanded, not reaching mouth. Labial furrows short, 1.5–2 times spiracle diameter. Teeth relatively large, tricuspidate; central cusps greatly enlarged (almost equal to spiracle diameter). Stomach not inflatable. Skin velvety; denticles very closely spaced, overlapping, tricuspidate; no ridge of enlarged denticles along dorsal caudal-fin margin. First dorsal-fin origin just in advance of pelvic-fin insertion; second dorsal-fin origin almost over middle of anal-fin base. First dorsal-fin apex distinctly rounded; base about 1.5 in interdorsal distance, longer than half anal-fin base; dorsal fins about equal in size. Pectoral fins small, short and broad with rounded outer tips. Pelvic fins connected by a membrane, forming a partial apron.
- **Colour:** Upper surface and fins greyish brown with several irregular, rusty-brown, saddle-like blotches; variably peppered with small black spots (most heavily concentrated on sides but which may extend to the undersurface of the head); sides with bluish grey blotches. Belly and undersurface of head pale.
- **Size:** Attains at least 43 cm.
- **Distribution:** Endemic to Australian waters, it has been collected from southern Western Australia between the Recherche Archipelago and Cape Naturaliste to a depth of 150 m. Possibly more widely distributed in the Great Australian Bight.
- **Remarks:** Reported as inhabiting caves and ledges off southwestern Australia but rarely seen due to its nocturnal habits. Little is known about its biology.
- **Local synonymy:** *Asymbolus* sp.: Hutchins & Swainston, 1986.
- **Reference:** Hutchins and Swainston (1986).

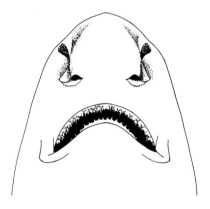

undersurface of head

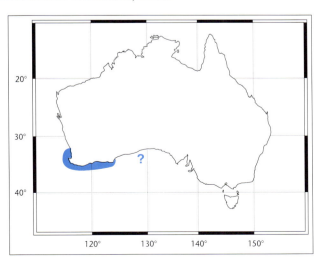

Orange Spotted Catshark — *Asymbolus* sp. D

26.12 *Asymbolus* sp. D (Plate 21)   **Fish Code:** 00 015024

- **Alternative names:** Rusty catshark, spotted catshark.
- **Field characters:** A small, pale brown catshark with numerous dark brown spots (with orange-brown borders), usually with an indistinct brownish blotch beneath the eye and a cluster of spots on the dorsal midline between each dark saddle, anterior nasal flaps enlarged but not reaching the mouth, relatively small labial furrows, and minute teeth. The first dorsal-fin origin is situated over or slightly in front of pelvic-fin insertion.
- **Distinctive features:** Body slender, subcylindrical; head relatively narrow, slightly flattened; snout short, rounded. Anterior nasal flaps expanded, not reaching mouth; posterior nasal flaps reduced. Labial furrows short, 1–2 times spiracle diameter. Teeth minute, with 3 or 5 cusps. Stomach not inflatable. Skin velvety; denticles dense, tricuspidate, with elongated central cusp; no ridge of enlarged denticles along dorsal caudal-fin margin. First dorsal-fin origin over or slightly in front of pelvic-fin insertion; second dorsal-fin origin over posterior third of anal-fin base. Dorsal fins about equal in size. First dorsal-fin base 1.75–2 in interdorsal distance, about half anal-fin base. Pectoral fins small, broad with rounded margins. Pelvic fins connected by a membrane, forming a broad partial apron over claspers of adult males; mature claspers extending about half their length past pelvic-fin tips.
- **Colour:** Light brown dorsally; faint greyish brown saddles and blotches located above gill area, beneath dorsal fins, and on sides in pectoral–pelvic interspace. Spots dark brown with orange-brown borders, larger than spiracle diameter; scattered irregularly and numerous (although variable in number). Upper surface of pectoral fins mostly with several spots; each posterior and anterior dorsal-fin margin usually with a single dark mark. Ventral surface, and pelvic and anal fins pale.
- **Size:** Attains at least 53 cm; males are mature at 37 cm.
- **Distribution:** Eastern Australia from Moreton Island (Queensland) to South East Cape (Tasmania) in depths of 80–290 m. The extent of its westerly distribution is unknown.
- **Remarks:** This species belongs to a group of four spotted catsharks which have all been confused in the past with the grey spotted catshark (26.15). It overlaps in range with the latter but occurs in deeper water. Little is known of its biology. Presently of no commercial value.
- **Local synonymy:** *Asymbolus analis*: Last *et al.*, 1983 (misidentification).
- **Reference:** Last *et al.* (1983).

undersurface of head

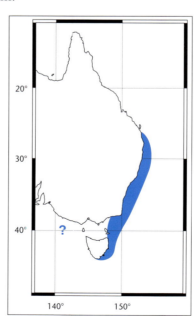

Pale Spotted Catshark — *Asymbolus* sp. E

26.13 *Asymbolus* sp. E (Plate 21)   Fish Code: 00 015025

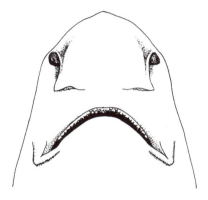
undersurface of head

- **Field characters:** A very small, pale yellowish catshark covered with dark brown spots (regularly distributed and sharply defined), lacking a dark spot or blotch beneath the eye, the anterior nasal flaps enlarged but not reaching the mouth, relatively short labial furrows, and minute teeth. The first dorsal-fin origin is over or just behind the pelvic-fin insertion.

- **Distinctive features:** Body slender, subcylindrical; head slightly depressed; snout short, broadly rounded. Anterior nasal flaps expanded, their tips slightly pointed, not reaching mouth; posterior nasal flaps rather short. Labial furrows short, 1.5–2 times spiracle diameter. Teeth minute, mostly with 5 cusps, lateral cusps reduced in size. Stomach not inflatable. Skin velvety; denticles very dense, tricuspidate, lateral cusps often barely distinguishable; no ridge of enlarged denticles along dorsal caudal-fin margin. First dorsal-fin origin over or just behind pelvic-fin insertion; second dorsal-fin origin slightly in advance of anal-fin insertion. Dorsal fins about equal in size. First dorsal-fin base about 2 in interdorsal distance, 0.5–0.75 of anal-fin base. Pectoral fins short, broad with broadly rounded posterior margins. Pelvic fins connected by a membrane, forming a partial apron over claspers of adult males; mature claspers extending about half their length past pelvic-fin tips.

- **Colour:** Dorsal surface pale yellowish brown with numerous rather regularly distributed, well-defined, dark brown spots (each about equal to spiracle diameter) on sides and back; lacking distinct saddles and blotches, and lacking spots and blotches below eye. Upper surfaces of pectoral, dorsal and caudal fins mostly with spots. Pale ventrally.

- **Size:** Reaches at least 46 cm; young hatch at about 19 cm; males mature at about 32 cm.

- **Distribution:** Continental slope off northeastern Australia between the Swain Reefs and Cairns in 270–400 m.

- **Remarks:** This small tropical species, which lives deeper than other members of the genus, is common throughout its range and may be more widespread along the continental slope off Queensland. It more closely resembles spotted catsharks from southern Australia than its immediate relative, the grey spotted catshark (26.15), which occurs off the New South Wales coastline. Little is known of its biology.

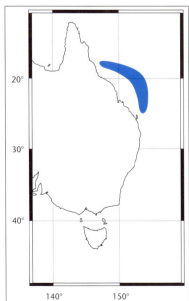

Western Spotted Catshark — *Asymbolus* sp. F

26.14 *Asymbolus* sp. F (Plate 21) Fish Code: 00 015026

- **Alternative name:** Spotted catshark.
- **Field characters:** A medium-sized, pale yellowish green catshark with 8 or 9 pale brownish saddles and an array of sharply defined, brownish black spots on the upper surface and sides (usually also with a prominent dark spot beneath each eye and a single spot on dorsal midline between each saddle). It also has enlarged anterior nasal flaps that do not reach the mouth, relatively small labial furrows, and minute teeth.
- **Distinctive features:** Body very slender, subcylindrical; head short, only slightly depressed; snout short, rounded. Anterior nasal flaps expanded, not reaching mouth; posterior nasal flaps relatively short. Labial furrows short, about 1.5–2 times spiracle diameter. Teeth minute, mostly with 3 cusps (sometimes with 5); central cusps rather broad, lateral cusps relatively small. Stomach not inflatable. Skin velvety; denticles very dense, overlapping, tricuspidate; lateral cusps very reduced, sometimes barely detectable; no ridge of enlarged denticles along dorsal caudal-fin margin. First dorsal-fin origin over or behind pelvic-fin insertion (almost an eye diameter behind in some males); second dorsal-fin origin over posterior third of anal-fin base. First dorsal-fin base less than half anal-fin base; dorsal fins about equal in size. Pectoral fins short, broad with broadly rounded posterior margins. Pelvic fins connected by a membrane, forming a partial apron over claspers of adult males; mature claspers extending about half their length past pelvic-fin tips.

undersurface of head

- **Colour:** Bright yellowish green above with dark spots and 8–9 pale brownish saddles with whitish borders (3 saddles predorsally, 1 beneath each dorsal fin, 1 interdorsally, and 2–3 on postdorsal tail); spots brownish black, large (each equal in size or larger than spiracle), most numerous in saddles; usually only a single dark spot on dorsal midline between each saddle; prominent dark spot usually located beneath each eye. Fins similar to body but with white margins; dorsal fins mostly with a dark mark on their leading edge and a black spot above their insertion; upper surfaces of other fins lightly spotted. Ventral surface uniformly pale. Eyes uniformly yellowish green. Juvenile brownish and white; pattern with more pronounced saddles and less well-defined spots.
- **Size:** Attains at least 60 cm; males are mature at 58 cm.
- **Distribution:** Recorded from south and southwestern Australia from Fowlers Bay (South Australia) to Perth (Western Australia) in 98–250 m. Most abundant on the outer continental shelf.
- **Remarks:** This rather large, spotted catshark is common off the Western Australian coast. The juvenile colour pattern differs dramatically from the adult and the association of the two forms was made only recently.
- **Local synonymy:** *Halaelurus analis*: Scott et al., 1980 (misidentification).
- **References:** Scott *et al.* (1980); May and Maxwell (1986).

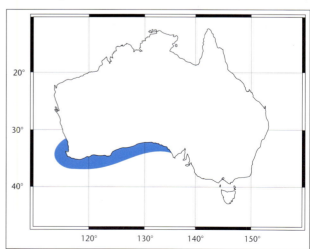

Grey Spotted Catshark — *Asymbolus analis* (Ogilby, 1885)

26.15 *Asymbolus analis* (Plate 21)　　　　　　　　　　**Fish Code:** 00 015027

- **Alternative names:** Australian spotted catshark, spotted dogfish.
- **Field characters:** A medium-sized, spotted catshark with a greyish background coloration, no dark spot beneath the eye, anterior nasal flaps enlarged but not reaching the mouth, relatively small labial furrows, and small teeth.
- **Distinctive features:** Body slender, subcylindrical; head relatively short, flattened; snout short, rounded. Anterior nasal flaps expanded into prominent fleshy lobes; flaps not reaching mouth. Labial furrows short, equal to or only slightly longer than spiracle diameter. Teeth very small, with 3 or 5 cusps. Stomach not inflatable. Skin rather rough; denticles dense, tricuspidate; no ridge of enlarged denticles along dorsal caudal-fin margin. First dorsal-fin origin over or just behind pelvic-fin insertion (more posterior in males); second dorsal-fin origin over posterior third of anal-fin base. Dorsal fins about equal in size. First dorsal-fin base 0.5–0.75 times anal-fin base length. Pectoral fins broad with broadly rounded posterior margins. Pelvic fins connected by a membrane, forming a partial apron over claspers of adult males; mature claspers extending less than half their length past pelvic-fin tips.

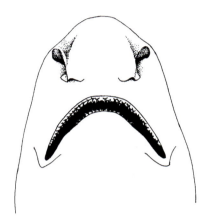

undersurface of head

- **Colour:** Greyish brown dorsally and on sides, with a pattern of faint, darker brown, saddle-like blotches, whitish flecks, and dark brownish spots (each usually larger than spiracle diameter); spots widely spaced and most heavily concentrated within blotches. Upper surface of pectoral fins sometimes with indistinct spots; dorsal fins with up to 3 indistinct spots. Uniformly pale only over small area below level of pelvic fins.
- **Size:** Attains at least 60 cm; males are mature at 52 cm.
- **Distribution:** Confined to eastern Australian waters from Port Macquarie to Green Cape (New South Wales) in 40–79 m. May also occur off eastern Victoria.
- **Remarks:** This species appears to be less common than other closely related, dark-spotted catsharks. It is oviparous, but little else is known of its biology. Caught occasionally by bottom trawlers but not utilised commercially.
- **Local synonymy:** *Halaelurus analis* (Ogilby, 1885).
- **Reference:** Springer (1979).

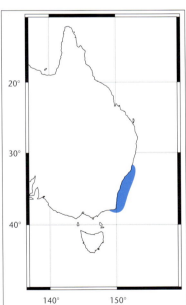

Gulf Catshark — *Asymbolus vincenti* (Zietz, 1908)

26.16 *Asymbolus vincenti* (Plate 22)   **Fish Code:** 00 015003

- **Field characters:** A medium-sized spotted catshark with a mottled greyish brown and white-spotted colour pattern, anterior nasal flaps enlarged but not reaching the mouth, relatively small labial furrows, and minute teeth.
- **Distinctive features:** Body slender, slightly compressed; head relatively deep, only slightly flattened; snout short, rounded distally. Anterior nasal flaps expanded into prominent fleshy lobes; not reaching mouth. Labial furrows short, 1–2 times spiracle diameter. Teeth minute, similar in both jaws, mostly with 3 cusps. Stomach not inflatable. Skin rough; denticles extremely dense, tricuspidate, lacking distinct ridges; no ridge of enlarged denticles along dorsal caudal-fin margin. First dorsal-fin origin over or just behind pelvic-fin insertion; second dorsal-fin origin over posterior end of anal-fin base. First dorsal-fin base about half length of anal-fin base; dorsal fins about equal in size. Pectoral fins small and broad with rounded tips. Pelvic fins connected by a membrane, forming a prominent apron over claspers of adult males; mature claspers extending half their length past pelvic-fin tips.

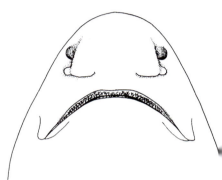

undersurface of head

- **Colour:** Heavily mottled; greyish brown with 7–8 faint, darker, saddle-like blotches and a scattering of numerous small (slightly less than spiracle diameter), whitish spots extending almost to ventral surface (darker coloration extending to just above mouth). Unpaired fins, and upper surfaces of pectoral and pelvic fins, similar to dorsal colour. Ventral surface pale. Pale spots less distinct in juveniles.
- **Size:** Attains at least 56 cm; males are mature at 38 cm.
- **Distribution:** Endemic to southern Australia; appears to be most common in the Great Australian Bight (west to Cape Leeuwin) where it has been reported from depths of 130–220 m. In the east, collected from Bass Strait and off western Tasmania, mostly in less than 100 m. Isolated records from New South Wales require confirmation.
- **Remarks:** This attractive catshark was first collected from South Australia. In Bass Strait, it is frequently found in seagrass beds near the coast.
- **Local synonymy:** *Juncrus vincenti*: Whitley, 1940; *Halaelurus vincenti*: Fowler, 1941.
- **References:** Zietz (1908a); Fowler (1941a); Springer (1979); Last *et al.* (1983).

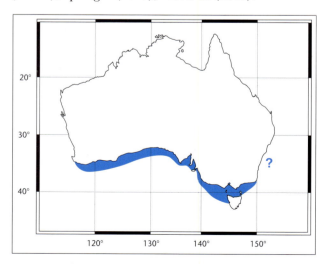

Banded Catshark — *Atelomycterus* sp. A

26.17 *Atelomycterus* sp. A (Plate 22)   Fish Code: 00 015005

- **Alternative name:** Longlip spotted catshark.
- **Field characters:** A slender catshark with very large anterior nasal flaps that extend posteriorly to overlap the mouth, long labial furrows (more than 4 times the spiracle diameter), and with 3–4 dark, predorsal saddle markings and relatively few, small black spots.
- **Distinctive features:** Body slender, subcylindrical; snout short, rounded. Anterior nasal flaps very large, sub-rectangular, extending posteriorly to overlap the mouth. Upper labial furrows long, extending forward to level of upper jaw symphysis; length equal to or slightly longer than lower furrows. Teeth small, similar in both jaws; tricuspid, with greatly enlarged central cusp. Skin velvety, denticles minute, dense, tricuspidate; no ridge of enlarged denticles along the dorsal caudal-fin margin. First dorsal-fin origin over posterior third of pelvic-fin bases; second dorsal-fin origin over or just anterior to midbase of anal fin. Dorsal fins about equal in size; rather low; posterior margins straight, sloping backward. Anal fin much smaller than second dorsal fin; its base about equal in length to first dorsal-fin base. Pectoral fins small with broadly rounded apices. Caudal fin very short, weakly developed (with a terminal lobe and long, low ventral lobe). Claspers of adult males moderately stout and elongate. Tooth count 56–73/50–59. Total vertebrae 149–161; precaudal 100–110.

undersurface of head

- **Colour:** Light greyish brown above with 3–4 darker saddles in the predorsal area; about 7 distinct saddles and bands behind first dorsal fin; additional narrow, dark bands variably developed and situated between main saddles; light scattering of small black spots on the head, body and precaudal fins. Belly and most of lower flanks pale. Specimens from the Torres Strait and Arafura Sea are darker with small white spots.

teeth from near symphysis (upper and lower jaw)

- **Size:** Attains about 45 cm. Males mature at about 32 cm and females at about 35 cm.
- **Distribution:** Confined to Australia where it has been recorded from the continental shelf off northern Western Australia (south to Exmouth Gulf), and off Melville Island (Northern Territory), the southeastern Gulf of Carpentaria, and Cape York (Queensland). Taken from depths of 27–122 m.
- **Remarks:** This catshark is commonly taken by demersal trawlers over sandy bottoms off Western Australia. A disjunct, and less well collected form found in the Torres Strait and Arafura Sea, has a different colour pattern and may be a separate species (Plate 22). It is oviparous but little else is known about the biology of this species.
- **Local synonymy:** *Halaelurus* sp. 1: Sainsbury *et al.*, 1985; *Atelomycterus macleayi*: Allen & Swainston, 1988 (misidentification).
- **Reference:** Sainsbury *et al.* (1985).

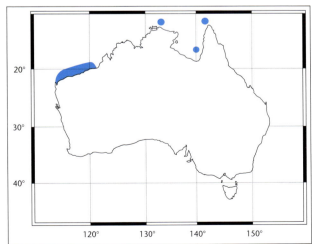

Marbled Catshark — *Atelomycterus macleayi* Whitley, 1939.

26.18 *Atelomycterus macleayi* (Plate 22) **Fish Code:** 00 015028

- **Alternative name:** Australian marbled catshark.
- **Field characters:** A slender catshark with very large anterior nasal flaps that partly overlap the mouth, long labial furrows (more than 4 times the spiracle diameter), and 7 dark, predorsal saddles and numerous, large black spots (saddles sometimes indistinct when spots are dense).
- **Distinctive features:** Body slender, subcylindrical; snout very short, rounded. Anterior nasal flaps very large, subrectangular, extending posteriorly to overlap the mouth. Upper labial furrows long, level with upper jaw symphysis; lower furrows almost joined. Teeth small, similar in both jaws; tricuspid, with greatly enlarged central cusp. Skin velvety; denticles minute, dense, feebly tricuspidate or with a single cusp; no ridge of enlarged denticles along the dorsal caudal-fin margin. First dorsal-fin origin over or slightly behind pelvic-fin insertion; second dorsal-fin origin over or just anterior to midbase of anal fin. Dorsal fins about equal in size, rather tall and upright, posterior margins concave. Anal fin much smaller than second dorsal fin, its base longer than first dorsal-fin base. Pectoral fins small with broadly rounded apices. Caudal fin short, weakly developed (with a terminal lobe and long, low ventral lobe). Claspers of adult males short and very stout. Tooth count 70/70*. Total vertebrae 167–183*; precaudal 112–132*.

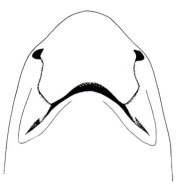

undersurface of head

- **Colour:** Upper surface medium brown with about 7 dark greyish saddle markings and numerous large, black spots and lines (saddles sometimes indistinct); spots extending well down sides to at least eye level and frequently onto ventral surface of tail; all fins normally spotted, including the ventral surfaces of the pectoral fins. Remainder of belly and undersurface of head pale.
- **Size:** Hatches at about 10 cm and attains at least 60 cm. Males are mature at 48 cm and females at 51 cm.
- **Distribution:** Confined to tropical Australia, from shallow water off northwestern Australia between Port Hedland (Western Australia) and Melville Island (Northern Territory). Records from Queensland require confirmation.
- **Remarks:** Appears to be most abundant inshore along the Northern Territory coast. Occurs on both sandy and rocky bottoms. Little else is known of its biology other than that it is oviparous.
- **References:** Whitley (1939b); Grant (1978); Springer (1979).

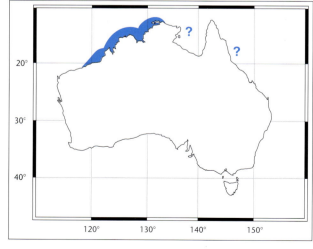

Black-spotted Catshark — *Aulohalaelurus labiosus* (Waite, 1905)

26.19 *Aulohalaelurus labiosus* (Plate 24)   **Fish Code:** 00 015029

- **Field characters:** A medium-sized, densely spotted catshark with the anterior nasal flaps enlarged but not reaching the mouth, very long upper labial furrows (almost reaching nasal flaps), the dorsal fins about equal in size, and the second dorsal-fin origin above the anterior half of the anal-fin base.

- **Distinctive features:** Body firm, rather slender, subcylindrical; head narrow; snout moderately flattened, short, rounded. Anterior nasal flaps expanded, nearly reaching mouth. Labial furrows very long; upper furrows reaching forward to front of mouth; lower furrows almost joined. Teeth small, similar in both jaws; mostly tricuspid, middle cusp elongate. Stomach not inflatable. Skin velvety, denticles dense, tricuspidate (lateral cusps sometimes indistinct); no ridge of enlarged denticles along the upper caudal-fin margin. First dorsal-fin origin over or just anterior to pelvic-fin insertion; second dorsal-fin origin over front quarter of anal-fin base. Dorsal fins about equal in size or second dorsal slightly larger; interdorsal space only slightly longer than first dorsal-fin base. Pectoral fins small, broad, with rounded tips. Inner margins of pelvic fins not forming an apron over claspers in mature males. Tooth count 50/45*.

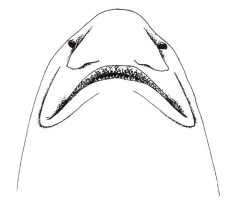

undersurface of head

- **Colour:** Upper surface greyish brown to yellowish brown with faint dusky saddles (5 predorsally) and large black spots. Spots numerous, sharply defined, randomly arranged, mostly larger than spiracle diameter; also with a scattering of small white spots; alternating light and dark spots usually on leading edges of dorsal and caudal fins. Pale ventrally.

- **Size:** Attains 67 cm; males mature at about 54 cm.

- **Distribution:** Confined to shallow coastal habitats and offshore reefs of southwestern Australia (Recherche Archipelago to the Houtman Abrolhos).

- **Remarks:** Reported to be common on coastal reefs but not often seen because of its nocturnal habits. Probably oviparous. Little is known of the biology of this species.

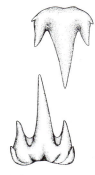

teeth from near symphysis
(upper and lower jaw)

- **Local synonymy:** *Catulus labiosus* Waite, 1905; *Halaelurus labiosus*: Garman, 1913.

- **References:** McKay (1966); Springer (1979); Hutchins and Swainston (1986).

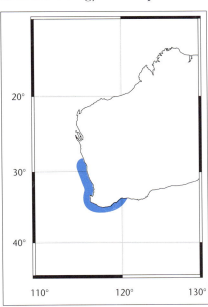

Whitefin Swell Shark — *Cephaloscyllium* sp. A

26.20 *Cephaloscyllium* sp. A (Plate 23)   **Fish Code:** 00 015013

- **Alternative names:** Draughtboard shark, swell shark.
- **Field characters:** A large, stocky catshark with a very broad head, rough skin, a highly inflatable stomach, no labial furrows, and a pattern of broad, dark blotches and saddles on the sides and back.
- **Distinctive features:** Body very robust; head short and very broad (width more than 4.2 times longer than preoral snout). Snout moderately flattened, short, broadly rounded. Anterior nasal flaps expanded but not reaching mouth. Labial furrows absent. Teeth very small, multicuspid, similar in both jaws. Stomach highly inflatable. Denticles arrowhead-shaped, relatively enlarged and widely spaced dorsally; no ridge of enlarged denticles along dorsal caudal-fin margin. First dorsal-fin origin usually over anterior half of pelvic-fin base; second dorsal-fin origin mostly behind anal-fin origin; pelvic–anal interspace less than or equal to horizontal length of anal fin in adults. First dorsal fin considerably larger than second dorsal fin. Pectoral fins relatively large and broad. Egg case flask-shaped, without transverse ridges. Tooth count 62/66*. Precaudal vertebrae 79–81*; monospondylous 45–47*.
- **Colour:** Medium brownish or greyish dorsally; broad saddles on back and large blotches on sides are darker; predorsal space usually with 5 bars; fins mostly dark with pale margins; mostly uniformly pale ventrally. Blotches and bars faint, but distinct in juveniles.
- **Size:** To at least 94 cm; males mature by 70 cm. Egg cases about 11 cm long, 5 cm wide.
- **Distribution:** Southeastern Australia from Port Stephens (New South Wales) to at least Port Lincoln (South Australia), including Tasmania. Upper continental slope in 240–550 m.
- **Remarks:** This catshark is common among the trawl bycatch off southern Australia. Compared to other similar members of the genus, the skin is rough, the head is very broad, and the first dorsal fin is located further forward (above the front half of the pelvic-fin base). It has been confused with the draughtboard shark (26.26), a shallower species with an overlapping range, and the closely related, New Zealand swellshark (*Cephaloscyllium isabella*). Like other members of the genus, it is capable of swelling enormously by swallowing water or air.
- **Local synonymy:** *Cephaloscyllium nascione*: May & Maxwell, 1986 (misidentification).

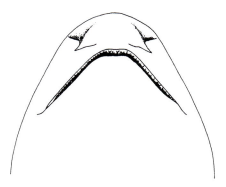

undersurface of head

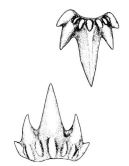

third teeth from symphysis (upper and lower jaw)

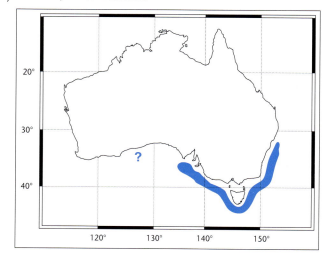

Saddled Swell Shark — *Cephaloscyllium* sp. **B**

26.21 *Cephaloscyllium* sp. **B** (Plate 24)　　　　　　　　　　Fish Code:　00 015030

- **Field characters:** A medium-sized, elongate catshark with a slender head, relatively smooth skin, an inflatable stomach, no labial furrows, and a pronounced pattern of dark saddles on the back.

- **Distinctive features:** Body comparatively slender, head short and not especially broad (width less than 4 times longer than preoral snout). Snout moderately flattened, short, broadly rounded. Anterior nasal flaps expanded but not reaching mouth. Labial furrows absent. Teeth very small, multicuspid, similar in both jaws. Stomach inflatable. Denticles arrowhead-shaped, relatively small and dense dorsally; no ridge of enlarged denticles along the dorsal caudal-fin margin. First dorsal-fin origin near level of insertion of pelvic fin; second dorsal-fin origin well behind anal-fin origin; pelvic–anal interspace greater than horizontal length of anal fin in adults. First dorsal fin considerably larger than second dorsal fin. Pectoral fins relatively large and broad. Precaudal vertebrae 78–81*; monospondylous 46*.

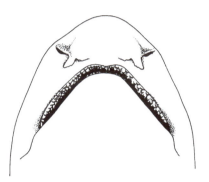

undersurface of head

- **Colour:** Medium brownish or greyish dorsally with dark saddles of variable width on back; predorsal space usually with 5 saddles; blotches usually absent on sides; fin margins sometimes pale; pale ventrally.

- **Size:** To at least 70 cm. Males mature at about 55 cm.

- **Distribution:** Continental slope off northeastern Australia between Townsville and the Saumarez Reef in 380–590 m. Possibly also on the Britannia Seamount (off Brisbane).

- **Remarks:** This tropical catshark has a similar colour pattern to the temperate whitefin swell shark (26.20) but is smaller and has a more slender head, finer denticle structure, and a more posteriorly located first dorsal fin than the latter. More specimens are required for research.

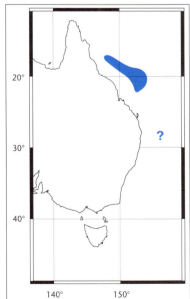

Northern Draughtboard Shark — *Cephaloscyllium* sp. C

26.22 *Cephaloscyllium* sp. C (Plate 24) **Fish Code:** 00 015031

- **Field characters:** A medium-sized, slender catshark with an inflatable stomach, no labial furrows, small and closely spaced denticles, dark saddles and blotches on the upper surface, and no dark median dark stripe on the belly of adults.

- **Distinctive features:** Body relatively slender, head short and broad. Snout moderately flattened, short, broadly rounded. Anterior nasal flaps expanded but not reaching mouth. Labial furrows absent. Teeth very small, multicuspid, similar in both jaws. Stomach inflatable. Denticles arrowhead-shaped, minute and dense dorsally; no ridge of enlarged denticles along the dorsal caudal-fin margin. First dorsal-fin origin over posterior half of pelvic-fin base; second dorsal-fin origin slightly behind anal-fin origin; pelvic–anal interspace about equal to horizontal length of anal fin. First dorsal fin considerably larger than second dorsal fin. Pectoral fins relatively large and broad. Egg case flask-shaped without transverse ridges. Tooth count 80/80*. Precaudal vertebrae 78*; monospondylous 46*.

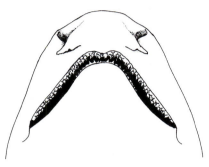

undersurface of head

- **Colour:** Upper surface dark greyish brown with indistinct darker bands (5 predorsally) suffused with irregular white blotches and flecks which extend onto the fins; belly pale, lacking a dark belly stripe. Fin margins not distinctly pale edged. Juveniles possibly paler with small, widely spaced spots.

- **Size:** To at least 65 cm. Egg cases about 7 cm long and 4 cm wide.

- **Distribution:** Outer continental shelf from Noosa (Queensland) to Wollongong (New South Wales) in depths of 90–140 m.

- **Remarks:** The relationship of this species to other eastern Australian swell sharks needs to be investigated further. It is known from only a few adults, although juvenile specimens and two egg cases collected from northern New South Wales are likely to belong to this species. Compared with the draughtboard shark (26.26) from southern waters, to which it is most similar, the skin is very smooth and the denticles are smaller and denser. More specimens are required.

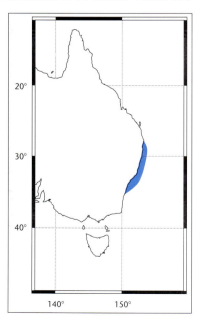

Narrowbar Swell Shark — *Cephaloscyllium* sp. D

26.23 *Cephaloscyllium* sp. **D** (Plate 23)   **Fish Code:** 00 015032

- **Field characters:** A small, stocky catshark with a broad head, an inflatable stomach, relatively smooth skin, no labial furrows, and a distinctive colour pattern of narrow, closely spaced, dark bars on a pale background.

- **Distinctive features:** Body rather stocky, head short and broad. Snout moderately flattened, short, broadly rounded. Anterior nasal flaps expanded but not reaching mouth. No labial furrows. Teeth small, multicuspid, similar in both jaws; slightly enlarged in mature males. Stomach inflatable. Denticles arrowhead-shaped, relatively fine, dense; no ridge of enlarged denticles along the upper caudal-fin margin. First dorsal-fin origin near middle of pelvic-fin base; second dorsal-fin origin well behind anal-fin origin; pelvic–anal interspace equal to or slightly shorter than horizontal length of anal fin in adults. First dorsal fin considerably larger than second dorsal fin. Pectoral fins medium-sized and rather broad. Tooth count 60–62/61–63*. Precaudal vertebrae 72*; monospondylous 40–41*.

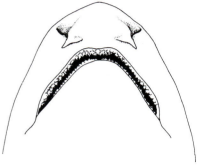

undersurface of head

- **Colour:** Dark brownish to cream with narrow, closely spaced bars on the dorsal surface; 17–18 bars predorsally (bars not united to form rings or open saddles); irregular lines present on snout; uniformly pale ventrally and on fins.

- **Size:** To at least 43 cm.

- **Distribution:** Based on only a few specimens trawled in 440 m near Flinders Reefs (Queensland).

- **Remarks:** The narrowbar swell shark is possibly more widely distributed along the continental slope off northeastern Australia. It is similar in size and general form to the reticulate swell shark (26.25) but the pattern of fine bars over the upper surface never become connected as in the latter. Specimens should be taken to a museum.

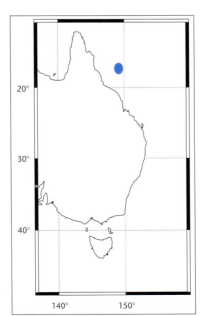

Speckled Swell Shark — *Cephaloscyllium* sp. E

26.24 *Cephaloscyllium* sp. E (Plate 23)   **Fish Code:** 00 015033

- **Field characters:** A medium-sized, rather robust catshark with an inflatable stomach, no labial furrows, lacking a dark median stripe on the belly, and with a densely speckled pattern on the upper surface.

- **Distinctive features:** Body moderately robust, head short and broad. Snout moderately flattened, short, broadly rounded. Anterior nasal flaps expanded but not reaching mouth. Labial furrows absent. Teeth very small, multicuspid, similar in both jaws. Stomach inflatable. Denticles arrowhead-shaped, small and dense dorsally; no ridge of enlarged denticles along the upper caudal-fin margin. First dorsal-fin origin over posterior half of pelvic-fin base; second dorsal-fin origin slightly behind anal-fin origin; pelvic–anal interspace subequal to horizontal length of anal fin in adults. First dorsal fin considerably larger than second dorsal fin. Pectoral fins relatively small. Precaudal vertebrae 74–76*; monospondylous 42*.

undersurface of head

- **Colour:** Upper surface pale greyish; intensely mottled with small dark blotches, and larger blotches and saddles bearing small white spots; a pair of rounded blotches behind eyes followed by 3 short saddles in predorsal area; a similar white-spotted blotch beneath each eye. Fins mostly pale with darker spots and blotches. Western forms white ventrally (eastern forms greyish); no median dark stripe on belly.

- **Size:** Reaches at least 68 cm. Males mature at about 64 cm.

- **Distribution:** Known from small pockets along the continental slopes of tropical Australia; off the Rowley Shoals (Western Australia) in 390–440 m, and near the Lihou Reef, off Innisfail (Queensland) in 600–700 m.

- **Remarks:** This rather attractive species is known only from a few specimens and more are required for research purposes.

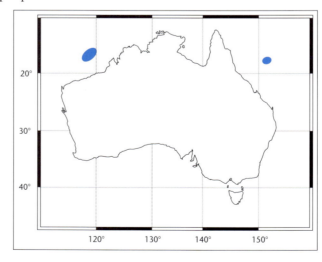

Reticulate Swell Shark — *Cephaloscyllium fasciatum* Chan, 1966

26.25 *Cephaloscyllium fasciatum* (Plate 23) Fish Code: 00 015007

- **Field characters:** A small, stocky catshark with an inflatable stomach, no labial furrows, and a striking colour pattern of dark transverse lines that form open-centred saddles and blotches on a pale background.

- **Distinctive features:** Body rather stocky, head short and broad. Snout moderately flattened, short, broadly rounded. Anterior nasal flaps expanded but not reaching mouth. No labial furrows. Teeth small, multicuspid, similar in both jaws. Stomach inflatable. Denticles arrowhead-shaped, relatively fine, dense; no ridge of enlarged denticles along the upper caudal-fin margin. First dorsal-fin origin over posterior half of pelvic-fin base; second dorsal-fin origin just behind anal-fin origin; pelvic–anal interspace longer than horizontal length of anal fin in adults. First dorsal fin considerably larger than second dorsal fin. Pectoral fins relatively small and narrow. Tooth count 48–60/47–58. Precaudal vertebrae 69–75; monospondylous 37–39.

undersurface of head

- **Colour:** Dorsal surface pale greyish or brownish with darker brown lines forming a striking pattern of open saddle-like markings and blotches; upper surface of pectoral fin some-times with a large dark ring; pale ventrally.

- **Size:** Hatching at about 12 cm and attaining at least 42 cm. Males mature at about 36 cm.

- **Distribution:** Indo–West Pacific; off Vietnam, China and northwestern Australia (from Geraldton to Broome). Continental slope in 220–450 m.

- **Remarks:** Little is known of the biology of this species other than that it is oviparous. The hatchlings do not appear to have the enlarged dorsal denticles that are used by other members of the genus to break out of the egg case. Differences exist between northern and southern hemisphere populations of this species and relationships between these forms needs further investigation.

- **Reference:** Springer (1979).

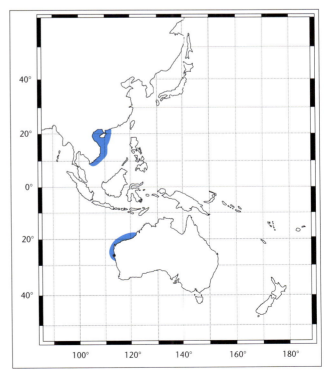

Draughtboard Shark — *Cephaloscyllium laticeps* (Duméril, 1853)

26.26 *Cephaloscyllium laticeps* (Plate 23) **Fish Code:** 00 015001

undersurface of head

- **Alternative names:** Australian swellshark, sleepy joe, nutcracker shark.
- **Field characters:** A large, stocky catshark with an inflatable stomach, no labial furrows, large and widely spaced denticles, a dark, mottled and blotched colour pattern, and mostly with a dark median stripe on the belly.
- **Distinctive features:** Body robust, head short and very broad. Snout moderately flattened, short, broadly rounded. Anterior nasal flaps expanded but not reaching mouth. No labial furrows. Teeth very small, multicuspid, similar in both jaws. Stomach inflatable. Denticles arrowhead-shaped; relatively enlarged and very widely spaced dorsally in immatures and juveniles (denser in adults); no ridge of enlarged denticles along the dorsal caudal-fin margin. First dorsal-fin origin between middle and insertion of pelvic-fin base; second dorsal-fin origin just behind anal-fin origin; pelvic–anal interspace about equal to horizontal length of anal fin in adults. First dorsal fin considerably larger than second dorsal fin. Pectoral fins relatively large and broad. Egg case flask-shaped, with 19–27* transverse ridges. Precaudal vertebrae 76–80*; monospondylous 43–45*.
- **Colour:** Upper surface medium brownish to greyish; ornamented with a dense array of irregular, dark blotches (mostly also with a few pale flecks); a broad dark saddle between eye and pectoral origin; area below eye dark. Fins similar to upper surface; margins of pectoral, pelvic and anal fins mostly pale. Ventral surface mostly pale (occasionally with greyish areas); adults normally with a dark median stripe on midline of belly.
- **Size:** Reported to reach 150 cm but rarely over 100 cm; males mature at about 82 cm. Egg cases about 13 cm long and 5 cm wide; hatching at about 14 cm.
- **Distribution:** Inshore on the continental shelf of southern Australia from the Recherche Archipelago (Western Australia) to Jervis Bay (New South Wales) to at least 60 m.
- **Remarks:** The most common catshark in coastal areas of southern Australia. It is a nuisance to lobster fishermen in Bass Strait where it frequently enters lobster pots in search of food. Other prey include small reef fishes, crustaceans and squid. Unlike other members of the genus, it has prominent transverse ridges across the egg case. The cream coloured egg cases, which are attached to bottom-dwelling invertebrates and seaweed, are often washed ashore after storms. Compared with other similarly pigmented adult members of the genus, the skin is relatively rough with larger, more widely spaced denticles.
- **Local synonymy:** *Cephaloscyllium isabella laticeps* Whitley, 1932; *Cephaloscyllium isabella nascione* Whitley, 1932.
- **References:** Scott (1963); Springer (1979); Coleman (1980).

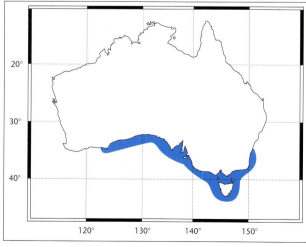

Slender Sawtail Shark — *Galeus* sp. A

26.27 *Galeus* sp. **A** (Plate 18) Fish Code: 00 015008

- **Field characters:** A very small, slender catshark with well developed labial furrows, the first dorsal fin slightly larger than the second dorsal fin, enlarged denticles on the anterior upper margin of the caudal fin (but absent from along the midline of the caudal peduncle), and with four short dark saddles on the tail and no markings predorsally.

- **Distinctive features:** Body firm, slender, subcylindrical; head short, narrow. Gill openings short; eyes directed laterally, suborbital ridges narrow. Mouth broadly arched; labial furrows short, not extending forward to level of anterior margin of eyes; teeth small, closely spaced, multicuspid; central cusps much longer than those adjacent. Stomach not inflatable. Denticles small, extremely dense, partially overlapping; crown broad, mostly tricuspidate; distinct crests of enlarged denticles along upper margin of tail, absent on lower edge of caudal peduncle; no naked areas (gill membranes and fin tips scaled). First dorsal-fin origin just in advance of pelvic-fin insertion; first dorsal slightly larger than second dorsal fin; interdorsal space less than 2.5 times first dorsal-fin base. Pectoral fins short and broad. Pelvic fins elongate, inner margin long; anal fin subtriangular. Caudal fin short, lower lobe distinct. Precaudal vertebrae 74–78*; monospondylous 33–36*.

- **Colour:** Pale greyish with 4 short, dusky saddles on the tail (beneath each dorsal fin and two on caudal fin); last saddle extending to ventral surface; saddles without pale margins. Pale ventrally.

- **Size:** To at least 34 cm.

- **Distribution:** Isolated records off Cape Cuvier and Port Hedland (Western Australia), off Melville Island (Northern Territory) and off Cape York (Queensland). Continental slope in 290–470 m.

- **Remarks:** This undescribed species has been taken from tropical latitudes, in the scampi grounds off northwestern Australia and off northern Queensland. It is small and of no commercial value.

undersurface of head

teeth from near symphysis
(upper and lower jaw)

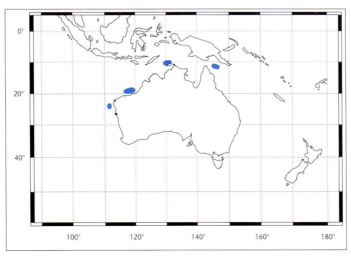

Northern Sawtail Shark — *Galeus* sp. **B**

26.28 *Galeus* sp. **B** (Plate 18)   **Fish Code:** 00 015034

- **Field characters:** A very small, banded catshark with the second dorsal fin only slightly larger than the first dorsal fin, and with a prominent crest of enlarged denticles on the front upper margin of the caudal fin and along the ventral midline of the caudal peduncle. Predorsal space with 10–16 narrow bands (some enlarged but all narrower than eye diameter).

- **Distinctive features:** Body firm, relatively slender, subcylindrical; head short, narrow. Gill openings short; eyes directed laterally with narrow suborbital ridges. Mouth broadly arched; labial furrows short, not extending forward to level of anterior margin of eyes; teeth small, closely spaced, multicuspid; central cusps rather long. Stomach not inflatable. Denticles small, extremely dense, partially overlapping; crown broad, mostly tricuspidate; distinct crests of enlarged denticles along upper margin of tail and on lower edge of caudal peduncle; no naked areas (gill membranes and fin tips scaled). First dorsal-fin origin over insertion of pelvic fin, first dorsal only slightly smaller than second dorsal fin; interdorsal space 2–3 times first dorsal-fin base. Pectoral fin relatively small, shorter than distance from snout tip to hind margin of eye. Pelvic fins elongate, inner margin long; anal fin subtriangular. Caudal fin short, lower lobe distinct. Precaudal vertebrae 86–88*; monospondylous 36–37*.

- **Colour:** Upper surface pale greyish brown, adorned with dark saddles and bars; predorsal space with 10–16 narrow saddles and bars (about 4 distinctly enlarged but only slightly broader than those adjacent); bars beneath dorsal fins with pale centres; saddles and bars pale-edged. Ventral surface and much of sides uniformly pale. Dorsal-fin bases pale greyish, paler distally; pectoral and pelvic fins, and lower caudal-fin lobe mostly pale.

- **Size:** To at least 41 cm; males mature at about 38 cm.

- **Distribution:** Northeastern Australia from Rockhampton to Townsville in depths of 310–420 m.

- **Remarks:** Similar to the sawtail shark (26.29) but is smaller, paler, more heavily banded, and has a relatively smaller pectoral fin than the latter. Little is known of its biology.

undersurface of head

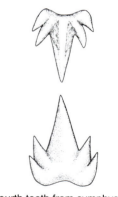

fourth teeth from symphysis (upper and lower jaw)

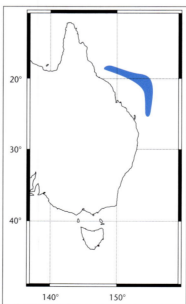

Sawtail Shark — *Galeus boardmani* (Whitley, 1928)

26.29 *Galeus boardmani* (Plate 18) Fish Code: 00 015009

- **Alternative names:** Australian sawtail catshark, banded shark.
- **Field characters:** A small, banded catshark with the second dorsal fin only slightly larger than the first dorsal fin, and with a crest of enlarged denticles on the anterior upper margin of the caudal fin and along the ventral midline of the caudal peduncle. Predorsal space with 3 enlarged saddles, each equivalent in width to eye diameter.
- **Distinctive features:** Body firm, relatively slender, subcylindrical; head short, narrow. Gill openings short; eyes directed laterally with narrow suborbital ridges. Mouth broadly arched; labial furrows short, not extending forward to level of anterior margin of eyes; teeth small, closely spaced, multicuspid; central cusps rather long. Stomach not inflatable. Denticles small, extremely dense, partially overlapping; crown broad, mostly tricuspidate; distinct crests of enlarged denticles along upper margin of tail and on lower edge of caudal peduncle; no naked areas (gill membranes and fin tips scaled). First dorsal-fin origin over insertion of pelvic fin, only slightly smaller than second dorsal fin; interdorsal space 2–3 times first dorsal-fin base. Pectoral fin medium-sized, mostly longer than distance from snout tip to hind margin of eye. Pelvic fins elongate, inner margin long; anal fin subtriangular. Caudal fin short, lower lobe distinct. Precaudal vertebrae 89–93; monospondylous 36–39.

undersurface of head

- **Colour:** Upper surface greyish with a variegated pattern of dark greyish brown saddles and bars; predorsal space with 3 broad saddles (each about equal to or slightly wider than the eye diameter) with a narrower (often less distinct) band between each; a broad saddle interdorsally, beneath each dorsal fin, and about 3 broad bands postdorsally (the last caudal fin band often forming a dark ring); most large bands and saddles pale-edged and sometimes with white flecks. Dorsal and pectoral fins with dark bases and pale margins. Ventral surface uniformly pale.
- **Size:** To 61 cm. Males mature at about 40 cm and females at about 43 cm.
- **Distribution:** Widespread and common off the southern coast of Australia (including Tasmania) from Carnarvon (Western Australia) to Noosa (Queensland). Outer continental shelf and upper slope in 150–640 m.
- **Remarks:** Common on the upper continental shelf, this catshark is frequently found in the demersal trawl bycatch. Like many other sharks, it appears to sometimes aggregate by sex. The diet consists mainly of fish, crustaceans and cephalopods, otherwise little is known of its biology. Presently of no commercial value.
- **Local synonymy:** *Pristiurus (Figaro) boardmani* Whitley, 1928; *Figaro boardmani socius* Whitley, 1939; *Figaro boardmani*: Whitley, 1940.
- **References:** Whitley (1928, 1939b).

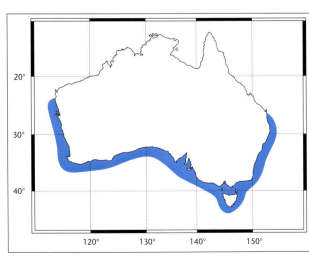

Dusky Catshark — *Halaelurus* sp. A

26.30 *Halaelurus* sp. A (Plate 18)   **Fish Code:** 00 015035

- **Field characters:** A uniformly dark, soft-bodied catshark with bristly skin, a very broad and flattened head, a short rounded snout, very short labial furrows (confined to corner of mouth), a relatively short-based and subtriangular anal fin, and lacking a crest of enlarged denticles along the anterior upper margin of the caudal fin.

- **Distinctive features:** Body relatively flattened anteriorly, head very broad; abdomen long. Snout short, extremely flattened, apex rounded. Eyes large (about twice length of first gill opening), directed sideways. Nostrils large. Mouth large and broadly arched, extending forward to front margin of eyes; labial furrows minute, confined to mouth corners. Teeth minute, multicuspid; central cusps barely longer than adjacent cusps. Skin rather rough (bristle-like); denticles slender, erect, widely spaced, tricuspidate; no naked areas on gill membranes or around fins. First dorsal-fin origin slightly behind origin of pelvic fins; dorsal fins tall, second only slightly larger than first; interdorsal space less than twice first dorsal-fin base. Pectoral fins widely separated, rounded. Anal fin tall, subtriangular, its base slightly shorter than interdorsal space. Caudal fin of medium length, lower lobe poorly developed.

- **Colour:** Body and fins uniform dark greyish brown; a few pale blotches on belly.

- **Size:** Known from a 44 cm immature male specimen.

- **Distribution:** Continental slope off the Ashmore Reef, northwestern Australia, in 900 m depth.

- **Remarks:** This species is more closely resembles the plainly coloured, deepwater catsharks, than the other shallow water member of the genus that occurs in this region (26.31).

undersurface of head

teeth from near symphysis (upper and lower jaw)

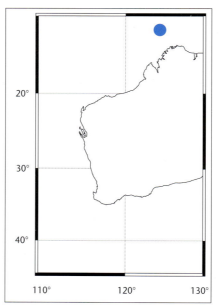

Speckled Catshark — *Halaelurus boesemani* Springer & D'Aubrey, 1972

26.31 *Halaelurus boesemani* (Plate 18)　　　　Fish Code: 00 015004

- **Alternative name:** Shortlip spotted catshark.
- **Field characters:** A small, banded catshark with short labial furrows (uppers barely evident), nasal flaps not extending to the mouth, widely spaced dorsal fins, the first dorsal-fin origin over the mid-pelvic base, and lacking crests of enlarged denticles on the caudal fin and caudal peduncle.
- **Distinctive features:** Body firm, elongate; head short, rather broad, slightly flattened. Gill openings short; eyes directed laterally; suborbital ridges rather broad. Mouth narrow, broadly arched; labial furrows very short, lower furrow much longer than upper; teeth minute, broad, multicuspid, with reduced lateral cusps. Stomach not inflatable. Denticles minute, extremely dense, tricuspidate; no crests of enlarged denticles along upper margin of tail or on lower edge of caudal peduncle; no naked areas on body. First dorsal-fin origin over middle of pelvic-fin base; first dorsal subequal to or slightly larger than second dorsal fin; interdorsal space more than twice length of first dorsal-fin base. Pectoral fins short and broad. Pelvic fins elongate, inner margins long and not joined over claspers in adult male; anal fin short-based, subtriangular. Caudal fin short, lower lobe distinct.
- **Colour:** Upper surface yellowish brown, ornamented with darker saddles, bands and spots; 8 broad, precaudal bars and saddles with narrower bars between; bars and saddles mostly with borders of fine brownish black spots; 4 broad predorsal bars (first between eyes, third widest); small dark spots present on snout; dorsal and caudal fins usually with spots and blotches. Upper pectoral-fin base brownish (sometimes with spots); broad, pale band on posterior margin of fin. Undersurface, and anal and pelvic fins pale.
- **Size:** To 42 cm. Males mature at about 35 cm.
- **Distribution:** Arabian Sea, Vietnam, Philippines, Indonesia and Australia. Locally, on the continental shelf off Western Australia from off Gantheaume Bay to the Rowley Shoals in 110–250 m.

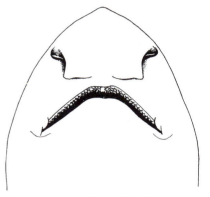

undersurface of head

teeth from near symphysis
(upper and lower jaw)

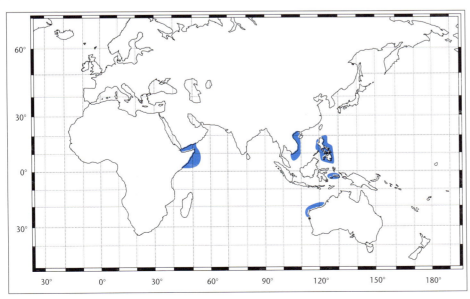

- **Remarks:** The colour pattern of this species closely resembles that of the sympatric banded catshark (26.17). Unlike the latter, however, the nasal flaps of the speckled catshark do not reach the mouth. Australian populations differ in colour pattern from those in other parts of the Indian Ocean and may well be another species.
- **Local synonymy:** *Halaelurus* sp. 1: Sainsbury *et al.*, 1985.
- **Reference:** Sainsbury *et al.* (1985).

Short-tail Catshark — *Parmaturus* sp. A

26.32 *Parmaturus* sp. A (Plate 20)   **Fish Code:** 00 015036

- **Field characters:** A slender, plain brownish catshark with velvety skin, a soft body, a conical snout, very short pectoral and caudal fins, crests of enlarged denticles on the lower caudal peduncle and upper caudal-fin lobe, and with the second dorsal fin inserted well behind the anal-fin insertion.
- **Distinctive features:** Body and head slender; snout medium length, almost conical. Gill openings small; eyes rather large, lateral. Nostrils large. Mouth broadly arched, extending forward well in advance of eyes; labial furrows well developed (but not extending to anterior margin of eyes); teeth small, mostly tricuspidate, central cusps much longer than adjacent cusps. Skin velvety on midflank, no dark naked areas on body. Denticles minute, dense, erect; crowns slender, tricuspidate. First dorsal-fin origin over pelvic-fin insertion; second dorsal fin tall, much larger than first dorsal fin and inserted well behind insertion of anal fin; interdorsal space slightly less than twice first dorsal-fin base. Pectoral fins rounded, very small, anterior margin about 3.4 in pectoral–pelvic distance. Anal fin triangular, base about twice first dorsal-fin base. Caudal fin very short, about 1.5 times caudal peduncle length.

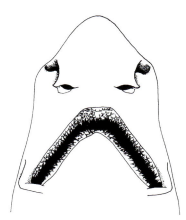

undersurface of head

- **Colour:** Uniform pale yellowish brown, somewhat paler below; fins similar to body but with paler margins.
- **Size:** Known only from a 71 cm adult female (with egg case).
- **Distribution:** Single record from the Saumarez Plateau, off northeastern Australia, in 590 m.
- **Remarks:** It differs from a New Zealand species, *Parmaturus macmillani*, in having a much longer snout and more posteriorly located dorsal fins. Most members of this genus are rare in collections. Specimens are required.

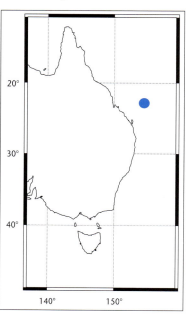

third teeth from symphysis (upper and lower jaw)

27

Family Triakidae

HOUND SHARKS

Hound sharks are small to medium-sized sharks (attaining 0.8–1.7 m) with fusiform bodies, ventrally placed mouths extending well past the eyes (which are oval and have a nictitating membrane), spiracles, five pairs of gill slits, two spineless dorsal fins (usually of a similar size), an anal fin, a spiral intestinal valve, and a caudal fin that is not lunate. They lack nasoral grooves (except in the South African *Scylliogaleus*), nasal barbels (except in *Furgaleus*), precaudal pits and caudal keels. The mid-base of the first dorsal fin is always forward of the pelvic-fin origin. The whalers (Carcharhinidae) and weasel sharks (Hemigaleidae) also have most of these features. Hound sharks, however, unlike members of these groups, have a spiral-type intestinal valve and lack precaudal pits.

Of the 34 species and 9 genera that comprise this family, 9 species and 6 genera are found in Australia. Most of these have restricted distributions, although one species occurs worldwide. Hound sharks generally live near the bottom on continental and insular shelves and upper slopes, feeding on small teleost fish and invertebrates. Reproduction is ovoviviparous or viviparous with a yolk-sac placenta. Some members of this family form the basis of important commercial fisheries, both in Australia and elsewhere.

Key to triakids

1 Nasal barbels present (fig. 1) .... ***Furgaleus macki*** (fig. 3; p. 207)

   Nasal barbels absent (fig. 2) 2

2 Jaws with blade-like cutting teeth (fig. 5) 3

   Jaws with pavement-like crushing teeth (fig. 6) 7

3 Second dorsal fin much smaller than first dorsal fin (half its area or less) (fig. 4); ventral lobe of caudal fin well developed (fig. 4) 4

   Second dorsal fin as large, or nearly as large, as first dorsal fin (two-thirds to about equal its area) (fig. 8); ventral lobe of caudal fin absent or weakly developed (fig. 8) 5

4 Second dorsal fin about as large as anal fin (fig. 4); subterminal lobe of caudal fin about half the length of upper caudal-fin margin (fig. 4) .... ***Galeorhinus galeus*** (fig. 4; p. 208)

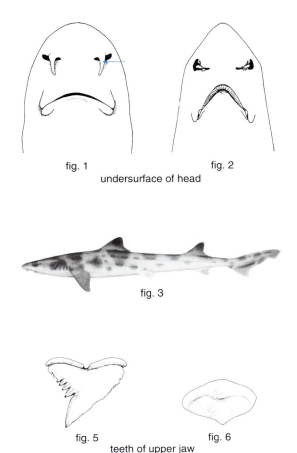

fig. 1   fig. 2
undersurface of head

fig. 3

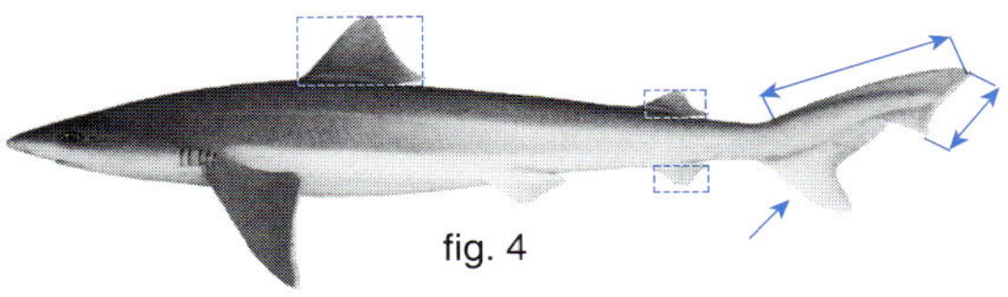

fig. 5   fig. 6
teeth of upper jaw

fig. 4

Second dorsal fin much larger than anal fin (fig. 7); subterminal lobe of caudal fin about one-third the length of upper caudal-fin margin (fig. 7) .... **Hypogaleus hyugaensis (fig. 7; p. 211)**

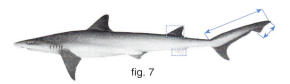
fig. 7

5 Origin of first dorsal fin over anterior half of pectoral-fin inner margins (fig. 8); back distinctly humped below first dorsal fin (fig. 8) .... ***Iago garricki*** **(fig. 8; p. 212)**

Origin of first dorsal fin about over pectoral-fin free rear tips (fig. 9); back not humped below first dorsal fin (fig. 9) **6**

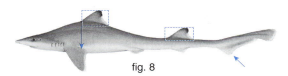
fig. 8

6 No dark stripe on the ventral surface of the snout; precaudal vertebrae 101–109 .... ***Hemitriakis*** **sp. A (fig. 9; p. 209)**

A dark stripe on the ventral surface of the snout (fig. 10); precaudal vertebrae 126–130 .... ***Hemitriakis*** **sp. B (fig. 11; p. 210)**

fig. 9

7 No white spots on body; lower labial furrows longer than uppers (fig. 12) .... ***Mustelus*** **sp. A (fig. 14; p. 213)**

Body with white spots; upper labial furrows longer than lowers (fig. 13) **4**

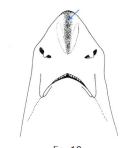
fig. 10
undersurface of head

8 Total vertebrae 125 or more; monospondylous vertebrae 34–37; total precaudal vertebrae 87 or fewer; southern Australia, usually at depths less than 100 m .... ***Mustelus antarcticus*** **(fig. 15; p. 215)**

Total vertebrae 130 or more; monospondylous vertebrae 37–39; total precaudal vertebrae more than 87; tropical eastern Australia, usually at depths greater than 120 m .... ***Mustelus*** **sp. B (eastern form) (fig. 16; p. 214)**

Total vertebrae mostly less than 125; monospondylous vertebrae 33–35; total precaudal vertebrae 81 or fewer; tropical western Australia, usually at depths greater than 120 m .... ***Mustelus*** **sp. B (western form) (fig. 16; p. 214)**

fig. 11

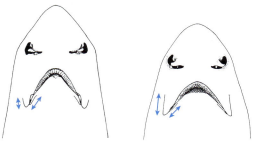
fig. 12   fig. 13
undersurface of head

fig. 14

fig. 15

fig. 16

Whiskery Shark — *Furgaleus macki* (Whitley, 1943)

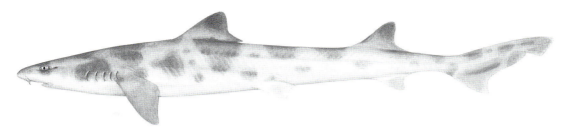

27.1 *Furgaleus macki* (Plate 27) **Fish Code:** 00 017003

- **Field characters:** A hound shark with anterior nasal flaps modified into slender barbels, two dorsal fins of a similar size and each larger than the anal fin. Juveniles have dark blotches on the body and fins.

- **Distinctive features:** Body moderately elongate and slender. Snout short and broadly rounded; eyes oval; anterior nasal flaps modified into slender barbels that are well separated from mouth. Upper labial furrows extending to level of upper jaw symphysis, lower labial furrows shorter than uppers. Upper teeth blade-like, compressed, with an oblique cusp and lateral cusplets; lower teeth erect, without cusplets. First dorsal-fin origin behind free rear tips of pectoral fin. Second dorsal-fin origin ahead of anal-fin origin. Dorsal fins about equal in size, both larger than anal fin. Subterminal lobe of caudal fin short, its length about 2.5–4 in dorsal caudal-fin margin. Tooth count 24–32/36–42

- **Colour:** Brownish grey dorsally, pale ventrally. Juveniles creamy white with dark blotches on body and fins; blotches fade with increasing size and may be absent in adults.

- **Size:** Born at about 25 cm and attains 160 cm. Both sexes mature at about 120 cm.

- **Distribution:** Temperate continental shelf waters from Exmouth Gulf (Western Australia) to Wynyard (Bass Strait). Rare off Victoria and Tasmania. Lives on or near the bottom to a depth of 220 m.

- **Remarks:** Little is known of the biology of this shark. It is ovoviviparous with litter sizes ranging from 5–24. It probably gives birth in spring or early summer. The diet consists mainly of cephalopods (particularly octopus), although teleost fish and crustaceans are also taken. The whiskery shark is a major species in the Western Australian shark fishery (total catch 1600 tonnes in 1989/90 with a value of $6 million) of which it comprises about 22% of the total catch by weight. The fishery is now thought to be fully exploited, with signs of over-fishing. Information is urgently needed on the biology of this species.

- **Local synonymy:** *Fur macki* Whitley 1943; *Fur ventralis* Whitley, 1943; *Furgaleus ventralis*: Whitley, 1951.

- **References:** Whitley (1943a, 1943b, 1951a); Scott (1980); Heald (1987).

undersurface of head

teeth from near symphysis (upper and lower jaw)

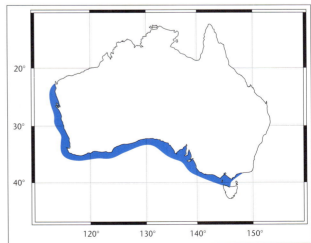

School Shark — *Galeorhinus galeus* (Linnaeus, 1758)

27.2 *Galeorhinus galeus* (Plate 27)   **Fish Code:** 00 017008

- **Alternative names:** Snapper shark, tope, soupfin shark.
- **Field characters:** A moderately slender, bronzy grey hound shark with a very large subterminal lobe on the caudal fin giving it a 'double-tailed' appearance, a small second dorsal fin (about equal in size to the anal fin), and sub-triangular teeth with oblique cusps and lateral cusplets.
- **Distinctive features:** Body fusiform, moderately slender. Snout relatively long, preoral length about equal to mouth width; anterior nasal flaps very small. Upper labial furrows not extending forward to level of upper jaw symphysis; lower labial furrows shorter than uppers. Teeth similar in both jaws, almost triangular with oblique cusps and lateral cusplets. First dorsal-fin origin over or slightly behind free rear tips of pectoral fins. Second dorsal fin much smaller than first dorsal fin, similar in size to anal fin. Subterminal lobe of caudal fin very large, almost as long as lower caudal-fin lobe. Tooth count 37/33* [30–40/31–39]. Total vertebrae 127* [129–139], precaudal 79* [81–87].
- **Colour:** Bronze to greyish brown dorsally, pale ventrally; ventral surface near snout tip often translucent.
- **Size:** In Australia, born at about 30 cm and attains 175 cm. Males mature at 120 cm and females at 130 cm. Elsewhere the species may reach a larger adult size (to 195 cm in the eastern North Pacific) and have a larger size of maturity.
- **Distribution:** Widespread in temperate waters of the eastern North Atlantic, western South Atlantic, eastern North and South Pacific, and off South Africa, New Zealand and southern Australia. Locally from Moreton Bay (southern Queensland) to Perth (Western Australia), including Lord Howe Island and Tasmania. Mainly demersal on the continental and insular shelves, but also on the upper slopes, at depths from the nearshore zone to 550 m.

undersurface of head

fifth teeth from symphysis
(upper and lower jaw)

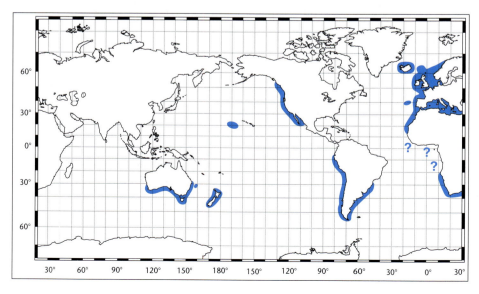

- **Remarks:** This shark often occurs in small schools composed predominantly of one sex and size group. It makes long migrations, apparently associated with reproduction. Movements of up to 2500 km have been recorded in the north-east Atlantic and up to 1400 km in southern Australia. It is ovoviviparous, producing litters of 15–43 pups in December and January (off southern Australia) after a gestation of about 12 months. Tagged adult individuals have been recaptured after 41 years, indicating a lifespan of at least 55 years. Females take 8–10 years to reach maturity. From feeding studies of southern Australian sharks, it eats mainly teleost fish and cephalopods. The school shark, which has been exploited since the mid-1920s and is now over-fished, is a major component of the southern Australian shark fishery (current annual production about 5000 tonnes and valued at $20 million to the fishermen). The meat is used for human consumption and is extremely popular in Victoria. This shark also supports (or has supported) large fisheries off South America, California, South Africa, New Zealand and northern Europe, where it is taken for its meat, fins and liver-oil.
- **Local synonymy:** *Notogaleus rhinophanes* Péron, 1807; *Galeus australis* Macleay, 1881; *Galeorhinus australis*: Ogilby 1898; *Carcharhinus cyrano* Whitley, 1930; *Notogaleus australis*: Whitley, 1931.
- **References:** Macleay (1881a); Whitley (1931a); Olsen (1954, 1959, 1984); Walker (1988).

Sicklefin Hound Shark — *Hemitriakis* sp. A

27.3 *Hemitriakis* sp. A (Plate 27)   **Fish Code:** 00 017009

- **Field characters:** A slender, greyish brown hound shark with distinct white tips to its similar-sized dorsal fins, with falcate dorsal, pectoral and anal fins in adults, and lacking stripes on the ventral surface of the snout.
- **Distinctive features:** Body moderately slender, with low interdorsal and postdorsal ridges. Head somewhat depressed; preoral snout long (1.1–1.3 times mouth width), tip bluntly pointed. Eyes large, oval, dorsolateral on head with a posterior notch. Upper labial furrows extending forward almost to level of upper jaw symphysis; lower labial furrows shorter than uppers. Teeth similar in both jaws; antero-laterals compressed, with oblique cusps and lateral cusplets; cusplets absent in small juveniles; posterior teeth with low keel-like cusps. First dorsal-fin origin almost over free rear tips of pectoral fins. Second dorsal-fin origin in front of anal-fin origin. Second dorsal fin about three-quarters size of first dorsal fin and larger than anal fin. Caudal fin with a short subterminal lobe and prominent ventral lobe. Pectoral, dorsal, anal and lower caudal fins falcate in adults. Tooth count 29–35/26–33. Total vertebrae 158–167; precaudal 101–109.
- **Colour:** Greyish brown dorsally, paler ventrally; prominent white dorsal-fin tips; thin white tips on the pectoral, pelvic and caudal fins; upper caudal-fin lobe with a dark tip.
- **Size:** Born at 20–25 cm; matures at about 65 cm; attains at least 80 cm.

undersurface of head

- **Distribution:** Currently known from tropical Australia on the outer continental shelf off Port Hedland (Western Australia) at depths of 145–200 m.
- **Remarks:** This species is known from only 10 specimens; nothing is known of its biology.

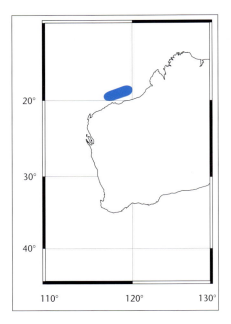

third teeth from symphysis (upper and lower jaw)

Darksnout Hound Shark — *Hemitriakis* sp. B

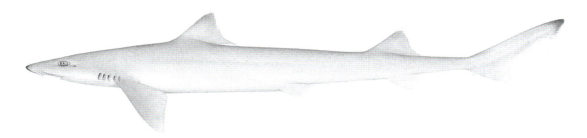

27.4 *Hemitriakis* sp. **B** (Plate 27) **Fish Code:** 00 017010

- **Field characters:** A slender, greyish brown hound shark with a dark stripe on the ventral surface of the snout, distinct white tips to its similar-sized dorsal fins, and falcate dorsal, pectoral and anal fins in adults.
- **Distinctive features:** Body moderately slender, with low interdorsal and postdorsal ridges. Head somewhat depressed; preoral snout long (1.2–1.3 times mouth width), tip bluntly pointed. Eyes large, oval, dorsolateral on head with a posterior notch. Upper labial furrows extending almost to level of upper jaw symphysis; lower labial furrows shorter than uppers. Teeth similar in both jaws; antero-laterals compressed with oblique cusps and lateral cusplets (absent in small juveniles); posterior teeth with low keel-like cusps. First dorsal-fin origin almost over free rear tips of pectoral fins. Second dorsal-fin origin in front of anal-fin origin. Second dorsal fin about three-quarters size of first dorsal fin and larger than anal fin. Caudal fin with a short subterminal lobe and prominent ventral lobe. Pectoral, dorsal, anal and lower caudal fins falcate in adults. Tooth count 34–37/28–33. Total vertebrae 186–193; precaudal 126–130.
- **Colour:** Greyish brown dorsally, paler ventrally; dorsal fins with white tips, otherwise darker than the upper body; pectoral, pelvic and pectoral fins with thin white tips and posterior margins; tip of the upper caudal-fin lobe dark. Dark tip of the snout extends ventrally as a longitudinal stripe to the level of the nostrils
- **Size:** Born at 20–25 cm, matures at about 65 cm and attains at least 80 cm.

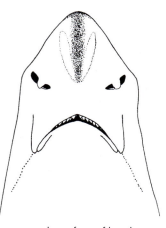

undersurface of head

fourth tooth from symphysis (upper jaw)

- **Distribution:** Known from the upper continental slope off Cairns (Queensland), and possibly off New Caledonia, at depths of 225–400 m.
- **Remarks:** Only six specimens of this sharks are held in museum collections. Nothing is known of its biology.

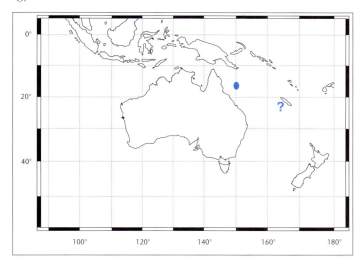

Pencil Shark — *Hypogaleus hyugaensis* (Miyosi, 1939)

27.5 *Hypogaleus hyugaensis* (Plate 27) **Fish Code:** 00 017006

- **Alternative names:** Western school shark, blacktip tope, blacktip houndshark, lesser soupfin shark.
- **Field characters:** A moderately slender, bronzy hound shark with a rather small subterminal caudal-fin lobe, a relatively large second dorsal fin (about two-thirds the size of the first dorsal fin and larger than the anal fin), and subtriangular teeth with oblique cusps and lateral cusplets.
- **Distinctive features:** Body fusiform and moderately slender. Snout relatively long, preoral length about equal to interorbital distance; anterior nasal flaps very small. Upper labial furrows not reaching forward to level of lower jaw symphysis; lower labial furrows shorter than uppers. Teeth similar in both jaws; almost triangular with oblique cusps and lateral cusplets. First dorsal-fin origin over or slightly behind free rear tips of pectoral fins. Second dorsal fin about two-thirds size of first dorsal fin and larger than anal fin. Second dorsal-fin origin slightly ahead of anal-fin origin. Subterminal lobe of caudal fin relatively small; length much less than half that of dorsal caudal-fin margin. Tooth count [48–50/43–46]*. Total vertebrae 160* [154–160]*; precaudal 97–98* [93–97]*.
- **Colour:** Bronze to grey-brown dorsally, paler ventrally; dorsal and upper caudal-fin tips dusky (more prominent in juveniles); pale posterior margins to most fins.
- **Size:** Born at about 35 cm; matures at about 95 cm; attains at least 130 cm.

undersurface of head

- **Distribution:** Disjunct in tropical and warm temperate waters of the Indian Ocean and western Pacific, including Australia. Locally also patchy, reported from Port Hedland (Western Australia) to Cairns (Queensland), including Bass Strait. Rarely caught, however, between Ceduna (South Australia) and Coffs Harbour (New South Wales). Demersal on the continental shelf in depths of 40–230 m.

- **Remarks:** Viviparous with a yolk-sac placenta, producing litters of 8–11. Limited information from Western Australia indicates that young sharks are born around February after a gestation period longer than 12 months. The few stomachs examined have contained teleost fish and cephalopods. Pencil sharks form part of the bycatch of the Western Australian shark fishery.

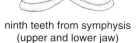

ninth teeth from symphysis (upper and lower jaw)

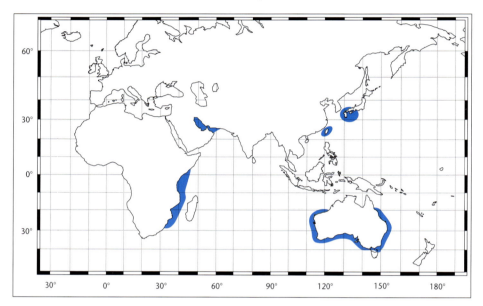

Longnose Hound Shark — *Iago garricki* (Fourmanoir & Rivaton, 1979)

27.6 *Iago garricki* (Plate 28)   **Fish Code:** 00 017007

- **Field characters:** A small hound shark in which the first dorsal fin is situated well forward (origin over the anterior half of the pectoral-fin inner margins), and the dorsal fins have conspicuous black tips.

- **Distinctive features:** Body moderately slender, larger individuals with distinctive 'humpbacked' appearance. Snout relatively long, bluntly pointed, preoral length 6.6–8.1% of total length; eyes relatively large and oval; upper labial furrows not extending to level of upper jaw symphysis; lower labial furrows shorter than uppers. Teeth similar in both jaws; blade-like cusps, erect near jaw symphysis, becoming oblique laterally; lateral cusplets variably developed. First dorsal-fin origin over anterior half of pectoral-fin inner margins. Second dorsal-fin origin just in front of anal-fin origin. Second dorsal fin about two-thirds size of first dorsal fin,

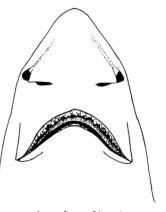

undersurface of head

considerably larger than anal fin. Ventral lobe of caudal fin poorly developed (particularly in juveniles). Total vertebrae 149–160; precaudal 95–100.

- **Colour:** Greyish brown dorsally, paler ventrally. Conspicuous black tips to the dorsal fins; upper caudal-fin margin dusky. Dorsal-fin free rear tips, and tips and posterior margins of pectoral and caudal fins pale.
- **Size:** Born at about 25 cm; attains at least 75 cm. Males mature at about 45 cm.
- **Distribution:** Off the New Hebrides and tropical Australia. Locally on the upper continental slope between Townsville and Cairns (Queensland), and Shark Bay and Darwin (northwestern Australia) at depths of 250–475 m.
- **Remarks:** A poorly known, viviparous shark with a yolk-sac placenta and which produces litters of 4–5 young. The few stomachs examined have contained cephalopods.

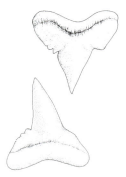
teeth from near symphysis (upper and lower jaw)

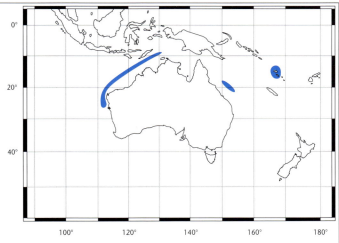

Grey Gummy Shark — *Mustelus* sp. **A**

27.7 *Mustelus* sp. **A** (Plate 28)   **Fish Code:** 00 017005

- **Field characters:** A slender, uniform bronzy hound shark with pavement-like crushing teeth, lower labial furrows longer than the upper labial furrows, and lacking white spots on the body.
- **Distinctive features:** Body slender; predorsal, interdorsal and postdorsal ridges present (predorsal ridge not always distinct). Snout relatively long (preoral length 6.0–6.3% of total length), tip relatively narrow and pointed; internarial space 2.4–2.7% of total length; eyes oval, dorsolateral on head; interorbital space relatively narrow, 5.6–6.0% total length; upper labial furrows 0.8–1.1% of total length, shorter than lower labial furrows; mouth relatively narrow (length 1.1–1.3 in width). Teeth in both jaws flattened, arranged in a pavement-like pattern; cusps rather high. First dorsal-fin origin over or behind free rear tips of pectoral fins. Second dorsal-fin

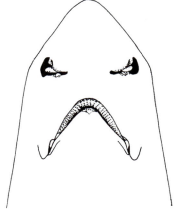
undersurface of head

origin in front of anal-fin origin. Second dorsal fin about three-quarters size of first dorsal fin; considerably larger than anal fin. Pectoral fins moderately broad, weakly falcate; apices pointed. Caudal fin with deep subterminal notch. Total vertebrae 138–146; precaudal 90–92; monospondylous 34–37.

- **Colour:** Uniformly bronze to greyish brown dorsally; pale ventrally. Second dorsal and upper caudal-fin tips usually with dark margins; pectoral and caudal fins with pale posterior margins.
- **Size:** Attains at least 73 cm. Males mature at about 58 cm.
- **Distribution:** Isolated records from tropical Australia; Dampier to Darwin (northwestern Australia) and possibly off Townsville (Queensland). Appears to be demersal at depths of 100–300 m.
- **Remarks:** This shark is likely to be much more widespread than the current records imply. More southerly records of the gummy shark (27.9) may sometimes be of this species or the related white-spotted gummy (27.8). Nothing is known of its biology.
- **Reference:** Sainsbury *et al.* (1985).

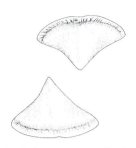
teeth from near symphysis (upper and lower jaw)

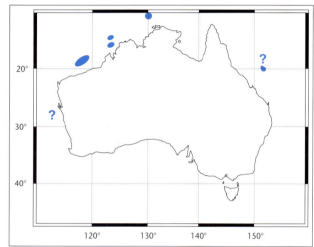

White-spotted Gummy Shark — *Mustelus* sp. B

27.8 *Mustelus* sp. B (Plate 28) **Fish Code:** 00 017004

- **Field characters:** A slender, white-spotted hound shark with pavement-like crushing teeth and upper labial furrows slightly longer than the lowers; tropical Australia.
- **Distinctive features:** Body slender; predorsal, interdorsal and postdorsal ridges present. Snout relatively long (preoral length 5.0–7.3% of total length), tip broad and rounded; internarial space 2.6–3.3% of total length; eyes oval, dorsolateral on head; interorbital space relatively broad, 6.1–7.1% of total length; upper labial furrows 1.9–2.4% of total length, slightly longer than lower labial furrows; mouth rather broad (length 1.4–1.7 in width). Teeth in both jaws flattened, arranged in a pavement-like pattern; cusps low. First dorsal-fin origin varying from slightly in front of pectoral-fin free rear tips to about half way along pectoral-fin inner margins. Second dorsal-fin origin in front of anal-fin origin. Second dorsal fin about three-quarters size of first dorsal fin; considerably larger than anal fin. Pectoral fins broad,

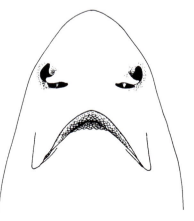
undersurface of head

apices rounded. Caudal fin with deep subterminal notch. Tooth count 68–72/67–79*. Total vertebrae (Western Australia) 119–128; precaudal 76–80; monospondylous 33–35. Total vertebrae (Queensland) 135–143; precaudal 88–95; monospondylous 37–39.

teeth from near symphysis
(upper and lower jaw)

- **Colour:** Bronze to greyish brown dorsally; usually covered with numerous small white spots; pale ventrally. Both dorsal fins sometimes with dusky or black tips. Juveniles with dusky tips and pale posterior margins to the dorsal fins; tips and posterior margins of pectoral, pelvic, anal and caudal fins pale; adults with less distinct fin markings.
- **Size:** Western Australian individuals born at about 27 cm and attain 103 cm; both males and females mature at about 60 cm. Off Queensland, attains 117 cm; males (and possibly females) mature at greater than 70 cm.
- **Distribution:** Tropical Australia off Dampier (Western Australia) and from Cairns to Bowen (Queensland), possibly also further north off both coasts. Unconfirmed records from Shark Bay (Western Australia) and Coffs Harbour (New South Wales). Demersal on the outer continental shelf and upper slope at depths of 120–400 m.
- **Remarks:** The two tropical forms of this shark are almost indistinguishable from the gummy shark (27.9). More research is required to determine whether Queensland and Western Australian populations are just variations of the gummy shark or represent one or two separate species. Both tropical forms are abundant along the upper continental shelf. Males are caught more frequently than females. Litters of 4–17 pups are produced throughout the year. The diet consists mainly of crustaceans (principally crabs), together with fish and cephalopods.
- **Local synonymy:** *Mustelus manazo*: Sainsbury *et al.* 1985.
- **Reference:** Stevens and McLoughlin (1991).

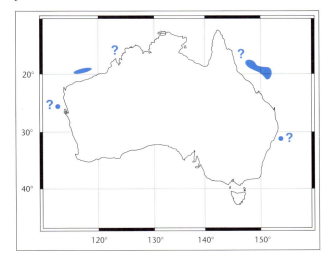

Gummy Shark — *Mustelus antarcticus* Günther, 1870

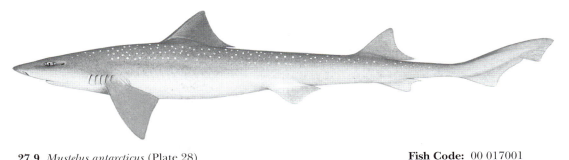

27.9 *Mustelus antarcticus* (Plate 28) **Fish Code:** 00 017001

- **Alternative names:** Australian smooth hound, Sweet William, flake
- **Field characters:** A slender, white-spotted hound shark with pavement-like crushing teeth and upper labial furrows slightly longer than the lower labial furrows; temperate Australia.

- **Distinctive features:** Body slender; predorsal, interdorsal and postdorsal ridges present. Snout relatively long (preoral length 5.7–7.4% of total length), tip broad and rounded; internarial space 2.6–3.2% of total length; eyes oval, dorsolateral on head; interorbital space relatively broad, 6.5–6.9% of total length; upper labial furrows 2.0–2.8% of total length, slightly longer than lower labial furrows; mouth relatively broad (length 1.6–1.8 in width). Teeth in both jaws flattened, arranged in a pavement-like pattern; cusps low. First dorsal-fin origin varying from slightly in front to slightly behind free rear tips of pectoral fins. Second dorsal-fin origin in front of anal-fin origin. Second dorsal fin about three-quarters size of first dorsal fin, considerably larger than anal fin. Pectoral fins broad; apices rounded. Caudal fin with deep subterminal notch. Tooth count 62–69/75–84*. Total vertebrae 125–133; precaudal 79–86; monospondylous 34–37.
- **Colour:** Bronze to greyish brown dorsally; usually with numerous small white spots (and rarely some black spots); pale ventrally.
- **Size:** Born at 30–35 cm and attains 175 cm. Males mature at about 80 cm and females at about 85 cm.
- **Distribution:** Temperate waters of Australia, at least from Port Stephens (New South Wales) to Geraldton (Western Australia). Possibly also ventures into southern Queensland. Demersal on the continental shelf from nearshore to about 80 m, although sometimes as deep as 350 m.
- **Remarks:** The gummy shark often occurs in small schools composed predominantly of one sex and size group. Tagging of sharks in Bass Strait and off eastern Tasmania has shown that it is capable of long migrations. Some tagged females were recaptured in South Australia and Western Australia. Similar movements have not been shown for males. It is ovoviviparous, producing litters of mostly about 14 pups (although the number may range from 1–38) in December after a gestation of 11–12 months. Ageing studies suggest that males mature at about 4 and females at about 5 years of age. The diet consists of cephalopods, crustaceans and, to a lesser extent, teleost fish. This species is a major component of the southern Australian shark fishery (current annual production about 5000 tonnes, valued at $20 million to the fishermen). It has been exploited heavily since the 1970s and is currently over-fished. The meat, which is used for human consumption mostly under the name of "flake", is very popular in Victoria and Tasmania. This gummy shark is difficult to distinguish from relatives off tropical Australia (27.8) and New Zealand and its distributional range is not well known. More taxonomic work is require on members of this genus in Australasian waters.
- **Local synonymy:** *Emissola maugeana* Whitley, 1939; *Emissola ganearum* Whitley, 1945; *Mustelus lenticulatus*: Munro, 1956 (misidentification).
- **References:** Whitley (1939b, 1945a); Walker (1988); Anon (1989); Lenanton *et al.* (1990).

undersurface of head

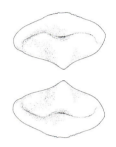

teeth from near symphysis (upper and lower jaw)

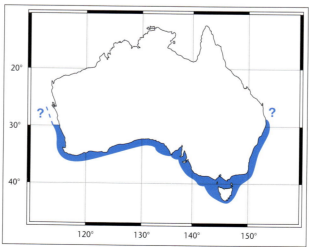

28 Family Hemigaleidae
WEASEL SHARKS

Weasel sharks are medium to large sharks (attaining 1.1–2.3 m) with fusiform bodies, ventrally placed mouths that extend well past the eyes, small spiracles, oval eyes with a nictitating membrane, five pairs of gill slits, two spineless dorsal fins (the second about two-thirds as large as the first), an anal fin, a strongly heterocercal caudal fin, and a spiral-type intestinal valve. They lack nasoral grooves, nasal barbels and caudal keels, but have upper and lower precaudal pits. The mid-base of the first dorsal fin is always forward of the pelvic-fin origin. The closely related whalers (Carcharhinidae) and hound sharks (Triakidae) also have most of these features. The weasel sharks, however, can be distinguished from the hound sharks in having precaudal pits, and from whalers in having a spiral-type intestinal valve (rather than a scroll-type).

Members of this family, which contains four genera and about seven species, have often been placed within the family Carcharhinidae. Two widespread, tropical Indo–West Pacific species occur off Australia. Reproduction is viviparous, with a yolk-sac placenta.

Key to hemigaleids

1 Teeth not noticeably protruding from mouth; upper teeth broad and short, with smooth inner edges and coarsely serrated outer edges (fig. 1) .... *Hemigaleus microstoma* (fig. 3; p. 218)

Teeth noticeably protruding from mouth; upper teeth narrower and more slender, both edges serrated (fig. 2) .... *Hemipristis elongata* (fig. 4; p. 219)

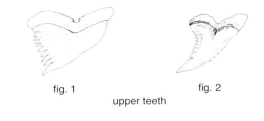

fig. 1 fig. 2
upper teeth

fig. 3

fig. 4

Weasel Shark — *Hemigaleus microstoma* Bleeker, 1852

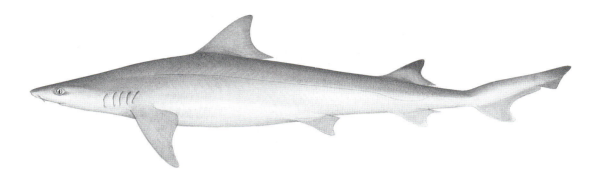

28.1 *Hemigaleus microstoma* (Plate 28)   Fish Code: 00 018020

- **Alternative name:** Sicklefin weasel shark.

- **Field characters:** A small, slender weasel shark with moderately long nasal lobes, oblique upper teeth with smooth inner edges and coarsely serrated outer edges, falcate fins, and a second dorsal fin about two-thirds the size of the first dorsal fin.

- **Distinctive features:** Body fusiform and slender; lateral line distinct, with a pronounced dip below second dorsal fin. Snout depressed, moderately rounded; orbit oval, with posterior notch; nostril lobes rather long and triangular. Upper teeth oblique, subtriangular, with smooth medial margins and coarsely serrated lateral margins. Lower teeth slender, erect, smooth-edged. First dorsal-fin origin behind free rear tips of pectoral fins. Second dorsal-fin origin ahead of anal-fin origin. Dorsal, anal, pelvic, pectoral and lower caudal fins falcate. Tooth count 28–34/43–54 [25–33/37–43]. Total vertebrae 111–119 [137–163]; precaudal 63–69 [79–92].

- **Colour:** Light bronze to greyish above, pale ventrally. Second dorsal and upper caudal-fin tips dark, fading in large specimens.

- **Size:** Born at about 30 cm and attains 110 cm. Males mature at about 60 cm and females at about 65 cm.

- **Distribution:** Tropical Indo–West Pacific, including tropical Australia from Shark Bay (Western Australia) to Brunswick Heads (New South Wales). On or near the bottom on continental and insular shelves down to depths of 170 m.

undersurface of head

fourth tooth from symphysis (upper and lower jaw)

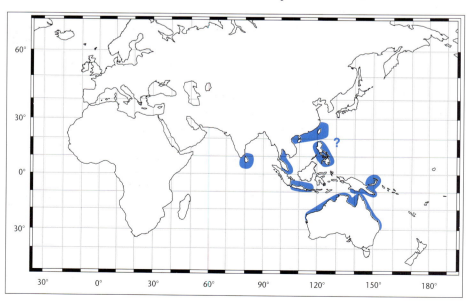

- **Remarks:** This shark is commonly trawled off northwestern Australia. It is also taken, although less frequently, by gillnet or hook and line. It is viviparous, and in northern Australia produces two litters of 1–19 pups each year (mainly in February and September) after a gestation of about 6 months. The weasel shark is a highly selective feeder, eating almost exclusively cephalopods, particularly octopus. Although caught elsewhere for its meat, it is of no commercial importance locally. Australian specimens have an unusually low vertebral count and need to be compared more closely with other Indo–Pacific populations.

- **Local synonymy:** *Negogaleus microstoma* Whitley, 1940.

- **References:** Stevens and Cuthbert (1983); Stevens and McLoughlin (1991).

Fossil Shark — *Hemipristis elongata* (Klunzinger, 1871)

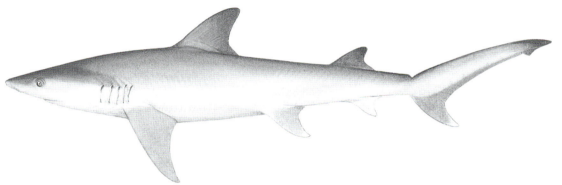

28.2 *Hemipristis elongata* (Plate 29)   **Fish Code:** 00 018011

- **Alternative name:** Snaggletooth shark.

- **Field characters:** A moderately large weasel shark with noticeably protruding teeth (the uppers long, curved and serrated on both edges), falcate fins, and a second dorsal fin about two-thirds the size of the first dorsal fin.

- **Distinctive features:** Body fusiform and moderately slender. Snout relatively long, broadly rounded; eyes moderately large; gill slits long. Upper teeth coarsely serrated on both margins, laterally curved and subtriangular (narrower near the symphysis). Lower teeth long, curved, awl-shaped near symphysis, with variably developed basal cusplets or serrations; posterior teeth serrated and more similar in shape to uppers. First dorsal-fin origin over or just behind free rear tips of pectoral fins. Second dorsal-fin origin ahead of anal-fin origin. All fins falcate, with markedly concave margins to second dorsal, anal, pelvic and pectoral fins. Tooth count 26–28/33–36 [26–30/30–36]. Total vertebrae 190–195 [190–194]*; precaudal 104–107 [103–104]*.

- **Colour:** Bronze to greyish brown dorsally, pale ventrally. Second dorsal and upper caudal fins with a dark blotch (less distinct in larger specimens).

- **Size:** Born at about 52 cm and attains 230 cm. Males mature at about 110 cm and females at about 120 cm.

undersurface of head

third tooth from symphysis (upper jaw)

- **Distribution:** Tropical Indo–West Pacific, including the Red Sea, south-eastern Africa and northern Australia. Locally, known from Lizard Island (Queensland) to Bunbury (Western Australia); in the west, rarely south of Shark Bay. Lives on the continental and insular shelves at depths down to at least 130 m.
- **Remarks:** The fossil shark is viviparous, producing 2–11 pups in April in northern Australia, after a gestation of 7–8 months. Australian sharks eat cephalopods, mainly squid and cuttlefish, together with fish (including some elasmobranchs). Elsewhere the diet is reported to consist of a variety of fish species. A minor component of the northern Australian gillnet fishery, its meat is used for human consumption. In India, where the fins and liver oil are also used, it is considered to be among the best sharks to eat.
- **References:** Bass (1979); Stevens and McLoughlin (1991).

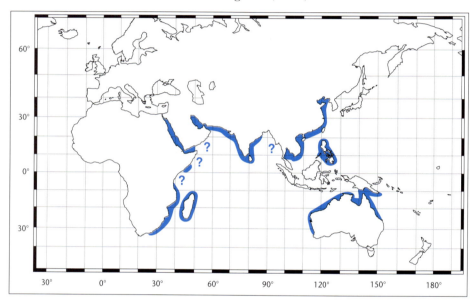

29

Family Carcharhinidae
WHALER SHARKS

Whalers are small to large sharks (attaining 0.7–6.0 m) with fusiform bodies, ventrally placed mouths extending well past the eyes, five pairs of gill slits, circular eyes with nictitating membranes, two spineless dorsal fins (the first much larger than the second, except in *Glyphis, Negaprion, Lamiopsis* and *Triaenodon*), an anal fin, a non-lunate caudal fin and a scroll-type intestinal valve. They lack nasoral grooves and nasal barbels, usually also lack spiracles and caudal keels, but they have upper and lower precaudal pits. The first dorsal fin is relatively high and short, and is inserted forward of the pelvic fin. The second dorsal fin is usually above the anal-fin origin. The hound sharks (Triakidae) and weasel sharks (Hemigaleidae) also have most of these characteristics, but the whalers can be distinguished from them by the presence of upper and lower precaudal pits on the caudal peduncle, and a scroll type intestinal valve.

Of the 49 species in this family, 30 are found in Australia. Most are pelagic in tropical and warm temperate areas, a few are oceanic, while at least one can penetrate far into fresh water. Many of the species are very similar and identification can be difficult. Identifying features are often subtle and the most important of these are tooth shape and numbers, position of the dorsal fins, colour, and the presence or absence of an interdorsal ridge. Whalers are typically viviparous, with a yolk-sac placenta (the ovoviviparous tiger shark is an exception). The family contains three of the four shark species most dangerous to humans.

Key to carcharhinids

1 Second dorsal fin half or more height of first dorsal fin (fig. 1) **2**

   Second dorsal fin less than half height of first dorsal fin (fig. 2) **4**

2 Distinct white tips to first dorsal and upper caudal fins (fig. 5); teeth with a single cusp and lateral cusplets (fig. 3) ....
.... ***Triaenodon obesus*** (fig. 5; p. 268)

   No white tips to first dorsal and upper caudal fins; teeth with a single cusp and no lateral cusplets (fig. 4) **3**

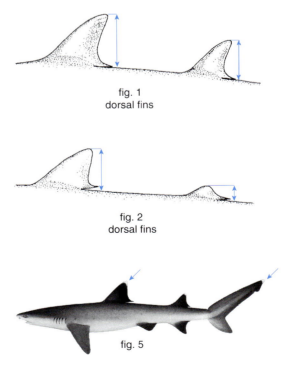

fig. 1
dorsal fins

fig. 2
dorsal fins

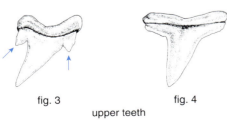

fig. 3   fig. 4
upper teeth

fig. 5

3   Upper teeth with broad, triangular, serrated cusps (fig. 6); second dorsal fin about half height of first dorsal fin (fig. 8) .... **Glyphis sp. A (fig. 8; p. 259)**

Upper teeth with narrow, non-serrated cusps (fig. 7); second dorsal fin almost or as high as first dorsal fin (fig. 9) .... **Negaprion acutidens (fig. 9; p. 262)**

4   Caudal peduncle with lateral keels (fig. 10); spiracles present (fig. 11); upper labial furrows very long, reaching forward to front of eyes (fig. 11); teeth in both jaws cockscomb-shaped (fig. 12) ....**Galeocerdo cuvier (fig. 13; p. 258)**

Caudal peduncle without lateral keels (weak ones present in *Prionace*); spiracles absent; upper labial furrows shorter, never reaching forward to front of eyes; teeth not cockscomb-shaped **5**

5   Mid-base of first dorsal fin closer to pelvic-fin origin than to pectoral-fin insertion (fig. 14); live dorsal coloration vivid blue .... **Prionace glauca (fig. 14; p. 263)**

Mid-base of first dorsal fin closer to pectoral-fin insertion than to pelvic-fin origin (fig. 13) (about midway in *Loxodon*, fig. 21); live dorsal coloration never blue, although sometimes greyish **6**

6   Anal-fin origin well in advance of second dorsal-fin origin (second dorsal-fin origin never in advance of middle of anal-fin base) (fig. 15) **7**

Anal-fin origin under or behind second dorsal-fin origin (second dorsal-fin origin always in advance of middle of anal-fin base) (fig. 16) **11**

7   Posterior rim of orbit with a distinct notch (fig. 17); first dorsal-fin origin posterior to adpressed

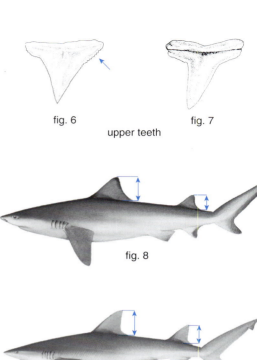

fig. 6   fig. 7
upper teeth

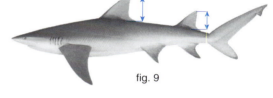

fig. 8

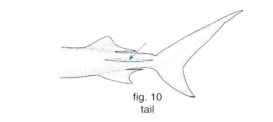

fig. 9

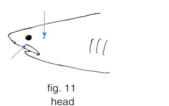

fig. 10
tail

fig. 11
head

fig. 12
upper tooth

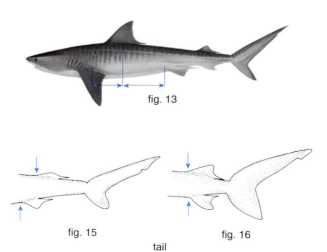

fig. 13   fig. 14

fig. 15   fig. 16
tail

fig. 17   fig. 18
eye shape

pectoral-fin free rear tips by distance greater than length of fourth gill slit (fig. 19) .... **Loxodon macrorhinus (fig. 21; p. 260)**

Posterior rim of orbit without a notch (fig. 18); first dorsal-fin origin rarely posterior to adpressed pectoral-fin free rear tips by distance equivalent to length of fourth gill slit (fig. 20) **8**

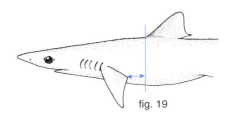
fig. 19

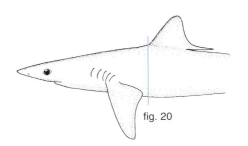
fig. 20

8 Upper teeth with relatively erect cusps and with basal cusplets (fig. 22); first dorsal-fin base more than 2 times in distance between pectoral-fin insertion and pelvic-fin origin (fig. 24) .... **Carcharhinus macloti (fig. 26; p. 249)**

Upper teeth with oblique cusps and no basal cusplets (fig. 23); first dorsal-fin base usually less than 2 times in distance between pectoral-fin insertion and pelvic-fin origin (fig. 25) **9**

9 Upper labial furrows relatively long and prominent, not confined to mouth corners (fig. 28) .... **Rhizoprionodon acutus (fig. 27; p. 265)**

Upper labial furrows very short, virtually confined to mouth corners (fig. 29) **10**

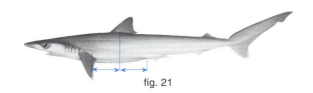
fig. 21

fig. 22   fig. 23
upper teeth

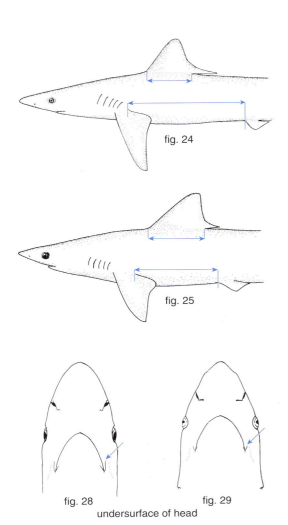

fig. 24

fig. 25

fig. 28   fig. 29
undersurface of head

fig. 26

fig. 27

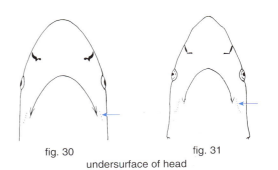
fig. 30   fig. 31
undersurface of head

10 Combined number of enlarged hyomandibular pores (lateral line pores just behind mouth corners) on both sides of the head usually less than 14 (7–16) (fig. 30); precaudal vertebrae 84–91 .... *Rhizoprionodon oligolinx* (fig. 32; p. 266)

Combined number of enlarged hyomandibular pores on both sides of the head more than 14 (15–22) (fig. 31); precaudal vertebrae 73–80 .... *Rhizoprionodon taylori* (fig. 33; p. 267)

11 Pectoral and first dorsal fins very broad distally and with broadly rounded tips, tapering only slightly towards their apices (fig. 34); most fin tips mottled white (fig. 36), some with black markings (more prominent in juveniles) .... *Carcharhinus longimanus* (fig. 36; p. 247)

Pectoral and first dorsal fins tapering distally, apices usually pointed or narrowly rounded (fig. 35) 12

12 First dorsal, pectoral, pelvic and caudal fins with distinct white tips (fig. 37) .... *Carcharhinus albimarginatus* (fig. 37; p. 228)

First dorsal, pectoral, pelvic and caudal fins without white tips, (the first dorsal fin of *C. amblyrhynchos* may have an indistinct white tip) 13

13 Second dorsal fin with a conspicuous black tip (fig. 38), other fins plain .... *Carcharhinus dussumieri* (fig. 38; p. 239)

Second dorsal fin never the only fin with a black tip 14

14 Caudal fin with a distinct wide black margin along entire posterior margin (fig. 40), anterior margins of caudal fin without black edging; first dorsal fin never black-tipped .... *Carcharhinus amblyrhynchos* (fig. 39; p. 232)

Caudal fin either plain, or if black-edged then either anterior margin also black-edged or first dorsal fin black-tipped 15

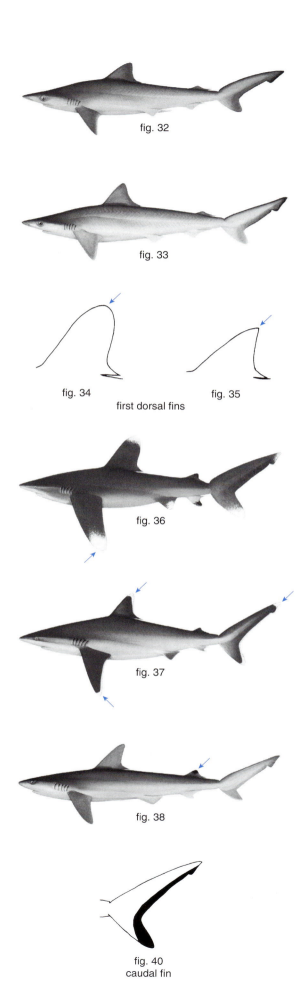

15 Upper anterolateral teeth with bent, hooked, narrow cusps (fig. 41) .... ***Carcharhinus brachyurus*** (fig. 42; p. 235)

Upper anterolateral teeth variably shaped but never with bent, hooked cusps **16**

16 Interdorsal ridge present (fig. 43) **17**

Interdorsal ridge absent **22**

17 Second dorsal, pectoral and lower caudal fins with striking black tips (sharply demarcated) (fig. 44) .... ***Carcharhinus sorrah*** (fig. 44; p. 255)

Fins plain or with dusky tips, but not strikingly black-tipped or sharply demarcated **18**

18 First dorsal-fin origin well behind pectoral-fin free rear tips (fig. 45); second dorsal-fin margin very long, usually more than second dorsal-fin height (fig. 46) .... ***Carcharhinus falciformis*** (fig. 45; p. 241)

First dorsal-fin origin over or anterior to pectoral-fin free rear tips (figs 49, 50); second dorsal-fin margin shorter, usually less than second dorsal-fin height (fig. 47) **19**

19 First dorsal fin situated forward, its origin closer to pectoral-fin insertions than to pectoral-fin free rear tips (fig. 49) **20**

First dorsal fin situated more posteriorly, its origin closer to pectoral-fin free rear tips than to pectoral-fin insertions (fig. 50) **21**

20 First dorsal fin relatively low, its height much less than half predorsal body length (fig. 48); upper anterolateral teeth very high (fig. 51), usually 15 rows of upper teeth; distance from nostrils to mouth less than 2.4 times in mouth width (fig. 53); precaudal vertebrae 101 or more .... ***Carcharhinus altimus*** (fig. 48; p. 230)

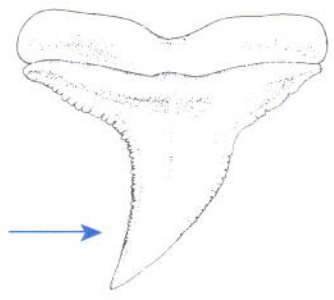

fig. 41
upper tooth

fig. 42

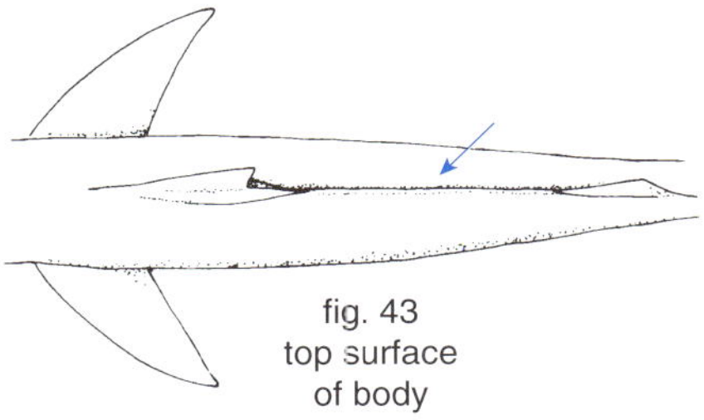

fig. 43
top surface
of body

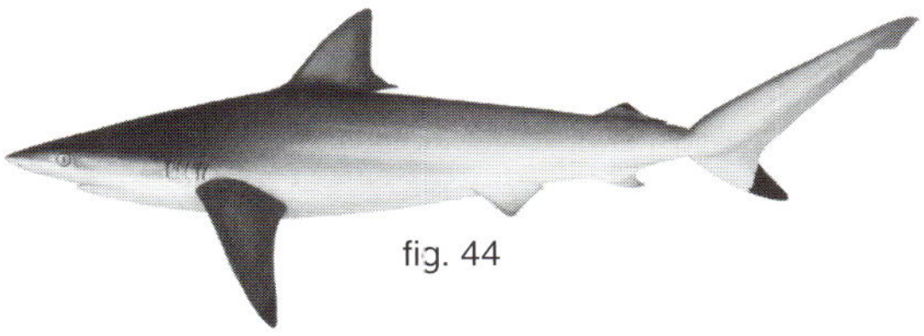

fig. 44

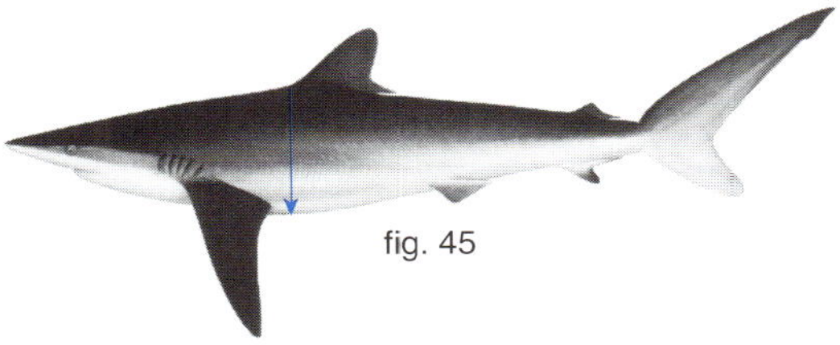

fig. 45

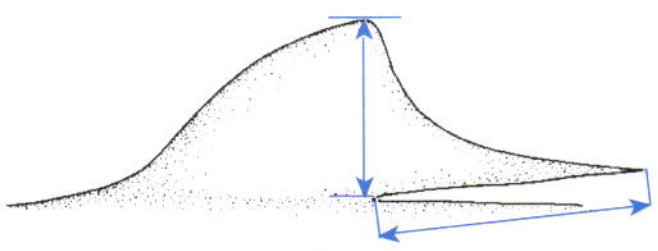

fig. 46 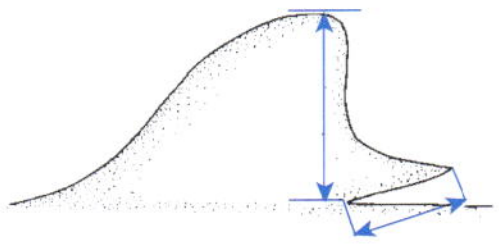 fig. 47
second dorsal fins

fig. 48

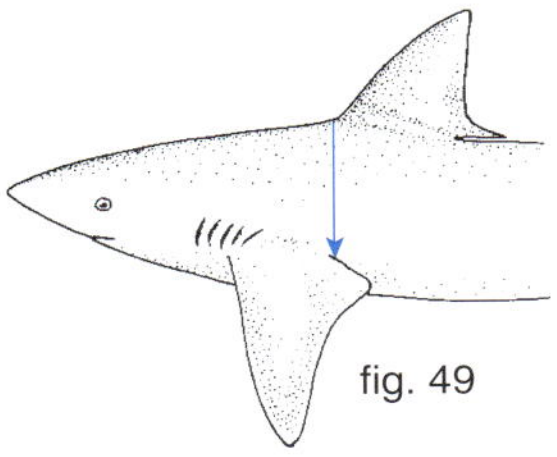

fig. 49 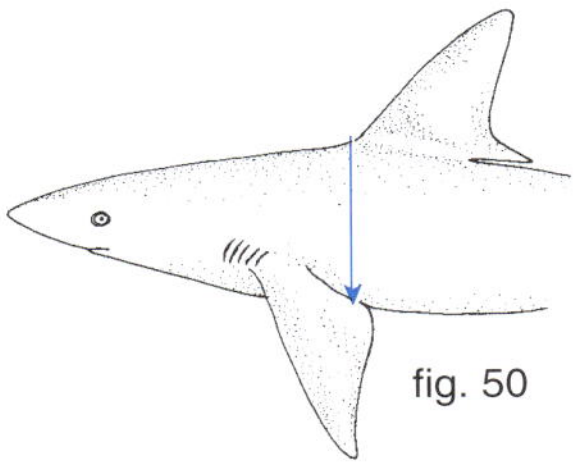 fig. 50

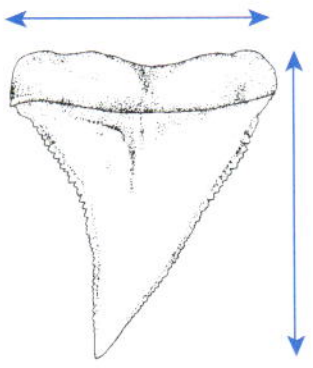

fig. 51 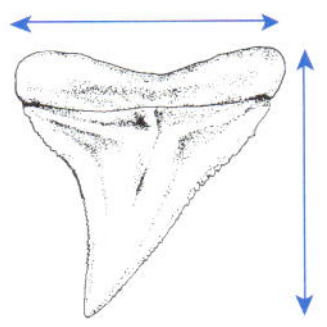 fig. 52
upper teeth

First dorsal fin very high, its height about half predorsal body length (fig. 55); upper anterolateral teeth only moderately high (fig. 52), usually 14 rows of upper teeth; distance from nostrils to mouth more than 2.4 times in mouth width (fig. 54); precaudal vertebrae 97 or less .... *Carcharhinus plumbeus* (**fig. 55; p. 253**)

21 Precaudal vertebrae 89–95; second dorsal-fin height usually less than 2.3% (1.5–2.3%) of total length (fig. 56); pectoral fins with nearly straight posterior margins (fig. 58); oceanic islands .... *Carcharhinus galapagensis* (**fig. 56; p. 243**)

Precaudal vertebrae 103–109; second dorsal-fin height usually more than 2.3% (2.1–3.3%) of total length (fig. 57); pectoral fins with distinctly concave posterior margins (fig. 59); continental coastlines ..... *Carcharhinus obscurus* (**fig. 57; p. 252**)

22 Upper anterolateral teeth broad, triangular and serrated (fig. 60) **23**

Upper anterolateral teeth not broad, triangular and serrated **24**

23 First dorsal-fin height more than 3.1 times second dorsal-fin height (fig. 61); usually 11 rows of teeth in lower jaw; angle of notch in anal-fin posterior margin usually less than a right angle (fig. 62); precaudal vertebrae 89–95 .... *Carcharhinus amboinensis* (**fig. 61; p. 234**)

First dorsal-fin height 3.1 times or less the second dorsal-fin height (fig. 64); usually 12 rows of teeth in lower jaw; angle of notch in anal-fin posterior margin usually a right angle or more (fig. 63); precaudal vertebrae 101–123 .... *Carcharhinus leucas* (**fig. 64; p. 244**)

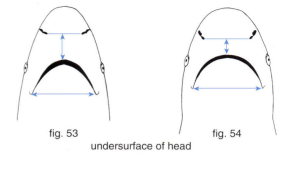

fig. 53   fig. 54

undersurface of head

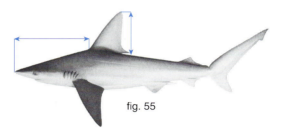

fig. 55

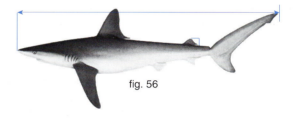

fig. 56

fig. 57

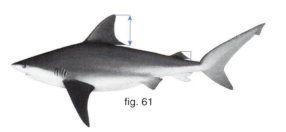

fig. 61

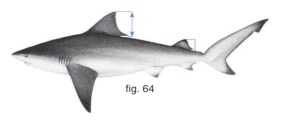

fig. 64

fig. 58   fig. 59   fig. 60

pectoral fins   upper tooth

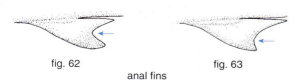

fig. 62   fig. 63

anal fins

24 Upper anterolateral teeth with basal cusplets (fig. 65) **25**

Upper anterolateral teeth without basal cusplets (fig. 66) **27**

25 First dorsal fin with a striking broad black tip (fig. 67) ....
.... ***Carcharhinus melanopterus*** (**fig. 67; p. 250**)

First dorsal fin without a black tip (anterior margin may be black-edged) **26**

26 Pectoral fins noticeably broad and triangular, their lengths 1.5 times or less in their anterior margin lengths (fig. 68); no black edging to dorsal or caudal fins ....
.... ***Carcharhinus fitzroyensis*** (**fig. 70; p. 242**)

Pectoral fins not noticeably broad and triangular, their lengths more than 1.5 times in their anterior margin lengths (fig. 69); black edging to dorsal and caudal fins ...... ***Carcharhinus cautus*** (**fig. 71; p. 238**)

27 Teeth very short compared to other members of the genus, distances between tips of teeth in rows 4–6 from symphysis usually longer than tooth heights (fig. 72); first dorsal-fin origin over or just behind pectoral-fin free rear tips (fig. 73); first dorsal fin relatively low, its height over 2.2 times in the interdorsal space (fig. 73) ....
.... ***Carcharhinus brevipinna*** (**fig. 73; p. 237**)

Teeth normal with respect to the genus, distances between tips of teeth in rows 4–6 from symphysis usually shorter than tooth heights (fig. 74); first dorsal-fin origin over or just behind pectoral-fin insertions (fig. 75); first dorsal fin relatively high, its height 2.2 times or less in the interdorsal space (fig. 75) **28**

28 Snout relatively short, internarial space 1.2 times or less (1.0–1.2) in preoral snout (fig. 76); second dorsal-fin height mostly less than 1.2 times (1.0–1.2) in inner margin length (fig. 78); precaudal vertebrae usually less than 82 ....
.... ***Carcharhinus amblyrhynchoides*** (**fig. 75; p. 231**)

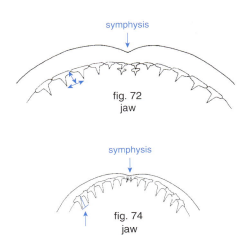

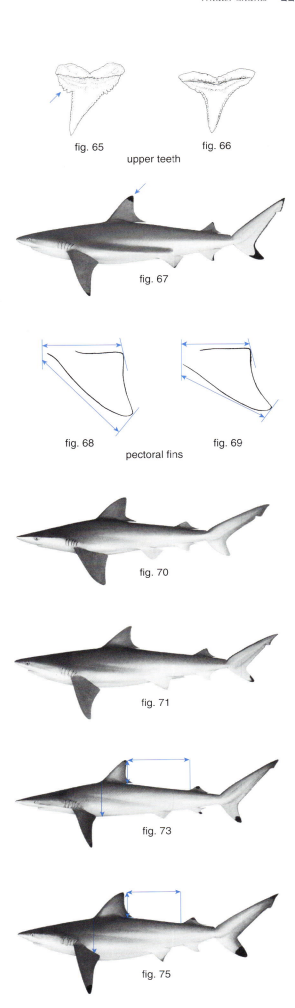

Snout longer, internarial space 1.3 or more times (1.3–1.7) in preoral snout (fig. 77); second dorsal-fin height mostly more than 1.2 times (1.1–1.6) in inner margin length; precaudal vertebrae more than 83 .... **29**

29 Precaudal vertebrae 94–102 ....
.... ***Carcharhinus limbatus* (fig. 79; p. 246)**
    Precaudal vertebrae 84–91 ....
.... ***Carcharhinus tilstoni* (fig. 80; p. 256)**

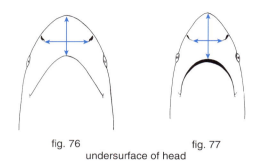

fig. 76   fig. 77
undersurface of head

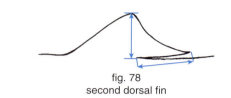

fig. 78
second dorsal fin

fig. 79

fig. 80

Silvertip Shark — *Carcharhinus albimarginatus* (Rüppell, 1837)

29.1 *Carcharhinus albimarginatus* (Plate 31)   **Fish Code:** 00 018027

- **Field characters:** A medium-sized whaler shark with conspicuous white tips to the first dorsal, pectoral and upper caudal fins, an interdorsal ridge, a small second dorsal fin, and a tapered first dorsal fin that is narrowly rounded or pointed (not broadly rounded) at its apex.

- **Distinctive features:** Body fusiform; interdorsal ridge present. Labial furrows small and inconspicuous; upper teeth triangular, oblique and serrated (somewhat coarser basally), outer margins distinctly notched, inner margins with a less distinct notch; lower teeth serrated, more slender and erect than uppers. First dorsal-fin origin over, or just anterior to, pectoral-fin free rear tips. First dorsal fin tapering to a pointed or narrowly rounded tip (not broadly rounded); second dorsal fin small, much less than half the height of first dorsal fin. Tooth count [26–30/25–30]. Total vertebrae 225* [216–231]; precaudal 116–118* [115–125].

- **Colour:** Dorsal surfaces bronze, fading to brown or grey after death or in preservative. Pale ventrally with an indistinct pale stripe on the flank extending anteriorly from above the pelvic fins. Tips of first dorsal, pectoral and caudal fins with conspicuous white tips. Second dorsal fin, anal fin and dorsal surfaces of pelvic fins sometimes dusky; pelvic fins sometimes with a white posterior margin.

- **Size:** Born at 55–80 cm and attains 275 cm. Males and females mature at about 170 cm and about 195 cm respectively.

- **Distribution:** Tropical Indo–Pacific, including northern Australian waters from Carnarvon (Western Australia) to Bundaberg (Queensland), with the exception of the Gulf of Carpentaria and Arafura Sea. Over or adjacent to continental or insular shelves and offshore banks, from close inshore to well offshore (but not mid-oceanic). Occurs throughout the water column from the surface to depths of about 800 m.

- **Remarks:** A tagging study on immature silvertips in the Indian Ocean showed relatively localised movements, with nearly 70% of the sharks recaptured within 2 km of the tagging site. The greatest distance travelled was 7 km and 34% of those tagged were recaptured. Growth appears to be fairly slow (about 9 cm a year for juveniles). The silvertip produces 1–11 pups in summer after a gestation period of about 12 months. Its diet includes a variety of both pelagic and demersal fishes. The silvertip is relatively aggressive and is prone to make sudden dashes at a food source. It is also strongly attracted to certain, artificially produced, low-frequency sounds. Not exploited commercially in Australia; in the Seychelles used for its meat which is dried and salted, and doubtless taken in other artisanal fisheries. Potentially dangerous to humans.

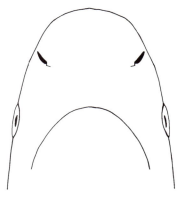

undersurface of head

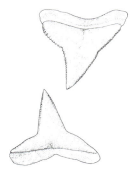

fourth teeth from symphysis
(upper and lower jaw)

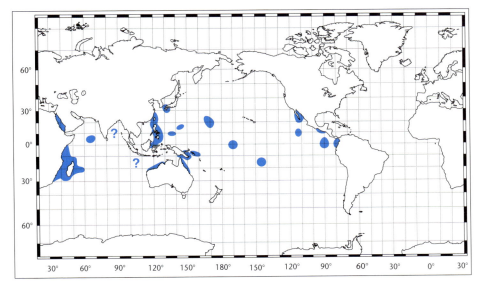

Bignose Shark — *Carcharhinus altimus* (Springer, 1950)

29.2 *Carcharhinus altimus* (Plate 34)

Fish Code: 00 018012

- **Field characters:** A medium-sized whaler shark with a tall first dorsal fin originating almost over the pectoral-fin insertion (or at least nearer to it than to the pectoral-fin free rear tips), a prominent interdorsal ridge, bronze to grey above, no distinctive fin markings (except in some small juveniles), and long, triangular, serrated upper teeth.

- **Distinctive features:** Body fusiform; interdorsal ridge conspicuous. Snout relatively long (preoral length 7.5–11.1% of total length); nasal flaps prominent; labial furrows very short, confined to mouth corners. Upper teeth triangular, serrated, erect to slightly oblique, noticeably long and pointed; lower teeth erect, narrow and serrated. First dorsal-fin origin varying from over pectoral-fin insertion to about half way along pectoral-fin inner margin. First dorsal fin relatively high, 8.3–13.3% of total length. Pectoral fins long, anterior margins 16.5–24% of total length. Tooth count [29–34/29–31]. Total vertebrae 195* [194–206]; precaudal 104–105* [101–110].

- **Colour:** Dorsal surfaces bronze to grey (grey after death and in preservative); ventral surfaces whitish. Fin tips sometimes dusky (more pronounced in small juveniles).

- **Size:** Born at 60–75 cm and attains 285 cm. Males and females mature at about 190 cm and about 225 cm respectively.

- **Distribution:** Records are patchy, but probably circumglobal in tropical and warm temperate seas. In Australia, currently recorded from northern New South Wales and Western Australia (the North-West Shelf region and off Rottnest Island, near

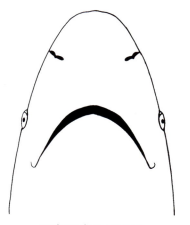

undersurface of head

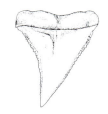

fourth tooth from symphysis (upper jaw)

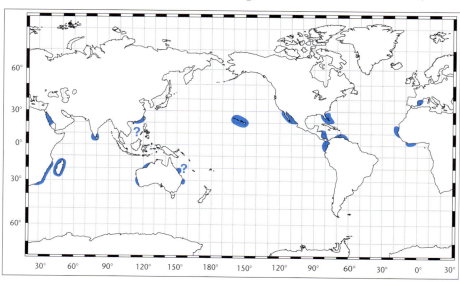

Perth), but almost certainly occurs in other tropical areas. Found mainly close to the bottom, near the edge of continental and insular shelves and on the upper slopes, in depths from about 80 to 430 m (occasionally shallower).

- **Remarks:** A mainly bottom-dwelling, deepwater shark (although five of six specimens caught adjacent to the North-West Shelf were taken near the surface over water depths of 400–810 m). It has a seasonal reproductive cycle producing litters of 3–15 pups, and a diet consisting mainly of bottom-living fish. Not commercial in Australia; elsewhere used for fishmeal, liver oil and shagreen.
- **Reference:** Springer (1950a).

Graceful Shark — *Carcharhinus amblyrhynchoides* (Whitley, 1934)

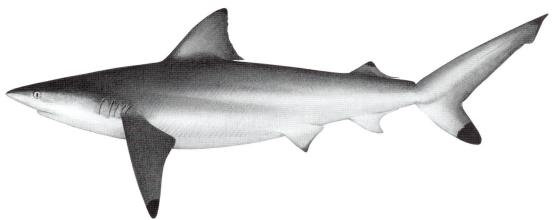

29.3 *Carcharhinus amblyrhynchoides* (Plate 33)   **Fish Code:** 00 018033

- **Alternative name:** Queensland shark.
- **Field characters:** A medium-sized, fairly short-snouted whaler shark with black tips to most fins, bronze to greyish dorsal coloration, lacking an interdorsal ridge, and with slender, erect, serrated upper teeth.
- **Distinctive features:** Body fusiform, moderately stout; interdorsal ridge absent. Snout moderately short, internarial space 0.9–1.2 in preoral length; labial furrows small and inconspicuous. Upper teeth with erect to slightly oblique, slender, finely serrated cusps (with somewhat coarser serrations basally); lower teeth erect, more slender, either with fine serrations or smooth-edged. First dorsal-fin origin over or just behind pectoral-fin insertions. Tooth count 32/31* [31–33/29–33]. Total vertebrae 168–170* [168–193]; precaudal 78–81* [78–96].
- **Colour:** Dorsal surfaces bronze, fading to grey after death or in preservative. Pale ventrally with a pale stripe extending along the mid-flank from the pelvic fin to below the first dorsal fin. All fin tips usually black or dusky (anal fin sometimes uniformly pale).
- **Size:** Born at 50–60 cm and attains 170 cm. Both sexes mature at 110–115 cm.
- **Distribution:** Tropical Indo–West Pacific; currently recorded from northern Australian waters between Townsville (Queensland) and Broome (Western Australia), but probably occurs throughout tropical northern waters. Mainly pelagic

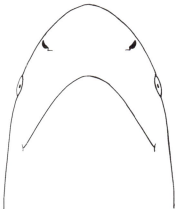

undersurface of head

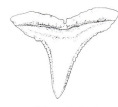

fifth tooth from symphysis (upper jaw)

in midwater over the continental and insular shelves from close inshore to a depth of at least 50 m.

- **Remarks:** The graceful shark is more stockily built than the very similar Australian blacktip (29.21) and common blacktip (29.14) sharks. It gives birth to 3 pups in January or February after a gestation period of 9–10 months. It eats mainly fish, with smaller numbers of crustaceans and cephalopods. This species is a minor component of the northern Australian gillnet fishery, where it is caught for its meat and fins.
- **Local synonymy:** *Gillisqualus amblyrhynchoides* Whitley, 1934.
- **References:** Whitley (1934a); Garrick (1982); Lyle and Timms (1984); Stevens and McLoughlin (1991).

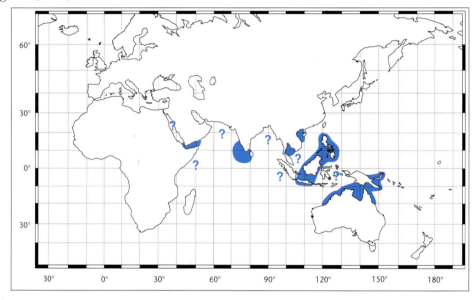

Grey Reef Shark — *Carcharhinus amblyrhynchos* (Bleeker, 1856)

29.4 *Carcharhinus amblyrhynchos* (Plate 32)   **Fish Code:** 00 018030

- **Alternative names:** Black-vee whaler, longnose blacktail shark.
- **Field characters:** A medium-sized, bronze to grey whaler shark with the first dorsal-fin origin about over the pectoral-fin free rear tips, usually no interdorsal ridge, and a prominent broad, black posterior margin to the caudal fin (less distinct in large specimens).

- **Distinctive features:** Body fusiform; interdorsal ridge usually absent (some specimens with a slight ridge). Labial furrows short and inconspicuous. Upper teeth triangular, erect to oblique, coarsely serrated; outer margins notched; inner margins slightly concave or sinuous. Lower teeth narrow, more erect, smooth-edged or weakly serrated. First dorsal-fin origin varying from slightly in front to slightly behind pectoral-fin free rear tips. Tooth count 28–30/27–29 [27–30/27–29]. Total vertebrae 217–221* [211–221]; precaudal 114–121* [110–119].
- **Colour:** Dorsal surfaces bronze or grey (grey after death or in preservative); ventral surfaces pale. Flank with an indistinct pale stripe extending anteriorly from above the pelvic fins. Caudal fin with a broad, black trailing edge (less distinct in large specimens), widening ventrally to cover the entire distal portion of the ventral lobe. First dorsal fin pale grey, sometimes with a small white tip and white trailing margin; remaining fin tips dusky.
- **Size:** Born at 50–60 cm and has been reported to attain 255 cm, but rarely exceeds 180 cm. Both sexes mature at 130–140 cm.
- **Distribution:** Tropical Indo–West and Central Pacific. In Australia, recorded in northern waters from Carnarvon (Western Australia) to Bundaberg (Queensland), including Lord Howe Island (off New South Wales). Continental and insular shelves, nearshore from the surface to a depth of about 280 m.
- **Remarks:** The grey reef shark is one of the most common sharks inhabiting coral reefs, where it usually lives near deep drop-offs or in atoll passes. It may also live over shallow reef flats, the normal habitat of the blacktip reef shark (29.17), when the latter is absent. Telemetry studies in the Pacific have shown that individuals living near reef drop-offs tend to be more nomadic than more site-attached individuals from lagoons. This gregarious species is known for its aggressive behaviour under baited conditions. It tends to be curious, investigating disturbances or unusual objects, and responding to low-frequency, artificially produced sounds. When approached too closely or startled by divers, it performs what appears to be a threat display, which may be intended to intimidate potential predators. The agitated shark wags its head and tail in lateral sweeps, arches its back, raises its head, depresses its pectoral fins and sometimes swims in a horizontal spiral. Displays may precede attack or flight by the shark. The grey reef shark matures at about 7 years, and produces 1–6 pups after a 12 month gestation period. Its diet consists of small fish and, to a lesser extent, cephalopods and crustaceans. Small quantities of this species have recently been marketed for food in Australia; elsewhere they are caught mainly for the flesh and fins. We currently recognise the very similar *C. wheeleri* Garrick, 1982, from the western Indian Ocean, as a valid species, although we caution that it may prove to be indistinct from *C. amblyrhynchos*.
- **Local synonymy:** *Galeolamna fowleri* Whitley, 1944; *Galeolamna tufiensis* Whitley, 1949; *Galeolamna coongoola* Whitley, 1964.
- **References:** Whitley (1944a, 1964a); Garrick (1982); Stevens and McLoughlin (1991).

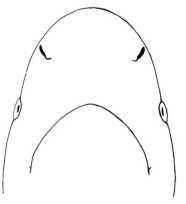

undersurface of head

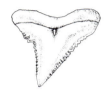

third tooth from symphysis (upper jaw)

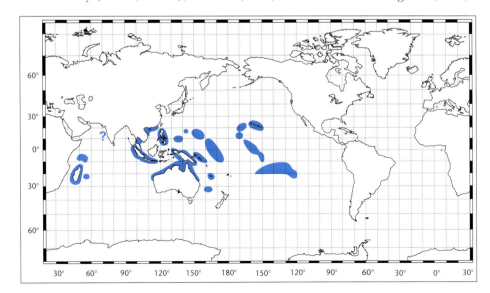

Pigeye Shark — *Carcharhinus amboinensis* (Müller & Henle, 1839)

29.5 *Carcharhinus amboinensis* (Plate 34) **Fish Code:** 00 018026

- **Alternative name:** Java shark.
- **Field characters:** A medium-sized, stout-bodied whaler shark with a short, blunt snout, a relatively small second dorsal fin (less than a third of first dorsal-fin height), no interdorsal ridge or distinctive fin markings, and with broad, triangular, serrated upper teeth.
- **Distinctive features:** Body fusiform, noticeably stout; interdorsal ridge absent. Snout short, blunt (internarial space 0.9–1.1 in preoral length); labial furrows short and inconspicuous; eye small, 0.7–1.7% of total length. Upper teeth erect to slightly oblique, broadly triangular, serrated; lower teeth similar but more slender. First dorsal-fin origin varying from slightly in front to slightly behind pectoral-fin insertions. First dorsal-fin height more than 3.1 times second dorsal-fin height. Tooth count 23–26/23–25* [23–27/21–25] (usually 11 teeth on each side of the lower jaw). Total vertebrae 190* [185–195]; precaudal 90–92* [89–95].
- **Colour:** Grey above, pale below; flank with an indistinct pale stripe extending anteriorly from above the pelvic fins. Fin tips dusky in juveniles, becoming indistinct in adults.
- **Size:** Born at 60–65 cm and attains 280 cm. Males mature at about 210 and females at about 215 cm.
- **Distribution:** Eastern North Atlantic (Nigeria) and Indo–West Pacific (South Africa, Madagascar, Gulf of Aden, Pakistan, Sri Lanka, Indonesia, Papua New Guinea, Bouganville and northern Australia). Locally, northern waters from about Carnarvon (Western Australia) to Bundaberg (Queensland). Inshore areas of the continental and insular shelves from the surface to depths of 100 m. Lives throughout the water column, but predominantly near the bottom. Occasionally enters brackish water.
- **Remarks:** Very similar to the bull shark (29.13), from which it differs in the relative heights of the first and second dorsal fins, the number of teeth on each side of the lower jaw and, most reliably, the number of precaudal vertebrae. Tag returns from juvenile pigeye sharks in Australia indicate that their movements are relatively localised (up to 60 km), while two larger individuals were recaptured 240 and 1080 km from the tagging site. Parturition in Australia is around November or December, with litters of 6–13 pups. The diet consists mainly of bottom fishes, including elasmobranchs, and crustaceans, cephalopods and other molluscs. Although this species is potentially dangerous, there are no proven attacks on humans. Caught for its meat and fins by a Taiwanese longline fishery off northern Australia (which ceased in 1991). Taken incidentally in the northern Australian gillnet shark fishery and processed for its meat; elsewhere also caught for its meat.

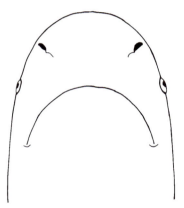

undersurface of head

third tooth from symphysis (upper jaw)

- **Local synonymy:** *Galeolamna (Lamnarius) spenceri*: Whitley, 1943 (misidentification).
- **References:** Whitley (1943c); Garrick (1982); Lyle and Timms (1984); Stevens and McLoughlin (1991).

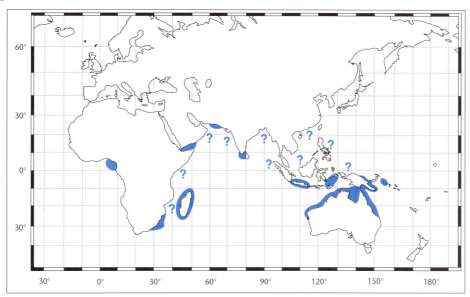

Bronze Whaler — *Carcharhinus brachyurus* (Günther, 1870)

29.6 *Carcharhinus brachyurus* (Plate 35) **Fish Code:** 00 018001

- **Alternative names:** Copper shark, cocktail shark.
- **Field characters:** A rather large whaler shark with narrowly triangular, hook-shaped upper teeth, usually no interdorsal ridge, a bronze to greyish dorsal coloration, and without distinctive fin markings.
- **Distinctive features:** Body fusiform, interdorsal ridge usually absent. Labial furrows small and inconspicuous. Upper teeth narrowly triangular, finely serrate and oblique; lateral margins concave; mesial margins slightly concave; tips narrowly pointed and hook-shaped (more noticeable in adult males). Lower teeth narrower, more erect and finely serrated. First dorsal-fin origin over, or just ahead of free rear tips of pectoral fin. Tooth count 32–33/31–32 [29–35/29–33]. Total vertebrae 193–197* [179–203]; precaudal 104* [96–110].

undersurface of head

- **Colour:** Bronze to greyish brown dorsally, fading to grey or greyish brown after death or in preservative. Ventral surfaces creamy white; a pale stripe extending anteriorly on the flank from above the pelvic fins. Fins without distinctive markings, but margins and tips sometimes dusky.
- **Size:** Born at 60–70 cm and attains 295 cm. Males mature at about 235 cm (somewhat smaller off South Africa); females mature at about 245 cm.
- **Distribution:** Cosmopolitan (but patchy) in warm temperate, and some tropical areas. Occurs throughout southern Australia from Jurien Bay (Western Australia) to Coffs Harbour (New South Wales); reports from off eastern Tasmania need to be verified. An inshore, continental shelf species found near the surf zone to at least 100 m depth.
- **Remarks:** The upper teeth of the bronze whaler are sexually dimorphic (those of adult males are proportionately longer, narrower and more hook-shaped than those of adult females or juveniles). Litter sizes vary from 7–20, with the smallest litters recorded from Californian specimens. Feeds on a variety of fishes (including those living near the bottom) and on cephalopods. The bronze whaler also feeds on schools of pelagic fish, such as Australian salmon (*Arripis* spp.). It often occurs in pairs, and may migrate in association with prey aggregations. It is regarded as potentially dangerous to humans, and there are numerous accounts of it menacing spear fishermen. Small specimens are marketed for their flesh in the Western Australian shark fishery.
- **Local synonymy:** *Galeolamna greyi* Owen, 1853; *Eulamia ahenea* Stead, 1938; *Galeolamna ahenea*: Whitley, 1940; *Galeolamna (Galeolamnoides) eblis*: Whitley, 1944 (misidentification).
- **References:** Whitley (1944a); Garrick (1982).

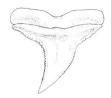

fifth tooth from symphysis (upper jaw)

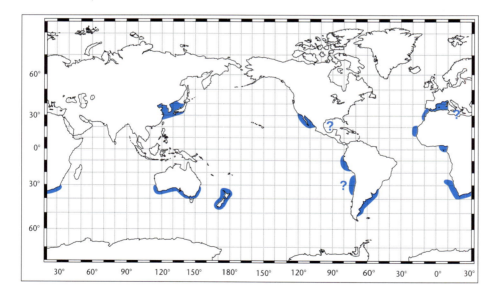

Spinner Shark — *Carcharhinus brevipinna* (Müller & Henle, 1839)

29.7 *Carcharhinus brevipinna* (Plate 33)   **Fish Code:** 00 018023

- **Alternative names:** Longnose grey shark, inkytail shark, smoothfang shark.
- **Field characters:** A medium-sized whaler shark with the first dorsal-fin origin over or slightly behind the pectoral-fin free rear tips, noticeably small upper teeth with erect, very slender and finely serrated cusps, a long and pointed snout, a bronze to greyish upper surface, and usually with black tips to all fins except the pelvics (specimens less than 100 cm total length have some or all fins uniformly coloured).
- **Distinctive features:** Body fusiform, moderately slender; interdorsal ridge absent. Snout long and pointed, preoral length 6.6–12.0% of total length; internarial space 1.5–1.9 in preoral length; upper labial furrows conspicuous, extending beyond mouth corners. Teeth relatively small, distance between tips of 4–6th teeth from symphysis usually greater than tooth height; uppers with erect to slightly oblique, very slender, finely serrated cusps; lowers erect, even more slender (almost needle-like in specimens up to 180 cm) and smooth-edged. First dorsal-fin origin over, or slightly behind, free rear tips of pectoral fins. Tooth count 34–36/33–35 [32–39/29–37]. Total vertebrae 178–181 [155–185]; precaudal 84–88 [76–91].
- **Colour:** Dorsal surfaces bronze to greyish, fading to grey after death or in preservative; pale ventrally. Fins plain at birth; distinct black tips arising on the second dorsal, anal and lower caudal fins between 80–100 cm; by 130 cm, all, except the pelvic fins, usually have black tips.
- **Size:** Born at 60–80 cm and attains 280 cm. In Australia, both sexes mature at about 190–200 cm.
- **Distribution:** Warm temperate and tropical areas of the Atlantic, Indian and Western Pacific Oceans. Tropical, throughout northern Australia, moving south in

undersurface of head

sixth tooth from symphysis (upper jaw)

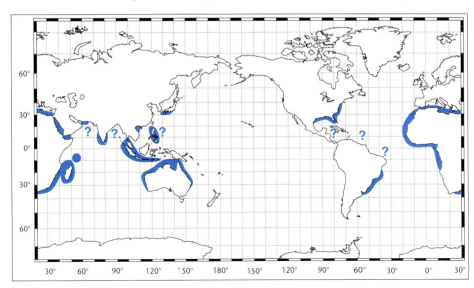

summer to about Geographe Bay (Western Australia) and Jervis Bay (New South Wales). Continental and insular shelves, nearshore to at least 75 m depth.

- **Remarks:** The seasonal migration of spinner sharks into New South Wales waters may be associated with reproduction, as near-term pregnant females and new-born young have been caught around March and April in these waters. Litter sizes vary from 3–15. Size at maturity varies considerably between regions (specimens from Brazil mature as small as 150–170 cm). The spinner shark is an active schooling species which frequently spins out of the water after making feeding runs through schools of small fishes. This behaviour has been responsible for its common name, but such jumps have not been reported in Australian sharks. The diet consists mainly of small fishes, including many pelagic species, and cephalopods. It is a minor component of the northern Australian gillnet and Western Australian shark fisheries, where its meat is used for human consumption; elsewhere used for its meat, hide, fins and liver oil.

- **Local synonymy:** *Uranga nasuta* Whitley, 1943; *Galeolamna fowleri* Whitley, 1944; *Longmania calamaria* Whitley, 1944.

- **References:** Whitley (1943c, 1944a); Garrick (1982); Stevens and McLoughlin (1991).

Nervous Shark — *Carcharhinus cautus* (Whitley, 1945)

29.8 *Carcharhinus cautus* (Plate 32) **Fish Code:** 00 018034

- **Field characters:** A small whaler shark with a bronze to greyish upper surface, narrowly triangular upper teeth that are deeply notched laterally and have coarse basal serrations, no interdorsal ridge, and with black tips to the lower caudal and pectoral-fin undersurfaces (usually also with fine black borders to the anterior margins of the dorsal and pectoral fins, and the upper anterior and entire posterior margins of the caudal fins).

- **Distinctive features:** Body fusiform; interdorsal ridge absent. Snout relatively short and blunt (internarial space 1.1–1.2 in preoral length); nasal lobes relatively long; labial furrows short and inconspicuous. Upper teeth narrowly triangular, irregularly serrated (coarser basally) and oblique; lateral margins deeply notched; medial margins relatively straight; lower teeth more slender, erect to oblique and serrated. Origin of first dorsal fin over, or just anterior to, free rear tips of pectoral fins. Tooth count 25–30/23–28. Total vertebrae 160–171; precaudal 86–91.

- **Colour:** Bronze to greyish dorsally, white ventrally; flanks with a white stripe extending anteriorly from above the pelvic fins; black tips to the lower caudal and pectoral fins (sometimes absent on the pectorals); usually dark along the anterior

undersurface of head

margins of the dorsal, pectoral and upper caudal fins, and along the entire posterior margin of the caudal fin.

- **Size:** Born at 35–40 cm and attains 150 cm. Both sexes mature at 80–85 cm.
- **Distribution:** Known only from the Solomon Islands, Papua New Guinea and tropical Australia, between Carnarvon (Western Australia) and Bundaberg (Queensland). Continental and insular shelves in shallow water.
- **Remarks:** A little-known species which probably has a wider geographical range than the records suggest. Very similar to the blacktip reef shark (29.17), but without conspicuous black tips to the first dorsal and anal fins. Females give birth to 1–5 pups in October or November after a gestation period of 8–9 months. The diet consists primarily of teleost fish, and to a lesser extent crustaceans and cephalopods; some 6% of stomachs examined from northern Australian sharks also contained terrestrial snakes. Small numbers are taken for food by the inshore component of the northern Australian gillnet fishery.
- **Local synonymy:** *Galeolamna greyi cauta* Whitley, 1945.
- **References:** Whitley (1945a); Garrick (1982); Lyle (1987a).

fourth tooth from symphysis (upper jaw)

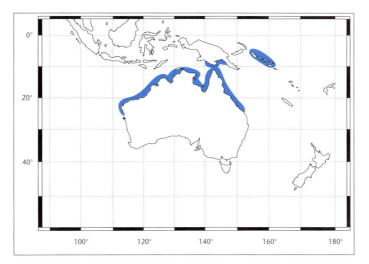

Whitecheek Shark — *Carcharhinus dussumieri* (Valenciennes, 1839)

29.9 *Carcharhinus dussumieri* (Plate 36)   **Fish Code:** 00 018009

- **Alternative name:** Widemouth blackspot shark.
- **Field characters:** A small whaler shark with a bronze upper surface, the first dorsal fin originating more or less over the free rear tips of the pectoral fins, a low

interdorsal ridge, and a conspicuous black tip to the second dorsal fin (all other fins without distinct markings).

- **Distinctive features:** Body fusiform; low interdorsal ridge usually present. Nasal lobes relatively long; labial furrows short, inconspicuous. Upper teeth triangular, serrated, oblique; lateral margins deeply notched, with coarse basal serrations. Lower teeth narrower, erect to oblique, serrated; sometimes noticeably narrower and more oblique in mature males. First dorsal-fin origin over, or somewhat anterior to, pectoral-fin free rear tips. Origin of second dorsal fin over, or usually slightly behind, anal-fin origin. Tooth count 25–28/25–28 [24–31/22–32]. Total vertebrae 130–150 [109–150]; precaudal 64–77 [54–74].

- **Colour:** Dorsal surfaces bronze (greyish or brownish after death or in preservative), paler ventrally. Second dorsal fin with a conspicuous black tip; other fins without markings.

- **Size:** Born at 35–40 cm and attains 90 cm. Both sexes mature at about 70 cm.

- **Distribution:** Tropical Indo–West Pacific, including northern Australia from about Dirk Hartog Island (Western Australia) to Fraser Island (Queensland). Inshore areas of continental and insular shelves down to about 170 m depth; usually near the bottom.

- **Remarks:** One of the most common sharks taken as bycatch by demersal trawlers off northern Australia. This species has often been confused with the very similar *Carcharhinus sealei*, which does not appear to occur in Australia. The usual litter size is 2; females breed every year but there is no reproductive seasonality. The diet consists primarily of fish, and to a lesser extent crustaceans and cephalopods. Caught by Taiwanese and Thai trawlers fishing in Australian waters and retained for its flesh and fins; elsewhere also marketed for its flesh.

- **Local synonymy:** *Carcharias (Prionodon) dussumieri* Valenciennes, *in* Müller & Henle, 1839; *Platypodon coatesi* Whitley, 1939.

- **References:** Whitley (1939b); Garrick (1982); Lyle and Timms (1984); Stevens and McLoughlin (1991).

undersurface of head

fifth tooth from symphysis (upper jaw)

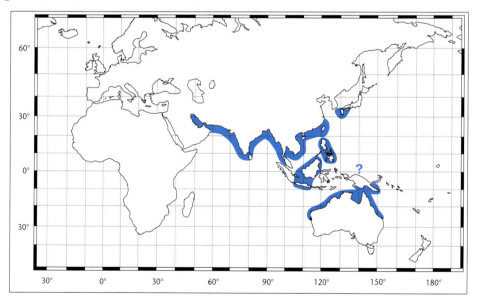

Silky Shark — *Carcharhinus falciformis* (Bibron, 1839)

29.10 *Carcharhinus falciformis* (Plate 35) **Fish Code:** 00 018008

- **Field characters:** A large, uniformly dark brownish to greyish whaler shark with the first dorsal fin originating behind the pectoral-fin free rear tips, an interdorsal ridge, and with very long inner margins and free rear tips to the second dorsal and anal fins.

- **Distinctive features:** Body fusiform, moderately slender; interdorsal ridge present. Labial furrows small and inconspicuous. Upper teeth erect to slightly oblique, serrated, triangular and with a notch about half way down the tooth on each side; lower teeth slender, erect, smooth-edged. First dorsal-fin origin behind pectoral-fin free rear tips. Second dorsal and anal fins with very long inner margins and rear tips (second dorsal-fin inner margin 1.6–3.0 times its height). Tooth count 34/33–34* [29–35/27–37]. Total vertebrae 210–212* [199–215]; precaudal 103–107* [98–106].

- **Colour:** Dorsal surfaces uniform dark brown to dark grey (grey in preservative); white ventrally. First dorsal fin plain; other fins (especially pectoral, second dorsal and anal fins) sometimes with dusky tips.

- **Size:** Born at 70–85 cm and attains 330 cm. Both sexes mature at 200–210 cm.

- **Distribution:** Circumtropical, may make seasonal incursions into warm temperate areas. Locally, off northern Australia (rare in the Gulf of Carpentaria), south to Sydney (New South Wales) and Lancelin (Western Australia). Oceanic and pelagic, but most abundant in offshore areas close to land masses and along the edges of continental and insular shelves. Occurs from the surface to a depth of at least 500 m.

- **Remarks:** The silky shark is found in water temperatures above 23°C. It breeds throughout the year producing litters of 2–15 pups after an unknown gestation

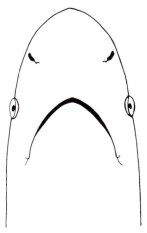

undersurface of head

fourth tooth from symphysis (upper jaw)

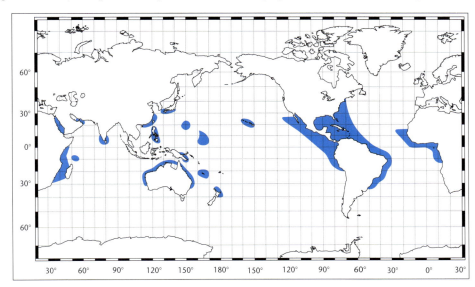

period. Its diet consists mainly of teleost fishes, but also includes cephalopods and pelagic crabs; it is often associated with schools of tuna. Taken as a bycatch of Japanese tuna longliners in northern Australian waters and retained for its fins. Not eaten in Australia; elsewhere caught for its meat, hide, fins and liver.

- **Local synonymy:** *Carcharias (Prionodon) falciformis* Bibron, *in* Müller & Henle, 1839.
- **References:** Stevens (1984); Stevens and McLoughlin (1991).

Creek Whaler — *Carcharhinus fitzroyensis* (Whitley, 1943)

29.11 *Carcharhinus fitzroyensis* (Plate 36) **Fish Code:** 00 018035

- **Field characters:** A small, bronze to greyish brown whaler shark with large, triangular pectoral fins, the first dorsal fin originating more or less over the pectoral-fin free rear tips, no interdorsal ridge, and no conspicuous fin markings.
- **Distinctive features:** Body fusiform; interdorsal ridge absent. Labial furrows small. Upper teeth relatively long, narrowly triangular, erect to oblique, serrated (coarser basally); lower teeth more slender, erect, serrated. First dorsal-fin origin over, or usually somewhat ahead of pectoral-fin free rear tips. Pectoral fins large and broad, maximum width 1.2–1.5 in anterior margin. Tooth count 30/28–30*. Total vertebrae 125*; precaudal 58*.
- **Colour:** Bronze dorsally, fading to greyish brown after death and in preservative; pale ventrally and with no distinctive fin markings.

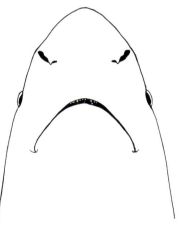

undersurface of head

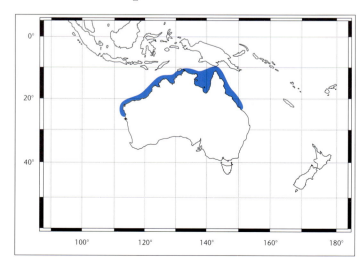

fifth tooth from symphysis (upper jaw)

- **Size:** Born at about 50 cm and attains 135 cm. Males and females mature at about 80 cm and about 90 cm respectively.
- **Distribution:** Currently known only off northern Australia from Gladstone (Queensland) to Cape Cuvier (Western Australia). Occurs mainly in inshore areas of the continental shelf from the intertidal zone to a depth of at least 40 m.
- **Remarks:** The creek whaler has an extended, seasonal reproductive cycle with females giving birth to 1–7 pups every year between February and May. The diet consists principally of teleost fishes, and to a lesser extent crustaceans. Small numbers are taken by the inshore component of the northern Australian gillnet fishery and utilised for their meat.
- **Local synonymy:** *Galeolamna (Uranganops) fitzroyensis* Whitley, 1943.
- **References:** Whitley (1943c); Garrick (1982); Lyle (1987a).

Galapagos Shark — *Carcharhinus galapagensis* (Snodgrass & Heller, 1905)

29.12 *Carcharhinus galapagensis* (Plate 35)　　　　　　　　**Fish Code:** 00 018040

- **Field characters:** A large, greyish whaler shark with a low interdorsal ridge, indistinct dusky tips to most fins (more distinct in juveniles), the first dorsal-fin origin over the posterior third of the pectoral-fin inner margin, and relatively long, broadly triangular, serrated upper teeth.
- **Distinctive features:** Body fusiform; interdorsal ridge low. Labial furrows short and inconspicuous. Upper teeth with relatively long, broad, triangular, serrated cusps; cusps erect to oblique, lateral margins concave, mesial margins somewhat sinuous. Lower teeth more slender, erect, serrated. First dorsal-fin height 9.1–12.1% of total length; origin over posterior third of pectoral-fin inner margin. Second dorsal-fin height 2.1–3.1% of total length. Tooth count 29/29* [27–33/27–33]. Total vertebrae 210* [200–215]; precaudal 106–108* [103–109].
- **Colour:** Dorsal surfaces brownish grey to dark grey; pale ventrally, with an indistinct pale stripe extending anteriorly on the flanks from the pelvic region. Most fins with dusky tips.
- **Size:** Born at 60–80 cm and attains 300 cm. Males mature at 210–230 cm and females at about 250 cm.

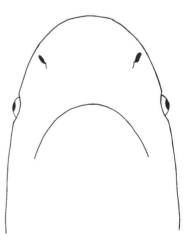

undersurface of head

- **Distribution:** Cosmopolitan in tropical and temperate seas, but generally associated with oceanic islands. In Australian waters, known only from Lord Howe Island and possibly Middleton and Elizabeth Reefs (off Queensland). Inshore to well offshore on insular and continental shelves from nearshore to the open ocean (adjacent islands); pelagic from the surface to at least 180 m.
- **Remarks:** Although habitat-limited, this shark can be very abundant in areas where it occurs (e.g. the Galapagos Islands and Clipperton Island in the eastern Pacific). It apparently prefers clear water and, although it will readily come to the surface, is often found swimming in aggregations near the bottom. Litter sizes range from 6–16 and the young are often found in shallower water (25 m or less) than the adults. A dangerous species and aggressive towards divers, it has been reported to make threat displays similar to those of the grey reef shark (29.4).

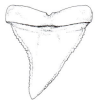

fourth tooth from symphysis (upper jaw)

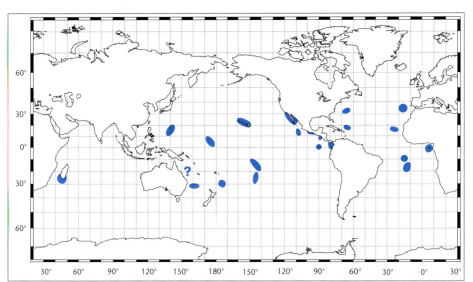

Bull Shark — *Carcharhinus leucas* (Valenciennes, 1839)

29.13 *Carcharhinus leucas* (Plate 34) **Fish Code:** 00 018021

- **Alternative names:** River whaler, freshwater whaler, Swan River whaler.
- **Field characters:** A large, stout-bodied whaler shark with a short, blunt snout, a rather large second dorsal fin (mostly more than a third of first dorsal-fin height), no interdorsal ridge or distinctive fin markings, and with broadly triangular, serrated upper teeth.

- **Distinctive features:** Body fusiform, stout; interdorsal ridge absent. Snout short and blunt (internarial space 0.8–1.0 in preoral length); labial furrows short and inconspicuous; eye small, 0.6–2.0% of total length. Upper teeth erect to slightly oblique, broadly triangular, serrated; lower teeth similar but more slender. First dorsal-fin origin over, or just behind, pectoral-fin insertion (rarely nearer their free rear tips). First dorsal-fin height up to 3.1 times second dorsal-fin height. Tooth count 25–27/25 [25–29/25–28 (usually 12 teeth on each side of the lower jaw)]. Total vertebrae 217–219* [198–227]; precaudal 113–120* [101–123].

- **Colour:** Dorsal surfaces grey; ventral surfaces pale; an indistinct pale stripe on each flank. Juveniles with dusky to black fin tips (particularly the caudal, pectoral, second dorsal and anal fins); upper caudal fin with a thin dusky posterior margin. Adults with indistinct fin markings.

- **Size:** Born at 55–80 cm and attains about 340 cm. There are no data on size at maturity in Australia. Elsewhere, different populations show considerable variation in maturation size, with males maturing between 160–225 cm and females between 180–230 cm.

- **Distribution:** Cosmopolitan in tropical and warm temperate seas, between Sydney (New South Wales) and Perth (Western Australia). A coastal, estuarine, riverine and lacustrine shark which, in the marine environment, occurs near the bottom from the surf zone down to a depth of at least 150 m.

- **Remarks:** The bull shark is the only widely distributed shark that penetrates far into fresh water for extended periods, where it occasionally even breeds (as in Lake Nicaragua). It has been found nearly 4000 km from the sea in the Amazon system. It can tolerate hypersaline conditions (up to 53 parts per thousand in the St Lucia lake system in South Africa) and is often found in turbid waters. It has been reported from numerous freshwater systems in warm temperate and tropical Australia, including the Adelaide, Daly, East Alligator, Herbert, Brisbane, Clarence and Swan Rivers, and Lake Macquarie. Whaler sharks reported from the Ord, Victoria and Pentecost Rivers are also likely to be this species. It has rarely been recorded in the sea off Australia, perhaps in part because of confusion with other whaler sharks. Females normally give birth in estuaries and river mouths; litter sizes range from 1–13 and the gestation period is 10–11 months. Ageing studies in South Africa indicate that sexual maturity is reached at about 6 years (at 250 cm) and that it lives for about 14 years. This shark lives well in captivity (one specimen was kept for over 15 years). It is markedly omnivorous (eating turtles, birds, dolphins, terrestrial mammals, crustaceans, molluscs and echinoderms) but favours bony fishes and elasmobranchs. It is a very dangerous shark, perhaps more so than the tiger (29.22) or even the white shark (25.1), because of its extremely aggressive nature, powerful jaws, broad diet, abundance, and its preference for shallow inshore habitats.

undersurface of head

third tooth from symphysis (upper jaw)

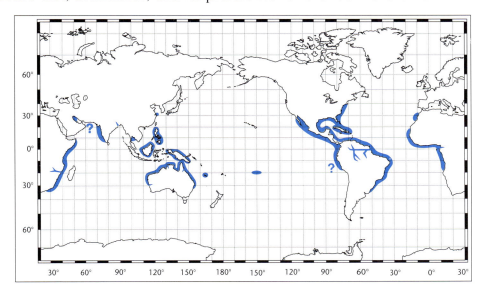

Probably responsible for many of the attacks in and around the Sydney Harbour area. Not eaten in Australia; elsewhere used for its meat, hide, fins, liver oil, and as fishmeal.

- **Local synonymy:** *Carcharias (Prionodon) leucas* Valenciennes, *in* Müller & Henle, 1839; *Carcharias brachyurus*: Waite, 1906 (misidentification); *Carcharias spenceri* Ogilby, 1910; *Galeolamna stevensi*: Whitley, 1940 (misidentification); *Galeolamna (Lamnarius) spenceri*: Whitley, 1943 (misidentification); *Galeolamna (Bogimba) bogimba* Whitley, 1943; *Galeolamna greyi mckaili* Whitley, 1945; *Galeolamna mckaili*: Whitley, 1951.
- **References:** Ogilby (1910a); Whitley (1943c, 1945a, 1951b); Garrick (1982).

Common Blacktip Shark — *Carcharhinus limbatus* (Valenciennes, 1839)

29.14 *Carcharhinus limbatus* (Plate 33) **Fish Code:** 00 018039

- **Alternative name:** Blacktip whaler.
- **Field characters:** A medium-sized whaler shark with a rather long snout, a bronze upper surface, no interdorsal ridge, black tips to most fins (some adults may lack distinctive fin markings), and slender, erect, serrated upper teeth.
- **Distinctive features:** Body fusiform; interdorsal ridge absent. Snout moderately long, internarial space 1.2–1.4 in preoral length; labial furrows small and inconspicuous. Upper teeth with erect to slightly oblique, slender, finely serrated cusps (somewhat coarser serrations basally); lower teeth erect, more slender, either with fine serrations or smooth-edged. First dorsal-fin origin usually over or just behind pectoral-fin insertion, exceptionally just in front of pectoral-fin free rear tip. Tooth counts 32–35/30–31 [30–35/27–34]. Total vertebrae 191–202 [182–203]; precaudal 94–101 [94–102].
- **Colour:** Dorsal surfaces bronze, fading to grey after death or in preservative. Ventral surfaces pale. Flanks with a pale stripe extending from the pelvic fins to below the first dorsal fin. Fin tips (except anal) usually black in juveniles; sometimes pelvic fin plain and anal fin black-tipped; distinctive fin markings lacking in some adults.
- **Size:** Born at 40–70 cm and attains 250 cm. Size at maturity shows considerable geographic variation with males maturing between 135–180 cm and females between 120–190 cm.
- **Distribution:** Cosmopolitan in tropical and warm temperate areas. Tropical Australia south to Sydney on the east coast; its southern limit on the west coast is uncertain. It is essentially pelagic over continental and insular shelves; it is commonly found close inshore, but is caught occasionally far offshore.

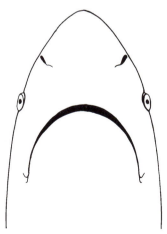

undersurface of head

third tooth from symphysis (upper jaw)

- **Remarks:** This species is very similar to the Australian black-tip shark (29.21) (see remarks for that species). An active, fast-swimming shark that occasionally leaps out of the water, apparently while feeding on small fish. It has been reported to occur in large aggregations, although it is not very abundant in Australian waters. A minor component in the catch of the northern Australian gillnet fishery; elsewhere used for its meat, hide and liver oil.
- **Local synonymy:** *Carcharias (Prionodon) limbatus* Valenciennes, *in* Müller & Henle, 1839.
- **References:** Stevens (1984); Stevens and Wiley (1986).

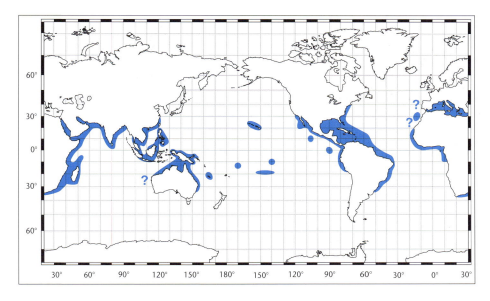

Oceanic Whitetip Shark — *Carcharhinus longimanus* (Poey, 1861)

29.15 *Carcharhinus longimanus* (Plate 31) **Fish Code:** 00 018032

- **Alternative name:** Whitetip whaler.
- **Field characters:** A large whaler shark with a greatly enlarged first dorsal fin (tapering only slightly towards the apex) and pectoral fins (each with broadly rounded tips), an interdorsal ridge, broadly triangular and serrated upper teeth, and mottled white tips to the first dorsal, pectoral and caudal fins (usually absent in specimens below 130 cm).

- **Distinctive features:** Body fusiform; low interdorsal ridge usually present. Labial furrows short, confined to mouth corners. Upper teeth serrated, broadly triangular and erect; lowers more slender, erect and serrated. First dorsal-fin origin just anterior to pectoral-fin free rear tips. Pectoral and first dorsal fins very large with broadly rounded tips; pectoral-fin anterior margin 20–31% of total length, maximum width 1.9–2.3 in anterior margin; first dorsal-fin height 9–17% of total length. Tooth count 30–31/27–29* [27–32/27–33]. Total vertebrae 230–232* [228–244]; precaudal 122–128* [123–131].

- **Colour:** Dorsal surfaces bronzy grey, ventral surfaces paler. New-born young and small juveniles with black tips to most fins (particularly the pelvic, second dorsal, anal and lower caudal fins), usually black saddle markings on the dorsal surface at the origin of the second dorsal and caudal fins (sometimes also at the second dorsal-fin insertion), and usually a black patch on the caudal fin above the subterminal notch. Black markings fade in specimens above 130 cm, which develop mottled white tips to the first dorsal, pectoral, pelvic and caudal fins.

- **Size:** Born at 60–65 cm and usually attains about 300 cm (although there is a record of a 350 cm specimen). Males mature at 175–195 cm and females at 180–200.

- **Distribution:** Cosmopolitan in tropical and warm temperate seas. Northern Australia (except for the Torres Strait, Gulf of Carpentaria and Arafura Sea) south to southern New South Wales. The southern limit off Western Australia is uncertain, but it would be expected at least as far south as Perth. One specimen was recorded south-west of Port Lincoln, South Australia. Oceanic and pelagic from the surface to a depth of at least 150 m. May occur close inshore where there is a narrow continental shelf.

- **Remarks:** Occupies a similar habitat to the blue shark (29.26), but prefers water above 20°C. Produces litters of 1–15 pups after about a 12 month gestation. Oceanic whitetips feed mainly on a variety of bony fishes and cephalopods. The mottled white fins may mimic schools of baitfish and attract such prey as scombrids (tunas and mackerels). This is one of the four most dangerous species to humans, and is probably responsible for many open-ocean attacks after air or sea disasters. Taken as a bycatch of Japanese tuna longliners in northern Australian waters and retained for its fins. Not used in Australia, but elsewhere caught for its meat, fins, hide and liver.

- **References:** Glover (1974); Stevens (1984).

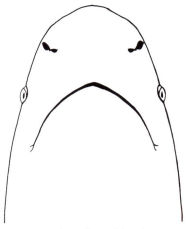

undersurface of head

fifth tooth from symphysis (upper jaw)

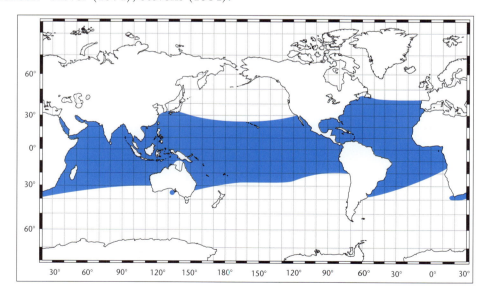

Hardnose Shark — *Carcharhinus macloti* (Müller & Henle, 1839)

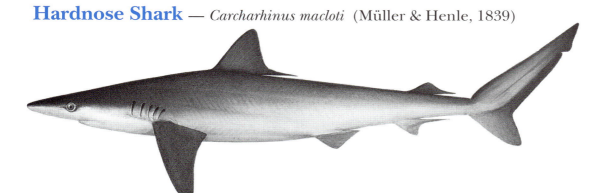

29.16 *Carcharhinus macloti* (Plate 36) Fish Code: 00 018025

- **Field characters:** A small whaler shark with a relatively long, narrow and noticeably stiff snout, the anal-fin origin forward of the second dorsal-fin origin, a bronze dorsal coloration, no distinct fin markings, and erect to oblique upper teeth with a narrow, smooth-edged central cusp and large basal serrations.

- **Distinctive features:** Body fusiform, slender; low interdorsal ridge sometimes present. Snout very firm, relatively long and narrow; preoral snout 8.1–10.4% of total length; internarial space 1.6–1.9 in preoral length; labial furrows short and inconspicuous; hyomandibular pores enlarged (forming discrete series extending anteriorly from outside mouth corners). Upper teeth narrow, erect to oblique, smooth-edged distally but with large basal serrations; lower teeth more slender, erect and smooth-edged. First dorsal-fin origin over, or somewhat anterior to, pectoral-fin free rear tips. Second dorsal-fin origin over middle of anal-fin base. Tooth count 29–31/26–27* [30–32/27–29]. Total vertebrae 135–141 [149–156]; precaudal 59–64 [68–71].

- **Colour:** Bronze dorsally fading to grey-brown after death and in preservative; ventral surfaces pale with an indistinct pale stripe extending anteriorly on the flanks from above the pelvic region. Distinct fin markings absent; second dorsal-fin anterior margin and both margins of the upper caudal fin sometimes dark-edged; posterior margins of some fins, in particular the pectoral and lower caudal fins, sometimes pale-edged.

- **Size:** Born at about 45 cm and attains 110 cm. Both sexes mature at 70–75 cm.

- **Distribution:** Tropical Indo–West Pacific; northern Australian waters from Bundaberg (Queensland) to Carnarvon (Western Australia). Continental and insular shelves from close inshore down to a depth of 170 m.

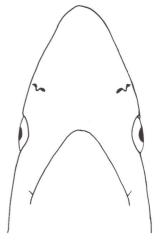

undersurface of head

fourth tooth from symphysis (upper jaw)

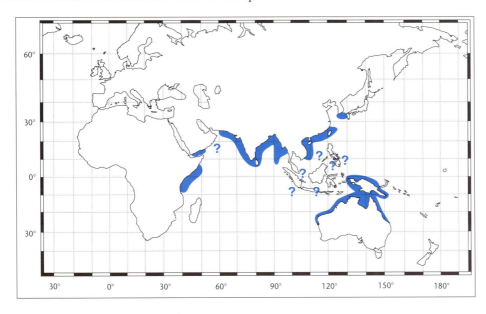

- **Remarks:** The hardnose shark often occurs in large aggregations off northern Australia where, together with the Australian blacktip shark (29.21) and the spot-tail shark (29.20), it is one of the most abundant sharks taken in gill-nets. It usually gives birth to two pups in July, after a gestation period of about 12 months. The diet consists mainly of fish, and to a lesser extent cephalopods and crustaceans. Although abundant in gillnet catches off northern Australia, it is of limited commercial value because of its small size; elsewhere its meat is used for human consumption.
- **References:** Lyle and Timms (1984); Stevens and Church (1984); Stevens and McLoughlin (1991).

Blacktip Reef Shark — *Carcharhinus melanopterus* (Quoy & Gaimard, 1824)

29.17 Carcharhinus melanopterus (Plate 32) Fish Code: 00 018036

- **Alternative names:** Blacktip shark, guliman.
- **Field characters:** A small whaler shark with a short and bluntly rounded snout, a yellowish brown to greyish upper surface, no interdorsal ridge, and with very distinct black tips to the first dorsal and lower caudal fins (as well as relatively smaller black tips to all other fins).
- **Distinctive features:** Body fusiform; interdorsal ridge absent. Snout short and bluntly rounded; preoral length 4.5–7.1% of total length; internarial space 0.8–1.1% in preoral length; nasal lobes relatively long; labial furrows short and inconspicuous. Upper teeth narrowly triangular, serrated (coarser basally), erect to oblique; lateral margins notched; medial margins relatively straight. Lower teeth more slender, erect and serrated. First dorsal-fin origin over free rear tips of pectoral fins. Tooth count 26–28/25* [23–28/21–27]. Total vertebrae 193–203* [196–214]; precaudal 113–114* [111–122].
- **Colour:** Dorsal surfaces varying from yellowish brown to grey; ventral surfaces white. A distinct pale stripe on each flank extends from above the pelvic area to level with the first dorsal-fin origin. First dorsal and lower caudal-fin tips distinctly black; all other fins with smaller black tips (second dorsal fin sometimes plain); upper caudal fin with black margins.
- **Size:** Locally, usually born at about 50 cm and can grow to 140 cm; size at birth in the Marshall Islands is smaller (about 35 cm) and adults may attain 180 cm. Both sexes mature between 95–110 cm.

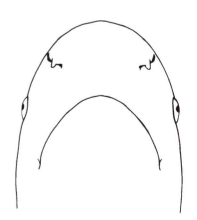

undersurface of head

fifth tooth from symphysis (upper jaw)

- **Distribution:** Tropical Indo–West and Central Pacific (including the eastern Mediterranean as an invader through the Suez Canal). Northern Australian waters from Moreton Bay (Queensland) to Shark Bay (Western Australia). Shallow areas of continental and insular shelves; usually in water only a few metres deep. Occasionally penetrates into brackish areas.

- **Remarks:** This whaler, along with the grey reef (29.4) and whitetip reef (29.30) sharks, are the most common sharks found on coral reefs in the Indo–Pacific region. The blacktip reef shark prefers water only a few metres deep and is often seen swimming on reef and sand flats with its dorsal and upper caudal fin exposed. It is also common in mangrove areas, moving in and out with the tide. Tagging studies at Aldabra atoll, Indian Ocean, showed that they move within an area of only a few square kilometres, with some individuals recaptured up to 7 times. Growth was slow (less than 4 cm per year even in juveniles). Reproduction is seasonal in northern Australia, where females give birth to 3–4 pups each year in November after an 8–9 month gestation period. At Aldabra, where the population density is high and food may be scarcer, females only breed every other year. The diet consists principally of fish, but also includes crustaceans, cephalopods, other molluscs and miscellaneous items. In a dietary study in northern Australia, almost a quarter of stomachs examined contained terrestrial snakes. The blacktip reef shark is not regarded as dangerous because of its small size. It can, however, be aggressive to spear fishermen and has been known to bite people's feet and legs while they were wading in shallow water. Rarely taken by the northern Australian gillnet fishery because of its shallow-water habitat. Eaten by the Australian Aborigines (mainly during the wet season) as buunhdhaarr, in which the liver and flesh are boiled separately and then minced and mixed together. Elsewhere used fresh and dry-salted for human consumption, and for its liver-oil.

- **Local synonymy:** *Mapolamia spallanzani*: Whitley, 1934 (misidentification).

- **References:** Whitley (1934a); Grant (1978); Lyle (1987a).

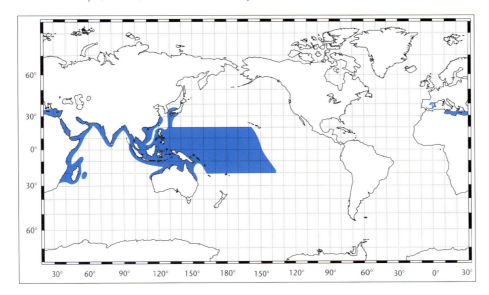

Dusky Shark — *Carcharhinus obscurus* (Lesueur, 1818)

29.18 *Carcharhinus obscurus* (Plate 35) **Fish Code:** 00 018003

- **Alternative names:** Black whaler, bronze whaler.
- **Field characters:** A large, greyish whaler shark with a low interdorsal ridge, the first dorsal-fin origin over or near the pectoral-fin free rear tips, indistinct dusky tips to most fins (more distinct in juveniles), and broadly triangular, serrated upper teeth.
- **Distinctive features:** Body fusiform; interdorsal ridge low. Labial furrows short and inconspicuous. Upper teeth with broad, triangular, serrated cusps; cusps erect to oblique, lateral margins concave, medial margins somewhat sinuous. Lower teeth more slender, erect, serrated. First dorsal-fin origin varying from slightly in front to slightly behind the pectoral-fin free rear tips. Tooth count 29–32/29–31 [29–33/27–32]. Total vertebrae 184–188* [173–194]; precaudal 89–92 [86–97].
- **Colour:** Dorsal surfaces bronzy grey to dark grey, pale ventrally. An indistinct pale stripe extends anteriorly along the flanks from the pelvic region. Fin tips (particularly lower caudal-fin lobe and ventral surfaces of pectoral fins) mostly dusky in juveniles; fin markings less distinct in adults.
- **Size:** Born at 70–100 cm (mostly at about 95 cm in Australia) and attains 365 cm. Both sexes mature at about 280 cm.
- **Distribution:** Cosmopolitan (but patchy) in tropical and warm temperate seas. Occurs throughout Australian waters (rare off southern Tasmania). Continental and insular shelves from the surf zone to adjacent oceanic waters and from the surface down to a depth of 400 m.
- **Remarks:** Tagging studies have shown that this species makes distinct seasonal migrations over parts of its range; little is known of its movements in Australian

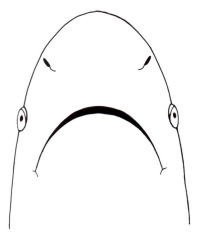

undersurface of head

fourth tooth from symphysis (upper jaw)

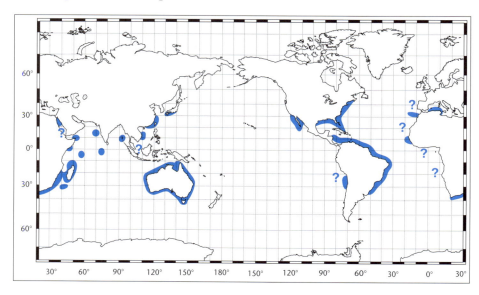

waters other than that adolescents and adults appear to move inshore into shallower water (less than 80 m) off Western Australia during summer and autumn. Litter sizes vary from 3–14 and the newborn young may occupy distinct nursery areas away from the rest of the population (e.g. in inshore areas off Western Australia). The diet consists principally of teleost and elasmobranch fishes, as well as crustaceans and cephalopods. Dusky sharks feed throughout the water column, but are more frequently found near the bottom than on the surface. The juveniles are preyed upon by other sharks and, off South Africa, the reduction in the numbers of the larger species of sharks inshore through beach protection netting operations may have been responsible for an increase in the numbers of young dusky sharks. No such effect has yet been reported in Australia, which has a similar meshing program. Large adults are potentially dangerous, although few attacks on humans can be directly attributed to this species. Dusky sharks are an important component of the Western Australian shark fishery (total catch 1600 tonnes in 1989/90 with a value of $6 million) comprising about a third of the catch (about 530 tonnes a year). The dusky shark catch is based essentially on newly-born fish in the 100 cm length class from the nursery areas. Sold fresh and frozen for human consumption; elsewhere used for its meat, hide, fins and liver-oil.

- **Local synonymy:** *Galeolamna macrurus* Ramsay & Ogilby, 1887; *Galeolamna (Galeolamnoides) eblis* Whitley, 1944.
- **References:** Whitley (1944a); Stevens (1984); Heald (1987).

Sandbar Shark — *Carcharhinus plumbeus* (Nardo, 1827)

29.19 *Carcharhinus plumbeus* (Plate 34)   **Fish Code:** 00 018007

- **Alternative name:** Thickskin shark.
- **Field characters:** A medium-sized whaler shark with a very tall first dorsal fin originating over or anterior to the pectoral-fin insertions, an interdorsal ridge, pale bronze to greyish upper surface, no distinctive fin markings, and with broadly triangular, serrated upper teeth.
- **Distinctive features:** Body fusiform; interdorsal ridge present. Labial furrows short and inconspicuous. Upper teeth broadly triangular, serrated, erect to oblique; posterior lateral margin concave; medial margins more or less straight. Lower teeth erect, more slender and serrated. First dorsal fin high (8.4–16.7% of total length),

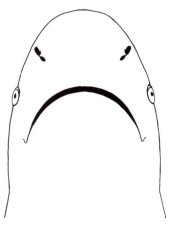

undersurface of head

particularly in adults; origin over, or slightly in front of, pectoral-fin insertions. Tooth count 29/27* [27–32/25–32]. Total vertebrae 176–177* [152–189]; precaudal 89–92* [82–97].

- **Colour:** Dorsal surfaces pale bronze to greyish brown; ventral surfaces pale. An inconspicuous pale stripe extends anteriorly along the flanks from above the pelvic area. Dorsal and upper caudal-fin margins sometimes with dusky edges; pectoral, pelvic, caudal and sometimes anal fins with pale tips and posterior margins (more pronounced in juveniles).

- **Size:** Born at 55–65 cm (up to 75 cm in some localities) and attains 240 cm. Size at maturity regionally variable; in Australia, both sexes mature at about 155 cm, elsewhere males mature between 130–180 cm and females between 145–185 cm.

- **Distribution:** Cosmopolitan (but patchy) in tropical and warm temperate areas. Locally off northern Australia, but extends south to Coffs Harbour (New South Wales) and Esperance (Western Australia). Continental and insular shelves and adjacent deep water from the intertidal to a depth of 280 m. Normally found near the bottom.

- **Remarks:** Tagging studies have shown that sandbar sharks in the western North Atlantic make extensive seasonal migrations of over 2700 km, and similar movements are known to occur off South Africa. Nothing is known about movements of this species in Australian waters. It is normally slow growing (western Atlantic specimens take 13 years to reach maturity and live for over 30 years). In northern Australia, females have 3–8 pups every other year in about February, after a gestation period of about 12 months. Juvenile sandbar sharks in the western Atlantic live in shallow-water nursery areas separate from the adult population. The diet consists principally of fish, as well as cephalopods and crustaceans. This species is not reported to be dangerous to humans. Once taken for its meat and fins by a Taiwanese longline fishery in northern Australia. A minor component of the Western Australian shark fishery, its meat being used for human consumption; elsewhere important for its meat, fins, hide and liver oil.

- **Local synonymy:** *Carcharias stevensi* Ogilby, 1911; *Galeolamna dorsalis* Whitley, 1944.

- **References:** Whitley (1944a); Garrick (1982); Stevens and McLoughlin (1991).

third tooth from symphysis (upper jaw)

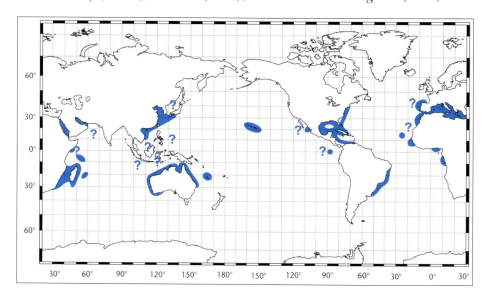

Spot-tail Shark — *Carcharhinus sorrah* (Valenciennes, 1839)

29.20 *Carcharhinus sorrah* (Plate 32) Fish Code: 00 018013

- **Alternative names:** School shark, sorrah shark.
- **Field characters:** A medium-sized whaler shark with a bronze to brownish grey upper surface, an interdorsal ridge, the first dorsal-fin origin more or less over the pectoral-fin free rear tips, conspicuous black tips to the pectoral, second dorsal and lower caudal fins, and with oblique, triangular, serrated upper teeth.
- **Distinctive features:** Body fusiform; interdorsal ridge present. Labial furrows small and inconspicuous. Upper teeth triangular, oblique, serrated (coarser basally); posterior lateral margin notched. Lower teeth similar to uppers, but narrower. First dorsal-fin origin varying from just anterior to just behind pectoral-fin free rear tips. Second dorsal fin with a long inner margin (2.0–2.6 times second dorsal-fin height). Tooth count 25/23–25 [23–30/23–30]. Total vertebrae 153–158 [153–175]; precaudal 69–72 [66–79].
- **Colour:** Dorsal surfaces bronze in life, fading to brownish grey or grey after death and in preservative. Ventral surfaces pale white. A pale stripe on the flanks extends anteriorly from above the pelvic fins. Pectoral, second dorsal and lower caudal fins (and often free rear tip of second dorsal fin) with distinct black tips. First dorsal and upper caudal fins with dusky margins.
- **Size:** Born at 50 cm and attains 160 cm. Both sexes mature at 90–95 cm.
- **Distribution:** Tropical Indo–West Pacific; northern Australian waters from Gladstone (Queensland) to Point Quobba (Western Australia). Found over shallow continental and insular shelves from the intertidal down to a depth of at least 80 m. Occurs throughout the water column, but mainly in midwater or near the surface.

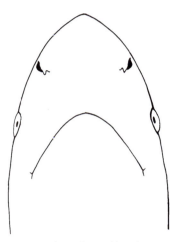

undersurface of head

fifth tooth from symphysis (upper jaw)

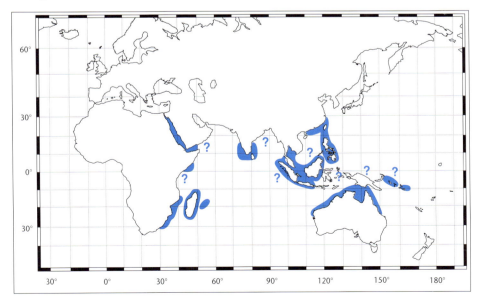

- **Remarks:** The spot-tail shark is common in open areas over muddy bottoms, but also occurs near coral reefs. Reproduction is seasonal, with 1–8 pups produced in January after a gestation period of 10 months. Growth is fairly rapid (about 25 cm in the first year, with sexual maturity attained in 2–3 years). The diet consists mainly of teleost fishes and, to a lesser extent, cephalopods and crustaceans. Tagging and genetic studies have shown that there is only one stock off northern Australia. About half of the tags recovered were taken within 50 km of the tagging site, but one shark was caught 1116 km away. This was the second most abundant shark species, after the Australian blacktip (29.21), taken by a Taiwanese gillnet fishery that operated from 1974–1986 off northern Australia. Until 1991, it was also caught there by Taiwanese longliners, who value the species for its meat, and to a lesser extent, its fins. The spot-tail shark and the Australian blacktip, form the basis of a small Australian gillnet fishery (up to 500 tonnes annually), which markets the flesh mainly in south-eastern Australia as 'flake'.
- **Local synonymy:** *Carcharias (Prionodon) sorrah* Valenciennes, *in* Müller & Henle, 1839; *Galeolamna (Galeolamnoides) isobel* Whitley, 1947.
- **References:** Stevens and Wiley (1986); Lyle (1987b); Lavery and Shaklee (1989).

Australian Blacktip Shark — *Carcharhinus tilstoni* (Whitley, 1950)

29.21 Carcharhinus tilstoni (Plate 33) **Fish Code:** 00 018014

- **Alternative name:** Blacktip whaler.
- **Field characters:** A medium-sized, long-snouted whaler shark with a bronzy to greyish dorsal coloration, the first dorsal-fin origin more or less over the pectoral-fin insertions, no interdorsal ridge, black tips to most fins, and slender, erect, serrated upper teeth.
- **Distinctive features:** Body fusiform; interdorsal ridge absent. Snout moderately long, internarial space 1.1–1.6 in preoral length; labial furrows small and inconspicuous. Upper teeth erect to slightly oblique, slender and finely serrated (somewhat coarser basally). Lower teeth erect, more slender, finely serrated or

smooth-edged. First dorsal-fin origin usually over or just behind pectoral-fin insertions, exceptionally just in front of pectoral-fin free rear tips. Tooth count 32–35/29–31. Total vertebrae 174–182; precaudal 84–91.

- **Colour:** Dorsal surfaces bronze, fading to grey after death and in preservative. Ventral surfaces pale. A pale stripe extends along each flank from the pelvic fin to below the first dorsal fin. All fins (except sometimes the pelvic and anal fins) black-tipped.
- **Size:** Born at 60 cm and attains 200 cm. Males and females mature at about 110 cm and about 115 cm respectively.
- **Distribution:** Currently known only from the continental shelf of tropical Australia. Occurs from close inshore to a depth of about 150 m. Found throughout the water column but mainly in midwater or near the surface.
- **Remarks:** This species is very similar to, and has only recently been separated from, the common blacktip shark (29.14). These two species can currently be reliably distinguished only on enzyme systems and vertebral counts. The Australian blacktip shark often occurs in large aggregations. It has a seasonal reproductive cycle producing, 1–6 pups in January after a 10 month gestation period. Ageing studies show that it grows relatively quickly (about 20 cm in the first year), and attains sexual maturity in 3–4 years. The diet consists of teleost fishes, and to a lesser extent, cephalopods. Tagging and genetic studies have shown that there is only one stock off northern Australia. About 60% of the tag recoveries were made within 50 km of the tagging site and the farthest recapture site was 1113 km away. This was the principal shark species taken by a Taiwanese gillnet fishery that operated from 1974–1986 off northern Australia. The species was caught for its meat, and to a lesser extent, its fins. Until 1991, it was taken by Taiwanese longliners in northern Australia and utilised in the same way. This blacktip also forms the basis of a small Australian gillnet fishery (up to 500 tonnes annually), which markets the flesh mainly in southeastern Australia. The flesh has a relatively high mercury concentration.
- **Local synonymy:** *Carcharhinus limbatus*: Garrick, 1982 (misidentification).
- **References:** Whitley (1950b); Stevens and Church (1984); Stevens and Wiley (1986); Lyle (1986); Davenport and Stevens (1988).

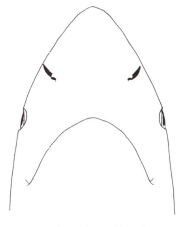

undersurface of head

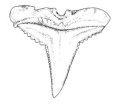

fifth tooth from symphysis (upper jaw)

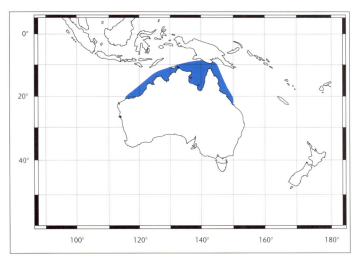

Tiger Shark — *Galeocerdo cuvier* (Péron & Lesueur, 1822)

29.22 *Galeocerdo cuvier* (Plate 29) **Fish Code:** 00 018022

- **Field characters:** A very large whaler shark with heavily serrated, cockscomb-shaped teeth, a broad and blunt head, long upper labial furrows, and a colour pattern of dark, vertical, flank bars (usually absent in specimens longer than 3 m).

- **Distinctive features:** Body fusiform, stout forward of the first dorsal fin; interdorsal ridge present; caudal peduncle relatively narrow; low lateral keels present. Head large and broad; snout short and blunt (internarial space 0.7–1.3 in preoral length); upper labial furrows long, extending level with front of eyes; mouth large; spiracles present. Teeth with single cusps and basal cusplets, similar in both jaws; curved, cockscomb-shaped and heavily serrated. Upper caudal-fin lobe tapering; tip thin and pointed. Tooth count 21–25/22–25 [18–26/18–26]. Total vertebrae 222–226* [216–233]; precaudal 102–109* [100–112].

- **Colour:** Dorsal surface grey with bold, dark reticulations in newly-born young; reticulations becoming vertical bars in specimens up to 300 cm; bars faint or missing in larger adults. Ventral surfaces white.

- **Size:** Born at 50–75 cm and attains approximately 600 cm. Males mature at about 300 cm and females at about 330 cm.

- **Distribution:** Cosmopolitan in tropical seas, with seasonal excursions into warm temperate areas. Off northern Australia, south to southern New South Wales and Perth (Western Australia). Close inshore to well off the continental shelf; from the surface to a depth of about 150 m.

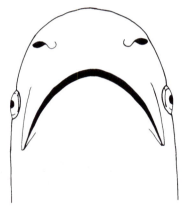

undersurface of head

third teeth from symphysis (upper and lower jaw)

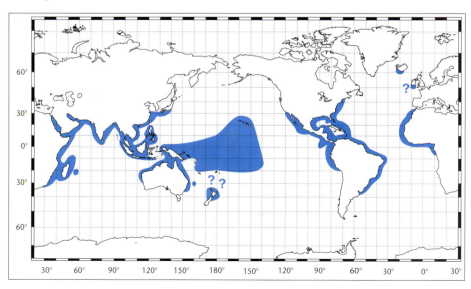

- **Remarks:** The tiger shark is the only ovoviviparous member of the family. It produces between 10 and 80 pups after a gestation period of about 12 months. It is most active at night, when it comes closer inshore or nearer the surface. This species is one of the few sharks that is a true scavenger, taking a wide range of marine prey such as fish, crustaceans and cephalopods as well as birds, mammals, reptiles and a variety of indigestible objects. It frequently preys on turtles and sea snakes in tropical areas. Its occurrence in shallow water, indiscriminate feeding habits and large size make it one of the most dangerous of sharks. Although common in Australia, it is not used commercially. In other parts of the world it is caught for its high-quality hide and for its fins, liver and flesh. Tiger sharks are hooked regularly by sport fishermen off eastern Australia (off Queensland, and off New South Wales during summer).
- **Local synonymy:** *Galeocerdo rayneri* McDonald & Barron, 1868.
- **References:** Stevens (1984); Stevens and McLoughlin (1991).

Speartooth Shark — *Glyphis* sp. A

29.23 *Glyphis* sp. A (Plate 29)   Fish Code: 00 018041

- **Field characters:** A medium-sized whaler shark with similar-sized dorsal fins, small eyes, no conspicuous colour pattern, a short and broadly rounded snout, and erect, broadly triangular, serrated upper teeth.
- **Distinctive features:** Body fusiform, moderately stout; interdorsal ridge absent. Snout short and broadly rounded; eyes small; labial furrows small and inconspicuous. Upper teeth erect, broadly triangular, serrated; lower teeth slender, not serrated; first few anterior teeth with cutting edges confined to slightly expanded spear-like tips. First dorsal-fin origin over front of pectoral-fin inner margin, its midbase much closer to the pectoral-fin bases than to the pelvic-fin bases. Second dorsal fin large, about 60% of the first dorsal-fin height; its origin slightly forward of anal-fin origin. Anal fin large, about same size as second dorsal fin. Pectoral fins relatively short and broad. Tooth count 33/32*. Total vertebrae 148–217*; precaudal 83–93*.
- **Colour:** Dorsal surfaces greyish; ventral surfaces pale. An inconspicuous pale stripe on the flanks extending anteriorly from above the pelvic area. No distinctive fin markings.
- **Size:** Possibly attaining 200–300 cm; the three Australian specimens ranged from 70 to 131 cm.

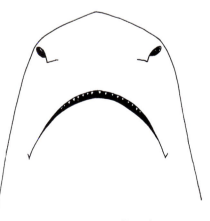

undersurface of head

- **Distribution:** Uncertain. Two Australian specimens were collected 17 km upstream in the Bizant River (Queensland), and more recently, a third was taken from the Adelaide River (Northern Territory). Very similar sharks occur in Borneo and New Guinea. Presumably inshore and often in fresh water.
- **Remarks:** Specimens of the genus *Glyphis* from the Northern Territory, Queensland, New Guinea and Borneo have been identified as *G. glyphis*. It is uncertain, however, whether they are all the same species, or whether any of them represent the true *G. glyphis* (the locality of the holotype is unknown). Even within Australia, there are large differences in vertebral counts between specimens from the Adelaide River (total 148; precaudal 83) and Bizant River (total 217; precaudal 93). More specimens from different localities are required to solve current taxonomic problems within this genus. The New Guinea form may reach up to 300 cm in length. Nothing is known about the biology of this shark.

teeth from near symphysis
(upper and lower jaw)

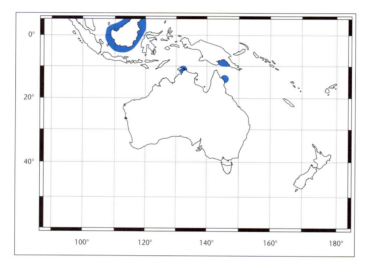

Sliteye Shark — *Loxodon macrorhinus* Müller & Henle, 1839

29.24 *Loxodon macrorhinus* (Plate 30)   **Fish Code:** 00 018005

- **Alternative names:** Jordans blue dogshark, slender dogshark.
- **Field characters:** A small, slender whaler shark with the second dorsal-fin originating almost over the anal-fin insertion, small and inconspicuous labial furrows, large eyes with a notch on the posterior orbital margin, and oblique, narrowly triangular, smooth-edged teeth.

- **Distinctive features:** Body fusiform, slender; interdorsal ridge usually absent. Snout relatively long and narrow, preoral length 8.1–9.8% of total length; internarial space 1.5–2.0 in preoral length; labial furrows small and inconspicuous; eye large, 2.3–3.6% of total length. Teeth similar in both jaws, oblique, narrowly triangular and smooth-edged. First dorsal-fin origin behind pectoral-fin free rear tips, its midbase about halfway between pectoral and pelvic-fin bases. Second dorsal-fin origin varying from slightly in front to slightly behind anal-fin insertion. Pectoral fins relatively short, 11.2–13.9% of total length. Tooth count 23–28/24–28 [24–28/24–29]. Total vertebrae 151–168 [148–191]; precaudal 82–95 [77–106].
- **Colour:** Dorsal surfaces bronze to greyish; ventral surfaces pale. Pectoral, pelvic and lower caudal-fin tips pale; pectoral fin posterior margins pale; first dorsal and caudal fins sometimes dark-edged.
- **Size:** Born at 40–45 cm and attains 90 cm. Both sexes mature at about 60 cm.
- **Distribution:** Tropical Indo–West Pacific; tropical Australia between Moreton Bay (Queensland) and North West Cape (Western Australia). Continental and insular shelves down to a depth of about 100 m; lives mainly near the bottom.
- **Remarks:** The sliteye shark is one of the most common sharks taken by bottom trawlers on the continental shelf off north-western Australia. There is no distinct reproductive seasonality in Australian waters; the usual litter size is two, with females breeding every year. The diet consists mainly of fish and crustaceans, and to a lesser extent cephalopods. Larger specimens taken by Taiwanese and Thai trawlers operating off northern Australia are retained for human consumption; elsewhere, also caught for its flesh.
- **Local synonymy:** *Scoliodon jordani* Ogilby, 1908; *Scoliodon affinis* Ogilby, 1912.
- **References:** Ogilby (1908b); Springer (1964); Stevens and McLoughlin (1991).

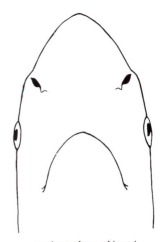

undersurface of head

fourth teeth from symphysis
(upper and lower jaw)

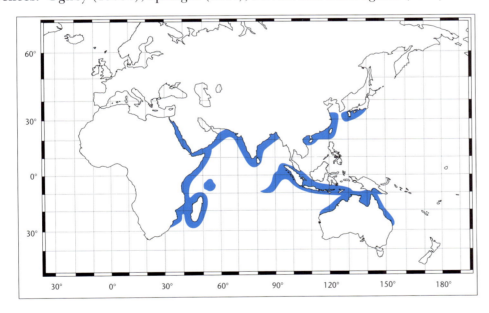

Lemon Shark — *Negaprion acutidens* (Rüppell, 1837)

29.25 *Negaprion acutidens* (Plate 31)

Fish Code: 00 018029

- **Alternative name:** Sharptooth shark.
- **Field characters:** A large, pale yellowish brown whaler shark with a blunt snout, similar-sized dorsal fins, and slender, erect, smooth-edged upper teeth.
- **Distinctive features:** Body fusiform, moderately stout; interdorsal ridge absent. Head broad; snout short and bluntly rounded (internarial space 0.8–1.2 in preoral length); labial furrows very short, confined to mouth corners; eyes relatively small. Cusps of upper teeth erect to slightly oblique, narrow and smooth-edged (fine serrations sometimes present on large specimens). Lower teeth somewhat more slender, never serrated. First dorsal-fin origin behind pectoral-fin free rear tips, its midbase closer to pelvic than pectoral-fin origin. Dorsal fins similar in size; second dorsal-fin height 0.8–1.1 of first dorsal-fin height; anal fin nearly as large as second dorsal fin. Pectoral fins broad (maximum width 1.2–1.5 in anterior margin) and rather strongly falcate. Tooth count 28–31/27–29 [29–32/28–30]. Total vertebrae 217–229 [223–227]*; precaudal 135–142 [136–140].
- **Colour:** Dorsal surfaces varying from a uniform pale yellow to light brown or grey. Ventral surfaces whitish. No conspicuous fin markings.
- **Size:** Born at 50–70 cm and attains 300 cm. Both sexes mature at about 220 cm.
- **Distribution:** Indian Ocean and central Pacific, including northern Australian waters from Moreton Bay (Queensland) to the Abrolhos Islands (Western Australia); rarely as far south as Perth (Western Australia). Continental and insular shelves from the intertidal zone to a depth of at least 30 m.

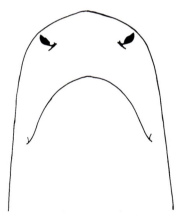

undersurface of head

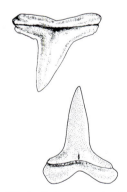

fifth teeth from symphysis
(upper and lower jaw)

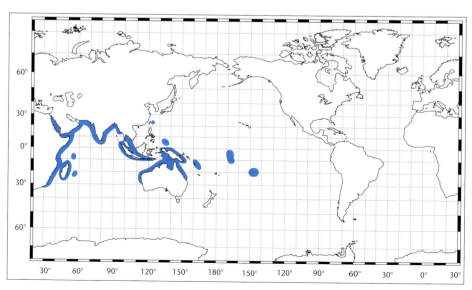

- **Remarks:** The lemon shark is commonly found close to the bottom in shallow, sandy lagoons and turbid, mangrove swamps. Juveniles often occur intertidally during the day, while adults are more active at night and remain in deeper channels. Lemon sharks produce litters of 1–14 pups after a 10–11 month gestation. Growth is relatively slow, in the order of 12 cm per year in juveniles. One tagging study demonstrated very restricted movements of juveniles, with 90% of recaptures made within 2 km of the tagging site. Juveniles are very inquisitive, while adults tend to be shy. If provoked, however, they sometimes respond aggressively. Not caught commercially in Australia; elsewhere used for their flesh, fins and liver.
- **Local synonymy:** *Aprionodon acutidens queenslandicus* Whitley, 1939; *Mystidens inominatus* Whitley, 1944; *Negaprion queenslandicus*: Munro, 1956.
- **Reference:** Whitley (1939b, 1944b).

Blue Shark — *Prionace glauca* (Linnaeus, 1758)

29.26 *Prionace glauca* (Plate 29)   Fish Code: 00 018004

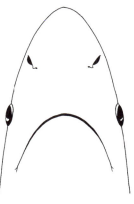

undersurface of head

- **Alternative names:** Blue whaler; great blue shark.
- **Field characters:** A large, slender-bodied whaler shark with a long, narrow snout, an indigo blue upper surface, the first dorsal fin originating well behind the pectoral-fin free rear tips, and long and scythe-like pectoral fins (except in specimens under 100 cm).
- **Distinctive features:** Body fusiform, slender; interdorsal ridge absent. Head narrow, snout long and narrow (internarial space 2.2–4.1 in preoral length); labial furrows very small, confined to mouth corners. Upper teeth relatively narrow, triangular, finely serrated and oblique (somewhat hook-shaped in large specimens); cusps of lower teeth more slender, erect, finely serrate. First dorsal-fin midbase closer to pelvic-fin origin than to pectoral-fin origin. Pectoral fins long and narrow in specimens over 150 cm; length about equal to head length (shorter than head length in small juveniles); maximum width 2.0–2.7 in anterior margin. Tooth count 29/27–31 [24–31/24–34]. Total vertebrae 241–253 [239–252]; precaudal 145–148 [142–150].

- **Colour:** Dorsal surfaces indigo blue (dull grey after death and in preservative); flanks grading from bright blue to silvery blue. Ventral surfaces pure white. Pectoral-fin tips dusky.
- **Size:** Born at 35–50 cm and attains 383 cm. Both sexes mature at about 220 cm.
- **Distribution:** Cosmopolitan in tropical and temperate seas. Found throughout Australian waters with the exception of the Arafura Sea, Gulf of Carpentaria and Torres Strait. Oceanic and pelagic from the surface to a depth of about 350 m. May occur close inshore where the continental shelf is narrow.
- **Remarks:** The blue shark is the most wide-ranging of all sharks; it prefers sea temperatures of 12–20°C and is found at greater depths in tropical waters. Courtship is a lively affair, with males biting the females (which are adapted by having skin more than twice as thick as that of males). The gestation period is 9–12 months and litter sizes average about 40 (although as many as 135 have been recorded). Growth is relatively rapid, with sexual maturity reached in 4–6 years. Tagging studies have demonstrated trans-Atlantic migrations, as well as movements from the Northern to the Southern Hemisphere. In Australia, a specimen tagged off Tasmania was recaught south of Java. The blue shark feeds mainly on small pelagic fish and cephalopods. It is potentially dangerous to humans, being very persistent, although not particularly aggressive. Although it is very abundant off southern Australia, the blue shark is not exploited by local commercial fishermen. The fins, however, are retained by Japanese tuna longliners who take a large bycatch of this species; elsewhere it is used for its flesh, fins and hide. Large numbers of this shark are caught by sport fishermen off New South Wales.
- **Local synonymy:** *Carcharhinus macki* Phillipps, 1935.
- **Reference:** Stevens (1984).

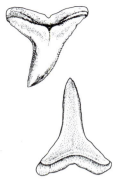

fourth teeth from symphysis
(upper and lower jaw)

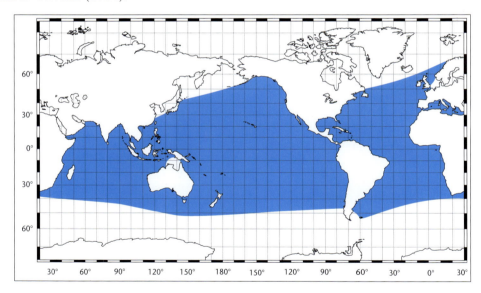

Milk Shark — *Rhizoprionodon acutus* (Rüppell, 1837)

29.27 *Rhizoprionodon acutus* (Plate 30)

Fish Code: 00 018006

- **Alternative names:** Longmans dogshark, fish shark, white-eye shark.
- **Field characters:** A small whaler shark with the second dorsal fin originating well behind the anal-fin origin, long labial furrows, usually more than 16 hyomandibular pores (total for both sides of the head), relatively large eyes, and oblique, narrowly triangular, smooth-edged teeth.
- **Distinctive features:** Body fusiform, moderately slender; interdorsal ridge usually absent. Snout relatively long and narrow, preoral length 6.7–11.4% of total length; labial furrows long (uppers 1.4–2.0% of total length); eye relatively large, 1.5–3.2% of total length; hyomandibular pores enlarged, in a distinct series, usually more than 16 in total for both sides of head. Teeth similar in both jaws, oblique, narrowly triangular, smooth-edged in juveniles, finely serrated in adults. First dorsal-fin origin varying from just anterior to just behind pectoral-fin free rear tips. Second dorsal-fin originating from over last third of anal-fin base to over anal-fin insertion. Tooth count 25/23–24 [23–27/22–26]. Total vertebrae 145–151 [121–162]; precaudal 67–76 [55–79].
- **Colour:** Dorsal surfaces bronze to greyish (grey after death and in preservative); ventral surfaces pale. Pectoral, pelvic, anal and lower caudal-fin tips pale; pectoral-fin posterior margins pale. Dorsal and upper caudal-fin tips dark in juveniles, sometimes dark-edged in adults.
- **Size:** Born at 35–40 cm and attains 100 cm (although a 178 cm specimen was recorded off Africa). Both sexes mature at about 75 cm.
- **Distribution:** Mainly tropical areas of the eastern Atlantic and Indo–West Pacific. Northern Australian waters from Fraser Island (Queensland) to Shark Bay (Western Australia). Continental and insular shelves from close inshore to a depth of about 200 m. Ranges through the water column but usually occurs near the bottom.

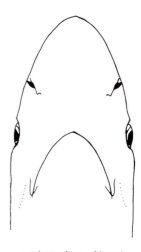

undersurface of head

sixth teeth from symphysis (upper and lower jaw)

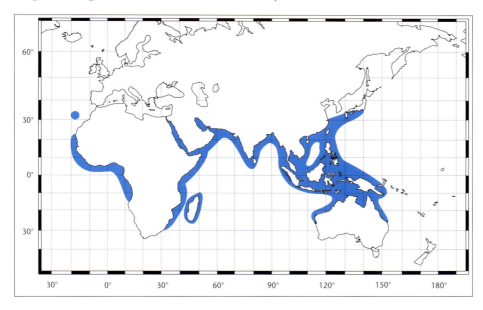

- **Remarks:** Populations of milk sharks from different areas vary considerably in size. Outside Australia, the size at birth may be as small as 25 cm and they may grow to almost 180 cm. Litter sizes vary from 1–8 and, in Australia, females give birth each year, but there is no reproductive seasonality. The diet consists principally of fish, with cephalopods and crustaceans of lesser importance. Off Natal, South Africa, the numbers of milk sharks have increased, possibly as a result of the shark meshing program, which has reduced the numbers of large inshore sharks that normally prey on them. This species is one of the most common sharks taken incidentally by Taiwanese and Thai demersal trawlers in northern Australia, who process larger specimens for human consumption; elsewhere an important species for food and fish meal.
- **Local synonymy:** *Scoliodon longmani* Ogilby, 1912.
- **References:** Springer (1964); Lyle and Timms (1984); Stevens and McLoughlin (1991).

Grey Sharpnose Shark — *Rhizoprionodon oligolinx* Springer, 1964

29.28 *Rhizoprionodon oligolinx* (Plate 30)   **Fish Code:** 00 018037

- **Field characters:** A small whaler shark with the second dorsal fin originating well behind the anal-fin origin, short labial furrows, usually less than 16 enlarged hyomandibular pores (total for both sides of the head), relatively large eyes, and oblique, narrowly triangular, smooth-edged teeth.
- **Distinctive features:** Body fusiform, moderately slender; interdorsal ridge occasionally present. Labial furrows short, uppers 0.2–1.3% of total length; hyomandibular pores enlarged, in a distinct series, usually less than 16 in total for both sides of head. Teeth similar in both jaws, oblique, narrowly triangular and smooth-edged; lower (and sometimes upper) anterior teeth of mature males more slender and oblique. First dorsal-fin origin over or just behind pectoral-fin free rear tips. Second dorsal fin origin over last third of anal-fin base. Tooth count [23–25/21–24]. Total vertebrae 161* [151–162]; precaudal 90* [84–91].
- **Colour:** Dorsal surfaces bronze to greyish (grey after death and in preservative); ventral surfaces pale. Pectoral-fin margins pale; upper caudal-fin margins dark-edged.
- **Size:** Born at 20–30 cm and attains 70 cm. Both sexes maturing at 35–40 cm (off India).
- **Distribution:** Continental and insular shelves of the tropical Indo–West Pacific. In the Australian region, recorded only from the Gulf of Carpentaria.
- **Remarks:** Only a single specimen, identified by its vertebral count, has been recorded from Australia. More collections are required to confirm the occurrence

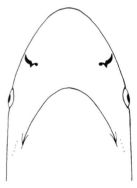

undersurface of head

tooth from near symphysis (upper jaw)

and distribution of this species locally. Off Bombay (India), litters of 3–5 pups are born mainly in January and February. The diet is not recorded, but probably consists mainly of fish, together with cephalopods and crustaceans. Its flesh is used for human consumption.

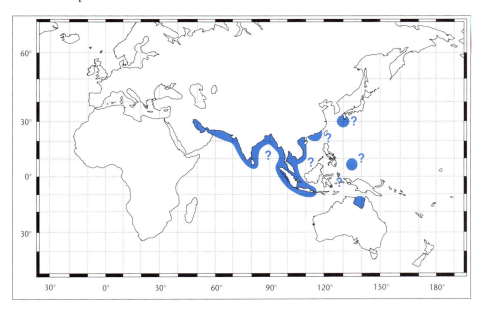

Australian Sharpnose Shark — *Rhizoprionodon taylori* (Ogilby, 1915)

29.29 *Rhizoprionodon taylori* (Plate 30)

Fish Code: 00 018024

- **Alternative name:** Taylors shark.
- **Field characters:** A small whaler shark with the second dorsal fin originating well behind the anal-fin origin, short labial furrows, usually more than 16 enlarged hyomandibular pores (total for both sides of the head), relatively large eyes, and oblique, narrowly triangular, smooth-edged teeth.
- **Distinctive features:** Body fusiform, moderately slender; interdorsal ridge occasionally present. Labial furrows short, uppers 0.7–1.1% of total length; hyomandibular pores enlarged (forming discrete series extending anteriorly from outside mouth corners), usually more than 16 in total for both sides of head. Teeth similar in both jaws, oblique, narrowly triangular and smooth-edged. First dorsal-fin origin varying from just anterior to just behind pectoral-fin free rear tips. Second

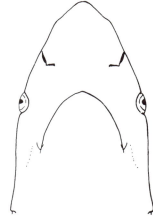

undersurface of head

dorsal-fin origin over last third of anal-fin base. Pectoral fins relatively short, 12.1–13.9% of total length. Tooth count 23–25/21–23. Total vertebrae 135–149; precaudal 73–81.

third tooth from symphysis (upper jaw)

- **Colour:** Dorsal surfaces bronze to greyish (grey after death and in preservative); ventral surfaces pale. Dorsal-fin anterior margins (and sometimes posterior margins) dark; upper caudal-fin tip dark and margins dark-edged; pectoral and lower caudal-fin posterior margins pale.
- **Size:** Born at 25–30 cm and attains 67 cm. Males mature at about 40 cm and females at about 45 cm.
- **Distribution:** Papua New Guinea and tropical Australia from the North-West Shelf (where it is uncommon) to southern Queensland. Continental shelf from close inshore down to a depth of at least 110 m. Ranges through the water column but more usually found near the bottom.
- **Remarks:** A locally common species taken incidentally in trawls and gillnets, but too small to be of any commercial importance. Reproduction appears to be seasonal, with females giving birth to 1–8 pups around January. Females ovulate in February but small embryos are not found until at least August, which suggests that the Australian sharpnose shark can suppress development of its eggs. The diet consists mainly of fish but also includes cephalopods and crustacea.
- **Local synonymy:** *Physodon taylori* Ogilby, 1915; *Protozygaena taylori*: Whitley, 1940.
- **References:** Ogilby (1915a); Springer (1964); Lyle and Timms (1984); Stevens and McLoughlin (1991).

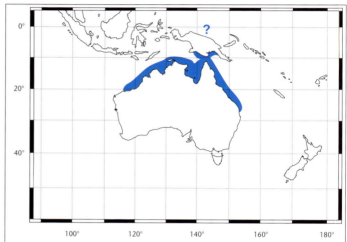

Whitetip Reef Shark — *Triaenodon obesus* (Rüppell, 1837)

29.30 *Triaenodon obesus* (Plate 31) **Fish Code:** 00 018038

- **Alternative names:** Whitetip shark, blunthead shark.
- **Field characters:** A whaler shark with distinct white tips to the first dorsal and upper caudal fin, small, smooth-edged teeth with one or more large basal cusplets on either side of the cusp, and a large second dorsal fin (half to three-quarters of the first dorsal-fin height).

- **Distinctive features:** Body fusiform, moderately slender; interdorsal ridge absent. Head relatively broad; snout extremely short and bluntly rounded (internarial space 0.7–0.8 in preoral length); labial furrows very short, restricted to mouth corners. Teeth small, similar in both jaws; erect to somewhat oblique, smooth-edged, with one or more large basal cusplets on either side of cusp. First dorsal-fin origin well behind pectoral-fin free rear tips; mid-base closer to pelvic-fin than pectoral-fin origin. Second dorsal fin about half to three-quarters of first dorsal-fin height, about equal in size to anal fin. Tooth count [42–50/42–48]. Total vertebrae [208–214]*; precaudal [128–129]*.

- **Colour:** Dorsal surfaces greyish brown, usually with a few scattered dark spots; ventral surfaces pale. First dorsal, upper caudal and sometimes second dorsal and lower caudal-fin tips distinctly white; sometimes distinctly darker beneath white tips on dorsal fins; other fins also sometimes dark-edged or darker than general body colour.

- **Size:** Born at 52–60 cm and attains 170 cm. Both sexes mature at about 105 cm.

- **Distribution:** Shallow waters of the Indo–Pacific, including tropical Australia from Point Quobba (Western Australia) to Gladstone (Queensland). Usually found in depths of 8–40 m, but recorded down to 300 m.

- **Remarks:** This reef-associated shark is usually found near the bottom, often resting in caves or under coral ledges during the day. Tagging and telemetry studies indicate that it is more active at night and during slack water. The whitetip reef shark appears to have a narrow home range, moving within a radius of only a few kilometres. It is adapted to feeding on the bottom, particularly in crevices and holes in the coral where it catches small fish, crustaceans and octopus. It appears inept at taking baits away from the bottom. The whitetip reef shark is one of the few shark species in which mating has been seen in the wild. The sharks observed copulated in a stationary parallel orientation with their heads on the bottom and their bodies angled upwards at about 45° into the water column. Litters of 1–5 pups are produced after an unknown gestation period. Although curious and will often approach divers, it is rarely aggressive. Not used commercially in Australia, but its flesh and liver are used elsewhere. This species has, exceptionally, been recorded as ciguatoxic.

- **Local synonymy:** *Triaenodon apicalis* Whitley, 1939.

- **Reference:** Whitley (1939b).

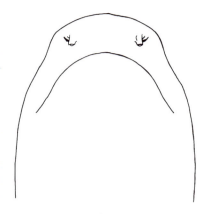

undersurface of head

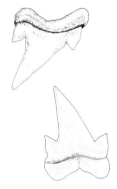

seventh teeth from symphysis
(upper and lower jaw)

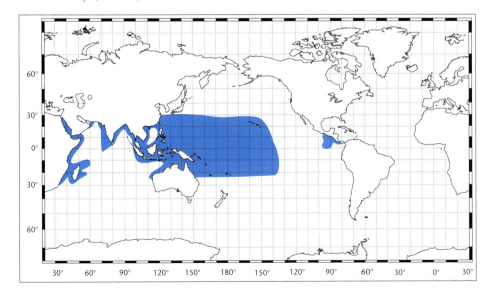

30 Family Sphyrnidae
HAMMERHEAD SHARKS

These unmistakable sharks have a unique head shape with the eyes located at the tips of laterally expanded blades which resemble a hammer. They also have fusiform bodies, ventrally placed mouths extending well beyond the eyes, five pairs of gill slits, circular eyes with a nictitating membrane, two spineless dorsal fins (the first considerably larger than the second), an anal fin, a non-lunate caudal fin, and a scroll-type intestinal valve. They lack nasoral grooves, nasal barbels, spiracles and caudal keels, but have both upper and lower precaudal pits. The first dorsal fin, which is relatively tall and short-based, is inserted ahead of the pelvic-fin origin. The anal-fin origin is usually slightly forward of the second dorsal-fin origin.

Hammerheads are small to large sharks (attaining 0.9–6.0 m) with worldwide or localised distributions in tropical and/or temperate seas. The group consists of 2 genera and 9 species, of which both genera and 4 species occur off Australia. Hammerheads are fast and active sharks, occurring throughout the water column. They range from inshore to offshore, but are never truly oceanic. Their bizarre head-shape acts as a bowplane, increasing manoeuverability and aiding in the capture of fast and elusive prey, and may also enhance their sensory capabilities. They are viviparous with a yolk-sac placenta. Despite a bad reputation, probably earned because of their strange shape, these sharks are not particularly dangerous to humans. However, there are numerous reports of large specimens bothering spear fishermen and they should be treated with caution.

Key to sphyrnids

1 Lateral blades of head very long, narrow and wing-like (fig. 1); head width 40–50% of total length .... **Eusphyra blochii** (fig. 4; p. 271)

   Lateral blades of head relatively broad, shorter and not wing-like (fig. 2); head width considerably less than 40% of total length 2

2 Anterior margin of head without a median indentation (fig. 2) .... **Sphyrna zygaena** (fig. 5; p. 275)
   Anterior margin of head with a median indentation (fig. 3) 3

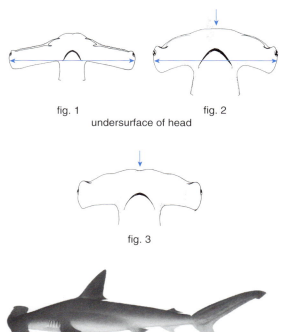

undersurface of head

fig. 1   fig. 2

fig. 3

fig. 4   fig. 5

3   First dorsal fin only moderately high and semi-falcate (fig. 8); anterior profile of head bulging forward near middle (fig. 8); teeth smooth-edged or weakly serrated (fig. 6); posterior margins of pelvic fins nearly straight (fig. 8); second dorsal-fin height less than or equal to length of third gill slit (fig. 8) .... *Sphyrna lewini* (**fig. 8; p. 272**)

First dorsal fin falcate, very tall in adults (fig. 9); anterior profile of head nearly straight (except in small juveniles) (fig. 9); teeth strongly serrated (fig. 7); posterior margins of pelvic fins markedly concave (fig. 9); second dorsal-fin height greater than length of third gill slit (fig. 9) ....*Sphyrna mokarran* (**fig. 9; p. 274**)

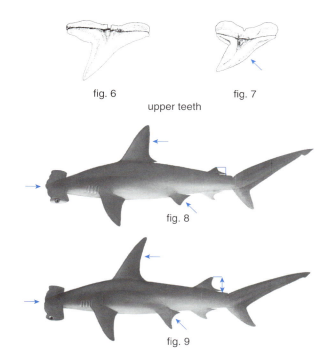

Winghead Shark — *Eusphyra blochii* (Cuvier, 1816)

30.1 *Eusphyra blochii* (Plate 37)   **Fish Code:** 00 019003

- **Alternative name:** Slender hammerhead.
- **Field characters:** A hammer-head shark with a very wide and slender 'hammer' that is equal to about half of the shark's total length.
- **Distinctive features:** Body fusiform, moderately slender; upper precaudal pit forming a narrow longitudinal groove. Head laterally expanded into two, very large, wing-like extensions; head width about 40–50% of total length. Nostrils greatly enlarged, their width about twice mouth width. Teeth similar in both jaws; relatively small, smooth-edged and oblique. First dorsal-fin origin above or slightly in front of pectoral-fin insertion. First dorsal fin relatively slender and falcate. Second dorsal-fin origin over posterior third of anal-fin base. Tooth count 30/29* [31/29]*. Total vertebrae 108* [116–124]*; precaudal 46–47* [51–54]*.
- **Colour:** Grey or greyish brown above, pale ventrally. No dark fin markings.
- **Size:** Australian specimens are born at 45–47 cm and attain 186 cm. Males mature at about 108 cm; females at about 120 cm.

undersurface of head

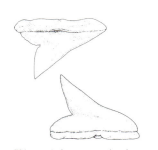
fifth teeth from symphysis (upper and lower jaw)

- **Distribution:** Tropical regions of the Indo–West Pacific Ocean, including northern Australia south to about 18°S between Ingham (Queensland) and Broome (Western Australia). Shallow areas of continental and insular shelves.
- **Remarks:** Viviparous, producing litters of 6–25 young after a gestation of 10–11 months. In Australia, the young are born in February and March. The diet consists mainly of small teleost fish, together with smaller quantities of crustaceans and some cephalopods (most of which are taken on or near the bottom). Not exploited in Australia (the flesh is relatively high in mercury); elsewhere used for its meat and liver oil.
- **Local synonymy:** *Sphyrna blochii* (Cuvier, 1817).
- **Reference:** Stevens and Lyle (1989).

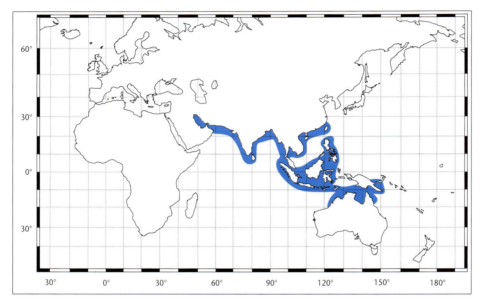

Scalloped Hammerhead — *Sphyrna lewini* (Griffith & Smith, 1834)

30.2 *Sphyrna lewini* (Plate 37) **Fish Code:** 00 019001

- **Alternative name:** Kidney-headed shark.
- **Field characters:** A hammerhead shark with the front margin of the head curved forward anteriorly (with a median indentation), the head width less than 40% of

total length, smooth-edged (or at most weakly serrated) teeth, straight posterior margins to the pelvic fins, and a rather low second dorsal fin (height less than or equal to length of third gill slit).

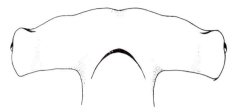
undersurface of head

- **Distinctive features:** Body fusiform, moderately slender; upper precaudal pit 'V' shaped. Head laterally expanded into prominent keels resembling a hammer; maximum width 24–30% of total length; anterior profile of 'hammer' curved anteriorly, with median and lateral indentations. Teeth relatively small, smooth-edged or occasionally weakly serrated; upper teeth narrowly triangular, first 3 nearly erect, remainder relatively oblique; lower teeth slightly more slender and erect than uppers. First dorsal-fin origin about over or slightly behind pectoral-fin insertion. First dorsal fin broad and relatively erect. Second dorsal-fin origin about over mid-point of anal-fin base; free rear tip of fin nearly reaching upper caudal-fin origin. Pelvic fin posterior margins nearly straight. Tooth count 32–33/31–33* [32–36/30–34]. Total vertebrae 195* [174–204]; precaudal 89–93 [89–96].

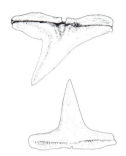
third teeth from symphysis (upper and lower jaw)

- **Colour:** Olive, bronze or brownish grey dorsally, pale ventrally. Ventral surface of pectoral-fin tips dusky in adults; no other fin markings. Pectoral, lower caudal and second dorsal-fin tips dark in juveniles.

- **Size:** Australian specimens are born at 45–50 cm and attain 350 cm. Males mature at 140–160 cm; females at about 200 cm.

- **Distribution:** Cosmopolitan in tropical and warm temperate seas. In Australia, recorded throughout the north to about 34°S on both coasts (Sydney to Geographe Bay). Occurs over the continental and insular shelves and adjacent deep water, from the surface to at least 275 m depth. Juveniles often occur close inshore.

- **Remarks:** The scalloped hammerhead sometimes forms large migratory schools. In Australia, little is known of its movements, although it occurs further south during the warmer months. Adult females are rarely caught inshore and may live in deeper water, only moving onto the continental shelf to mate and give birth. They are viviparous producing litters of 13–23 young between October and January (in Australian waters), after a gestation of 9–10 months. Their diet consists mainly of teleost fishes and cephalopods which, in adults at least, suggests a pelagic way of life. Tracking experiments on a Californian seamount showed that they disperse from their day-time aggregations and move into deep water at night to feed. This species was utilised for its meat and fins by a Taiwanese gillnet fishery which operated off northern Australia until 1986, but is rarely landed by domestic vessels because of its relatively high mercury content.

- **Reference:** Stevens and Lyle (1989).

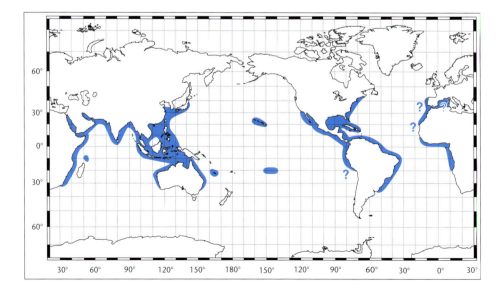

Great Hammerhead — *Sphyrna mokarran* (Rüppell, 1837)

30.3 *Sphyrna mokarran* (Plate 37) **Fish Code:** 00 019002

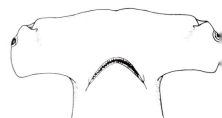

undersurface of head

fifth tooth from symphysis (upper jaw)

- **Field characters:** A hammer-head shark with the front margin of the head nearly straight (except in small juveniles) but with a median indentation, the head width less than 40% of total length, a very tall and falcate first dorsal fin, serrated teeth, concave posterior margins to the pelvic fins, and a relatively tall second dorsal fin (height greater than length of third gill slit).

- **Distinctive features:** Body fusiform, moderately slender; upper precaudal pit 'V' shaped. Head laterally expanded into prominent keels, resembling a hammer; maximum width 23–27% of total length; anterior profile of 'hammer' nearly straight, but with both median and lateral indentations. Teeth relatively small, with definite serrations; upper teeth triangular and oblique, lower teeth more erect. First dorsal-fin origin about over or slightly behind pectoral-fin insertion. First dorsal fin very tall, slender and falcate. Second dorsal-fin origin about over or slightly behind anal-fin origin; its free rear tip ending well forward of upper caudal-fin origin. Pelvic-fin posterior margins markedly concave. Tooth count 35/35–36* [36–39/34–38]. Total vertebrae 204–205* [197–212]; precaudal 93–97* [94–98].

- **Colour:** Bronzy to greyish brown dorsally, pale ventrally. No fin markings in adults; second dorsal-fin tip dark in juveniles.

- **Size:** Born at about 65 cm and attains 600 cm (although rarely exceeding 450 cm). In Australia, males mature at about 225 cm and females at about 210 cm.

- **Distribution:** Circumglobal in tropical and warm temperate seas. Recorded throughout northern Australia, south to Sydney (New South Wales) and the

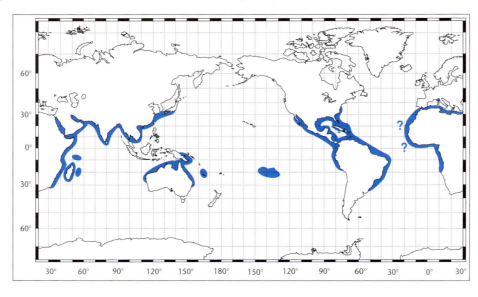

Abrolhos Islands (Western Australia). Continental and insular shelves, from the surface (and in very shallow water) to at least 80 m depth.

- **Remarks:** This species is widely distributed and, while normally occurring in the tropics, has been recorded as far south as Sydney (34°S). It is viviparous, producing litters of 6–33 young in December and January (in northern Australia), after a gestation of 11 months. Stomach contents of northern Australian specimens contained fish, smaller numbers of crustaceans and some cephalopods. The fish components were mainly demersal species (including numerous sharks and rays), suggesting that this hammerhead feeds close to the bottom. Similar diets have been reported from other areas. This species was utilised for its meat and fins by a Taiwanese gillnet fishery which operated in northern Australian waters until 1986. It is not, however, landed by domestic vessels because of its relatively high mercury content. A small number of attacks on humans have been attributed to large hammerheads but the species responsible has not been accurately identified. However, there have been numerous reports locally of large specimens bothering spear fishermen and it seems most probable that this species is the culprit. The diets, relatively small teeth and jaws, and unaggressive (sometimes timid) behaviour of both the scalloped (30.2) and smooth (30.4) hammerheads, suggest their bad reputation is based more on their strange appearance than on fact.

- **Reference:** Stevens and Lyle (1989).

Smooth Hammerhead — *Sphyrna zygaena* (Linnaeus, 1758)

30.4 *Sphyrna zygaena* (Plate 37) **Fish Code:** 00 019004

- **Alternative name:** Common hammerhead.
- **Field characters:** A hammerhead shark with the front border of the head curved forward anteriorly (but with no median indentation), the head width less than 40% of total length, and finely serrated (occasionally smooth-edged) teeth.
- **Distinctive features:** Body fusiform, moderately slender; upper precaudal pit 'V' shaped. Head laterally expanded into enlarged keels resembling a hammer; maximum width 26–29% of total length; anterior margin of 'hammer' curved anteriorly, with lateral but without median indentations. Teeth relatively small, finely serrated or occasionally smooth-edged; upper teeth triangular and oblique; lower teeth similar but slightly smaller. First dorsal-fin origin above pectoral-fin insertion. First dorsal fin broad and relatively erect. Second dorsal-fin origin about over mid-point of anal-fin base; its free rear tip ending well forward of upper caudal-fin origin. Pelvic fin posterior margins relatively straight. Tooth count 30–31/29–30 [29–31/25–31]. Total vertebrae 195–202 [193–206]; precaudal 98–100 [94–102].

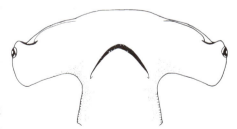

undersurface of head

fifth tooth from symphysis (upper jaw)

- **Colour:** Olive to dark greyish brown dorsally, white ventrally; ventral tips of pectoral fins dusky.
- **Size:** Born at 50–60 cm and attains about 350 cm. Males mature at about 250 cm; females at about 265 cm.
- **Distribution:** Widespread, temperate in both Hemispheres (also tropical in some regions). Southern Australia north to about 30°S between Coffs Harbour (New South Wales) and Jurien Bay (Western Australia). Occurs over the continental and insular shelves from the surface to at least 20 m depth.
- **Remarks:** Like the scalloped hammerhead (30.2), this species occasionally forms large schools. It is viviparous, producing litters of 20–50 young between January and March in Australia, after a gestation of 10–11 months. Stomach contents of New South Wales specimens consisted primarily of cephalopods and, to a lesser extent, teleost fishes. Elsewhere, they have been found to eat teleost and elasmobranch fishes, cephalopods and crustaceans. It is taken as bycatch in the Western Australian shark fishery and retained for its meat; elsewhere used for its meat, fins, liver oil and hide.
- **Reference:** Stevens (1984).

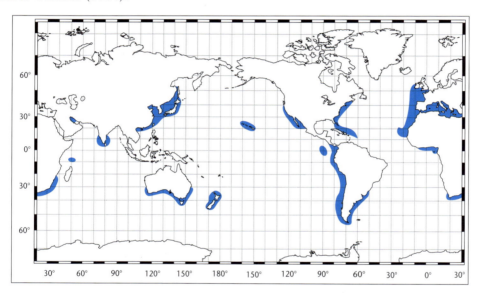

31 Family Squatinidae
ANGEL SHARKS

Angel sharks resemble rays in having dorsoventrally flattened bodies, extremely broad pectoral fins, and dorsally located eyes and spiracles. Unlike rays, however, the gill slits are located laterally on the head and forward of the pectoral-fin origins (although the pectoral fins extend forward of the gill slits as a lobe). A terminal mouth is ornamented with well-developed nasal barbels and labial folds. They also have well-developed dorsal fins without spines, a hypocercal caudal fin, a long and depressed tail, but lack an anal fin.

This family, represented worldwide in tropical and warm temperate waters by a single genus, contains about 14 species with four occurring in the Australian region. Most species live close inshore but one angel shark occurs to depths of 1300 m. The flesh is of high quality. Eastern Australian species are caught in moderate quantities for food and are sold under the recommended marketing name of 'angel shark', although 'monkfish' is another well-known local name for these fishes. Elsewhere, angel sharks are harvested for oil, fishmeal, leather, and shagreen for woodworking.

Key to squatinids

1 Body with 3 pairs of granular ocelli near bases of pectoral and pelvic fins (fig. 1) .... *Squatina tergocellata* (fig. 1; p. 281)

   No ocelli on upper surface 2

2 Interorbital space flat or convex (fig. 2); no orbital thorns (fig. 2); lower lobe of caudal fin with numerous dark spots (fig. 4) .... *Squatina australis* (fig. 4; p. 280)

   Interorbital space concave (fig. 3); orbital thorns normally present (fig. 3); caudal fin with more pale spots than dark spots (fig. 5) 3

3 Upper surface mainly covered with small white, dark-edged spots; no thorns along midline of trunk (fig. 5) .... *Squatina* sp. A (fig. 5; p. 278)

   Upper surface mainly covered in bluish spots (only single white spot on nuchal region); thorns mostly in single row along midline of trunk (fig. 6) .... *Squatina* sp. B (fig. 6; p. 279)

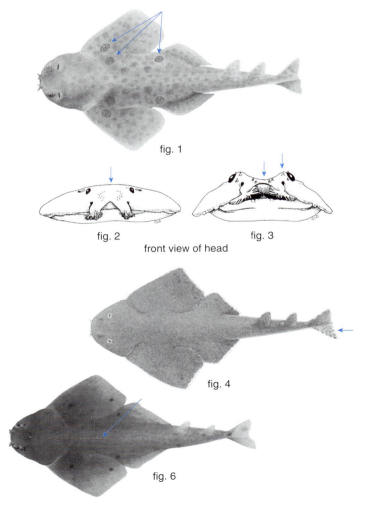

Eastern Angel Shark — *Squatina* sp. A

31.1 *Squatina* sp. A (Plate 38)

Fish Code: 00 024004

- **Field characters:** An angel shark with a concave interorbital region, a dense pattern of fine white spots on the upper surface but without distinctive ocelli, no dark spots on the caudal fin, and with orbital thorns but no median row of trunk thorns.

- **Distinctive features:** Head and body very broad and depressed; snout short. Eyes small, much less than an eye diameter from head margin; interorbital region broad, slightly concave, with widely-spaced denticles; spiracles less than an eye diameter behind eye. Mouth broad, terminal; anterior nasal barbels strongly fringed; posterior nasal barbels with slightly uneven margin; skin folds along sides of head slender, without triangular lobes. Teeth long, sharply pointed, similar in both jaws. Dorsal surface granular with short, very widely spaced denticles (becoming more dense in adults); central denticles with distinct ridges anteriorly. Thorns short, conical; mostly with 2 postorbital, 2 preorbital and 2 prenasal (somewhat irregular in number and size in adults); no rows of thorns along dorsal midline. Ventral surface mostly smooth, except for denticles and fine granular patches along outer half of pectoral and pelvic fins, and on mid-trunk and tail. Dorsal fins short, mostly separated by about 1.5 times first dorsal-fin base or less; first dorsal-fin origin ahead of pelvic-fin free rear tips. Pectoral-fin apex narrowly rounded, posterior margin slightly concave. Tail with prominent lateral keels.

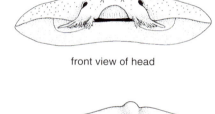
front view of head

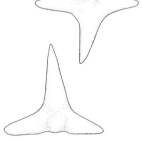
third teeth from symphysis (upper and lower jaw)

- **Colour:** Dorsal surface yellowish brown to chocolate brown with numerous sharply defined, small, dark-edged white spots (as dense on body as on pectoral and pelvic fins); several lighter or darker brownish blotches scattered more or less regularly; margins of pectoral and pelvic fins pale; dorsal and caudal fins with dark bases, becoming paler posteriorly. Ventral surface white.

- **Size:** To at least 63 cm (immature at that size).

- **Distribution:** Outer continental shelf and upper slope off eastern Australia between Cairns (Queensland) and Lakes Entrance (Victoria) in 130 to 315 m.

- **Remarks:** This species has been confused in the past with its shallower inshore relative, the Australian angel shark (31.3). It is good to eat but quantities reaching markets are currently small. Little is known of its biology.

Western Angel Shark — *Squatina* sp. **B**

31.2 *Squatina* sp. **B** (Plate 38)   **Fish Code:** 00 024005

- **Field characters:** An angelshark with a concave interorbital space, a pattern of bluish spots and brownish blotches, no dark granular ocelli, no dark spots on the caudal fin, and mostly with orbital thorns and a median row of trunk thorns.

- **Distinctive features:** Head and body very broad and depressed; snout short. Eyes small, less than an eye diameter from head margin; interorbital region broad, distinctly concave (with prominent ridges above eyes), with relatively widely spaced denticles; spiracles less than an eye diameter behind eye. Mouth broad, terminal; anterior nasal barbels strongly fringed; posterior nasal barbels with distinctly crenulated margins; skin folds along sides of head slender, without triangular lobes. Teeth long, sharply pointed, similar in both jaws. Dorsal surface of adults granular with short, broad, widely spaced denticles; denticles with distinct ridges anteriorly; smaller specimens with few denticles near margins of pectoral and pelvic fins. Irregular numbers of short thorns before and after eye and in prenasal area; thorns persisting with age; adults mostly with single row of short thorns along midline of back. Ventral surface mostly smooth, except for denticles and fine granular patches along outer half of pectoral and pelvic fins, and on mid-trunk and tail. Dorsal fins short, separated by about 1.5 times first dorsal-fin base or more; first dorsal-fin origin well ahead of pelvic-fin free rear tips (often more posterior in juveniles); pectoral-fin apex narrowly rounded, posterior margin straight to slightly concave. Tail with prominent lateral keels.

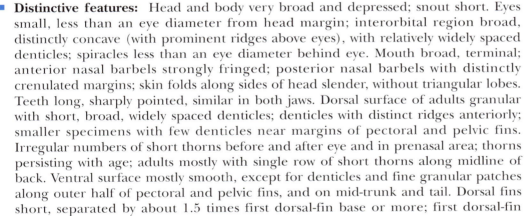

front view of head

- **Colour:** Dorsal surface medium to pale brownish or greyish; covered with widely spaced, bluish spots and darker brownish blotches; dorsal and caudal fins mostly uniformly coloured; no distinct 'mitotic' ocelli; upper lobe of caudal fin sometimes with a few bluish spots and 1–2 brownish blotches. Ventral surface uniformly pale.

- **Size:** Attains at least 64 cm.

- **Distribution:** Demersal on the outer continental shelf and upper slope from Broome to Shark Bay (Western Australia) in 150–310 m.

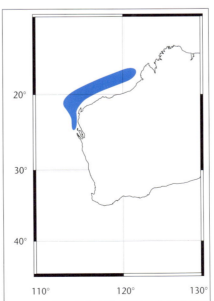

- **Remarks:** The earliest picture of an Australian elasmobranch, published after William Dampier's voyage to Western Australia in 1699, may have been this undescribed species. It has been confused since with the ornate angel shark (31.4), with which it may be sympatric in the southernmost part of its distribution. The western angel shark lacks the unusual ocelli, resembling cells in the process of mitotic division, that characterize its sister species.
- **Local synonymy:** *Squatina tergocellata*: Sainsbury *et al.*, 1985.
- **Reference:** Sainsbury *et al.* (1985).

Australian Angel Shark — *Squatina australis* Regan, 1906

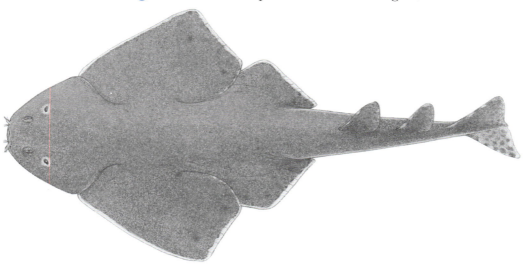

31.3 *Squatina australis* (Plate 39)　　　　　　　　　　　　　　**Fish Code:** 00 024001

- **Alternative names:** Angelshark, monkfish.
- **Field characters:** An angelshark with a flat to slightly convex interorbital space, irregular white flecks and spots on the upper surface, dark spots on the caudal fin, median trunk thorns (juveniles mostly with a central patch of enlarged denticles) and without orbital thorns.

front view of head

- **Distinctive features:** Head and body very broad and depressed; snout short. Eyes very small, more than an eye diameter from head margin; interorbital region broad, flat to slightly convex, densely granulated; spiracles about an eye diameter behind eye. Mouth broad, terminal; anterior nasal barbels strongly fringed; posterior nasal barbels with distinctly crenulated margins; skin folds along sides of head slender, without triangular lobes. Teeth long, sharply pointed, similar in both jaws. Dorsal surface densely granulated with minute, sharp denticles; central denticles without anterior ridges; young specimens with multiple rows of enlarged predorsal denticles (some persisting in adults); a few enlarged denticles on snout. No enlarged thorns. Ventral surface mostly smooth, except for denticles and fine granular patches along outer half of pectoral and pelvic fins, and on mid-trunk and tail. Dorsal fins short, separated by about 1.5 times first dorsal-fin base; first dorsal-fin origin slightly behind pelvic-fin free rear tips; pectoral-fin apex angular, posterior margin slightly concave. Tail with prominent lateral keels.
- **Colour:** Dorsal surface dull grey to greyish brown with dense pattern of small, irregular white spots and flecks (less pronounced in juveniles); flecks most densely concentrated on pectoral and pelvic fins; dorsal and caudal fins paler, latter with small black spots on the ventral lobe; margins of paired fins with white edges. Ventral surface uniformly pale; pectoral-fin margins slightly darker.

- **Size:** Attains at least 152 cm.
- **Distribution:** Continental shelf of southern Australia from Rottnest Island (Western Australia) to at least Sydney (New South Wales). Mainly coastal but down to 130 m.
- **Remarks:** This shark is common in Bass Strait and off surf beaches along the eastern seaboard. It feeds mainly on small fishes and crustaceans. The largest Australian squatinid, its flesh makes excellent eating. It is marketed, mainly in Sydney and Melbourne, as angel shark or flake.
- **References:** Regan (1906a); Coleman (1980); Edgar *et al.* (1982).

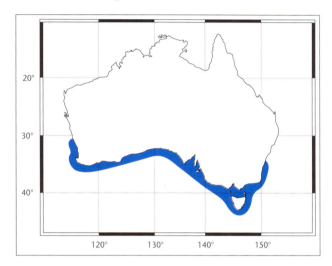

Ornate Angel Shark — *Squatina tergocellata* McCulloch, 1914

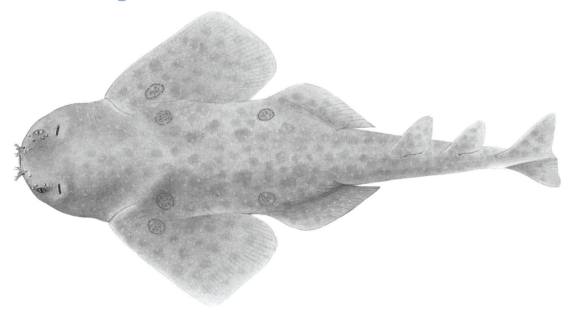

31.4 *Squatina tergocellata* (Plate 38)   Fish Code: 00 024002

- **Alternative name:** Archbishop.
- **Field characters:** An angelshark with three pairs of dark granular (mitotic) ocelli and a dense pattern of bluish spots on the upper surface, a concave interorbital region, orbital thorns but no median row of trunk thorns, and without dark spots on the caudal fin.

- **Distinctive features:** Head and body very broad and depressed; snout short. Eyes small, less than an eye diameter from head margin; interorbital region broad, distinctly concave (with prominent ridges above eyes), with relatively widely spaced denticles (almost smooth in small specimens); spiracles less than an eye diameter behind eye. Mouth broad, terminal; anterior nasal barbels strongly fringed; posterior nasal barbels with distinctly crenulated margins; skin folds along sides of head slender, without triangular lobes. Teeth long, sharply pointed, similar in both jaws. Dorsal surface of adults granular, with short, broad, widely spaced denticles; central denticles with distinct ridges anteriorly; adult males with small thorns along outer margins of pectoral and pelvic fins (denticles and thorns lacking in young). Thorns on head conical, mostly with 2 postorbital, 2 preorbital and 2 prenasal (less distinct in adults); no well-developed row of stellate thorns along midline of back. Ventral surface mostly smooth, except for denticles and fine granular patches along outer half of pectoral and pelvic fins, and on mid-trunk and on tail. Dorsal fins short, separated by about 1.5 times first dorsal-fin base or more; first dorsal-fin origin mostly ahead of pelvic-fin free rear tips; pectoral-fin apex narrowly rounded, posterior margin straight to slightly concave. Tail with prominent lateral keels.

front view of head

- **Colour:** Dorsal surface pale yellowish brown with numerous fine greyish blue or white spots; 3 prominent pairs of dark brown 'mitotic' ocelli on body (2 near each pectoral-fin base and 1 above each pelvic-fin origin); ocelli less intricate in mature specimens; dusky rings (often faint) on nape, outer pectoral and pelvic fins, and along mid-body; dorsal and caudal fins pale with brown and white blotches. Ventral surface uniformly pale.
- **Size:** To at least 100 cm.
- **Distribution:** Demersal on the continental shelf and upper slope from Geraldton (Western Australia) to Port Lincoln (South Australia) in 130–400 m.
- **Remarks:** This rather attractive shark is captured frequently in small quantities by trawlers when fishing the Great Australian Bight for flathead (*Platycephalus* spp.) and morwong (*Nemadactylus macropterus*). The distinctive ocelli on the upper surface, which resemble cells in the process of mitotic division, are unique within the family. This species appears to be most common in about 300 m depth. The flesh is tasty.
- **Reference:** Scott *et al.* (1980).

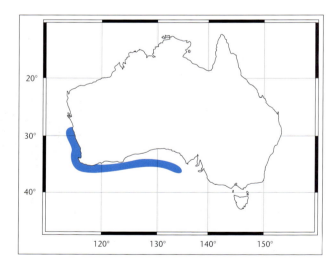

32 Family Rhinobatidae
SHOVELNOSE RAYS

Shovelnose rays (also referred to as guitarfishes) are small to medium-sized rays with a rather small, variably-shaped disc (mostly oval to shovel-shaped) and a long, broad and rather flattened tail without a stinging spine. They have two well-developed dorsal fins, with the first dorsal fin originating behind the pelvic fins. The upper caudal lobe is well developed, but the posterior margin of the tail is straight (rather than concave as in the closely related family Rhynchobatidae) and the lower lobe is small. The mouth is armed with numerous, small, blunt teeth, and the nostrils may or may not be covered with an internasal flap. The spiracles are well developed and frequently have small (but prominent) skin folds along their hind margins.

This family, represented worldwide by four genera and about 40 species, is most diverse in the Indo–West Pacific. Three genera and at least 8 species are represented in the Australian region. As some of these rays are not well known locally, the true number of species may be higher. All Australian shovelnose rays live primarily on soft substrates and feed mainly on shellfish and other bottom-dwelling invertebrates.

Key to rhinobatids

1 Internasal flap present (fig. 1); disc oval, snout broadly rounded (fig. 3) 2

   No internasal flap (fig. 2); disc not oval, snout triangular (fig. 13) 4

2 Upper surface bluish black with pale disc margin (fig. 3); no thorns around eye ....
   .... *Trygonorrhina melaleuca* (fig. 3; p. 294)

   Upper surface brownish with elaborate pattern of dark-edged bands (figs 6, 7); thorns normally present around eye 3

3 Distinctive dark triangular or diamond-shaped marking behind eyes (fig. 4) ....
   .... *Trygonorrhina* sp. A (fig. 6; p. 291)

   No dark triangular or diamond-shaped marking behind eyes (3 pale parallel stripes instead) (fig. 5)
   .... *Trygonorrhina fasciata* (fig. 7; p. 292)

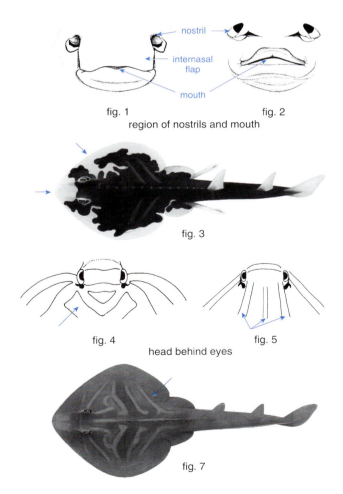

4 Nostrils oblique, much wider than internasal space (figs 8, 9); mouth horizontal (fig. 8); spiracles with 2 distinct skin folds (fig. 11) **5**

Nostrils almost transverse, their width equal to or narrower than internasal space (fig. 10); mouth strongly curved forward (fig. 10); spiracles without distinct skin folds (sometimes with low bump) (fig. 12) **6**

5 Eye small, snout length more than 5 times orbit diameter (fig. 13); thorns and enlarged denticles present along centre of disc (fig. 13); anterior nasal aperture almost rectangular (fig. 8) .... ***Rhinobatos typus* (fig. 13; p. 290)**

Eye relatively large, snout length less than 5 times orbit diameter (fig. 14); no thorns or enlarged denticles along centre of disc; anterior nasal aperture circular (fig. 9) .... ***Rhinobatos* sp. A (fig. 14; p. 289)**

6 Upper surface covered with white spots (fig. 15); snout apex very slender (fig. 15) .... ***Aptychotrema* sp. A (fig. 15; p. 285)**

Upper surface not covered with white spots (plain or with dark blotches); snout apex rather broad, tip normally more rounded (fig. 19) **7**

7 Snout very long, mostly more than 6 times longer than orbit diameter (fig. 17); front skin fold of anterior aperture of nostril relatively broad (less so in adults) (fig. 16); eastern Australia .... ***Aptychotrema rostrata* (fig. 17; p. 286)**

Snout long, mostly less than 6 times longer than orbit diameter (fig. 19); front skin fold of anterior aperture of nostril relatively narrow (fig. 18); western and southern Australia .... ***Aptychotrema vincentiana* (fig. 19; p. 287)**

fig. 16
left nostril

fig. 18
left nostril

fig. 8   fig. 9
region of nostrils and mouth

fig. 11

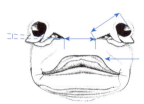

fig. 10   fig. 12
region of nostrils and mouth   region of eye and spiracle

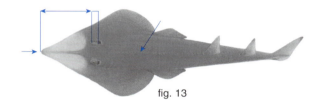

fig. 13

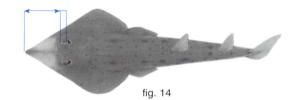

fig. 14

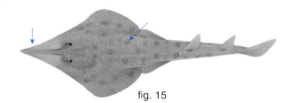

fig. 15

fig. 17

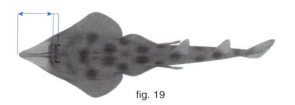

fig. 19

Spotted Shovelnose Ray — *Aptychotrema* sp. A

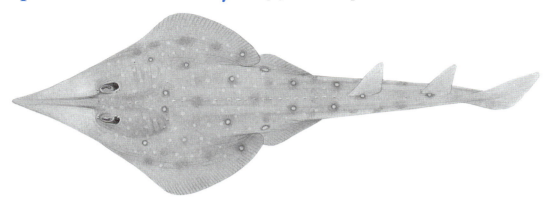

32.1 *Aptychotrema* sp. A (Plate 40) Fish Code: 00 027007

- **Field characters:** A small, white-spotted, shovelnose ray with a narrow triangular snout, curved mouth, small eyes, a short lower caudal-fin lobe, and relatively narrow and transverse nostrils with no internasal flap.

- **Distinctive features:** Disc wedge-shaped; snout long (preorbital snout about 5.4 times orbit length), narrow; tip rather slender; anterior margin slightly concave. Pectoral-fin tips rounded. Spiracle lacking skin folds. Mouth curved slightly; upper lip strongly arched. Eyes small. Nostrils nearly transverse, slightly narrower than internasal space. Dorsal surface uniformly granular; 2 short thorns before eye; 3–4 thorns near spiracle; about 27 widely-spaced thorns along midline of disc and tail before first dorsal fin; sometimes with 2 enlarged thorns on snout tip and a few on each shoulder; thorns strong, compressed slightly. Tail longer than disc. Dorsal fins widely spaced, subequal in size, tips rather angular; first dorsal fin behind pelvic-fin tips. Caudal fin without prominent lower lobe. Claspers of mature male long, very slender, slightly bulbous distally. Total vertebrae 165*; monospondylous 36*; caudal 38*.

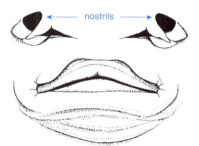

region of nostrils and mouth (adult)

upper and lower jaw teeth

- **Colour:** Brownish above with a peppering of light, widely-spaced spots from eye level to caudal fin (each spot smaller than pupil of eye); spots with dusky borders, sometimes paired (notably on shoulder, above pectoral and pelvic insertions, before first dorsal fin, near origins of pelvic and dorsal fins, and near free rear tip of pectoral fins). Ventral surface uniformly pale; snout tip pale rather than dark.

- **Size:** To at least 51 cm.

- **Distribution:** Known from a few specimens collected from the Timor Sea, off Melville Island (Northern Territory) in about 120 m.

- **Remarks:** This rather distinctive member of the genus is undescribed. It is known from only a few specimens and additional material is required.

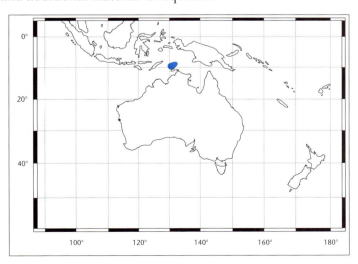

- **Local synonymy:** *Aptychotrema* sp. 2: Sainsbury *et al.*, 1985; *Aptychotrema* sp.: Gloerfelt-Tarp & Kailola, 1984; *Aptychotrema* sp.: Allen & Swainston, 1988.
- **Reference:** Sainsbury *et al.* (1985).

Eastern Shovelnose Ray — *Aptychotrema rostrata* (Shaw & Nodder, 1794)

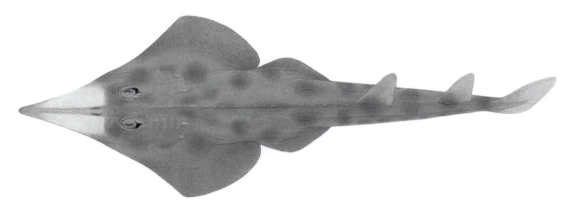

32.2 *Aptychotrema rostrata* (Plate 40)   **Fish Code:** 00 027009

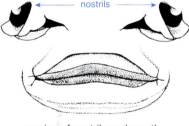

region of nostrils and mouth (juvenile)

- **Alternative names:** Banks shovelnose ray, eragoni, shovelnose shark.
- **Field characters:** A medium-sized, greyish brown shovelnose ray (either plain coloured or ornamented with dark blotches) with a very long and triangular snout, small eyes, curved mouth, a short lower caudal-fin lobe, and relatively narrow and transverse nostrils with no internasal flap. Juveniles have a relatively broad fleshy lobe on the margin of the anterior nasal aperture.
- **Distinctive features:** Disc wedge-shaped; snout very long (preorbital snout 5.7–7.6 times orbit length); tip rather broadly rounded; anterior margin weakly double concave. Pectoral-fin tips relatively angular. Spiracle without distinct skin folds, mostly with a low, fleshy lump. Mouth strongly curved; upper lip strongly arched. Eyes relatively small. Nostrils nearly transverse, length equal to or slightly shorter than internasal space; outer anterior lobe and inner posterior lobe relatively broad (less so in adults). Dorsal surface uniformly granular; 2–3 short thorns before eye; 1–2 thorns near spiracle; about 18–20 short, widely-spaced thorns along midline of disc and predorsal tail; 2 enlarged thorns on snout tip; 2 groups of thorns on each shoulder (barely distinguishable in adults); thorns small (sharper in juveniles), compressed slightly. Tail longer than disc. Dorsal fins widely spaced, subequal in size, apex narrowly rounded; first dorsal fin situated behind pelvic-fin tips. Caudal fin without prominent lower lobe. Claspers of mature male long, very slender, slightly bulbous distally. Total vertebrae 171–172*; monospondylous 39*; caudal 41–44*.
- **Colour:** Greyish brown above; central disc with or without dark or dusky blotches; paler beside rostral cartilage; blotches about equal in size (three or four on central disc, two before each dorsal fin, one at base of each dorsal fin); rarely dark around orbit. Ventral surface white with irregular dark flecks; snout of juveniles and some adults with broad black tip.
- **Size:** Reported to reach at least 120 cm (largest examined 85 cm); largest mature male 72 cm.

- **Distribution:** Continental slope off eastern Australia between Moreton Bay (Queensland) and Jervis Bay (New South Wales); mainly inshore but to depths of 50 m or more.
- **Remarks:** This ray is fairly common in lower parts of estuaries and off beaches. It is often caught in nets but will also take a hook. It is good to eat and is frequently marketed as skate. An active predator and scavenger, the eastern shovelnose ray feeds mainly on sand-dwelling crustaceans, molluscs and fish. Another shovelnose ray, *A. bougainvillii*, also thought to occur off New South Wales and southern Queensland, is doubtfully distinct from this species.
- **Local synonymy:** *Rhinobatus banksii* Müller & Henle, 1841; *Rhinobatus tuberculatus* Macleay, 1882.

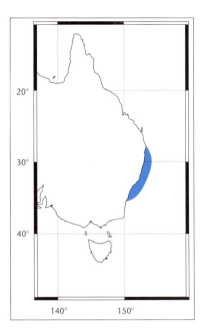

Western Shovelnose Ray — *Aptychotrema vincentiana* (Haake, 1885)

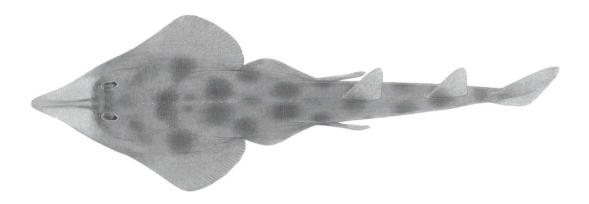

32.3 *Aptychotrema vincentiana* (Plate 40) Fish Code: 00 027001

- **Alternative name:** Southern shovelnose ray, yellow shovelnose ray.
- **Field characters:** A medium-sized, yellowish brown shovelnose ray (adults ornamented with dark blotches) with a long and triangular snout, small eyes, curved mouth, a short lower caudal-fin lobe, and relatively narrow and transverse nostrils with no internasal flap. Juveniles plain coloured with a narrow fleshy lobe on the margin of the anterior nasal aperture.
- **Distinctive features:** Disc wedge-shaped; snout long (preorbital snout 5.1–5.9 times orbit length); tip rather broadly rounded; anterior margin weakly double concave. Pectoral-fin tips relatively angular to narrowly rounded (less so in juveniles).

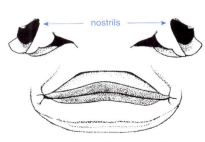

region of nostrils and mouth

Spiracle lacking distinct skin folds, mostly with a low fleshy lump. Mouth strongly curved (more so in adult males); upper lip strongly arched. Eyes rather small. Nostrils nearly transverse, about equal to internasal space; outer anterior lobe and inner posterior lobe relatively narrow. Dorsal surface uniformly granular; 2–4 short thorns before eye; 1–2 thorns near spiracle; 18–20 short, widely-spaced thorns along midline of disc and tail before first dorsal fin; sometimes with 2 enlarged thorns on snout tip and 2 small groups on each shoulder; thorns small (sharper in juveniles), compressed slightly. Tail longer than disc. Dorsal fins widely spaced, subequal in size, their apices narrowly rounded; first dorsal fin behind pelvic-fin tips. Caudal fin without prominent lower lobe. Claspers of mature male long, very slender, slightly bulbous distally. Total vertebrae 167–171*; monospondylous 35–36; caudal 44–47.

- **Colour:** Yellowish brown above with dark blotches on central disc and tail (mostly more distinct in adults); prominent black bands usually extending diagonally forward from eye in live specimens; similar narrower bands often on midsnout. Juveniles uniformly coloured (lacking blotches), paler beside rostral cartilage. Blotches about equal in size (2–3 on central disc, 2 before each dorsal fin, one at base of each dorsal fin); dark areas around orbit. Ventral surface white with irregular flecks; snout tip black in juveniles.
- **Size:** To at least 79 cm; males mature at about 65 cm.
- **Distribution:** Southern and western Australia from the Kent Islands (Bass Strait) to Port Hedland (Western Australia) to depths of 32 m.
- **Remarks:** Tropical northern populations occur mainly on the mid-continental shelf whereas along the south coast this species is more common near the shore. In the south-west, juveniles are often caught in beach seines as they migrate into the shallows with the rising tide. There are also subtle shape differences between the two populations suggesting that more than one species may be involved.
- **Local synonymy:** *Aptychotrema* sp. 1: Sainsbury *et al.*, 1985.
- **Reference:** Sainsbury *et al.* (1985).

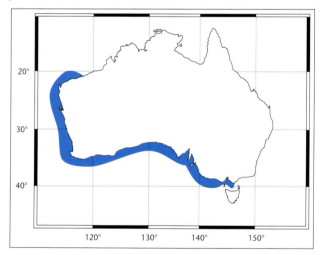

Goldeneye Shovelnose Ray — *Rhinobatos* sp. A

32.4 *Rhinobatos* sp. A (Plate 39)　　　　　　　　　　Fish Code: 00 027003

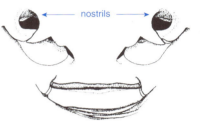

region of nostrils and mouth

- **Field characters:** A small, yellowish shovelnose ray with a relatively short, broadly triangular snout, expanded nostrils with enlarged nasal lobes (no internasal flap), and a circular anterior nasal aperture. It has no thorns or enlarged denticles on the dorsal midline and the lower caudal-fin lobe is short.

- **Distinctive features:** Disc wedge-shaped; snout moderate (preorbital snout 3.3–3.9 times orbit length); tip relatively broad; anterior margin almost straight. Pectoral-fin tips broadly rounded. Spiracle with 2 distinct skin folds (outer more prominent). Mouth and upper lip almost straight. Eyes moderately large. Nostrils oblique, longer than internasal space (1.4–1.6 times internasal space in adults, smaller in juveniles); anterior aperture almost circular; inner posterior lobe greatly enlarged (bigger than anterior aperture). Dorsal surface uniformly granular; thorns absent. Tail longer than disc. Dorsal fins widely spaced, subequal in size; first dorsal fin behind pelvic-fin tips. Caudal fin without prominent lower lobe. Claspers long, very slender, slightly bulbous distally. Total vertebrae 177–185; monospondylous 35–40; caudal 41–44.

- **Colour:** Mostly pale yellowish brown, sometimes with faint golden brown or dusky blotches on upper disc and tail (occasionally with prominent, small, dark brown spots); snout beside rostral cartilage, hind margin of disc and pelvic fins, and dorsal and caudal fins pale; eye golden. Ventral surface white.

- **Size:** To at least 56 cm; males mature at about 55 cm. Smallest specimen examined was 22 cm.

- **Distribution:** Common on the continental shelf off northwestern Australia between Dampier and Broome in 70–200 m.

- **Remarks:** There are two sympatric forms of this species which differ in their colour patterns (either intensely spotted or plain), but which appear to have almost identical vertebral counts and similar shapes. Their relationship to each other requires further study.

- **Local synonymy:** *Rhinobatos* sp. 1: Gloerfelt-Tarp & Kailola, 1984; *Rhinobatos* sp. 1: Sainsbury *et al.*, 1985; *Rhinobatos* sp. 2: Sainsbury *et al.*, 1985; *Rhinobatos* sp.: Allen & Swainston, 1988.

- **Reference:** Sainsbury *et al.* (1985).

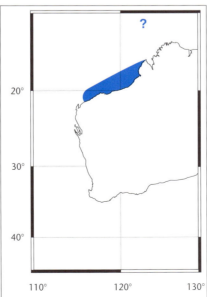

Giant Shovelnose Ray — *Rhinobatos typus* Bennett, 1830

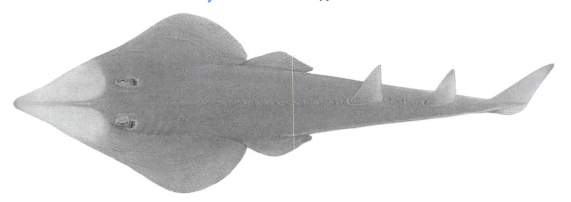

32.5 *Rhinobatos typus* (Plate 39)　　　　　　　　**Fish Code:** 00 027010

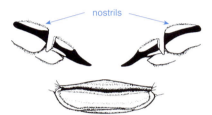

region of nostrils and mouth

- **Alternative name:** Common shovelnose ray.
- **Field characters:** A medium-sized, greyish brown shovelnose ray with a relatively short, broad triangular snout, very broad oblique nostrils with small nasal lobes (no internasal flap), and a rectangular anterior nasal aperture. It lacks a pronounced lower caudal-fin lobe, and has small thorns and a band of enlarged denticles along the dorsal midline.
- **Distinctive features:** Disc wedge-shaped; snout long (preorbital snout 6.2–7.1 times orbit length), broad; tip broadly rounded; anterior margin almost straight. Pectoral-fin tips broadly rounded. Spiracle with 2 low, widely-separated skin folds (outer frequently more prominent). Mouth and upper lip almost straight. Eyes very small. Nostrils oblique, much longer than internasal space (1.9–2.2 times longer); anterior aperture almost rectangular; inner posterior lobe rather small (much smaller than anterior aperture). Dorsal surface uniformly granular, denticles enlarged near midline (largest near median row of thorns); thorns small, only in median row along disc and tail, around eye and in 2 patches on shoulders (better developed in juveniles). Tail longer than disc. Dorsal fins widely spaced, subequal in size; first dorsal fin behind pelvic-fin tips. Caudal fin without prominent lower lobe. Claspers of mature male long, very slender, slightly bulbous distally.
- **Colour:** Upper surface greyish brown to olive along central disc and tail; distinctly paler and yellowish around pectoral, pelvic, dorsal and caudal fin margins; snout pale yellow with dark brown stripe over rostral cartilage, becoming abruptly greyish brown just before eyes. Ventral surface uniformly pale; snout tip sometimes greyish.

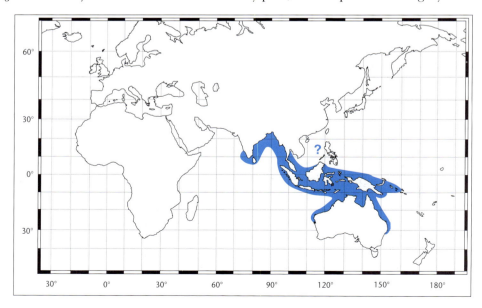

- **Size:** Reaches at least 270 cm.
- **Distribution:** Widely distributed in the Indo–Pacific from India, east to Australia (possibly also Melanesia). Locally in northern waters, from Shark Bay (Western Australia) to Forster (New South Wales). Juveniles occur inshore around atolls, in mangrove swamps and occasionally venture into estuaries; adults are found deeper on the continental shelf to about 100 m.
- **Remarks:** This large ray is common inshore, with juveniles moving over shallow sandflats during the incoming tide to feed on shellfish. Reported to be able to live and breed permanently in freshwater. Another rather rough-skinned species, *Rhinobatos granulatus*, co-occurs with this species at the western extremity of its range (off India) but has not been positively identified from the Australian region.
- **Local synonymy:** *Rhinobatus granulatus* Macleay, 1881 (misidentification); *Rhinobatus armatus* Gray, 1834; *Rhinobatos batillum* Whitley, 1939.
- **References:** Gray (1834); Macleay (1881b); Whitley (1939b); Grant (1978).

Eastern Fiddler Ray — *Trygonorrhina* sp. A

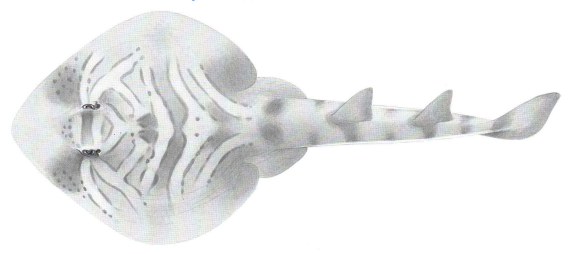

32.6 *Trygonorrhina* sp. **A** (Plate 41) **Fish Code:** 00 027006

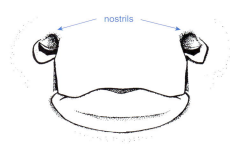

region of nostrils and mouth

- **Alternative names:** Fiddler ray, banjo shark.
- **Field characters:** A medium-sized shovelnose ray with an oval disc, nostrils partly covered by a fleshy internasal flap, a short lower caudal-fin lobe, and with an ornate pattern of dark-edged bands and a distinct triangular or diamond-shaped marking behind the interorbital space.
- **Distinctive features:** Disc almost oval, shorter than tail length; snout semicircular, rather short (preorbital snout about 4* times orbit length), obtusely pointed. Spiracle with single skin fold. Internasal flap present, broad; nostrils connected to mouth by groove; mouth broad, almost straight to slightly convex. Eyes very small. Dorsal surface uniformly granular; denticles minute, slightly enlarged on centre of disc and along upper tail; enlarged thorn-like denticles along dorsal midline of disc (about 18* predorsally); single preorbital and postorbital thorns; thorns in 2 rows on shoulder. Tail almost semi-spherical in cross-section, rather depressed, broad-based. Dorsal fins separated by horizontal length of first dorsal fin or less, fins subequal in size; first dorsal-fin origin behind pelvic fin. Caudal fin without prominent lower lobe.

- **Colour:** Upper surface light brown with distinctive pattern of transverse lilac bands (with dark brown margins); bands forming a distinctive dark triangular or diamond-shaped marking behind eyes; dark brown spots often present near disc margin and midline anterior to eyes; snout pale. Ventral surface white. Dorsal and caudal fins pale.
- **Size:** Reported to attain 120 cm; largest specimen observed 92 cm.
- **Distribution:** Eastern Australia from southern Queensland south to at least Twofold Bay (New South Wales) in depths to 100 m. Records further south and west have yet to be substantiated.
- **Remarks:** Fiddler rays were named because of their shape and scroll-like colour pattern resembling the holes of a violin. This species is fairly common in shallow water throughout its range. It is an active scavenger and will enter fish traps, often becoming ensnared by the disc. The young fiddlers develop in golden egg-cases (2–3 embryos in each) inside the mother and are born alive.
- **Local synonymy:** *Trygonorrhina fasciata*: Whitley, 1940 (misidentification).
- **Reference:** Grant (1978).

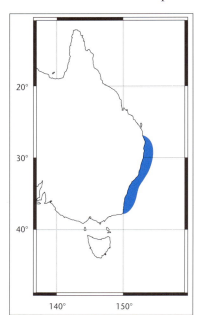

Southern Fiddler Ray — *Trygonorrhina fasciata* Müller & Henle, 1841

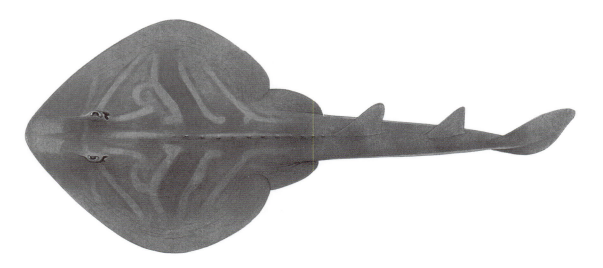

32.7 *Trygonorrhina fasciata* (Plate 41)   **Fish Code:** 00 027002

- **Alternative names:** Fiddler, fiddler ray, banjo shark, green skate, parrit, southern fiddler.

- **Field characters:** A medium-sized shovelnose ray with an oval disc, nostrils partly covered by a fleshy internasal flap, an ornate pattern of dark-edged bands but without a distinct triangular or diamond-shaped marking behind the interorbital space, and a poorly defined lower caudal-fin lobe.

- **Distinctive features:** Disc almost oval, shorter than tail length; snout semicircular, relatively short (preorbital snout 2.8–4.2 times orbit length), broadly rounded. Spiracle with one very enlarged skin fold. Internasal flap broad; nostrils connected to mouth by groove; mouth broad, almost straight to slightly convex. Eyes very small. Dorsal surface uniformly granular; denticles minute, slightly enlarged on centre of disc and along upper tail; enlarged thorn-like denticles along dorsal midline of disc (about 12–16 predorsally); single preorbital, postorbital and spiracular thorns; thorns in 2 rows on shoulder (outer row longest); denticles covering bases of thorns forming small mounds; thorn tips small, blunt. Tail almost semi-spherical in cross-section, rather depressed, broad-based. Dorsal fins separated by about horizontal length of first dorsal fin or slightly less, about equal in size; first dorsal-fin origin behind pelvic fin. Caudal fin without prominent lower lobe.

- **Colour:** Upper surface, and dorsal and caudal fins, yellowish to brown with distinct dark-edged bluish grey transverse bands; bands radiating posteriorly and laterally from eye, mid-disc and longitudinally behind eye (as 3 distinct stripes and not forming a triangular marking); darker band between eyes extending forward obliquely to disc margin (more prominent in juveniles); disc and pelvic-fin margins pale. Ventral surface and lateral skin folds uniformly pale.

- **Size:** Reported to attain 126 cm; largest specimen observed 97 cm.

- **Distribution:** Southern Australia from eastern Bass Strait (including the northern Tasmanian coast) to Lancelin (Western Australia) in shallow water.

- **Remarks:** Common near the coast throughout much of its range. It is frequently seen by divers around jetties and wharves or near seagrass meadows. The flesh is good to eat, although only a small quantity is sold in seafood outlets. A shovelnose ray, *Rhinobatus dumerilii*, described from Western Australia in the last century and which has not been collected since, is likely to be this species.

- **Local synonymy:** *Rhinobatus dumerilii* Castelnau, 1873; *Trygonorhina fasciata guanerius* Whitley, 1932; *Trygonorhina guanerius*: Last *et al.*, 1983.

- **References:** Castelnau (1873a); Scott *et al.* (1980).

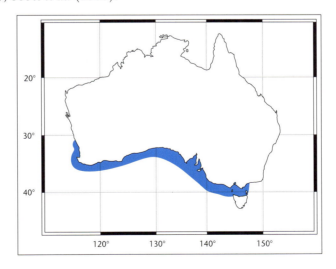

Magpie Fiddler Ray — *Trygonorrhina melaleuca* Scott, 1954

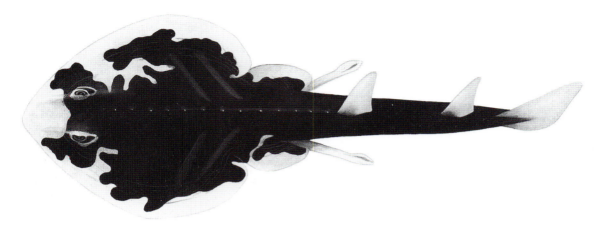

32.8 *Trygonorrhina melaleuca* (Plate 41)   **Fish Code:** 00 027008

- **Field characters:** A medium-sized shovelnose ray with an oval disc, nostrils partly covered by a fleshy internasal flap, a bluish black upper surface with a pale disc margin (not ornamented with dark-edged bands), and a short lower caudal-fin lobe.

- **Distinctive features:** Disc almost oval, shorter than tail length; snout semicircular, rather short (preorbital snout about 4* times orbit length), obtusely pointed. Spiracle large. Internasal flap broad; nostrils connected to mouth by groove; mouth broad. Dorsal surface uniformly granular; denticles minute, slightly enlarged on centre of disc and along upper tail; enlarged, thorn-like denticles along dorsal midline of disc (about 13 predorsally); orbital thorns absent; thorns in 2 rows on shoulder. Tail almost semi-spherical in cross-section, rather depressed, broad-based. Dorsal fins separated by at least length of first dorsal fin; first dorsal fin slightly taller than second dorsal; first dorsal-fin origin behind pelvic fin. Caudal fin without prominent lower lobe.

- **Colour:** Upper surface of disc and tail bluish black; pale, irregularly-shaped markings along disc and pelvic-fin margins; 2 rather faint, oblique, dark grey bars on each side of central disc. Ventral surface white. Dorsal and caudal fins pale.

- **Size:** Reaches at least 90 cm.

- **Distribution:** Known only from a few specimens taken from shallow water in St Vincents Gulf, South Australia.

- **Remarks:** No specimens are available for examination in museum collections. This species appears to be exceptionally rare and may be a mutant form of the southern fiddler ray (32.7). If collected, specimens should be donated to a museum.

- **References:** Scott (1954); Scott *et al.* (1980).

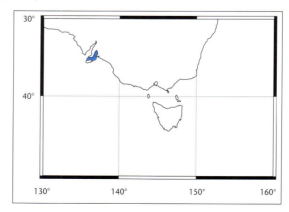

33

Family Rhynchobatidae

SHARKFIN GUITARFISHES

Sharkfin guitarfishes are medium to large, shark-like rays with the pectoral fins and head distinct from each other, and with a long, broad and rather flattened tail. They have two large dorsal fins, the first dorsal fin originating forward of the pelvic-fin insertions, and both fins lacking stinging spines. Both upper and lower lobes of the caudal fin are well developed and its posterior margin is deeply concave. The teeth are small, blunt and numerous. The nostrils are large, but are not covered with an internasal flap. Their spiracles are rather large, and frequently have small (but prominent) skin folds along their hind margin.

This family, represented worldwide by 2 genera and possibly 5 or more species, is most diverse in the Indo–West Pacific. Both genera and at least 2 species are represented in the Australian region but additional members may occur here. The taxonomy of the genus *Rhynchobatus* is in need of more research. Sharkfin Guitarfishes are amongst the bulkiest rays, and some species reach in excess of 3 m.

Key to rhynchobatids

1 Snout broadly rounded (fig. 1); back with ridges bearing large thorns (fig. 1); no spiracular folds (fig. 2) .... ***Rhina ancylostoma*** (fig. 1; p. 296)

   Snout triangular (fig. 4); back with small thorns (fig. 4); two spiracular folds (fig. 3) .... ***Rhynchobatus djiddensis*** (fig. 4; p. 297)

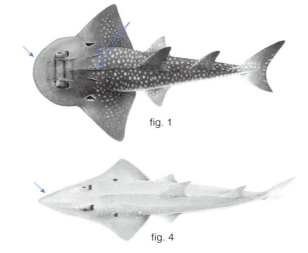
fig. 1

fig. 2 fig. 3 fig. 4
region of eye and spiracle

Shark Ray — *Rhina ancylostoma* Bloch & Schneider, 1801

33.1 *Rhina ancylostoma*   (Plate 42)   **Fish Code:** 00 026002

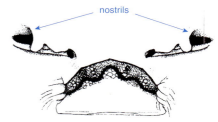

region of nostrils and mouth

teeth from near symphysis (upper and lower jaw)

- **Alternative names:** Bowmouth guitarfish, mud skate.
- **Field characters:** A very deep-bodied guitarfish with its head distinctly demarcated from its pectoral fins, a broadly rounded snout, large thorns on the back located on horny ridges, and no spiracular folds.
- **Distinctive features:** Broadly rounded anteriorly; head clearly distinct from pectoral fins; body greatly thickened centrally; snout relatively short (preorbital length about 4* times orbit diameter). Spiracle very large, without skin folds. Lower jaw distinctly trilobed; lobes recessing into concavities in upper jaw; anterior labial furrows poorly developed. Eyes moderately large. Nostrils elongate, almost transverse, curved slightly posteriorly towards mouth, equal to or slightly less than internasal space; skin folds well developed; anterior aperture oval, inner posterior lobe rather well developed. Anterior gill openings very widely separated, only about an eye diameter from disc margin. Dorsal surface uniformly granular, denticles minute. Thorns greatly enlarged, very broad-based, somewhat compressed, subtriangular, with sharp tips; located on prominent horny ridges on midline of back (finishing well before first dorsal fin), above eye (extending to beyond spiracle), in smaller patch anterior to eye and in 2 slender patches on shoulder (inner patch longer). Tail much longer than disc, rather deep, broad-based. Dorsal fins very tall, distinctly falcate with slightly concave posterior margins; relatively close together (horizontal length of first dorsal fin exceeding interdorsal space); first dorsal fin distinctly larger than second dorsal; first dorsal-fin origin about over pelvic-fin origin. Pectoral fins triangular, apices pointed; posterior margin slightly longer than anterior margin. Caudal fin greatly enlarged, with lower lobe more than half length of upper lobe; posterior margin slightly concave.
- **Colour:** Bluish grey on raised portion of disc and tail, dotted with large white spots; prominent white-edged black blotch above pectoral-fin base and dark bands between eyes; paler greyish to white around margin of pectoral fin and snout. Ventral surface mostly pale. Dorsal and caudal fins similar to body, occasionally with white spots. Colour markings more pronounced in juveniles; adults frequently brownish with faint spots and lines.
- **Size:** Reported to reach 270 cm.
- **Distribution:** Widely distributed in the Indo–Pacific west to Natal (South Africa). Locally, off northern Australia from Exmouth Gulf (Western Australia) to Forster (New South Wales) in coastal areas and on offshore reefs.

- **Remarks:** This most unusual ray is caught commercially off Asia, but is rarely marketed locally. Large specimens, which may weigh in excess of 125 kg, are a nuisance to trawl fisherman. In the confined working spaces of a small prawn trawler they are difficult to handle and may also damage the commercial catch. The shark ray feeds on crabs and shellfish.
- **Reference:** Gordon (1992).

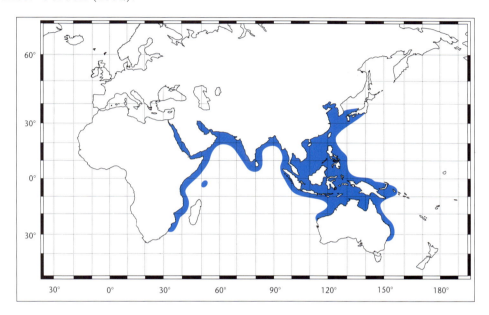

White-spotted Guitarfish — *Rhynchobatus djiddensis* (Forsskål, 1775)

33.2 *Rhynchobatus djiddensis* (Plate 42)
Fish Code: 00 026001

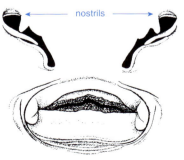

region of nostrils and mouth

- **Alternative names:** Giant guitarfish, sandshark, whitespot ray, whitespot shovelnose ray.
- **Field characters:** A very large guitarfish with its head more or less distinct from its pectoral fins, an almost triangular snout, small thorns on the back (not located on raised ridges), and with 2 large spiracular folds.
- **Distinctive features:** Disc wedge-shaped; centre of disc raised evenly; snout moderately long (preorbital length about 4.7–6.3 times orbit diameter), broadly triangular, tip bluntly rounded; disc margin concave beside eye. Spiracle with 2 prominent, equal-sized skin folds. Middle of lower jaw forming a prominent hump, recessing into similarly-shaped concavity in upper jaw; upper labial furrows present. Eyes rather large. Nostrils very elongate (longer than internasal space), oblique,

curved posteriorly towards mouth; skin folds poorly developed; anterior aperture almost oval, inner posterior lobe reduced. Anterior gill openings very widely separated, only about an eye diameter from disc margin. Surface uniformly granular, denticles minute. Thorns small, short, blunt; present along midline, behind eye, on shoulder (in 2–3 short rows) and around upper margin of eye. Tail much longer than disc, rather deep and very broad-based; caudal peduncle broad, tapering, very depressed. Dorsal fins widely spaced, distinctly falcate with deeply concave posterior margins; first dorsal fin distinctly larger than second dorsal; first dorsal-fin origin over or just behind pelvic-fin origin. Pectoral-fin apices angular, posterior margin relatively short. Caudal fin lower lobe more than half length of upper lobe; posterior margin deeply concave. Claspers very slender.

upper and lower jaw teeth

- **Colour:** Upper surface, and dorsal and caudal fins, mostly greyish to yellowish brown; 10–30 distinctive white spots (mostly less than half eye diameter in size and sometimes dark-edged) extending from mid-pectoral fin along side of body to posterior tip of first dorsal fin; sometimes with additional white spots below first dorsal fin and usually with prominent white spot on snout midline just in advance of eye; 1–2 black spots above base of pectoral fin (sometimes with black spots and streaks around eye and on orbital membrane in juveniles); margin of disc and pelvic fins pale. Ventral surface uniformly pale with some irregularly placed darker blotches (always present on snout tip).
- **Size:** Attains at least 300 cm; males mature at about 110 cm.
- **Distribution:** Continental shelf of tropical and warm temperate Australia from Fremantle (Western Australia) to Coffs Harbour (New South Wales). Similar forms recorded from other parts of the Indo–Pacific from southern Africa to New Caledonia (including the Red Sea), and north to Japan.
- **Remarks:** This rather stocky ray, reaching more than 220 kg in weight, is very inquisitive and will approach divers underwater. It is characteristically ray-like in its habit of propping off the bottom on its pectoral fins, but is more shark-like in appearance when swimming. It has a reputation as an excellent game fish and is caught mainly with crayfish or pilchards as bait. A reasonable food fish, it is sometimes sold at fried fish outlets in Queensland and Western Australia. The fins command a high price on Asian markets. It is represented in this region by 2 forms that may be separate species. Smaller specimens, including some mature males, are pale in colour and have distinctive white spots along the side. Some large individuals are dark, almost black, and appear to lack white spots. Other forms which may be separate species exist outside Australia.
- **Local synonymy:** *Rhynchobatus djiddensis australiae* Whitley, 1939.
- **References:** Whitley (1939b); Grant (1978).

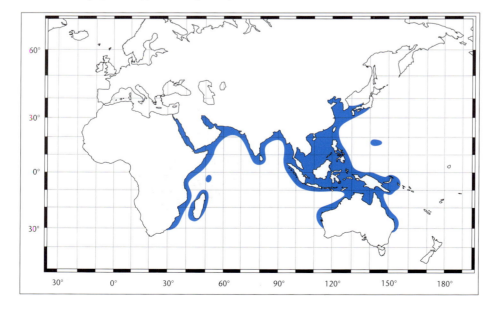

34

Family Rajidae

SKATES

Skates are small to large rays with discs varying in shape from almost circular to rhomboidal. The snout, which is variable in shape, is mostly supported by a central cartilage, the shape of which is important in distinguishing genera. They have a rather slender tail that usually has two (rarely one or none) small dorsal fins near its apex. The caudal fin is very small or barely distinguishable. The pelvic fins are generally deeply notched and the relative lengths of the anterior and posterior lobes vary within the group. Sharp thorns are present on most species and their distribution around the eyes, and on the shoulders, dorsal midline and tail are important characters for distinguishing between species.

This widely distributed and diverse family is found worldwide in all oceans except for insular areas of the western Pacific. Skates are primarily marine on the continental slopes to more than 2000 m. In some temperate areas they occur inshore and one Australian species is common in estuaries. The family contains at least seven genera and almost 200 species. Several of the genera contain subgenera which, in time, may receive generic status. The Australian fauna consists of 38 species from at least five genera. The subgeneric placements of these species are given in the key below where known. Unassigned species of the complex genera *Raja* and *Pavoraja* do not conform to current definitions and require further investigation.

Key to rajids

1   Snout supported by a firm cartilage; rostral cartilage broad and flattened (fig. 1), visible and/or detectable by touch from just in front of eye for more than three quarters of snout length **2**

    Snout flexible; rostral cartilage either absent, incomplete, rod-like or very narrow with its width not exceeding its depth (offering more substantial support in some *Notoraja* spp.) **24**

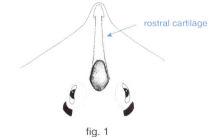

fig. 1
upper surface of snout

2   No orbital thorns; both upper and lower surfaces of disc uniformly covered with dense granulations in specimens exceeding 20 cm disc width ....
    .... ***Raja whitleyi*** **(fig. 2; p. 355)**

    Orbital thorns present; upper and lower surfaces of disc not uniformly covered with dense granulations (smooth on the ventral surface apart from on the anterior disc margin and snout) **3**

fig. 2

3 Thorns in a more or less triangular patch in nuchal area and/or with 2–3 thorns on each shoulder (fig. 3) 4

Thorns in nuchal area in a single median row or absent (not forming a triangular patch and absent from shoulders) 6

4 Pectoral-fin tips angular (fig. 3); tail less than three-quarters of precloacal length; dorsal surface with dark blotches (fig. 3) ....
.... ***Raja (Amblyraja) hyperborea*** **(fig. 3; p. 350)**

Pectoral-fin tips broadly rounded (fig. 4); tail longer than three-quarters of precloacal length; dorsal surface uniform in colour (sometimes irregular where skin is removed) 5

5 Ventral surface of disc dark; pale dorsally ....
.... ***Raja (Rajella)* sp. P (fig. 4; p. 345)**

Ventral surface of disc pale; pale or only slightly darker dorsally ....
.... ***Raja (Leucoraja)* sp. O (fig. 5; p. 344)**

6 Distance from first dorsal-fin origin to tail tip (when undamaged) exceeding distance from snout tip to spiracle (fig. 6); interdorsal distance greater than or equal to eye diameter ....
.... ***Raja (Okamejei)* sp. N (fig. 6; p. 342)**

Distance from first dorsal-fin origin to tail tip less than distance from snout tip to spiracle (fig. 9); interdorsal distance shorter than eye diameter 7

7 Snout very long and slender, preorbital length more than 5 times orbit diameter (occasionally slightly less in juveniles) (fig. 7); pectoral-fin tips rather angular (fig. 7) 8

Snout only moderately elongate, preorbital length less than 5 times orbit diameter (fig. 8); pectoral-fin tips narrowly rounded (may be more angular in *Raja* sp. C) (fig. 8) 15

8 Mucous pores not dark-edged and barely visible against blackish snout (fig. 9); undersurface of disc dark, often almost black (slightly darker than dorsal surface); granular portion of anterior ventral margin pale (fig. 9); snout not extremely elongated (fig. 9) .... ***Raja (Dipturus)* sp. I (fig. 9; p. 335)**

Mucous pores dark-edged and distinct on snout (fig. 10); colour not as above (if undersurface darker than dorsal surface then snout very long and pointed); snout extremely elongated (except in juveniles and in *Raja* sp. H) 9

9   Disc very short and broad, width more than 1.25 times length (fig. 11); anterior margin of disc undersurface darker than area adjacent (more obvious in small specimens) (fig. 11) .... ***Raja (Dipturus)* sp. H (fig. 11; p. 333)**

   Disc relatively long and narrow, width less than 1.25 times length (fig. 12); anterior margin of disc under-surface rarely darker than area adjacent .. **10**

10 No malar thorns in adult males; pectoral-fin apex not especially angular; no nuchal thorns **11**

   Malar thorns present in adult males; pectoral-fin apex normally sharply angular (fig. 17) (less so in *Raja* sp. L); nuchal thorn usually present (usually absent in *Raja* sp. F and rarely absent in *Raja* sp. K) **12**

11 Caudal fin and outer half of dorsal fins pale; rarely more than 3 thorns around eye and not forming rosette in adults (fig. 12); juveniles lacking dark blotches and adults greyish brown above .... ***Raja (Dipturus)* sp. G (fig. 12; p. 332)**

   Caudal and dorsal fins dark or dusky; normally 3 or more thorns around eye that form a rosette in adults (fig. 13); juveniles with dark blotches and adults greyish to greyish green above .... ***Raja (Dipturus) gudgeri* (fig. 13; p. 349)**

12 Upper disc almost black, darker than lower surface; thorns surrounded by denticles around eyes and along middle of disc and tail .... ***Raja* sp. L (fig. 14; p. 339)**

   Colour not as above, upper disc pale to medium and normally paler than undersurface; denticles sparse or absent from around eyes and along middle of disc and tail **13**

13 Caudal fin whitish, distinctly paler than dorsal fins (fig. 15); tropical eastern Australia .... ***Raja (Dipturus)* sp. K (fig. 15; p. 338)**

   Caudal fin dark, similar in colour to dorsal fins **14**

14 Outer half of pectoral fins dark (fig. 16) (black on juveniles); snout very extended in adults (fig. 16); nuchal thorn present; southern Australia .... ***Raja (Dipturus)* sp. J (fig. 16; p. 336)**

   Outer half of pectoral fins pale (fig. 17); snout relatively less extended (fig. 17); rarely with a nuchal thorn; tropical western Australia .... ***Raja (Dipturus)* sp. F (fig. 17; p. 330)**

fig. 11

fig. 12

fig. 13

fig. 14

fig. 15

fig. 16

fig. 17

15 A prominent dark ocellus on each pectoral fin (fig. 18) .... **Raja sp. E (fig. 18; p. 329)**

No prominent dark ocellus on each pectoral fin **16**

16 Upper surface with clusters of fine dark spots with pale borders (fig. 19) .... **Raja polyommata (fig. 19; p. 353)**

Upper surface lacking clusters of fine dark spots (plain coloured or with pale markings or dark blotches) **17**

17 Undersurface of snout pale with a prominent dark marking at tip (fig. 20); undersurface lacking dark-edged pores **18**

Undersurface of snout rarely with a prominent dark marking at tip; undersurface with dark-edged pores (fig. 22) **19**

18 Interorbital space broad, 2.7 or more in preorbital snout length; thorns mostly continuous along dorsal midline from nuchal area to tail (fig. 20); colour pattern dominated by reticulations and blotches; 45 predorsal diplospondylous vertebrae .... **Raja lemprieri (fig. 20; p. 352)**

Interorbital space extremely broad, less than 2.7 in preorbital snout length; nuchal thorns, when present, widely separated from those extending along midline of tail (fig. 21); colour plain or with fine white spots; 36–38 predorsal diplospondylous vertebrae .... **Raja sp. M (fig. 21; p. 341)**

19 Disc undersurface with denticles restricted to snout tip, anterior margin behind tip smooth to touch **20**

Disc undersurface with granular denticles widespread over preoral snout and/or along anterior margin (rough to touch); denticles often barely detectable in juveniles **21**

20 Dorsal surface yellowish with small white spots .... **Raja cerva (fig. 22; p. 347)**

Uniform greyish above .... **Raja sp. B (fig. 23; p. 325)**

21 Tail relatively long, more than 85% of precloacal length (fig. 24); dorsal fins widely separated (interdorsal distance more than three-quarters of eye diameter) (fig. 24) .. **Raja sp. D (fig. 24; p. 328)**

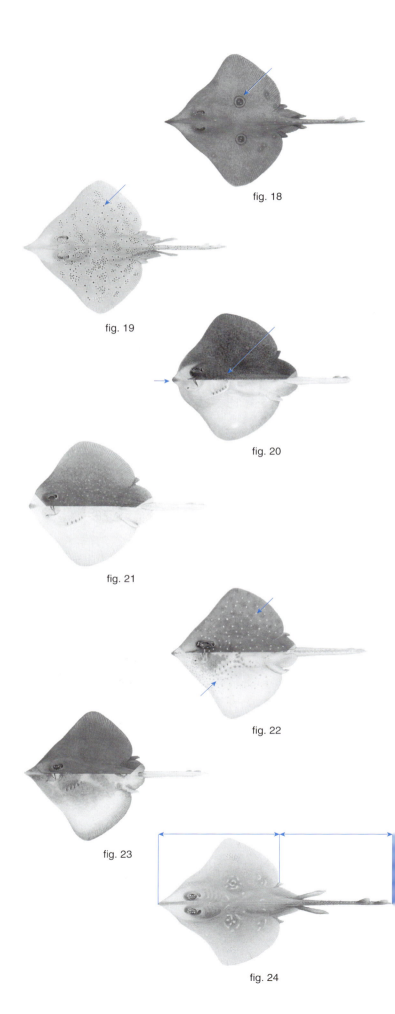

fig. 18

fig. 19

fig. 20

fig. 21

fig. 22

fig. 23

fig. 24

Tail relatively shorter, less than 85% of precloacal length (occasionally more in small juveniles) (fig. 25); interdorsal distance much less than three-quarters of eye diameter (fig. 25) **22**

22 Disc relatively broad and short, width more than 1.25 times length (fig. 25); tail slender, width less than 1.7 times height at first dorsal fin origin; uniform greyish above ....
.... ***Raja (Dipturus)* sp. C (fig. 25; p. 326)**

Disc quadrangular, width less than 1.25 times length (fig. 26); tail broad and flattened, width more than 1.7 times height at first dorsal fin origin; yellowish above with paler markings **23**

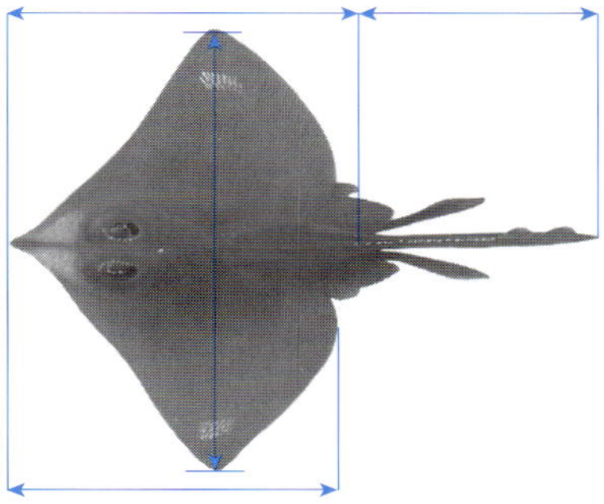

23 Two malar thorn patches on each side in mature males (fig. 26); tail of adults with 3 rows of thorns in males, 5 rows in females; only one nuchal thorn; upper surface with pale blotches ....
.... ***Raja australis* (fig. 26; p. 346)**

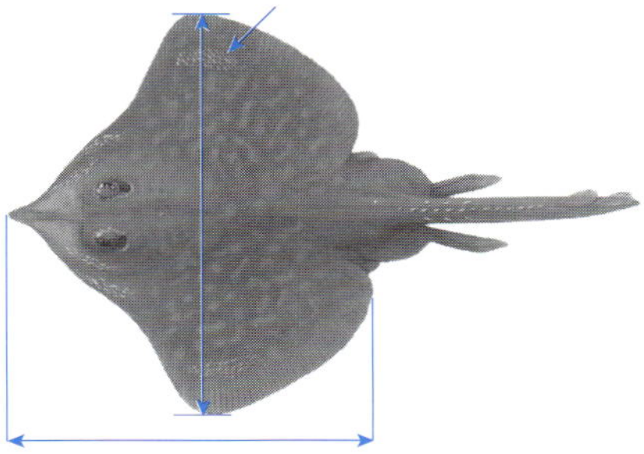

One malar thorn patch on each side in mature males (fig. 27); tail of adults with single row of thorns in males, 3 rows in females (lateral thorns, when present, never forming a continuous row); 0–7 nuchal thorns; upper surface mostly with pale spots, blotches and reticulations ....
.... ***Raja* sp. A (fig. 27; p. 323)**

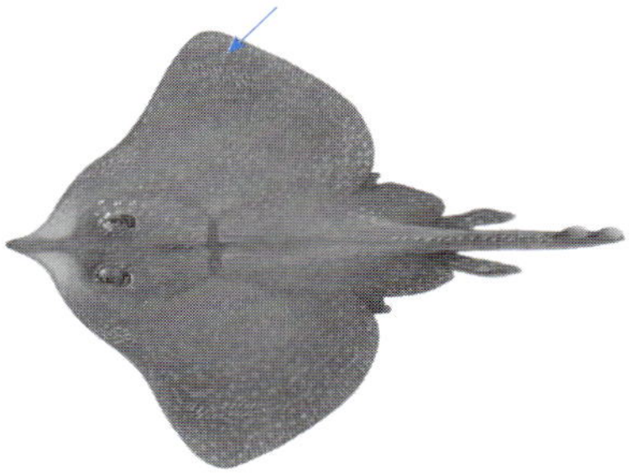

24 Disc quadrangular, broadly wedge-shaped anteriorly (fig. 28); preorbital snout more than 5.5 times orbit length (fig. 28) ....
.... ***Bathyraja* sp. A (fig. 28; p. 306)**

Disc almost circular (fig. 29) to heart-shaped (fig. 31); preorbital snout less than 5.5 times orbit length (fig. 29) **25**

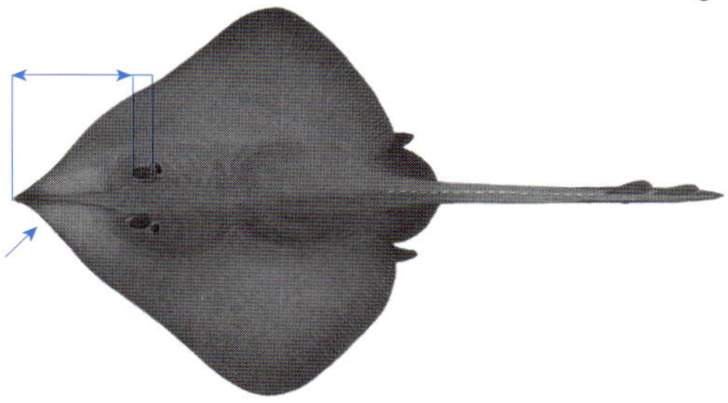

25 Disc circular (feebly angular at snout in mature males) (fig. 29); dorsal surface uniformly smooth; brownish with blue spots above, almost black below
.... **26**

Disc heart-shaped (fig. 31) to subcircular (fig. 33); dorsal surface covered with granular denticles (extensive smooth patches in adults of *Notoraja* spp. A & D); colour not as above **27**

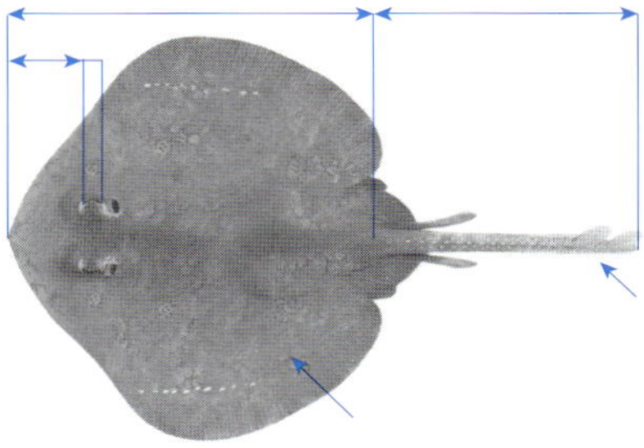

26 Tail relatively short and broad (shorter than or equal to precloacal length, fig. 29), not tapering distally; upper surface of disc with dense clusters of fine bluish spots .... ***Irolita waitii* (fig. 29; p. 308)**

Tail relatively longer and more slender (mostly longer than precloacal length, fig. 30), tapering distally; upper surface of disc with larger, more widely-spaced bluish spots ....
.... ***Irolita* sp. A (fig. 30; p. 307)**

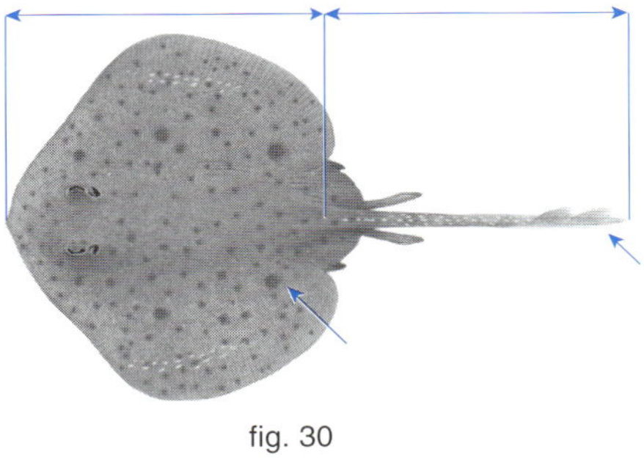

fig. 25

fig. 26

fig. 27

fig. 28

fig. 29

fig. 30

27 No thorns on tail; skin flabby on both dorsal and ventral surfaces **28**

Thorns present on tail; skin not flabby (looser on the undersurface of some species) **29**

28 No dorsal fins or orbital thorns .... ***Pavoraja* sp. A (fig. 31; p. 314)**

Dorsal fins present (fig. 32); a single preorbital thorn (fig. 32) .... ***Pavoraja* sp. B (fig. 32; p. 315)**

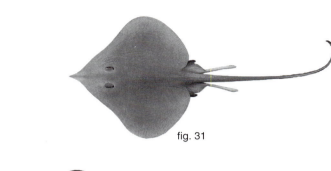
fig. 31

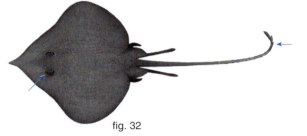
fig. 32

29 Both dorsal and ventral surfaces uniformly pale **30**

Colour not as above **31**

30 Rostral cartilage continuous to snout tip; no orbital thorns; tail thorns arranged in 1–3 indistinct rows (fig. 33), absent from ventral surface .... ***Notoraja* sp. B (fig. 33; p. 311)**

Rostral cartilage not continuous to snout tip; a single preorbital thorn (fig. 34); tail thorns bristle-like, not arranged in distinct rows, also on undersurface .... ***Notoraja* sp. C (fig. 34; p. 312)**

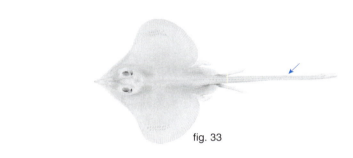
fig. 33

31 Rostral cartilage present, flexible, rod-like or slender (obvious to touch at midsnout and mostly visible when held to backlight) **32**

No rostral cartilage **33**

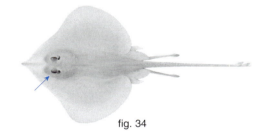
fig. 34

32 Upper surface uniform bluish or with bluish mottling, lower surface almost entirely dark .... ***Notoraja* sp. A (fig. 35; p. 309)**

Both upper and lower surfaces white with large greyish brown blotches and patches .... ***Notoraja* sp. D (fig. 36; p. 313)**

33 Upper surface of disc with a distinctive pattern dominated by small spots and/or reticulations **34**

Upper surface of disc mostly plain coloured (occasionally with a few large faint dusky blotches) **36**

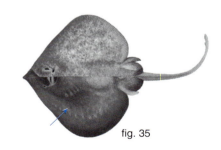
fig. 35

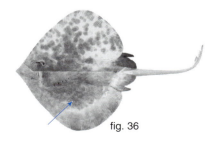
fig. 36

34 Spots dark, irregular in shape and size, densely arranged over upper disc to form a dark mosaic pattern; no nuchal thorns; preoral snout exceeding 13.5 % of total length (fig. 37) ....
.... ***Pavoraja* sp. D (fig. 37; p. 317)**

Spots pale, regular in size and shape, not forming a dark mosaic pattern; nuchal thorns mostly present; preoral snout mostly less than 13.5% of total length (fig. 38) **35**

35 Spots arranged into irregular clusters (fig. 38); tail relatively broad (midwidth rarely more than 27 in tail length) ....
.... ***Pavoraja (Pavoraja) nitida* (fig. 38; p. 322)**

Spots not arranged into clusters (fig. 39); tail relatively narrow (midwidth rarely less than 27 in tail length) ....
.... ***Pavoraja (Pavoraja) sp.* E (fig. 39; p. 318)**

36 Thorns on dorsal surface immediately preceding first dorsal fin similar in size to those more anteriorly on tail; tail relatively rounded in cross-section (sum of its height at pelvic-fin tip and first dorsal-fin origin more than 1.2 times orbit length) .... ***Pavoraja (Pavoraja) sp.* C (fig. 40; p. 316)**

Dorsal surface immediately preceding first dorsal fin lacking thorns (if present much smaller than those more anteriorly on tail); tail relatively depressed in cross-section (sum of its height at pelvic-fin tip and first dorsal-fin origin less than 1.2 times orbit length) **37**

37 Dorsal surface of disc yellowish brown (frequently with dusky blotches); median series of tail thorns extending forward onto disc above cloaca; fewer than 69 pectoral-fin radials ....
.... ***Pavoraja (Pavoraja) alleni* (fig. 41; p. 320)**

Dorsal surface of disc greyish brown (frequently with dusky blotches); median series of tail thorns not extending forward onto disc above cloaca (normally commencing above insertion of pelvic fins); more than 69 pectoral-fin radials ....
.... ***Pavoraja (Pavoraja) sp.* F (fig. 42; p. 319)**

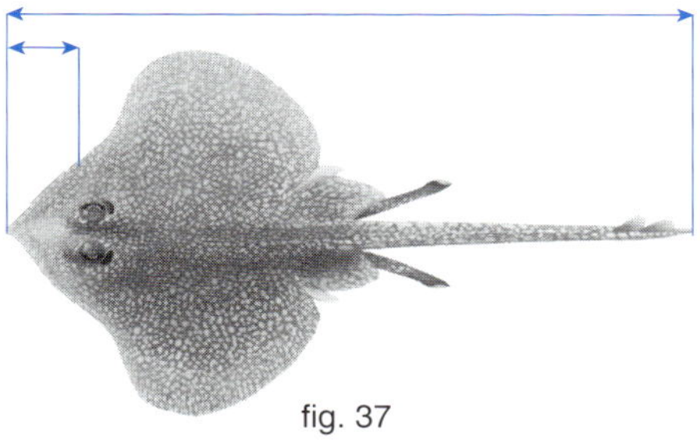

fig. 37

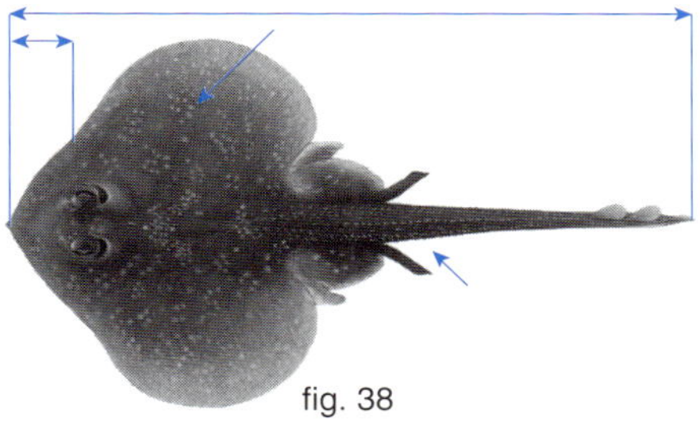

fig. 38

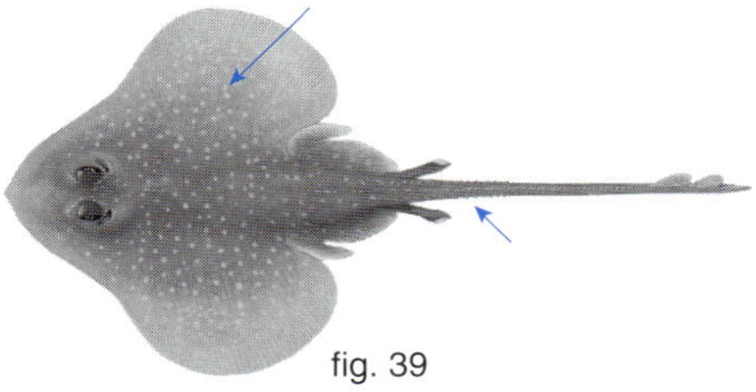

fig. 39

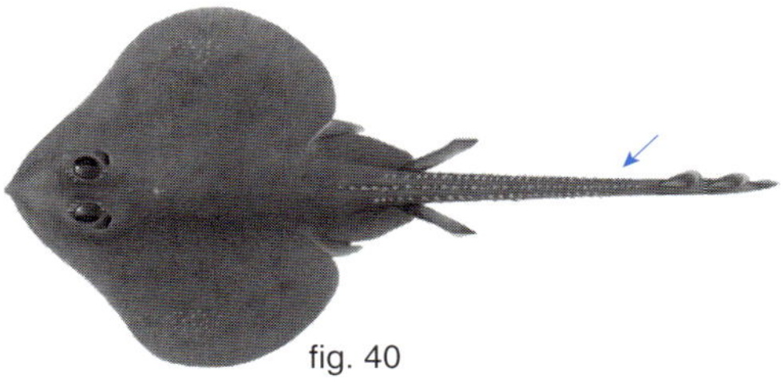

fig. 40

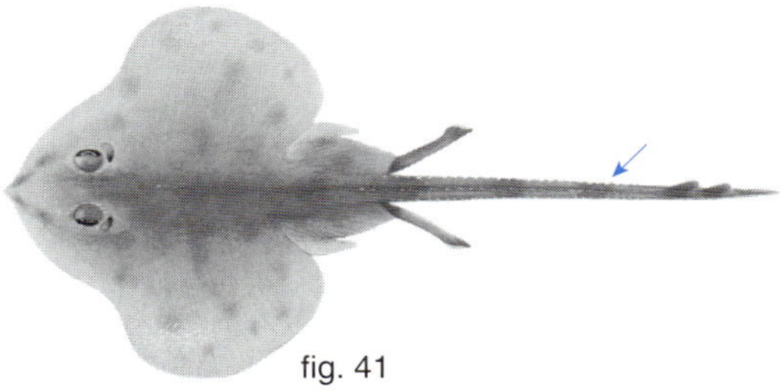

fig. 41

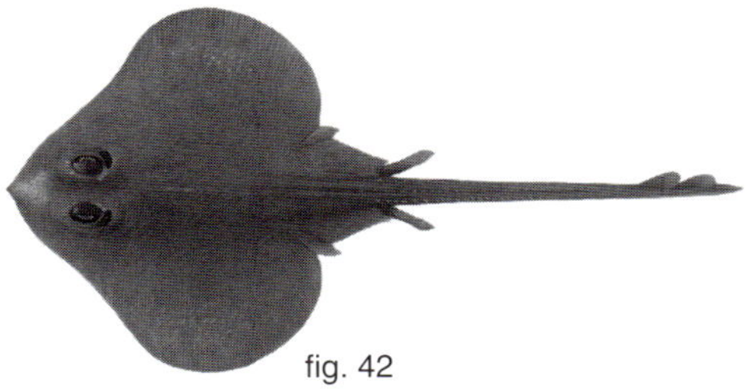

fig. 42

Abyssal Skate — *Bathyraja* sp. A

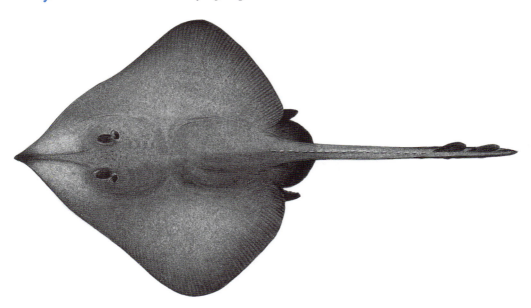

34.1 *Bathyraja* sp. A (Plate 53)   **Fish Code:** 00 031016

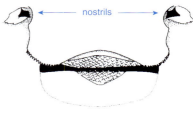

region of nostrils and mouth

- **Field characters:** A very deepwater skate with dark upper and lower surfaces, a flexible rostral cartilage, a long and broad snout, and a thornless back.

- **Distinctive features:** Disc quadrangular, broader than long; pectoral-fin apex broadly rounded; anterior margin of disc straight to slightly double concave. Snout relatively long, very broad (orbit about 6* in preorbital snout), tip pointed, supported by a flexible medial cartilage; orbit small, about 1.6* in interorbital space. Mouth moderately broad; internasal flap narrow-lobed, with dermal fringe. Tail of moderate length (about equal to precloacal length), slightly depressed, slender; lateral skin folds well developed posteriorly. No thorns on dorsal midline of disc before cloaca. Dorsal surface of disc smooth-skinned, except for a patch of fine denticles along anterior margin and around orbit. Ventral surface of disc and tail smooth. Tail with single median row of blunt thorns. Pelvic fins relatively small; anterior lobe slender, slightly less than length of posterior lobe. Dorsal fins small, oblique, about equal in height, separated by about a quarter of length of first dorsal-fin base; base of upper lobe of caudal fin slightly shorter than first dorsal-fin base, separated from second dorsal fin; lower lobe of caudal fin reduced to a narrow ridge.

- **Colour:** Dorsal surface coffee brown, darker near disc margins. Ventral surface dark brown with pale patches around mouth and gill openings. Tail mostly brownish; dorsal and caudal fins dark brown.

- **Size:** Mature males to at least 120 cm.

- **Distribution:** Known only from the Naturaliste Plateau, off Western Australia, in about 2300 m.

- **Remarks:** This very deepwater species was collected by Russian research vessels on the abyssal plain and is currently under investigation by foreign scientists. No specimens are held in Australian museums and more are required urgently. This is the only member of the very widely distributed genus, *Bathyraja*, yet recorded from the Australian region. Related species also occur off New Zealand and in Australia's Antarctic Territory.

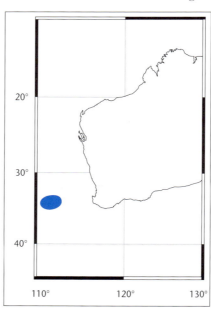

Western Round Skate — *Irolita* sp. A

34.2 *Irolita* sp. A (Plate 48)   Fish Code: 00 031017

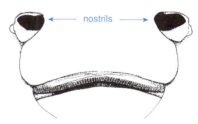

region of nostrils and mouth

tooth of upper jaw

- **Alternative name:** Round skate.
- **Field characters:** A small round skate with a relatively long tail, a flexible snout, a smooth upper disc usually without orbital thorns, a dense coverage of ventral pores, and rather widely spaced, bluish spots on the upper surface.
- **Distinctive features:** Disc almost circular, broadly heart-shaped in mature males, slightly broader than long; anterior margin of disc slightly convex. Snout flexible, very short (orbit 1.9–3.1 in preorbital snout), with a minute, fleshy lobe at tip; orbit small to moderate, 1.05–1.3 in interorbital space. Mouth broad; internasal flap broadly lobed, lacking dermal fringe. Tail relatively long (0.95–1.15 times precloacal length), almost oval in cross-section, tapering slightly from base, slender; lateral skin folds moderately well developed. Dorsal surface of disc smooth rather than granular; orbital thorns rarely present, no thorns on dorsal midline of disc before cloaca; alar thorns non-retractable, arranged in an elongate band extending posteriorly from anterior margin of disc; malar thorns absent. Ventral surface of disc and tail uniformly smooth. Tail thorns enlarged, widely spaced, in 3–5 irregular rows. Pelvic fins moderately large, margin between lobes deeply concave; anterior lobe moderately elongate, about 1.25 in posterior lobe. Dorsal fins small, slanting backward, about equal in size, separated slightly; upper lobe of caudal fin minute. Claspers long, slender, slightly bulbous distally.
- **Colour:** Upper surface brownish with medium-sized, bluish spots (not arranged in distinct clusters); usually with enlarged blue spots beside eye, on back and just in advance of pectoral-fin insertion; eye membrane usually with a single dark brown spot; 3–4 brownish bands on tail. Ventral surface greyish green to dark brown; pores with greyish centres. Dorsal fins uniformly brownish; claspers brownish above, pale below.
- **Size:** Attains at least 42 cm; males mature at about 35 cm.
- **Distribution:** Off northwestern Australia, from Shark Bay to Port Hedland, in 150–200 m.

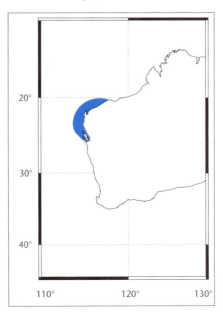

- **Remarks:** This second member of this endemic Australian genus was only recently distinguished from its southern sister species, the round skate (34.3). The two differ from each other subtly in shape and colour pattern.
- **Local synonymy:** *Irolita waitii*: Sainsbury *et al.*, 1985 (misidentification).
- **Reference:** Sainsbury *et al.* (1985).

Southern Round Skate — *Irolita waitii* (McCulloch, 1911)

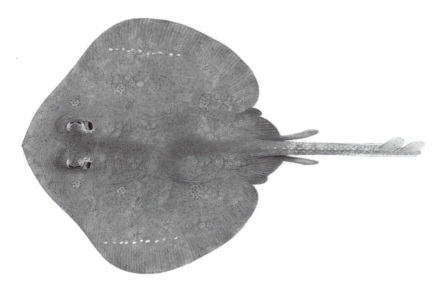

34.3 *Irolita waitii* (Plate 48) **Fish Code:** 00 031001

- **Alternative name:** Round skate.
- **Field characters:** A small round skate with a moderately long tail, a flexible snout, a smooth upper disc without orbital thorns, a dense coverage of ventral pores, and dense clusters of fine bluish spots on the upper surface.
- **Distinctive features:** Disc almost circular, slightly broader than long; anterior margin of disc straight to slightly convex. Snout flexible, very short (orbit 2.3–3 in preorbital snout), with a minute, fleshy lobe at tip; orbit moderate, 1.1–1.45 in interorbital space. Mouth broad; internasal flap broadly lobed, lacking dermal fringe. Tail relatively short (0.85–1 times precloacal length), semi-circular in cross-section, relatively broad, not tapering distally; lateral skin folds very well developed, extending beyond tip of tail. Dorsal surface of disc uniformly smooth rather than granular; no thorns on dorsal midline of disc before cloaca; usually with single anterior and posterior orbital thorns (occasionally with more); alar thorns non-retractable, arranged in an elongate band extending posteriorly from anterior margin of disc; malar thorns absent. Ventral surface of disc and tail uniformly smooth. Tail thorns short, usually in 3 rows along the dorsal midline. Pelvic fins moderately large, margin between lobes deeply concave; anterior lobe moderately long, about 1–1.3 in posterior lobe. Dorsal fins small, rather upright, subequal in size, separated slightly; upper lobe of caudal fin barely distinguishable. Claspers long, slender, slightly bulbous distally.
- **Colour:** Upper surface pale yellowish to brownish with dense clusters of fine blue spots interspersed with similar aggregations of minute brownish spots; bluish markings forming broken ocelli beside eye, on back and just in advance of pectoral

fin insertion; ocular membrane pale with brownish peppering; skin folds pale. Ventral surface medium greyish to black (normally darkest on head); pores distinctly white. Claspers pale with brown peppering dorsally. Dorsal fins similar to body.

- **Size:** Attains at least 52 cm; males mature at about 46 cm.
- **Distribution:** Great Australian Bight from Geraldton (Western Australia) to Port Lincoln (South Australia) in 50–200 m.
- **Remarks:** This very attractive skate was described some time ago but does not appear to be common. It is also rare in collections. Like most members of the family, little is known of its biology.
- **Local synonymy:** *Psammobatis waitii*: McCulloch, 1929.
- **References:** McCulloch (1911); Scott *et al.* (1980).

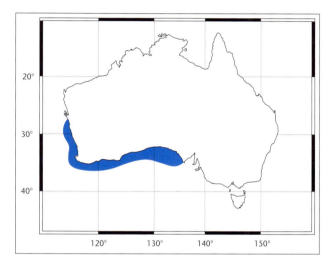

Blue Skate — *Notoraja* sp. **A**

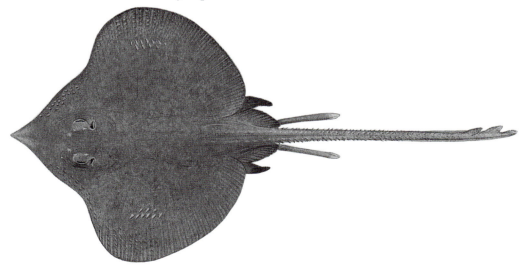

34.4 *Notoraja* sp. **A** (Plate 50)　　　　　　　　　　Fish Code: 00 031018

- **Field characters:** A small skate with a heart-shaped disc, a uniformly granular dorsal surface in juveniles (becoming smooth in adults), prominent thorns in the malar region of both males and females, and a long tail that has a smooth undersurface. Bluish above, rather dark below.

- **Distinctive features:** Disc heart-shaped in males, subcircular in females and juveniles, broader than long; pectoral-fin apex very broadly rounded; anterior margin of disc straight in females and juveniles, slightly concave in mature males. Snout flexible, short (orbit 3.3–4.15 in preorbital snout), with a short, fleshy lobe at tip; medial cartilage not firm; orbit small to moderate, 0.8–0.95 in interorbital space. Mouth rather narrow; internasal flap broadly lobed, lacking dermal fringe. Tail usually very long (1.15–1.35 times precloacal length in mature specimens); relatively longer in juveniles; oval, slightly depressed, very slender posteriorly; lateral skin folds well developed. No thorns on dorsal midline of disc before cloaca. Juveniles covered uniformly with fine denticles on dorsal surface; 1 anterior and 1 posterior orbital thorn; several irregular rows of enlarged thorns on tail. Adults with fewer denticles on disc (entirely smooth patches on large specimens); sometimes with persistent orbital thorns and interorbital denticles; large females with thorns in centre of disc and along anterior margin; males with extensive united alar and malar patches; alar thorns non-retractable. Ventral surface of disc uniformly smooth; occasionally with a few isolated denticles on middle of tail. Pelvic fins small, margin between lobes deeply concave; anterior lobe long, almost as long as posterior lobe. Dorsal fins small, lanceolate, subequal in height, usually connected; upper lobe of caudal fin low, base generally longer than base of first dorsal fin, usually separated from second dorsal fin. Claspers slender, not expanded distally.

- **Colour:** Dorsal surface usually uniform pale greyish blue to dark blue (sometimes with fine black spots, or pale with bluish grey mottling). Ventral surface mainly medium brown centrally, darker bluish black along disc and tail margins; pores sometimes pale; frequently pale around gill slits and mouth; ventral tip of anterior pelvic-fin lobe pale; dorsal fins variable.

- **Size:** Females to 60 cm; males to 53 cm and maturing at about 48 cm; free swimming by 20 cm.

- **Distribution:** Widely distributed through temperate Australia from Perth (Western Australia) to the Solitary Islands (New South Wales), including Tasmania, in 840–1120 m.

- **Remarks:** This species is represented by at least 3 regional forms which could be separate species. Apart from exhibiting colour differences, they can be distinguished on vertebral counts. Eastern populations are very dark with fine black spots, whereas populations in the Great Australian Bight are pale blue and speckled. Both of these forms have fewer monospondylous vertebrae (24–25) than the southern form (26–27). More taxonomic work is required to resolve this problem.

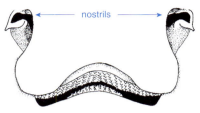

region of nostrils and mouth

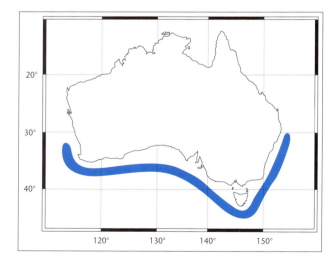

Pale Skate — *Notoraja* sp. **B**

34.5 *Notoraja* sp. **B** (Plate 50)　　　　　　　**Fish Code:** 00 031019

- **Field characters:** A small, uniformly pale skate with a firm rostral cartilage (flexible only at snout tip), a uniformly granular upper surface without thorns (non-retractable alar thorns only on males), very bulbous claspers, and a dark band on the tail through the second dorsal fin.

- **Distinctive features:** Disc heart-shaped, broader than long; pectoral-fin apex broadly rounded; anterior margin of disc straight to slightly concave. Snout short (orbit 3.3–3.95 in preorbital snout), with a short, fleshy lobe at tip; supported by a firm medial cartilage for most of its length (becoming flexible at apex); orbit of moderate size (0.8–1.2 in interorbital space). Mouth broad; internasal flap broadly lobed, lacking a dermal fringe. Tail long (about 1.3* times precloacal length in mature specimens, longer in juveniles), oval, slightly depressed, very slender; lateral skin folds indistinct. Dorsal surface of disc uniformly granular; no malar, nuchal, orbital or precloacal thorns; alar thorns non-retractable. Ventral surface of disc uniformly smooth. Tail thorns short, dense, in 1–3 indistinct rows (some thorns barely larger than adjacent denticles). Pelvic fins small, margin between lobes deeply concave; anterior lobe long, about 1.25 in posterior lobe. Dorsal fins small, almost lanceolate, subequal in height, situated close together or connected; upper lobe of caudal fin low, often barely distinguishable, usually connected to second dorsal fin. Claspers very slender, distal quarter greatly expanded.

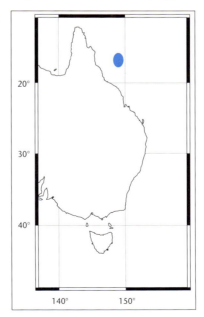

- **Colour:** Upper surface uniform pale yellow, almost translucent near edge of disc and on snout. Undersurface pale, skin translucent; pores not dark-edged. Tail with prominent dark band through base of second dorsal fin.

- **Size:** Attains at least 36 cm; males mature at about 35 cm.

- **Distribution:** Known from a few specimens taken from the continental slope off Cairns (Queensland) in 400–465 m.

- **Remarks:** First collected by a CSIRO research vessel, 'Soela', during a fishing survey for commercial crustaceans off northern Queensland. Its uniform paleness is unusual among Australian skates.

Ghost Skate — *Notoraja* sp. C

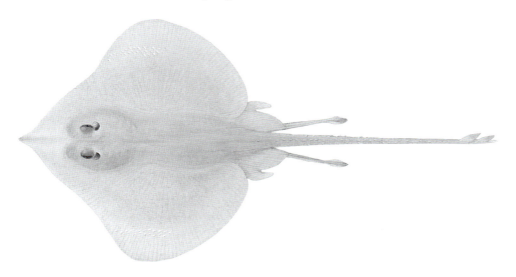

34.6 *Notoraja* sp. C (Plate 51)   Fish Code: 00 031015

- **Field characters:** A small, pale skate with a heart-shaped disc, a flexible snout, a uniformly granular upper surface with only a single preorbital thorn, and a very long, slender tail with fine denticles on its undersurface.

- **Distinctive features:** Disc heart-shaped in mature males, subcircular in juveniles, slightly broader than long; pectoral-fin apex broadly rounded; anterior margin of disc straight in females and juveniles, slightly concave in mature males. Snout short but slightly longer in mature males (orbit 2.3–4.2 in preorbital snout), with fleshy lobe at tip; flexible, medial cartilage not firm; orbit small to moderate, 0.7–0.95 in interorbital space. Mouth rather narrow; internasal flap broadly lobed, lacking dermal fringe. Tail very long (1.35–1.5 times precloacal length), oval, very slender; lateral skin folds indistinct. Dorsal surface of disc uniformly granular; thorns confined to orbit (1 preorbital thorn); alar thorns non-retractable; ventral surface uniformly smooth. Tail thorns bristle-like, regularly arranged on dorsal surface; less dense on ventral surface. Pelvic fins small, margin between lobes extremely concave; anterior lobe elongate, about equal in length to, or longer than, posterior lobe. Dorsal fins very small, lanceolate, equal in height, situated close together; upper lobe of caudal fin low, base usually longer than first dorsal-fin base, usually separated from second dorsal fin. Claspers very slender, slightly bulbous distally.

- **Colour:** Upper surface pale. Undersurface white or translucent; pores not dark-edged. Claspers, dorsal fins and anterior pelvic-fin lobes white.

- **Size:** Attains at least 45 cm.

- **Distribution:** Continental slope from Shark Bay to the Monte Bello Islands (Western Australia) in 590–760 m.

- **Remarks:** A distinctive skate known from only a few specimens; more are required.

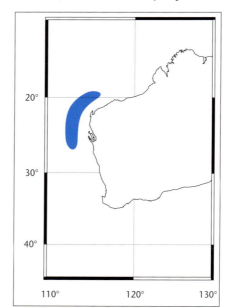

Blotched Skate — *Notoraja* sp. **D**

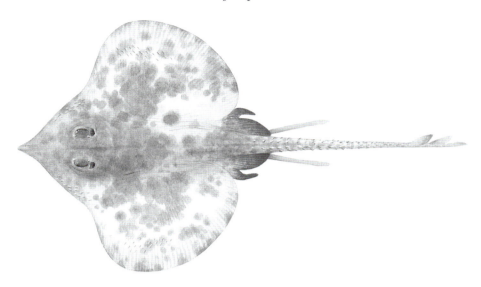

34.7 *Notoraja* sp. **D** (Plate 51)   **Fish Code:** 00 031020

- **Field characters:** A small skate with a heart-shaped disc with large greyish brown blotches on both surfaces, and a flexible, rod-like rostral cartilage. It has orbital and interorbital thorns (and malar thorns in mature males) but lacks thorns along the mid-disc.

- **Distinctive features:** Disc broadly heart-shaped in mature males, subcircular in females, broader than long; pectoral-fin apex very broadly rounded; anterior margin of disc extremely concave in mature males, straight in females. Snout short (orbit about 4.5* in preorbital snout), with a prominent fleshy lobe at tip; rostral cartilage rod-like, wavy, flexible; orbit moderate, about 0.85* in interorbital space. Mouth broad; internasal flap broadly lobed, lacking dermal fringe. Tail long (1.2–1.25* times precloacal length in mature specimens), oval to slightly depressed, tapering rapidly, slender posteriorly; lateral skin folds well developed posteriorly. Dorsal surface of disc mostly smooth, denticles present on anterior head of females; no thorns on dorsal midline before cloaca; orbital and interorbital regions with a few small thorns; alar thorns non-retractable, located on centre and along outer margin of pectoral fin; malar thorns prominent; ventral surface smooth. Tail with several regular rows of enlarged thorns; lacking denticles (except occasionally along ventral midline). Pelvic fins moderate, margin between lobes deeply concave; anterior lobe long, almost as long as posterior lobe. Dorsal fins small, lanceolate, equal in height, situated very close together; upper lobe of caudal fin low, base about equivalent to first dorsal-fin base, not connected to second dorsal fin. Claspers very slender, slightly bulbous distally.

- **Colour:** Upper surface white with large, greyish brown blotches (pattern individually variable but often more concentrated on central disc); similar on undersurface, except blotches cover more of disc. Tail mostly pale above, dark below; claspers white; dorsal fins uniform greyish brown or white.

- **Size:** Largest specimen a mature male of 53 cm.

- **Distribution:** Continental slope off Eucla, Western Australia, in 820–930 m.

- **Remarks:** Known from only a few specimens.

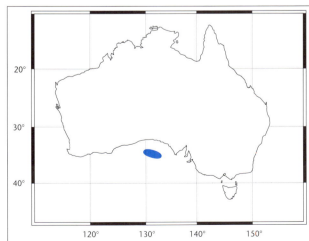

Eastern Looseskin Skate — *Pavoraja* sp. **A**

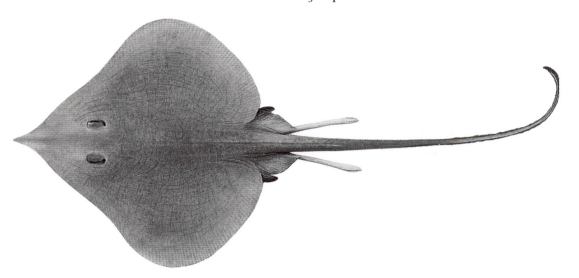

34.8 *Pavoraja* sp. **A** (Plate 49) Fish Code: 00 031021

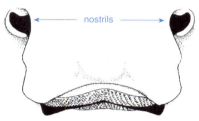

region of nostrils and mouth

- **Field characters:** A small, pale bluish skate with flabby skin, a uniformly granular upper surface that lacks thorns (alars present in mature males), a very long and thin tail with no dorsal fins, and a well-developed caudal fin.

- **Distinctive features:** Disc subcircular in females and juveniles, heart-shaped in mature males, broader than long; pectoral-fin apex very broadly rounded; anterior margin of disc straight to moderately double concave. Snout short to moderate (orbit about 3.7–5.3* in preorbital snout), with a prominent fleshy lobe at tip (larger in males); supported by a flexible, rod-like cartilage; orbit small, about 0.8–1.0* in interorbital space. Mouth moderately broad; internasal flap broadly lobed, lacking dermal fringe. Tail extremely long (1.2–1.35* times precloacal length), almost circular in cross-section, slightly depressed posteriorly, very slender; lateral skin folds well developed posteriorly. Dorsal surface of disc and tail uniformly granular (anterior lobe of pelvic fin naked); no nuchal, orbital or tail thorns; alar thorns non-retractable, extending forward to anterior margin of disc. Ventral surface of disc uniformly smooth, skin extremely loose; tail uniformly granular underneath. Pelvic fins small, margin between lobes very deeply concave; anterior lobe long, subequal to posterior lobe. Dorsal fins absent; upper lobe of caudal fin low, long (base almost twice orbit diameter); lower lobe of caudal fin rudimentary. Claspers long, very slender.

- **Colour:** Upper surface uniformly pale bluish; anterior margin of disc and snout tip pale; anterior lobe of pelvic fin greyish brown, darker distally; tail with irregular blue and white patches. Skin of ventral surface almost transparent; beneath skin greyish blue and white areas; tail pale or greyish blue. Caudal fin and lateral skin folds bluish to black; claspers pale dorsally, mostly greyish blue ventrally.

- **Size:** Reaches at least 57 cm.

- **Distribution:** Continental slope off Cairns, Queensland, in depths of 800–880 m.

- **Remarks:** Looseskin skates, as the name implies, have unusually flabby skin. Only two species are known and both of these are endemic locally. This species is unique among Australian skates in lacking dorsal fins. Known from only a few specimens.

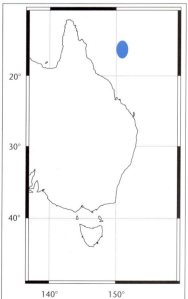

Western Looseskin Skate — *Pavoraja* sp. **B**

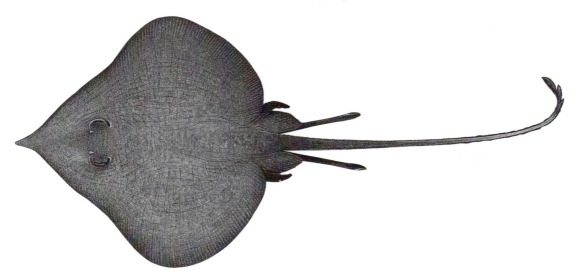

34.9 *Pavoraja* sp. **B** (Plate 49)   **Fish Code:** 00 031022

- **Field characters:** A small, dark bluish skate with flabby skin, a uniformly granular upper surface, single preorbital thorns but no other thorns on the body or tail (except in the alar region in mature males), a very long and thin tail, two dorsal fins, and a moderately elongate caudal fin.

- **Distinctive features:** Disc subcircular in females and juveniles, heart-shaped in mature males, broader than long; pectoral-fin apex very broadly rounded; anterior margin of disc straight to moderately double concave. Snout short to moderate (orbit 3.7–4.9 in preorbital snout), with a prominent fleshy lobe (larger in males) at tip; supported by a flexible, rod-like cartilage; orbit small, 1.0–1.1 in interorbital space. Mouth moderately broad; internasal flap broadly lobed, lacking dermal fringe. Tail extremely long (1.4–1.55 times precloacal length), almost circular in cross-section, slightly depressed anteriorly, very slender; lateral skin folds well developed posteriorly. Dorsal surface of disc and tail uniformly granular (anterior lobe of pelvic fin naked); single large preorbital thorns; alar thorns non-retractable, extending forward to anterior margin of disc; no nuchal or tail thorns. Ventral surface of disc uniformly smooth, skin loose; tail uniformly granular beneath. Pelvic fins small, margin between lobes very deeply concave; anterior lobe long, usually equal to or slightly longer than posterior lobe. Dorsal fins small, subequal in size, situated close together; upper lobe of caudal fin low, base about equal to first dorsal-fin base; lower lobe of caudal fin rudimentary. Claspers long, very slender.

- **Colour:** Upper surface medium to dark blue with irregular white flecks; anterior margin of disc and snout tip not paler than remainder of disc; tail and anterior pelvic-fin lobe bluish brown. Skin of ventral surface almost transparent; bluish brown beneath, darkest on outer disc; tail pale. Dorsal and caudal fins bluish black; lateral skin folds black. Claspers dark dorsally, mostly pale ventrally.

- **Size:** Attains at least 54 cm; males maturing at about 50 cm.

- **Distribution:** Continental slope off Western Australia between Lancelin and Ashmore Reef in 610–1200 m, but most abundant in 900–1100 m.

- **Remarks:** Although only discovered recently, this species is the most abundant skate found on the continental slope of tropical and warm temperate Western Australia. It has a rather broad depth range and is caught occasionally by trawlers fishing for scampi. Little is known of its biology.

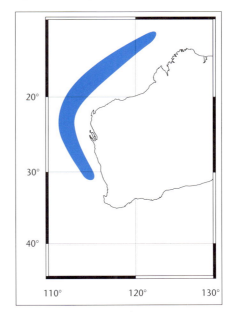

Sandy Skate — *Pavoraja* sp. C

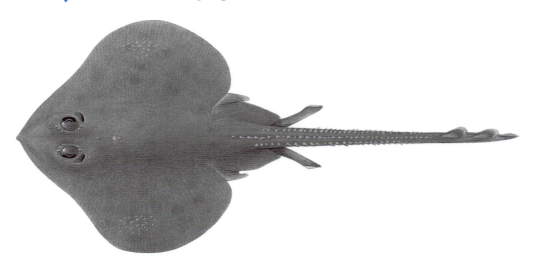

34.10 *Pavoraja* sp. C (Plate 45)
Fish Code: 00 031023

- **Field characters:** A very small, uniformly yellowish skate with a circular to heart-shaped disc, nuchal and orbital thorns, and lacking a firm supporting cartilage in the snout. The tail is long and relatively rounded with the predorsal thorns all about the same size.

- **Distinctive features:** Disc subcircular in females and juveniles, heart-shaped in mature males; slightly broader than long; pectoral-fin apex very broadly rounded; anterior margin of disc convex anteriorly, becoming concave or straight posteriorly. Snout relatively short (orbit about 2.2–3.1* in preorbital snout), flexible (no firm rostral cartilage), with a very small fleshy lobe at tip; orbit moderately large, about 1.4–1.65* in interorbital space. Mouth rather narrow; internasal flap broadly lobed, lacking dermal fringe. Tail moderately long (0.75–1.25* times precloacal length), slender based, slightly depressed; lateral skin folds well developed posteriorly. Dorsal surface of disc and tail granular (anterior lobe of pelvic fin naked); 2–3 small nuchal thorns; orbital thorns small (1–2 anterior, 1–2 posterior); tail thorns rather large, widely spaced, in three rows, not decreasing in size posteriorly (enlarged thorns present just before first dorsal fin); alar thorns small, non-retractable, usually in a single row; malar thorns very small, extending anteriorly to just behind greatest concavity of disc. Ventral surface of disc and tail uniformly smooth. Pelvic fins small, margin between lobes very deeply concave; anterior lobe moderately elongate (1–1.3 in posterior lobe). Dorsal fins small, separated by about the length of first dorsal-fin base; upper lobe of caudal fin shorter than first dorsal-fin base, low, its base separated from second dorsal-fin base; lower lobe of caudal fin rudimentary. Claspers moderately elongate, very slender, slightly enlarged near tip. Tooth rows in lower jaw 31–37*. Pectoral fin radials 71–79*. Predorsal diplospondylous vertebrae 66–70*.

- **Colour:** Upper surface uniformly pale yellow; anterior lobes of pelvic fins, median region of snout, and pored prenuchal area paler. Ventral surface mostly uniformly white; outer corners of disc and apical part of tail occasionally with greyish areas. Dorsal and caudal fins predominantly dusky.

- **Size:** Attains at least 33 cm; males mature at about 30 cm.

- **Distribution:** Known from the Great Australian Bight between Eucla (Western Australia) and Fowlers Bay (South Australia) in depths of 200–520 m.

- **Remarks:** This species closely resembles Allens skate (34.14) but has larger thorns preceding the first dorsal fin, a more rounded tail, and fewer vertebrae. More specimens are required.

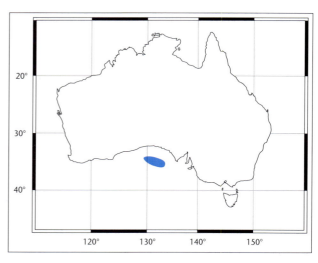

Mosaic Skate — *Pavoraja* sp. **D**

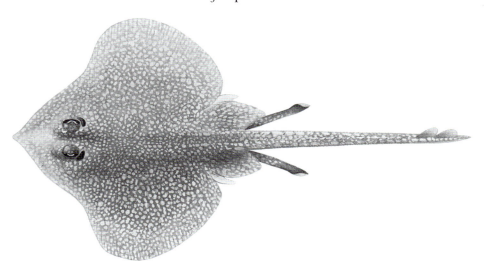

34.11 *Pavoraja* sp. **D** (Plate 46)   **Fish Code:** 00 031024

- **Field characters:** A very small skate with a distinctive pattern of fine reticulations and spots, a circular to heart-shaped disc, no nuchal and only small orbital thorns, and lacking a firm supporting cartilage in the snout.

- **Distinctive features:** Disc subcircular in females and juveniles, heart-shaped in mature males; slightly broader than long; pectoral-fin apex very broadly rounded; anterior margin of disc convex to slightly concave in females and immature males, usually deeply concave opposite spiracles in mature males. Snout relatively short (orbit about 2.8–3.1* in preorbital snout), flexible (no firm rostral cartilage), with a very small fleshy lobe at tip; orbit moderately large, about 1.25–1.65* in interorbital space. Mouth rather narrow; internasal flap broadly lobed, lacking dermal fringe. Tail moderately long (0.9–1.05* times precloacal length), slender based, slightly depressed; lateral skin folds well developed posteriorly. Dorsal surface of disc and tail granular (anterior lobe of pelvic fin naked); nuchal thorns absent; orbital thorns very small (1–3 anterior, 0–2 posterior); tail thorns very small, usually in three rows (lateral rows sometimes absent, particularly in juveniles); alar thorns small, non-retractable, in 2–3 irregular rows; malar thorns small, extending anteriorly to greatest concavity of anterior margin of disc. Ventral surface of disc and tail uniformly smooth. Pelvic fins small, margin between lobes very deeply concave; anterior lobe moderately elongate (1.2–1.5 in posterior lobe). Dorsal fins very small, connected; upper lobe of caudal fin slightly shorter than first dorsal-fin base, low, usually connected to second dorsal fin; lower lobe of caudal fin rudimentary. Claspers moderately elongate, very slender, slightly enlarged near tip. Tooth rows in lower jaw 38–44*. Pectoral fin radials 72–74*. Predorsal diplospondylous vertebrae 72–78*.

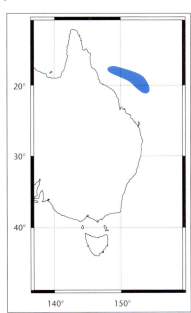

- **Colour:** Upper surface pale with a dense pattern of fine, medium to dark brown reticulations and spots; preorbital snout and posterior margin of disc paler. Ventral surface mostly uniformly white; outer corners of disc sometimes greyish. Dorsal fins light brown with pale margins; caudal fin and anterior lobe of pelvic fin white.

- **Size:** Attains about 30 cm; smallest mature male examined was 27 cm.

- **Distribution:** Known only from the continental slope off Queensland between Ingham and Mackay in depths of 300–400 m.

- **Remarks:** This very attractive little skate is sometimes caught in the same trawl as the false peacock skate (34.12). Known from only a few specimens.

False Peacock Skate — *Pavoraja* sp. E

34.12 *Pavoraja* sp. E (Plate 47)   Fish Code: 00 031025

- **Field characters:** A very small, yellowish brown skate with randomly arranged white spots (rather than in distinct clusters), a circular to heart-shaped disc, nuchal and orbital thorns, and lacking a firm supporting cartilage in the snout.

- **Distinctive features:** Disc subcircular in females and juveniles, heart-shaped in mature males; slightly broader than long; pectoral-fin apex very broadly rounded; anterior margin of disc convex anteriorly, becoming slightly concave or straight posteriorly in females and immatures, usually deeply concave in mature males. Snout relatively short (orbit about 2.3–3.1* in preorbital snout), flexible (no firm rostral cartilage), with a very small fleshy lobe at tip; orbit moderately large, about 1.4–2.0* in interorbital space. Mouth rather narrow; internasal flap broadly lobed, lacking dermal fringe. Tail moderately long (0.85–1.3* times precloacal length), slender based, slightly depressed; lateral skin folds well developed posteriorly. Dorsal surface of disc and tail granular (anterior lobe of pelvic fin naked); 0–5 nuchal thorns (often absent in juveniles, usually 2–3 in adults); scapular thorns rarely present; orbit with 3–10 thorns (1–5 anterior, 2–5 posterior), almost forming a rosette; interorbital thorn sometimes present. Tail thorns rather large, widely spaced, in three rows, decreasing in size and number posteriorly; lateral row usually absent from posterior half of tail (often absent in immediate predorsal area); alar thorns small, non-retractable, in 1–2 rows; malar thorns small, extending from immediately behind greatest concavity of anterior margin of disc to about cloaca level. Ventral surface of disc and tail uniformly smooth. Pelvic fins small, margin between lobes very deeply concave; anterior lobe moderately elongate, (1–1.5 in posterior lobe). Dorsal fins small, closely spaced but rarely connected; upper lobe of caudal fin variable in length, generally shorter than first dorsal-fin base, low, its base separated from second dorsal-fin base; lower lobe of caudal fin rudimentary. Claspers moderately elongate, very slender, slightly depressed near tip. Tooth rows in lower jaw 39–45*. Pectoral fin radials 70–76*. Predorsal diplospondylous vertebrae 75–82*.

- **Colour:** Upper surface mostly uniform yellowish brown to pale brown, densely covered with fine white spots; spots random, not aggregated into clusters; anterior lobes of pelvic fins pale. Ventral surface mostly uniformly pale; outer corners of disc often dusky. Dorsal fins with pale to white anterior margins, darker posteriorly and at base; caudal fin pale.

- **Size:** Attains at least 37 cm; males maturing at about 34 cm.

- **Distribution:** Known from the continental slope off Queensland between Rockhampton and Cairns in depths of 210–500 m.

- **Remarks:** A tropical relative of the temperate peacock skate (34.15). It differs from the latter in having fewer spots which are arranged randomly rather than in clusters, a relatively narrower tail and more vertebrae. Possibly the most abundant skate on the upper continental slope of tropical eastern Australia.

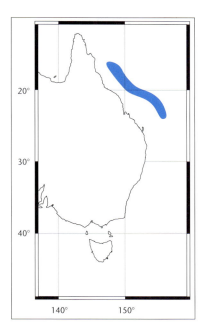

Dusky Skate — *Pavoraja* sp. F

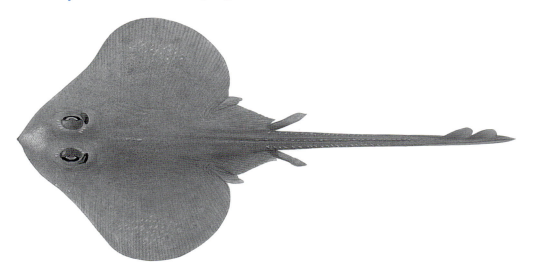

34.13 *Pavoraja* sp. F (Plate 46)   Fish Code: 00 031026

- **Field characters:** A very small, greyish skate with a circular to heart-shaped disc, nuchal and orbital thorns, a relatively large eye, and lacking a firm supporting cartilage in the snout.
- **Distinctive features:** Disc subcircular in females and juveniles, heart-shaped in mature males; broader than long; pectoral-fin apex broadly rounded; anterior margin of disc slightly convex in females and immature males, deeply concave opposite spiracles in mature males. Snout relatively short (orbit about 2.1–2.55* in preorbital snout), flexible (no firm rostral cartilage), with a very small fleshy lobe at tip; orbit moderately large, about 1.6–1.95* in interorbital space. Mouth rather narrow; internasal flap broadly lobed, lacking dermal fringe. Tail moderately long (0.95–1.3* times precloacal length), slightly depressed, very slender; lateral skin folds well developed posteriorly. Dorsal surface of disc and tail granular (anterior lobe of pelvic fin naked); 3–4 prominent nuchal thorns; orbital thorns large (2–4

anterior, 1–4 posterior); tail thorns rather large, widely spaced, in three rows (lateral rows variable, occasionally absent), poorly developed on posterior part of tail; alar thorns small, non-retractable, in 1–3 rows; malar thorns small, situated laterally to alar thorns, extending anteriorly to greatest concavity of anterior margin of disc. Ventral surface of disc and tail uniformly smooth. Pelvic fins small, margin between lobes very deeply concave; anterior lobe moderately elongate (1.2–1.5 in posterior lobe). Dorsal fins small, barely connected; upper lobe of caudal fin short, equal to or slightly shorter than first dorsal-fin base, low, its base usually separated from second dorsal-fin base; lower lobe of caudal fin rudimentary. Claspers moderately elongate, very slender, slightly enlarged near tip. Tooth rows in lower jaw 38–42*. Pectoral fin radials 70–73*. Predorsal diplospondylous vertebrae 74–82*.

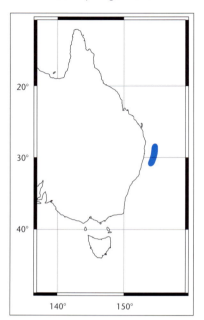

- **Colour:** Upper surface uniformly greyish to greyish brown, occasionally slightly mottled or with diffuse dark brown blotches. Ventral surface mostly uniformly pale (frequently darker greyish brown on central and outer corners of disc). Dorsal fins generally dusky; caudal fin pale.

- **Size:** Attains about 37 cm; males mature at about 32 cm.

- **Distribution:** Continental slope off eastern Australia between Coffs Harbour (New South Wales) and Moreton Island (Queensland) in depths of 360–730 m.

- **Remarks:** Little is known of the biology of this species.

Allens Skate — *Pavoraja alleni* McEachran and Fechhelm, 1982

34.14 *Pavoraja alleni* (Plate 45) **Fish Code:** 00 031027

- **Field characters:** A very small, uniform yellowish brown skate (sometimes with very faint dusky blotches) with a circular to heart-shaped disc, nuchal and orbital thorns, and lacking a firm supporting cartilage in the snout. The tail is long and relatively flattened with the thorns immediately before the dorsal fins distinctly smaller than those more anteriorly.

- **Distinctive features:** Disc subcircular in females and juveniles, heart-shaped in mature males; slightly broader than long; pectoral-fin apex broadly rounded; anterior margin of disc convex anteriorly, becoming concave or straight posteriorly in females and immature males, usually deeply concave in mature males. Snout relatively short (orbit about 2–3.25* in preorbital snout), flexible (no firm rostral cartilage), with a very small, fleshy lobe at tip; orbit moderately large, about 1.1–1.6* in interorbital space. Mouth rather narrow; internasal flap broadly lobed, lacking a dermal fringe. Tail moderately long (1.05–1.3* times precloacal length), slightly depressed, very slender; lateral skin folds well developed posteriorly. Dorsal surface of disc and tail granular (anterior lobe of pelvic fin naked); 1–6 prominent nuchal thorns; usually with 2–8 orbital thorns; mature specimens with 1–4 prominent scapular thorns; tail thorns moderately large, in three rows (absent or reduced just before first dorsal fin; lateral rows generally absent in juveniles); alar thorns small, non-retractable, in 1–3 (rarely 4–5) irregular rows; malar thorns small, situated anterolaterally to (and continuous with) alar thorns; an additional very small patch of scattered thorns adjacent to orbits. Ventral surface of disc and tail uniformly smooth. Pelvic fins small, margin between lobes very deeply concave; anterior lobe moderately elongate (0.9–1.6 in posterior lobe). Dorsal fins small, separated by about length of first dorsal-fin base or less; upper lobe of caudal fin short, base usually equal to or less than first dorsal-fin base, low, separated from second dorsal-fin base; lower lobe of caudal fin rudimentary. Claspers moderately elongate, very slender, enlarged slightly near tip. Tooth rows in lower jaw 32–37*. Pectoral fin radials 64–68* [63–67]. Predorsal diplospondylous vertebrae 70–74* [71–79].

- **Colour:** Upper surface mostly yellowish to pale brown with a scattering of large, dark brown blotches (forming a series of narrow bands on tail); blotches most prominent in immatures, diffuse and faint (or almost undetectable) in most adults. Ventral surface pale or medium greyish brown; outer corners of disc not conspicuously darker. Dorsal fins dusky; caudal fin pale to dusky; posterior skin folds dark.

- **Size:** Attains about 35 cm; males mature at about 32 cm; free swimming by about 14 cm.

- **Distribution:** Widely distributed along the continental slope off Western Australia between Perth and Broome in depths of 200–460 m.

- **Remarks:** The most abundant and widely distributed skate on the upper continental slope off Western Australia. Little is known of its biology.

- **Reference:** McEachran and Fechhelm (1982).

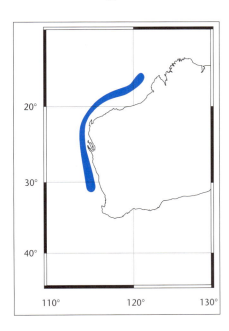

Peacock Skate — *Pavoraja nitida* (Günther, 1880)

34.15 *Pavoraja nitida* (Plate 47)　　　　　　　　　　　　　　　Fish Code: 00 031009

- **Alternative names:** Graceful skate, roughback skate, shining skate.

- **Field characters:** A very small, brownish skate with fine white spots arranged in distinct clusters, a circular to heart-shaped disc, nuchal and orbital thorns, and lacking a firm supporting cartilage in the snout.

- **Distinctive features:** Disc subcircular in females and juveniles, heart-shaped in mature males; broader than long; pectoral-fin apex very broadly rounded; anterior margin of disc convex to slightly concave in immatures and females, usually deeply concave opposite spiracles in mature males. Snout relatively short (orbit about 1.75–2.4* in preorbital snout), flexible (no firm rostral cartilage), with a very small fleshy lobe at tip; orbit moderately large, about 1.45–1.8* in interorbital space. Mouth rather narrow; internasal flap broadly lobed, lacking a dermal fringe. Tail moderately long (0.9–1.15* times precloacal length), slightly depressed, very slender; lateral skin folds well developed posteriorly. Dorsal surface of disc and tail granular (anterior lobe of pelvic fin naked); 1–5 nuchal thorns; orbital thorns usually enlarged (2–3 anterior, 0–1 median, 2–3 posterior); mature specimens generally with interspiracular thorns; small scapular thorns sometimes present; tail thorns rather large, widely spaced, in three rows (additional lateral rows near base of tail); alar thorns small, non-retractable, in 1–3 rows; malar thorns small, situated laterally to alar thorns, extending anteriorly to greatest concavity of anterior margin of disc. Ventral surface of disc and tail uniformly smooth. Pelvic fins small, margin between lobes very deeply concave; anterior lobe moderately elongate (1.4–1.8 in posterior lobe). Dorsal fins small, separated by less than half first dorsal-fin base; upper lobe of caudal fin short, base usually shorter than first dorsal-fin base, low, usually connected to second dorsal-fin base; lower lobe of caudal fin rudimentary. Claspers moderately elongate, very slender, enlarged slightly near tip. Tooth rows in lower jaw 28–33*. Pectoral fin radials 73–75* [62–73]. Predorsal diplospondylous vertebrae 63–69* [66–72].

- **Colour:** Upper surface medium to dark brown with poorly defined clusters of small pale spots; sometimes also with an irregular scattering of paler and darker blotches. Ventral surface mostly uniformly white; outer corners of disc occasionally dusky. Dorsal and caudal fins uniformly pale brownish or yellowish.

- **Size:** Attains at least 35 cm; males maturing at about 33 cm.

- **Distribution:** Southeastern Australia (including Tasmania), north to at least Newcastle (New South Wales) and west to Eyre (Great Australian Bight) in depths of 30–390 m.

- **Remarks:** The most widely distributed and first described member of the genus *Pavoraja*, this small skate is very abundant on the continental shelf of southern Australia.
- **References:** Günther (1880a); Last *et al.* (1983).

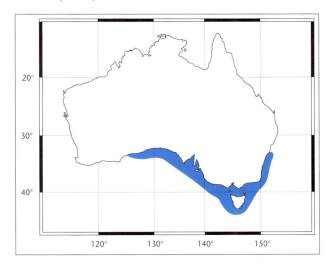

Longnose Skate — *Raja* sp. **A**

34.16 *Raja* sp. **A** (Plate 62) Fish Code: 00 031005

- **Field characters:** A medium-sized skate with a quadrangular disc, a moderately elongate snout with a firm rostral cartilage, usually with nuchal thorn(s) and malar thorns in males, a short and broad tail, and denticle patches on the anteroventral margins of the disc. The dorsal surface has a pattern of pale spots, blotches and reticulations, and there are dark pores over most of the ventral surface.
- **Distinctive features:** Disc quadrangular, broader than long; pectoral-fin apex broadly rounded; anterior margin of disc straight or weakly double concave in females and juveniles, strongly double concave in mature males. Snout moderately elongate (orbit 3.2–4.1 in preorbital snout), pointed, supported by a firm medial

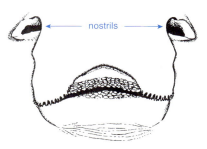

region of nostrils and mouth

cartilage; orbit moderate, about 1–1.25 in interorbital space. Mouth broad; internasal flap narrowly lobed with prominent dermal fringe. Tail short (0.65–0.75* times precloacal length in mature specimens, longer in juveniles), very depressed, rather broad; lateral skin folds well developed. Nuchal thorns present or absent, 0–7 (usually more than one), no other thorns on dorsal midline of disc before cloaca; orbital thorns forming a rosette around eye in mature specimens (2 anterior, 1 median, 1 posterior in juveniles). Dorsal surface of disc of males and juveniles mostly naked; females mostly granular along anterior margin of disc and on head, largest females often with weak granulations on remainder of disc and tail; alar and malar thorns well developed in mature males; alar thorns retractable; malar thorns in long single band beside eyes. Ventral surface granular along anterolateral margin of disc and on snout and ventral fins (often barely detectable in juveniles). Tail with double median rows of thorns in mature males; additional midlateral row (and occasionally with a few lateral thorns) in mature females. Pelvic fins large; anterior lobe short, about 1.5–2 in posterior lobe. Dorsal fins small to moderate, broadly rounded, equal in height, separated slightly or joined at base; upper lobe of caudal fin low, base much shorter than first dorsal-fin base, usually connected to second dorsal fin; lower lobe of caudal fin absent. Claspers rather broad, not bulbous distally.

- **Colour:** Dorsal surface brownish or yellowish (rarely greyish), usually with a complex pattern of paler spots, blotches and reticulations. Ventral surface white with grey areas (concentrated around head and central disc); ventral pores black-edged.

- **Size:** Reaches at least 70 cm, females larger than males; males mature at about 47–53 cm.

- **Distribution:** Confined to the continental shelf off southeastern Australia from Portland (Victoria) to Eden (New South Wales), including Tasmania, from 40–250 m.

- **Remarks:** The genus *Raja*, the largest of the skate genera, is comprised of several subgenera. This species, often mistaken for the New Zealand endemic skate, *Raja nasuta*, is one of the few true cool-temperate Australian skates. It is quite common in trawl catches, particularly off Tasmania, where it is sometimes taken with the white-spotted skate (34.33). The two species are similar in appearance but the longnose skate is rough along the anteroventral margin of the snout (rather than smooth as in the white-spotted).

- **Local synonymy:** *Raja nasuta*: Whitley, 1940 (misidentification).

- **Reference:** Last *et al.* (1983).

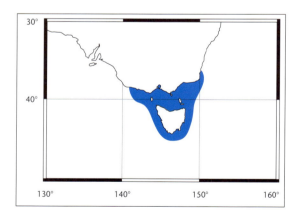

Grey Skate — *Raja* sp. **B**

34.17 *Raja* sp. **B** (Plate 60) **Fish Code:** 00 031028

- **Field characters:** A medium-sized skate with a quadrangular disc, a moderately elongate snout with a firm rostral cartilage, usually with nuchal thorn(s) and malar thorns in males, a relatively short and thin tail, and the anteroventral margins of the disc lacking denticles. The dorsal surface is uniformly greyish and the ventral surface is mostly pale with few dark pores.

- **Distinctive features:** Disc quadrangular, broader than long; pectoral-fin apex broadly rounded; anterior margin of disc straight to slightly double concave in females and juveniles, weakly double concave in mature males. Snout of moderate length (orbit 3.45–4.85 in preorbital snout), pointed, supported by a strong medial cartilage; orbit small to medium, 1.0–1.35 in interorbital space in mature specimens (relatively larger eye in juveniles). Mouth broad to very broad; internasal flap narrow-lobed with dermal fringe. Tail relatively short (0.8–0.9 times precloacal length in mature specimens, longer in juveniles), slightly depressed, not especially broad; lateral skin folds poorly developed. Nuchal thorns usually present (1–3), no other thorns on dorsal midline of disc before cloaca; orbital thorns forming a rosette around eye in mature specimens (mid-orbital thorns present in juveniles). Dorsal surface of disc mostly smooth, mature specimens granular on snout tip; males with granulations in malar region, females sometimes granular at base of tail; alar and malar thorns well developed in mature males; alar thorns retractable; malar thorns large, in a small patch at eye level. Ventral surface uniformly smooth, sometimes granular at snout apex in mature males. Tail usually with paired median rows of thorns, sometimes with a few lateral thorns on anterior half in males (tail thorns often lost); additional lateral row in females (tail thorns rarely missing). Pelvic fins moderately large; anterior lobe moderately short, about 1–1.5 in posterior lobe. Dorsal fins small to moderate, broadly rounded, equal in height, separated; upper lobe of caudal fin rather long, usually connected to second dorsal fin, its base usually about equal to first dorsal-fin base; lower lobe minute or absent. Claspers slender, slightly bulbous distally.

- **Colour:** Upper surface uniformly pale grey to greyish brown (white where deciduous skin has been removed). Ventral surface whitish, uniformly pale or with irregular dark greyish patches around inner angle of pelvic fin and around cloaca; ventral pores often indistinct, sometimes black-edged.

- **Size:** Reaches a total length of 90 cm; males mature at about 63–72 cm.

- **Distribution:** Occurs along the continental slope off southern Australia from Jervis Bay (New South Wales) to at least Eucla (Western Australia), including Tasmania. Most abundant on the slope in depths of 450–600 m, but distributed from 330–950 m.
- **Remarks:** This skate is possibly the most abundant *Raja* species living on the continental slope of southern Australia. It is often found amongst the bycatch of trawlers fishing for blue grenadier (*Macruronus*). Specimens from the Great Australian Bight tend to be paler on both surfaces, have less distinct pores, and lack pronounced markings around the pelvic and anal regions. Not presently retained as food but good eating and may become more valuable in the future. Little is known of its biology.
- **Local synonymy:** *Raja* sp. 2: Last *et al.*, 1983.
- **Reference:** Last *et al.* (1983).

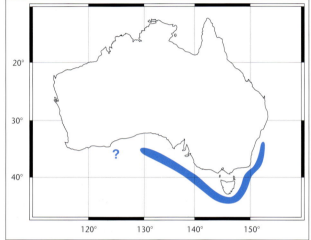

Grahams Skate — *Raja* sp. C

34.18 *Raja* sp. C (Plate 60)   Fish Code: 00 031029

- **Field characters:** A medium-sized skate with a quadrangular disc, a moderately elongate snout with a firm rostral cartilage, a nuchal thorn, no malar thorns in males, a short tail, and denticle patches on the anteroventral margins of the disc.

The dorsal surface is uniform dark brownish and the ventral surface is a slightly paler brown with dark-edged pores.

- **Distinctive features:** Disc quadrangular, much broader than long; pectoral-fin apex relatively angular; anterior margin of disc concave in females and juveniles, weakly double concave in mature males. Snout moderately elongate (orbit 4–4.8 in preorbital snout), rather broad at eye level, supported by a firm medial cartilage; orbit moderate, 1.05–1.45 in interorbital space. Mouth moderately broad; internasal flap narrow-lobed with dermal fringe. Tail short (0.75–0.8 times precloacal length), moderately depressed; lateral skin folds poorly developed. Nuchal thorn present, no other thorns on dorsal midline of disc before cloaca; rosette of short thorns distributed equally around orbit in mature specimens (medial thorns present in juveniles). Dorsal surface of disc smooth except for dense patch of denticles on snout tip and along anterior margin of disc in mature males; alar thorns retractable; malar thorns absent. Undersurface of preoral head mostly covered with denticles; denticle band on anteroventral surface narrow posteriorly and extending about halfway to two-thirds along anterior margin of disc. Tail with 1 row of regular-sized thorns along dorsal midline of male; females with additional lateral rows and short midlateral rows near base of tail. Pelvic fins moderately large; anterior lobe of moderate length, about 1.25–1.5 in posterior lobe. Dorsal fins medium-sized, rather elongate, equal in height, separated; base of upper lobe of caudal fin much shorter than first dorsal-fin base, usually connected to second dorsal fin; lower lobe of caudal fin reduced to a narrow ridge. Claspers slender, slightly bulbous distally.

- **Colour:** Upper surface uniformly dark chocolate brown, paler beside rostral cartilage. Ventral surface medium to dark greenish brown (paler than dorsal surface), no large white patches; sensory pores black-edged, without broad greyish borders. Tail similar to disc; basal half of dorsal fins dark with pale outer areas.

- **Size:** Reaches at least 63 cm; males mature at about 55 cm.

- **Distribution:** Continental slope off New South Wales between Wollongong and Coffs Harbour in 70–450 m, but most common in 200–380 m. Possibly also Western Australia.

- **Remarks:** This species replaces the grey skate (34.17) as the dominant *Raja* species on the continental slope off central eastern Australia. A single specimen of a skate collected recently off Western Australia may be this species, but more specimens are required.

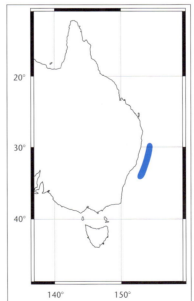

False Argus Skate — *Raja* sp. **D**

34.19 *Raja* sp. **D** (Plate 61) **Fish Code:** 00 031030

- **Alternative name:** Eye skate.
- **Field characters:** A small skate with a quadrangular disc, a moderately elongate snout with a firm rostral cartilage, a small nuchal thorn, a slender tail with widely separated dorsal fins, and denticle patches on the anteroventral margins of the disc. The mid-dorsal surface is yellowish with paler blotches, spots and streaks.
- **Distinctive features:** Disc quadrangular, broader than long; pectoral-fin apex broadly rounded; anterior margin of disc weakly double concave in females and juveniles. Snout short to moderate length (orbit 3.75–4.65 in preorbital snout), supported by a strong medial cartilage; orbit rather small, 0.95–1.2 in interorbital space. Mouth broad; internasal flap narrow-lobed with dermal fringe. Tail of medium length (0.9–0.95 times precloacal length), slightly depressed, slender; lateral skin folds well developed. Nuchal thorn small, no other thorns on dorsal midline of disc before cloaca; usually with 3 orbital thorns (2 anterior, 1 posterior). Dorsal surface of disc uniformly smooth in juveniles; females occasionally with a few sparse denticles on anterior margin adjacent to greatest concavity. Ventral surface granular on snout tip and anterior margin before mouth level in juveniles and some females (also on mid-belly in larger females). Tail with irregular row of thorns along midline and a few isolated midlateral thorns in immature males; midlateral row better developed and with short lateral rows near tail base in females. Pelvic fins rather small; anterior lobe long, about 1–1.25 in posterior lobe. Dorsal fins small to moderate, low, raked backward, equal in height, widely separated (often by almost an eye diameter); upper lobe of caudal fin extremely low, base much shorter than first dorsal-fin base, often separated from second dorsal fin; ventral lobe of caudal fin absent.
- **Colour:** Upper surface pale yellowish grey to medium brown; usually with an irregular arrangement of white spots, streaks and blotches on centre of disc behind scapular region (markings often indistinct, widely separated); paler beside rostral cartilage. Ventral surface mostly pale, sometimes with greyish areas around mouth and on central disc; pores dark-edged, sometimes encircled by grey areas. Dorsal fins brownish near base, pale distally.
- **Size:** Reaches about 48 cm; largest male examined was immature at 32 cm.
- **Distribution:** Continental shelf off Western Australia between Dampier and Port Hedland in 60–200 m.
- **Remarks:** Known from only a small sample of specimens.

- **Local synonymy:** *Raja* sp. 1: Sainsbury *et al.*, 1985; *Raja* sp.: Allen & Swainston, 1988.
- **Reference:** Sainsbury *et al.* (1985).

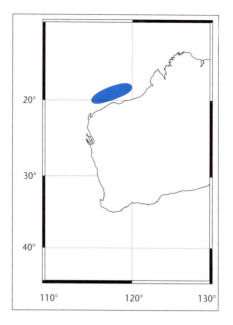

Oscellate Skate — *Raja* sp. E

34.20 *Raja* sp. E (Plate 63)   **Fish Code:** 00 031031

- **Field characters:** A medium-sized skate with a quadrangular disc, a moderately elongate snout with a firm rostral cartilage, a weak nuchal thorn, a short and broad tail, and denticle patches on the anteroventral margins of the disc. The dorsal surface is brownish with a large, dark ocellus on either side of the disc centre.
- **Distinctive features:** Disc quadrangular, broader than long; pectoral-fin apex broadly rounded; anterior margin of disc straight to weakly concave in females, double concave in maturing males. Snout moderate in length (orbit 3.7–4.6 in preorbital snout), moderately elongate and sharply pointed, supported by a strong medial cartilage; orbit moderate, 0.9–1.2* in interorbital space. Mouth broad; internasal flap narrow-lobed with dermal fringe. Tail short (0.7–0.9* times precloacal length), depressed, broad; lateral skin folds well developed. Nuchal thorn present, no other thorns on dorsal midline of disc before cloaca; orbital thorns short, 4–7. Dorsal surface of disc uniformly smooth, apart from short

denticles at snout tip. Ventral surface of disc and tail mostly smooth, granulations confined to snout tip and anterior third of disc margin. Tail with paired median row of thorns in males; females with additional lateral rows and short midlateral rows. Pelvic fins medium-sized; anterior lobe short, about 1.5 in posterior lobe. Dorsal fins short, broadly rounded, equal in height, separated; upper lobe of caudal fin small, base much shorter than first dorsal-fin base, usually connected to second dorsal fin; lower lobe of caudal fin reduced to a narrow ridge.

- **Colour:** Upper surface greenish grey with a pinkish margin, faint dusky blotches, and a single dark brown ocellus on each pectoral fin; ocelli with a large, dark spot, encircled by a darker ring; some irregular pale spots with broad, dark margins on posterior disc; head greyish pink around rostral cartilage. Ventral surface pale greyish pink with dark pores lacking dark borders; tail similar to body, but with about 6 olive saddles.

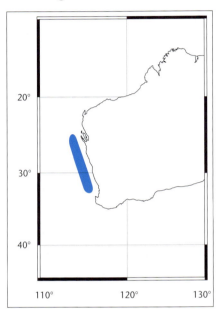

- **Size:** Reaches 58 cm; males mature at about 40–47 cm.
- **Distribution:** Off Western Australia between Shark Bay and Bunbury in 200–250 m.
- **Remarks:** This very distinctive skate has a narrow geographic range and depth preference. Like other medium-sized Australian skates, it is likely to be good to eat. Not presently exploited.

Leylands Skate — *Raja* sp. F

34.21 *Raja* sp. **F** (Plate 58) **Fish Code:** 00 031011

- **Field characters:** A medium-sized skate with a quadrangular disc (width rarely more than 1.2 times length), an elongate snout with a firm rostral cartilage, a rosette of thorns around eye in adults, rarely with nuchal thorns, malar thorns present in males, a slender and almost oval tail, and denticle patches on the anteroventral margins of the disc. The dorsal surface is uniform brownish and there are no dark-edged pores on the ventral surface.

- **Distinctive features:** Disc quadrangular, broader than long; pectoral-fin apex rather angular; anterior margin of disc double concave in females and juveniles, deeply double concave in mature males. Snout elongate (orbit about 5.5–7.6 in preorbital snout), extremely pointed, supported by a solid medial cartilage; orbit medium-sized, 1.2–1.6 in interorbital space. Mouth broad; internasal flap narrow-lobed, with dermal fringe. Tail moderately long (0.8–0.85 times precloacal length), rather oval in cross-section, slender; lateral skin folds usually poorly developed. Nuchal thorns rarely present, no other thorns on dorsal midline of disc before cloaca; orbital thorns short, forming a rosette around eye (2 anterior, 1 posterior at birth). Dorsal surface of disc with band of fine denticles on anterolateral margin of males; females smooth; alar and malar thorns well developed in mature males; alar thorns retractable; malar thorns large, beside eye. Ventral surface with large denticles near snout tip and fine denticles extending along anterior margin (evident on small juveniles). Tail with 1 irregular row of thorns in males and juveniles; females with additional and better developed lateral rows. Pelvic fins small to medium; anterior lobe long, about 1–1.3 in posterior lobe. Dorsal fins short, slender, raked backward, equal in height, connected or separated by less than length of first dorsal-fin base; upper lobe of caudal fin low, usually connected to second dorsal fin, its base frequently equal to length of first dorsal-fin base, extending onto ventral surface as a ridge. Claspers medium-sized, bulbous distally.

- **Colour:** Upper surface yellowish brown to greyish brown; darkest around eyes, on centre of disc and along tail; paling gradually toward disc margin; snout above rostral cartilage white. Ventral surface variable, pale with scattered darker patches or dark greyish brown centrally and mostly pale around outer disc; sensory pores without distinct dark edges. Dorsal and caudal fins dark in juveniles, mostly dusky or dark in mature specimens.

- **Size:** Attains at least 72 cm; males maturing at about 50 cm.

- **Distribution:** Western Australia on the upper continental slope between Ashmore Reef and Bunbury in 200–440 m.

- **Remarks:** Very common throughout its range. Little is known of its biology.

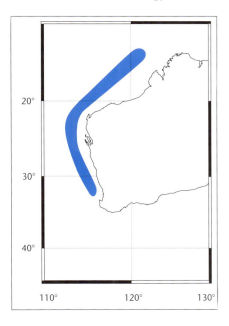

Pale Tropical Skate — *Raja* sp. **G**

34.22 *Raja* sp. **G** (Plate 57)   Fish Code: 00 031032

- **Field characters:** A medium-sized skate with a quadrangular disc (width rarely more than 1.2 times length), a very elongate snout with a firm rostral cartilage, no nuchal or malar thorns, no rosette of thorns around the eye, and a slender (almost oval) tail. The dorsal surface is uniform greyish brown, the ventral surface is dusky with dark-edged pores, and the upper caudal-fin lobe is pale.

- **Distinctive features:** Disc quadrangular, broader than long (usually by about 2–3 orbit diameters); pectoral-fin apex not particularly angular; anterior margin of disc double concave in females and juveniles, strongly concave in mature males. Snout very elongate (orbit about 4.85–7.55 in preorbital snout), extremely pointed, supported by a solid medial cartilage; orbit rather small, about 1.0–1.6 in interorbital space. Mouth of medium width; internasal flap narrow-lobed, with dermal fringe. Tail rather short (0.75–0.85* times precloacal length in mature specimens, longer in juveniles) oval in cross-section, slender, slightly depressed, sometimes bulging near middle; lateral skin folds poorly developed. Nuchal thorn absent, no thorns on dorsal midline of disc before cloaca; orbital thorns short (2 anterior, 1 posterior in juveniles), rarely with additional thorns in adults and not forming a rosette around eye. Dorsal surface of disc with denticles on snout and along anterolateral margin of males; females and juveniles smooth; alar thorns retractable; malar thorns absent. Ventral surface mostly smooth except for preoral area and along disc margin to greatest concavity. Tail with 1 row of widely spaced thorns in males and juveniles; females sometimes with short lateral row near base. Pelvic fins rather small; anterior lobe long, longer than posterior lobe in juveniles, becoming relatively shorter (1–1.5 in posterior lobe) in adults. Dorsal fins rather rounded, raked backward, subequal in height, well separated (by half an eye diameter or more); upper lobe of caudal fin low, base subequal to length of first dorsal-fin base, connected to second dorsal fin; ventral lobe barely detectable. Claspers long, slender, bulging slightly over distal half.

- **Colour:** Upper surface uniformly medium greyish brown; paler around disc margin and on snout. Ventral surface irregular; pale grey to dark brownish; snout mostly pale, except for dark stripe below rostral cartilage; margin of disc pale; pores black-edged; undersurface of tail mostly pale. Dorsal-fin bases brownish, outer half and upper caudal-fin lobe pale; claspers similar to body colour.

- **Size:** To at least 77 cm, adult females larger than males. Males mature from 55–65 cm; born at about 17 cm.

- **Distribution:** Continental slope off northeastern Australia, between Bundaberg and Cairns (Queensland), in depths of 225–550 m (mainly in 300–400 m).
- **Remarks:** Closely related to the larger temperate bight skate (34.34). Common throughout its range and one of the most abundant rays on the upper continental shelf off Queensland.

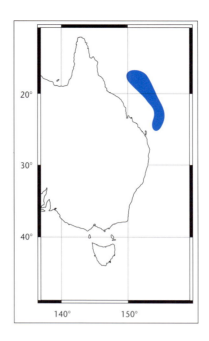

Blacktip Skate — *Raja* sp. **H**

34.23 *Raja* sp. **H** (Plate 56) **Fish Code:** 00 031033

- **Field characters:** A large skate with a very broad, quadrangular disc (width exceeding 1.2 times length), a very elongate snout with a firm rostral cartilage, no nuchal or malar thorns, a rosette of thorns around the eye of adults, and a slender tail. The dorsal surface is uniform yellowish to greyish, the ventral surface is mostly dusky with dark-edged pores, and the upper caudal-fin lobe is dark.
- **Distinctive features:** Disc broadly quadrangular, much broader than long (by almost preoral snout length); pectoral-fin apex extremely angular; anterior margin of disc double concave in females and juveniles, deeply double concave in mature

males. Snout elongate (orbit about 4.6–6.85 in preorbital snout), extremely pointed, supported by a solid medial cartilage; orbit small, 0.9–1.4 in interorbital space. Mouth relatively narrow; internasal flap narrow-lobed, with dermal fringe. Tail rather short (0.75–0.85 times precloacal length in mature specimens, longer in juveniles), quadrangular in cross-section, slender, with minor median bulge; lateral skin folds usually poorly developed. No nuchal thorns or thorns on dorsal midline of disc before cloaca; orbital thorns short usually up to 6), forming a rosette around eye (2 anterior, 1 posterior in juveniles). Dorsal surface of disc with denticles on snout tip and along anterior margin to level of alar patch in males; females and juveniles smooth; alar thorns retractable; malar thorns absent. Ventral surface of disc and tail usually smooth except on snout and along anterior margin to greatest concavity. Tail with 1 median row of thorns. Pelvic fins small; anterior lobe long, subequal in length to posterior lobe in juveniles, 1.25–1.5 in posterior lobe in mature males. Dorsal fins moderately well developed, raked backward, subequal in height, usually separated slightly; upper lobe of caudal fin low, usually connected to second dorsal fin, its base subequal to length of first dorsal-fin base; lower lobe of caudal fin very short, low. Claspers very long, bulging slightly distally.

- **Colour:** Upper surface pale yellowish to greyish, whitish above rostral cartilage. Ventral surface mostly pale in juveniles, becoming darker greyish brown in adults with definite pale patches on head, and around mouth, disc centre and pelvic fin; outer half of anterior disc with a distinctive black margin in juveniles, less distinct in adults (anterior half mostly with a distinct pale margin); pores dark-edged. Dorsal fins, and caudal fin and its base black in small individuals, often dusky in adults. Claspers similar to body.

- **Size:** Attains a length of at least 76 cm and males mature by 62 cm; free swimming at 20 cm.

- **Distribution:** Eastern Australia from Wooli (New South Wales) to Townsville (Queensland) on the continental slope in 240–650 m.

- **Remarks:** This particularly broad, short-bodied skate has a very angular disc with a characteristic dark front margin on its undersurface in young individuals. It is easily confused with other members of the subgenus *Dipturus*, the pale tropical (34.22) and Queensland deepwater skates (34.26), that occur in the same area; these species need to be identified carefully.

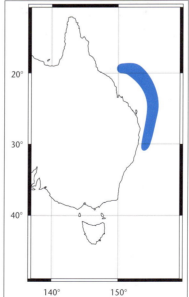

Wengs Skate — *Raja* sp. I

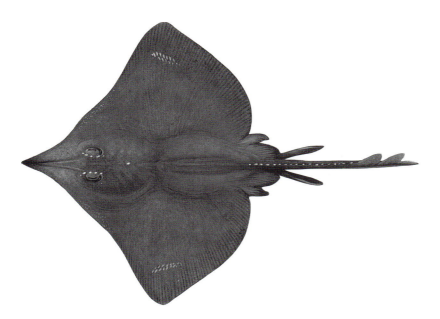

34.24 *Raja* sp. I (Plate 58)　　　　　　　　　　　　**Fish Code:** 00 031034

- **Field characters:** A large skate with a broad quadrangular disc (width exceeding 1.2 times length), a moderately elongate snout with a firm rostral cartilage, no malar thorns but often with a nuchal thorn, rarely with a rosette of thorns around the eye of adults, and a broad, almost circular tail. Both dorsal and ventral surfaces are dark.

- **Distinctive features:** Disc quadrangular, much broader than long (by more than 4 orbit diameters); pectoral-fin apex rather angular; anterior margin of disc weakly double concave in juveniles and small females, becoming moderately concave in mature individuals. Snout moderately elongate (orbit about 7–8.5* in preorbital snout in mature specimens, orbit relatively larger in juveniles), relatively broad, pointed, supported by a solid medial cartilage; orbit small, about 1.6–2.1* in interorbital space in mature males, relatively larger (about 1.2–1.4* in interorbital space) in juveniles. Mouth relatively narrow; internasal flap narrow-lobed, with dermal fringe. Tail short (0.7–0.75* times precloacal length in mature specimens, longer in juveniles), semi-spherical in cross-section in adults (almost rounded in juveniles), rather slender, becoming greatly enlarged centrally; lateral skin folds poorly developed. Nuchal thorns present in eastern Australian populations, usually absent in western Australian populations; no other thorns on dorsal midline of disc before cloaca; orbital thorns long, sharp (2 anterior, 1 posterior in juveniles), usually persisting but shorter in adults (sometimes with more thorns but rarely forming a rosette). Dorsal surface of disc uniformly smooth; alar thorns retractable; malar thorns absent. Ventral surface with large denticles confined to below rostral cartilage and along front half of anterior margin; very large females with widely spaced granulations on belly. Tail with 1 row of median thorns, becoming widely and evenly spaced and deciduous in adults. Pelvic fins small to medium; anterior lobe elongate (usually longer than posterior lobe, except in mature males). Dorsal fins rounded, raked backward, subequal in height, close together but rarely connected; upper lobe of caudal fin low, base usually greater than length of first dorsal-fin base, connected to second dorsal fin; ventral lobe very short. Claspers slender.

- **Colour:** Upper surface greyish brown to dark brown on mid-disc, greyish to black along disc margin and over rostral cartilage. Ventral surface dark, similar to dorsal surface; front half of anterior margin pale; sometimes with pale areas at tip of anterior lobe of pelvic fin and around cloaca. Dorsal and caudal fins mostly black in small specimens, sometimes dusky in adults. Claspers dark.

- **Size:** To at least 115 cm; males mature at about 102 cm; born at about 26 cm.
- **Distribution:** Widely distributed on the continental slope off tropical and temperate Australia in depths of 400–1030 m (mainly between 400–600 m). Along eastern Australia between Cairns (Queensland) and Cape Pillar (Tasmania); off western Australia between Port Hedland and Geraldton; possibly also off the south coast.
- **Remarks:** This widely distributed ray is one of the few skates with an eastern and western Australian distribution. It is most abundant in tropical waters and is seldom caught in cool temperate areas. Similar species occur in other regions.

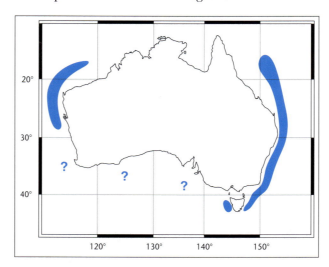

Deepwater Skate — *Raja* sp. J

34.25 *Raja* sp. J (Plate 59) **Fish Code:** 00 031035

- **Alternative name:** Brown bight skate.
- **Field characters:** A large skate with a quadrangular disc (width usually less than 1.2 times length), a nuchal thorn, malar thorns in males, a rosette of thorns around the eye of adults, a very elongate snout with a firm rostral cartilage, and a slender (almost oval) tail. The dorsal surface is distinctly paler than the ventral surface and the caudal fin is dark.

- **Distinctive features:** Disc quadrangular, broader than long (by about 2–3 orbit diameters); pectoral-fin apex rather angular; anterior margin of disc deeply concave. Snout very long (orbit 5.2–7.8 in preorbital snout), slender, pointed; supported by solid medial cartilage; orbit small, 0.95–1.3 in interorbital space. Mouth narrow; internasal flap narrow-lobed, with dermal fringe. Tail short (about 0.8* times precloacal length in mature specimens, usually longer in juveniles), almost oval in cross-section, very slender, slightly enlarged medially; lateral skin folds poorly developed. Usually with single nuchal thorn, no other thorns on dorsal midline of disc before cloaca; orbital thorns long, sharp in juveniles (usually 2 anterior, 1 posterior) but short, forming a rosette in adults. Dorsal surface of disc mostly smooth (except anterior margin of males), uneven coverage of granulations around eyes; alar and malar thorns well developed in mature males; alar thorns retractable; malar thorns large, lateral to eye. Ventral surface smooth except for snout and near anterior margin of disc. Tail with 1–5 irregular rows of large thorns (long, sharp in juveniles), lacking granulations. Pelvic fins small; anterior lobe long (usually longer than or equal to posterior lobe in adolescents and juveniles). Dorsal fins moderate-sized, rounded, relatively upright, subequal in height, separated by half an eye diameter or less; upper lobe of caudal fin low, connected at base to second dorsal fin, its base usually subequal to length of first dorsal-fin base; lower lobe of caudal fin minute. Claspers slender.
- **Colour:** Dorsal surface almost uniformly white (in western Australian specimens) to pale grey or greyish brown with light irregular patches where skin removed (in eastern Australian specimens); paler around rostral cartilage; posterior margin of disc dark in juveniles (area broader on ventral surface). Ventral surface mostly brownish black, with irregular lighter flecks; paler on snout; tail and pelvic-fin margins mostly black in small individuals; pores dark-edged. Post-dorsal tail black in juveniles; dorsal fins mostly black in adolescents, dusky in adults.
- **Size:** Attains a length of 133 cm, adult females much larger than males. Males mature at about 80 cm; born at about 20 cm.
- **Distribution:** Widely distributed along the mid-continental slope of southern Australia from Esperance (Western Australia) to Sydney (New South Wales), including Tasmania, in 800–1400 m (but mainly less than 1000 m).
- **Remarks:** The geographical range of this skate is likely to be extended when the continental slope of southwestern Australia has been more thoroughly explored. This is one of the largest deepwater skates, with females large compared to males. Common, but little is known of its biology.
- **Local synonymy:** *Raja gudgeri*: Last *et al.*, 1983 (misidentification).
- **References:** Last *et al.* (1983).

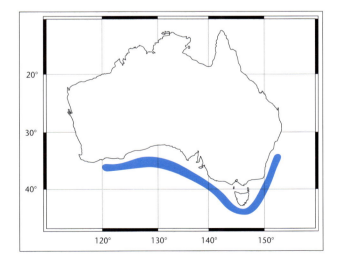

Queensland Deepwater Skate — *Raja* sp. K

34.26 *Raja* sp. K (Plate 59) Fish Code: 00 031036

- **Field characters:** A medium-sized skate with a quadrangular disc (width usually less than 1.2 times length), with or without a nuchal thorn, malar thorns in males, a rosette of thorns around the eye, a very elongate snout with a firm rostral cartilage, and a slender (almost oval) tail. The dorsal surface is not appreciably paler than the ventral surface and the caudal fin is pale.

- **Distinctive features:** Disc quadrangular, broader than long (usually by about 2–3 orbit diameters); pectoral-fin apex very angular; anterior margin of disc deeply concave to deeply double concave. Snout very elongate (orbit 5.65–7.35 in preorbital snout), extremely pointed, supported by a solid medial cartilage; orbit relatively small, about 1.1–1.3 in interorbital space. Mouth of moderate width; internasal flap narrow-lobed, with dermal fringe. Tail short (0.7–0.85* times precloacal length in mature specimens, longer in juveniles), oval in cross-section, very slender, bulging medially; lateral skin folds well developed posteriorly. Nuchal thorn present or absent, no other thorns on dorsal midline of disc before cloaca; orbital thorns elongate, sharp, forming a rosette around eye (well developed in small specimens). Dorsal surface of disc with denticles on snout tip and along anterior margin in males; females and juveniles smooth; alar thorns retractable; malar thorns well developed, in small patch beside eye. Ventral surface mostly smooth except along anterior margin and on snout (rarely extending much past mouth level). Tail thorns closely spaced; 1 row in mature males and juveniles; females with lateral rows (sometimes incomplete). Pelvic fins small to medium; anterior lobe long (slightly shorter than posterior lobe in juveniles and females, up to about 1.3 in posterior lobe in mature males). Dorsal fins upright, broadly rounded, subequal in size, separated slightly; upper lobe of caudal fin elongate, base usually longer than first dorsal-fin base, connected to second dorsal fin; ventral lobe of caudal fin rudimentary. Claspers slender, slightly bulbous distally.

- **Colour:** Upper surface uniformly pale grey to greyish brown; paler beside rostral cartilage and around posterior margin of disc. Ventral surface variable, pale greyish white; frequently covered with extensive dark brown areas and blotches (not confined to pores), a dark area behind mouth between gill slits consistently present; gill slits pale; margin of posterior lobe of pelvic fin pale; pores mostly greyish, occasionally with broad borders. Dorsal fins blackish or dusky; caudal fin mostly white. Claspers similar to body colour.

- **Size:** Attains at least 76 cm; males mature by 65 cm.

- **Distribution:** Off northeastern Australia on the continental slope, between the Swain Reefs and Townsville, in 440–650 m.
- **Remarks:** One of the the few Australian *Dipturus* species in which the nuchal thorn may be present or absent. Specimens lacking a nuchal thorn are sometimes confused with the blacktip skate (34.23).

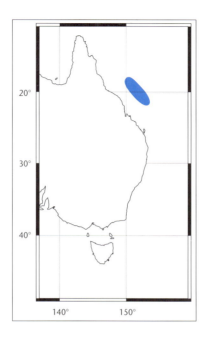

Maugean Skate — *Raja* sp. L

34.27 *Raja* sp. L (Plate 55) **Fish Code:** 00 031037

- **Field characters:** A medium-sized skate with a quadrangular disc (width less than 1.2 times length), a very elongate snout with a firm rostral cartilage, nuchal thorns, malar thorns present in males, a broad and slightly depressed tail, spatulate claspers, and the denticles on undersurface confined to the preoral head. The dorsal surface is greyish black and the ventral surface is also dark with dark-edged pores.

- **Distinctive features:** Disc quadrangular, slightly broader than long; pectoral-fin apex angular; anterior margin of disc deeply concave; snout elongate (orbit 5.7–5.9* in preorbital snout), narrow, pointed, supported by a firm medial cartilage; orbit moderately large, 1.4–1.55* in interorbital space. Mouth moderately broad; internasal flap narrow-lobed, with dermal fringe. Tail moderate (0.7–0.75* times precloacal length), slightly depressed, rather broad; lateral skin folds well developed. Two nuchal thorns, no other thorns on midline of disc before cloaca; orbital thorns 4–5, short, widely separated. Tail with a regular median row and midlateral rows of thorns in mature males. Dorsal surface of disc with fine denticles situated along anterior margin, around eyes, and along entire midline (including tail); pelvic fin smooth; alar and malar thorns present in mature males; alar thorns retractable; malar thorns in a small patch opposite eye. Ventral surface of preoral head mostly uniformly granular; remainder of ventral surface smooth, with a thick mucus covering. Pelvic fins large; anterior lobe short, about 1.5–1.75 in posterior lobe. Dorsal fins rather large, rounded, equal in height, separated; base of upper lobe of caudal fin shorter than first dorsal-fin base, connected to second dorsal fin; lower lobe of caudal fin reduced to narrow ridge. Claspers large, spatulate.
- **Colour:** Upper surface almost uniformly greyish black with faint (often barely detectable), pale spots. Undersurface dark (but paler than upper surface), darkest near centre of disc and paling laterally to greyish white; pores dark-edged; pelvic fins and claspers dark grey; white patch around cloaca; tail uniformly dark grey.
- **Size:** Known from two mature males of 64 and 67 cm.
- **Distribution:** Bathurst Harbour near Port Davey (south-western Tasmania) in 5 m.
- **Remarks:** Known only from Port Davey, a large estuary in Tasmania's south-west. Skates taken on fisheries surveys from Macquarie Harbour, a short distance to the north, may also have been this species. Its closest relative, the rough skate (*Raja nasuta*), occurs inshore off New Zealand.

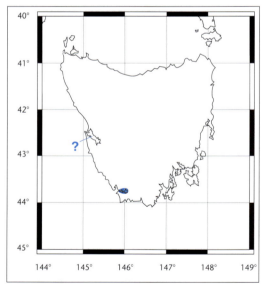

Pygmy Thornback Skate — *Raja* sp. **M**

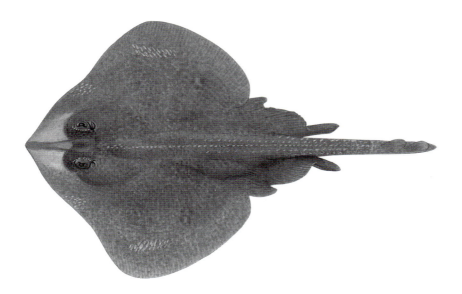

34.28 *Raja* sp. **M** (Plate 54)   **Fish Code:** 00 031038

- **Alternative name:** Thornback skate.

- **Field characters:** A very small skate with a circular to heart-shaped disc, a short snout with a firm rostral cartilage, a relatively wide interorbital space (2.2–2.65 in preorbital snout), small malar thorns, nuchal thorns (usually), no denticles on the ventral surface, and a very short, broad and extremely depressed tail. The undersurface is mostly pale with a prominent dark snout tip and no dark-edged pores.

- **Distinctive features:** Disc subcircular to broadly heart-shaped, broader than long; pectoral-fin apex broadly rounded; anterior margin of disc slightly concave (weakly double concave in mature males). Snout short (orbit 2.65–3.45 in preorbital snout), bluntly pointed, supported by a firm medial cartilage; orbit rather small, 1.15–1.4 in interorbital space. Mouth very broad; internasal flap narrow-lobed with dermal fringe. Tail very short (0.75–0.9 times precloacal length), very broad and depressed; lateral skin folds very well developed. Up to 4 nuchal thorns (occasionally absent); median thorns mostly minute and blunt; median row variable, rarely beginning in advance of cloaca, but extending onto tail; tail with additional lateral rows in mature males, often additional 2 midlateral rows in females; up to 6 orbital thorns, mostly small. Dorsal surface of disc lacking granulations in juveniles, denticles present along anterior margin of disc and on snout in mature males (frequently sparse, widespread over disc in mature females); alar thorns retractable; malar thorns small, in small patch beside eye. Ventral surface of disc and tail entirely smooth. Pelvic fins large; anterior lobe short, about 1.25–1.75 in posterior lobe. Dorsal fins moderately large, broadly rounded, first slightly larger than second, usually slightly separated; upper lobe of caudal fin minute, connected to second dorsal fin; lower lobe of caudal fin rudimentary. Claspers short, rather bulbous. Tooth rows in lower jaw 35–37*. Pectoral fin radials 74–75*. Predorsal diplospondylous vertebrae 36–38*.

- **Colour:** Upper surface pale yellowish brown, covered with dense pattern of fine white spots (sometimes plain); lacking well-developed reticulations and fine black spots (except sometimes in interorbital area). Ventral surface whitish; snout tip distinctly greyish or black; pores lacking dark borders. Dorsal fins and claspers similar to body colour.

- **Size:** To about 36 cm; males mature at about 27 cm.
- **Distribution:** Continental shelf off southern Australia from Albany (Western Australia) to Victor Harbour (South Australia) in 20–35 m.
- **Remarks:** A dwarf relative of the thornback skate (34.36). Earlier reports of the latter from South Australia are likely to be of this species.

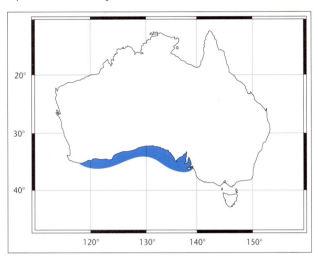

Thintail Skate — *Raja* sp. **N**

34.29 *Raja* sp. **N** (Plate 56) Fish Code: 00 031013

- **Field characters:** A small skate with a quadrangular disc, a moderately elongate snout with a firm rostral cartilage, no nuchal thorns, a very thin tail, and very forward and widely spaced dorsal fins. The disc is pale on both surfaces and the ventral pores are not dark-edged.

- **Distinctive features:** Disc quadrangular, much broader than long; pectoral-fin apex somewhat angular in adults; anterior margin of disc weakly double concave in females and juveniles, strongly double concave in mature males. Snout moderately elongate (orbit 4.0–5.1 in preorbital snout), pointed, supported by a firm medial cartilage; orbit moderate, 1.05–1.2 in interorbital space. Mouth broad; internasal flap narrow-lobed, with dermal fringe. Tail moderately elongate (0.95–1.2 times precloacal length), slender; lateral skin folds poorly developed. No nuchal thorns or thorns on dorsal midline of disc before cloaca; usually with more than 4 orbital thorns, forming rosette around eye in mature specimens (mainly with 3 anterior and 2 posterior in juveniles). Dorsal surface of disc smooth except for snout tip and narrow band along anterior margin; alar thorns retractable; malar thorns well developed, beside eye. Ventral surface of disc and tail smooth behind mouth; denticles confined to snout and along disc margin anterior to mouth level. Tail usually with single row of large, sharp thorns in males and juveniles; additional lateral rows (sometimes also midlaterals) in females. Pelvic fins large; anterior lobe short, about 1.25–1.5 in posterior lobe. Dorsal fins small, broadly rounded, equal in height, very widely separated (by an eye diameter or more); situated well forward on tail (tail length behind first dorsal-fin origin exceeding distance from snout tip to spiracle); base of upper lobe of caudal fin much longer than first dorsal-fin base, separated or connected to second dorsal fin by low membrane; lower lobe of caudal fin barely detectable. Claspers elongate, slightly bulbous distally.

- **Colour:** Upper surface uniformly greenish brown to yellowish centrally, paler around disc margin and on snout; no distinct pattern of spots and ocelli. Undersurface whitish; no dark greyish patches or dark-edged pores. Dorsal fins dark grey or black in juveniles, pale to dusky in adults. Claspers greyish brown above, pale below.

- **Size:** Attains about 56 cm; males mature from 41–45 cm; adult females larger than males.

- **Distribution:** Off Western Australia from Bunbury to Port Hedland in 400–735 m.

- **Remarks:** Superficially similar to some other species from the region, this skate can be identified by its very widely spaced dorsal fins and long postdorsal tail. A member of the subgenus *Okamejei*, its closest relatives appear to be found in other parts of the Indo-Pacific. It occurs in two disjunct populations (Bunbury to Shark Bay and from Dampier to Port Hedland) and needs closer examination.

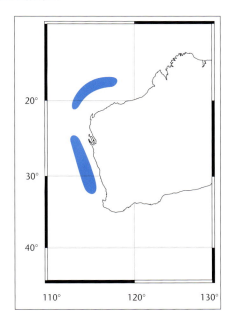

Sawback Skate — *Raja* sp. O

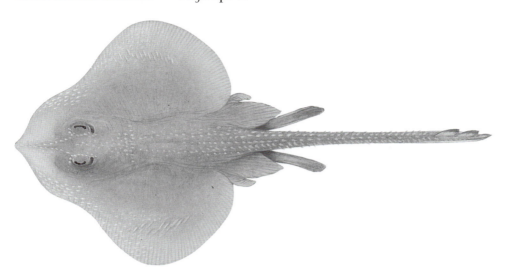

34.30 *Raja* sp. O (Plate 52)　　　　　　　　　　　　　　**Fish Code:** 00 031039

- **Field characters:** A small skate with a subcircular to heart-shaped disc, a short snout with a firm rostral cartilage, numerous enlarged thorns around the orbit, a triangular patch of thorns on the shoulder, malar thorns, no denticle patches on the anteroventral margins of the disc, and a moderately elongate, broad and very depressed tail. Both surfaces of the disc are pale and the ventral pores are not dark-edged.

- **Distinctive features:** Disc subcircular in females and juveniles, heart-shaped in mature males, broader than long; pectoral-fin apex broadly rounded; anterior margin of disc straight to slightly convex in females and juveniles, deeply double concave in mature males. Snout short (orbit 2.85–3.55 in preorbital snout), rather bluntly rounded, supported by a firm medial cartilage; orbit rather small, 0.9–1.05 in interorbital space. Mouth broad; internasal flap narrow-lobed, with dermal fringe. Tail moderately elongate (0.95–1.2* times precloacal length in mature specimens, longer in juveniles), very depressed, rather broad; lateral skin folds poorly developed. Triangular patch of thorns in nuchal area, usually 2 rows of thorns along back posterior to nuchal patch, continuing onto tail as 3–7 rows; dense patch of thorns around eye and on interorbital (less well developed in juveniles). Dorsal surface of disc uniformly granular in juveniles, denticles less dense in adults; additional thorns along snout in malar region of both sexes; alar thorns retractable. Ventral surface of disc uniformly smooth; tail sometimes granular. Pelvic fins large; anterior lobe short, about 1.25–1.5 in posterior lobe. Dorsal fins moderate, raked backward, subequal in height, usually connected; base of upper lobe of caudal fin much shorter than first dorsal-fin base, usually connected to second dorsal fin; lower lobe of caudal fin absent. Claspers stocky, elongate.

- **Colour:** Upper surface uniformly pale greyish or brownish, somewhat translucent around edge of disc and on pelvic fins. Undersurface uniformly pale or translucent; pores not dark-edged. Dorsal and caudal fins with pale bases and dark margins (less distinct in adults). Claspers similar to body coloration.

- **Size:** Attains about 40 cm; males mature at about 33 cm.

- **Distribution:** Continental slope off Western Australia between Ashmore Reef and Dampier in 350–420 m.

- **Remarks:** The upper surface of this short-snouted species is extremely thorny. It belongs to the subgenus *Leucoraja* and is the only member of this group recorded from the Australasian region.

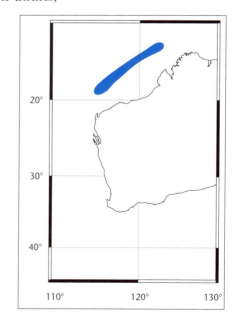

Challenger Skate — *Raja* sp. **P**

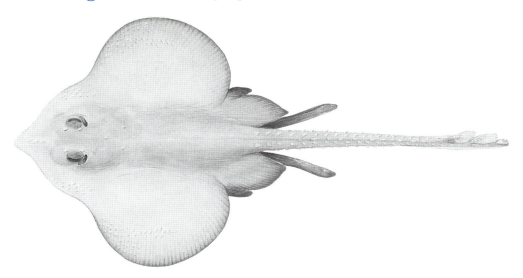

34.31 *Raja* sp. **P** (Plate 52)   **Fish Code:** 00 031040

- **Field characters:** A small skate with a subcircular to heart-shaped disc, a rather short snout with a firm rostral cartilage, malar thorns, numerous enlarged thorns around the orbit and along the mid-disc, no denticle patches on the anteroventral margins of the disc, and a rather elongate, oval tail. The upper disc is pale and the ventral surface is dark.

- **Distinctive features:** Disc subcircular in females and juveniles, heart-shaped in mature males, slightly broader than long; pectoral-fin apex very broadly rounded; anterior margin of disc straight to weakly double concave in females and juveniles, deeply concave in mature males. Snout short to moderate (orbit 3.15–3.65 in preorbital snout in adults, snout shorter in juveniles), bluntly rounded, supported by a firm medial cartilage; orbit small to moderate (0.95–1 in interorbital space). Mouth relatively narrow; internasal flap narrow-lobed, with dermal fringe. Tail long (1.15–1.25* times precloacal length in adults, much longer in juveniles), oval, tapering, broad basally; lateral skin folds poorly developed anteriorly. Adults usually lacking granulations; juveniles covered uniformly with large denticles. Usually with 1–3 rows of thorns extending posteriorly along dorsal midline behind a large triangular patch of nuchal thorns (sometimes absent); precloacal thorns extending onto tail as a single row in juveniles, 3–5 rows in adults; thorns greatly enlarged and sharp; usually 5 orbital thorns in juveniles (often more in adults); additional thorns on snout and in malar region of both sexes; alar thorns retractable. Ventral surface of disc and tail uniformly smooth. Pelvic fins large; anterior lobe short, about 1.25–1.5 in posterior lobe. Dorsal fins small, raked backward, equal in height, usually connected; base of upper lobe of caudal fin very short, usually connected to second dorsal fin; lower lobe of caudal fin minute. Claspers almost conical, rather spatulate posteriorly.

- **Colour:** Upper surface uniformly white to pale grey, margins of disc black; anterior lobes and margins of pelvic fins black. Ventral surface almost entirely dark brown or black; often with pale areas around mouth and cloaca, on anterior lobe of pelvic fin, and irregularly on disc; tail mostly dark. Claspers brownish; dorsal fins mostly pale.

- **Size:** Attains about 55 cm; males mature at about 51 cm; born at about 24 cm.

- **Distribution:** Southern Australia from Sydney (New South Wales) to Albany (Western Australia) in 860–1500 m. Abundant on the mid–lower continental slope off Tasmania in depths exceeding 1100 m.

- **Remarks:** This species was discovered by the Tasmanian fisheries research vessel 'Challenger' during an exploratory research cruise for orange roughy (*Hoplostethus* *atlanticus*) off the Tasmanian coast in the early 1980s. It is the only member of the subgenus *Rajella* found in Australasian waters. Like most skates from this region, little is known of its biology.

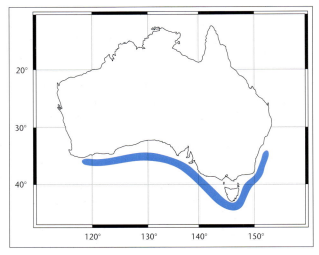

Sydney Skate — *Raja australis* Macleay, 1884

34.32 *Raja australis* (Plate 61) **Fish Code:** 00 031002

- **Alternative name:** Common skate.
- **Field characters:** A medium-sized skate with a quadrangular disc, a moderately elongate snout with a firm rostral cartilage, a nuchal thorn, two malar thorn patches in males, no denticles on the undersurface of the disc behind the mouth, and a short, broad and very depressed tail. The dorsal surface is brownish with paler yellowish blotches and the ventral pores are dark-edged.
- **Distinctive features:** Disc quadrangular, broader than long; pectoral-fin apex broadly rounded; anterior margin of disc straight to slightly concave in females and juveniles, double concave in mature males. Snout moderately elongate (orbit 3.25–3.8 in preorbital snout), pointed, supported by a firm medial cartilage; orbit moderate (0.8–1.0 in interorbital space). Mouth broad; internasal flap narrow-lobed with dermal fringe. Tail short (0.7–0.8 times precloacal length in adults), very depressed,

broad; lateral skin folds well developed. Dorsal surface of disc smooth except for a dense patch of denticles on snout tip and a narrow band along the mid-anterior margin. No nuchal thorns or thorns on dorsal midline of disc before cloaca; orbital thorns evenly spaced, forming rosette around eye in mature specimens (2 anterior, 2 posterior in juveniles); alar thorns retractable; malar thorns large, in 2 groups on each side of eye. Ventral surface of disc and tail smooth behind mouth; denticles distributed irregularly over snout in males, confined to tip and along snout margin in females. Tail with 3 rows of large, sharp thorns in males, 5 rows in females. Pelvic fins large; anterior lobe short, about half length of posterior lobe. Dorsal fins small to moderate, broadly rounded, equal in height, situated close together or connected; base of upper lobe of caudal fin much shorter than first dorsal-fin base, usually connected to second dorsal fin; lower lobe of caudal fin minute. Claspers rather elongate, broad, slightly bulbous distally.

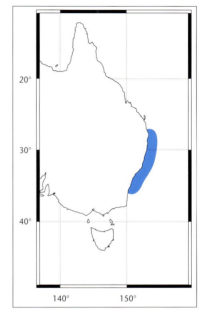

- **Colour:** Upper surface of disc dark yellowish brown with paler yellowish blotches; paler near margin and on mid-snout. Undersurface mostly whitish with irregular greyish patches near disc centre; mucous pores black-edged. Tail and dorsal fins similar to disc.
- **Size:** Reaches about 50 cm; adult males and females similar in size; males maturing from 43–48 cm.
- **Distribution:** Continental shelf off eastern Australia between Moreton Bay (Queensland) and Jervis Bay (New South Wales) in 50–180 m.
- **Remarks:** The most common skate on the continental shelf of central eastern Australia. Records of this species from prawn trawl catches from southern Queensland require validation. Compared to other related species, the tail is unusually broad and flattened.
- **Reference:** Macleay (1884a).

White-spotted Skate — *Raja cerva* Whitley, 1939

34.33 *Raja cerva* (Plate 62)   **Fish Code:** 00 031003

- **Field characters:** A medium-sized skate with a quadrangular disc, a moderately elongate snout with a firm rostral cartilage, one or more nuchal thorns, malar thorns present in males, a short and broad tail, and no denticle patches on the anteroventral margins of the disc. The upper surface is brownish with small white spots.

- **Distinctive features:** Disc quadrangular, broader than long; pectoral-fin apex broadly rounded; anterior margin of disc straight to slightly concave in females and juveniles, distinctly double concave in mature males. Snout short to moderate (orbit 3.35–3.55 in preorbital snout), shortest in juveniles, longest in mature males; supported by a strong medial cartilage; orbit moderate (0.9–1.15 in interorbital space). Mouth broad; internasal flap narrow-lobed with dermal fringe. Tail relatively short (0.75–0.85* times precloacal length in adults, longer in juveniles), depressed, slender; lateral skin folds well developed. Dorsal surface of disc uniformly smooth in juveniles; large females with granulations on snout tip, extending along anterolateral margin in mature males. Nuchal thorns present, 1–3 (usually one), no other thorns on dorsal midline of disc before cloaca; usually with a rosette of 4–9 prominent orbital thorns in adults (2 anterior, 1 posterior in juveniles); alar thorns retractable; malar thorns large, in a small single patch at eye level. Ventral surface normally smooth, mature males sometimes granular at snout tip. Tail with double median rows of sharp thorns (and sometimes with a few midlateral thorns anteriorly) in males; additional lateral rows (rarely in double rows) in females. Pelvic fins large; anterior lobe short, about 1.5–2 in posterior lobe. Dorsal fins small to moderate, broadly rounded, equal in height, separated by less than an eye diameter or bases connected by a low membrane in juveniles; upper lobe of caudal fin low, base much shorter than first dorsal-fin base, usually connected to second dorsal fin; lower lobe absent. Claspers relatively slender, not particularly bulbous.
- **Colour:** Upper surface yellowish brown to greyish brown with numerous, small, regular-sized, white spots and no pale reticulations; occasionally with large, irregular-sized, dusky blotches. Undersurface whitish, often with greyish areas; ventral pores usually black-edged, but frequently lacking encircling greyish areas. Juveniles similar to mature specimens, but with fewer and larger white spots and with dark tail bands through the dorsal-fin bases.
- **Size:** To about 60 cm; adult males and females similar in size; males mature from about 45 cm onwards; born at about 15 cm.
- **Distribution:** Demersal on the continental shelf and upper slope of southern Australia from Eden (New South Wales) to the Recherche Archipelago (Western Australia), including Tasmania, in depths of 20–470 m.
- **Remarks:** Distinguished from the closely related and sympatric longnose skate (34.16) by lacking granulations along the anteroventral margin of the disc. Possibly the most abundant skate on the continental shelf of the Great Australian Bight. It has very good quality flesh which can be spoiled by poor preparation. Once regarded as a trash species, increasing quantities of 'skate flaps' are now available in fish shops.
- **Reference:** Whitley (1939b).

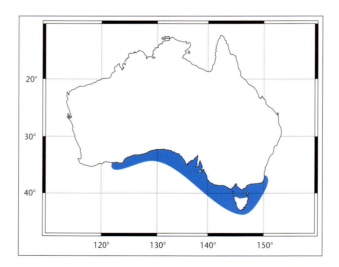

Bight Skate — *Raja gudgeri* (Whitley, 1940)

34.34 *Raja gudgeri* (Plate 57) Fish Code: 00 031010

- **Alternative name:** Greenback skate.

- **Field characters:** A large skate with a quadrangular disc (width rarely more than 1.2 times length), a very elongate snout with a firm rostral cartilage, a rosette of thorns around eye in adults, no nuchal or malar thorns, and a slender (almost oval) tail. The dorsal surface is normally uniform greyish green (juveniles may have blotches), the ventral surface is dusky with dark-edged pores, and the upper caudal-fin lobe is dark.

- **Distinctive features:** Disc quadrangular, broader than long (by 1.5–3 orbit diameters); pectoral-fin apex not especially angular; anterior margin of disc deeply concave. Snout very long (orbit 6.6–8.2 in preorbital snout), pointed, supported by a firm medial cartilage; orbit small, about 1.2–1.9 in interorbital space. Mouth moderately broad; internasal flap narrow-lobed, with dermal fringe. Tail short (0.65–0.7* times precloacal length in mature males, longer in juveniles), somewhat semi-circular, depressed, slender, with slight median bulge; lateral skin folds poorly developed. Dorsal surface of disc smooth in juveniles, sparsely granulated in females; granulations restricted to head, anterior margin of pectoral fins and above cloaca in large males. No nuchal thorns or thorns on dorsal midline of disc before cloaca; orbital thorns short, forming a rosette around eye in adults (2 anterior, 1 median, 1 posterior in juveniles); alar thorns retractable; malar thorns absent. Ventral surface of disc smooth (except for anterior margin and snout tip in juveniles); granular in mature specimens, although feeble over much of posterior region; tail granular, less so in large males. Tail with 1–3 rows of thorns in adults, single row in juveniles. Pelvic fins rather small; anterior lobe long (usually longer than posterior lobe in juveniles, about 1.75 in posterior lobe in large adult males). Dorsal fins rounded, raked backward, subequal in height, separated slightly (usually by about a quarter of an eye diameter or more); upper lobe of caudal fin low, base subequal to first dorsal-fin base, connected to second dorsal fin, extending onto ventral surface as minute lobe. Claspers slender, slightly spatulate distally and slightly bulbous.

- **Colour:** Upper surface uniformly greyish to greyish green in adults (sometimes brownish in immatures); darkest on posterior margins of pectoral and pelvic fins, slightly paler above rostral cartilage; juveniles often covered with large, dusky blotches. Ventral surface variable, greenish grey to greyish brown; posterior margin of disc normally much darker; pores mostly greyish to black. Dorsal and caudal fins dusky or black.

- **Size:** Attains at least 140 cm; males mature from 127 cm onwards.

- **Distribution:** Occurs along the outer continental shelf and upper slope of southern Australia from Shellharbour (New South Wales) to Geraldton (Western Australia), including Tasmania, in depths of 160–700 m (but most common in 400–550 m).
- **Remarks:** This large, long-snouted skate is frequently taken as bycatch by trawlers on the blue grenadier (*Macruronus novaezelandiae*) grounds off Tasmania and western Victoria. The pectoral-fin tips are less angular than in most other Australian skates with a protruding snout.
- **Local synonymy:** *Raja* sp. 1: Last *et al.*, 1983.
- **Reference:** Last *et al.* (1983).

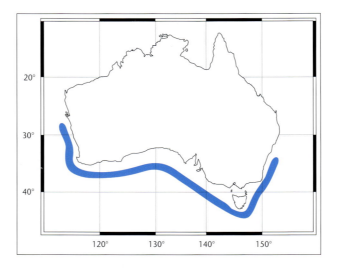

Boreal Skate — *Raja hyperborea* Collett, 1879

34.35 *Raja hyperborea* (Plate 55)   **Fish Code:** 00 031041

- **Field characters:** A large skate with a quadrangular disc, a broad snout with a firm rostral cartilage, no malar thorns, enlarged thorns around the orbit and along the mid-disc, a uniformly smooth undersurface, and a very short tail. The upper disc is greyish brown with darker blotches.

- **Distinctive features:** Disc quadrangular, much broader than long; pectoral-fin apex rather angular; anterior margin of disc double concave (more so in mature males). Snout moderately elongate and pointed (orbit about 5–6* in preorbital snout in mature males), supported by a firm medial cartilage; orbit rather small, 2.15–2.7* in interorbital space in mature males (larger in juveniles); interorbital space very broad. Mouth very broad; internasal flap narrow-lobed, with a dermal fringe. Tail very short (0.65–0.7 times precloacal length), semi-spherical, broad-based, tapering rapidly; lateral skin folds feeble anteriorly. Dorsal surface of disc, pelvic fins and upper half of tail uniformly covered in fine granulations. Median row of thorns along disc and tail commencing in nuchal area; thorns broad-based, sharp, upright, widely spaced (more prominent in juveniles); 2 orbital thorns, 1 interspiracular thorn; 3 enlarged scapular thorns (about 1–2 eye diameters from midline) on each side; juveniles with additional smaller thorns in malar and prepelvic regions; alar thorns retractable; no malar thorns in mature males. Ventral surface of disc and tail uniformly smooth. Pelvic fins large; anterior lobe short, about 1.25–1.5 in posterior lobe. Dorsal fins moderate, raked backward, equal in height, situated close together or connected; base of upper lobe of caudal fin much shorter than first dorsal-fin base, usually connected to second dorsal fin; lower lobe of caudal fin small. Claspers extremely robust, broad, very depressed.
- **Colour:** Upper surface greyish brown with large, widely spaced, dusky blotches; snout tip dark; tail with about 6 black saddles. Undersurface mostly dark greyish brown (rarely uniformly white); mouth, gill membranes, cloaca and snout frequently pale. Dorsal fins and claspers uniformly dark.
- **Size:** Attains at least 106 cm; males mature at about 94 cm.
- **Distribution:** This deepwater skate appears to be widely distributed in temperate parts of the Atlantic, Pacific and Indian oceans in both hemispheres. Moderately common on the lower continental slope off southern Australia (particularly off Tasmania) and New Zealand in 1300–1500 m. Elsewhere recorded to 2500 m.
- **Remarks:** This bulky, deep-water species is represented worldwide by variable colour forms which are currently being studied by skate taxonomists. Hence, this identification must be treated as somewhat provisional. The boreal skate belongs to the subgenus *Amblyraja*.

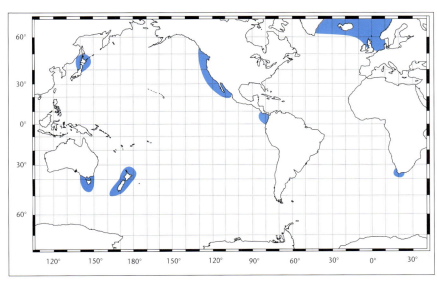

Thornback Skate — *Raja lemprieri* Richardson, 1845

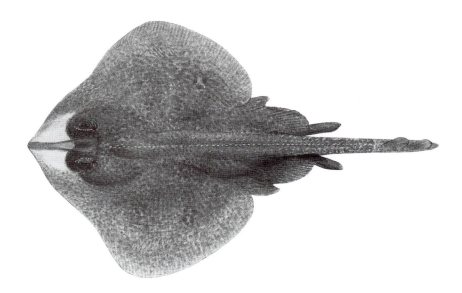

34.36 *Raja lemprieri* (Plate 54)　　　　　　　　　　　　　　　**Fish Code:** 00 031007

- **Alternative names:** Denticulate skate, Lemprieres skate.

- **Field characters:** A small skate with a subcircular to quadrangular disc, a short snout with a firm rostral cartilage, a relatively narrow interorbital space (2.7-3.2 in preorbital snout), prominent malar thorns and several nuchal thorns, no denticles on the ventral surface, and a very short, broad and extremely depressed tail. The undersurface is mostly pale with a prominent dark patch on the snout tip and no dark-edged pores.

- **Distinctive features:** Disc subcircular to quadrangular, broader than long; pectoral-fin apex broadly rounded; anterior margin of disc straight to slightly concave (weakly double concave in mature males). Snout short (orbit 2.85–3.4 in preorbital snout), bluntly pointed, supported by a firm medial cartilage; orbit rather small, 1.05–1.35 in interorbital space. Mouth broad (occasionally very broad in adults); internasal flap narrow-lobed with dermal fringe. Tail very short (0.7–0.9 times precloacal length), very broad and depressed; lateral skin folds well developed. Orbital ridge with 6–10 thorns (forming a rosette in juveniles); median disc thorns commencing just in advance of nuchal area, usually extending along disc midline and onto tail; median thorns with 2–6 additional midlateral rows extending posteriorly from just in advance of cloaca in most adults (juveniles usually with a single continuous median row). Dorsal surface of disc and tail with fine granulations (very dense in adults); alar thorns retractable; malar thorns well developed, beside eye. Ventral surface of disc almost smooth (no denticles along anterior margin); occasionally granular on pelvic fin and tail. Pelvic fins large; anterior lobe short, about 1.25–1.5 in posterior lobe. Dorsal fins moderately large, broadly rounded, subequal in height, separated slightly or connected; upper lobe of caudal fin minute, connected to second dorsal fin; lower lobe of caudal fin rudimentary. Claspers enlarged, rather elongate, somewhat bulbous. Tooth rows in lower jaw 30–33*. Pectoral fin radials 76–79*. Predorsal diplospondylous vertebrae 45*.

- **Colour:** Upper surface greyish black or brownish, mostly covered with darker blotches and fine reticulations. Ventral surface creamish or white, occasionally with irregular greyish blotches near margin of disc and on tail; snout tip distinctly darker, greyish or black; pores lacking dark borders. Dorsal fins and claspers uniformly greyish to brownish.

- **Size:** To at least 52 cm, with males maturing at about 39 cm.

- **Distribution:** Occurs on the continental shelf off southern Australia, from Jervis Bay (New South Wales) to Beachport (South Australia), including Tasmania, from the shore to about 170 m but usually shallower (less than 40 m).
- **Remarks:** The thornback skate is the most common inshore skate of southern waters. It is frequently caught in graball nets set over muddy bottoms in shallow bays and estuaries of Tasmania. Along with the closely related pygmy thornback skate (34.28), it differs from other Australian skates in having a prominent blackish patch on the undersurface of the snout tip. It feeds mainly on small bottom-dwelling fish and crustaceans.
- **Local synonymy:** *Raja dentata* Klunzinger, 1872.
- **References:** Whitley (1939b); Last *et al.* (1983).

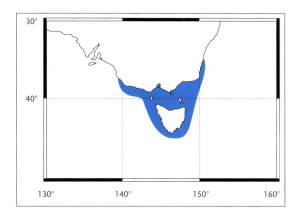

Argus Skate — *Raja polyommata* Ogilby, 1910

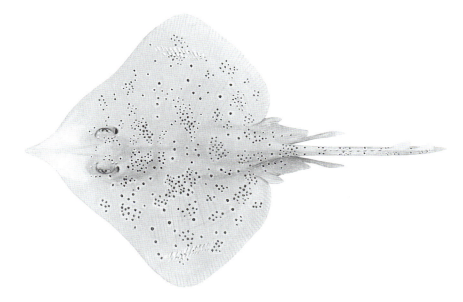

34.37 *Raja polyommata* (Plate 63)   **Fish Code:** 00 031042

- **Field characters:** A small skate with a quadrangular disc, a rather short snout with a firm rostral cartilage, a small nuchal thorn, malar thorns present in males, a slender tail, and no denticles on the undersurface of the disc behind the mouth. The dorsal surface is brownish with distinctive clusters of small dark spots.

- **Distinctive features:** Disc quadrangular, broader than long; pectoral-fin apex broadly rounded; anterior margin of disc straight to slightly concave in females and juveniles, slightly double concave in mature males. Snout short to moderate (orbit 3.75–4.6 in preorbital snout), supported by a strong medial cartilage; orbit moderate in size, 0.95–1.1 in interorbital space. Mouth broad; internasal flap narrow-lobed with dermal fringe. Tail of medium length (0.75–0.95 times precloacal length), depressed, slender; lateral skin folds well developed. Nuchal thorn small, no other thorns on dorsal midline of disc before cloaca; orbital thorns usually absent from mid-eye, usually less than 5 in mature specimens (2 anterior, 1 posterior in juveniles). Dorsal surface of disc uniformly smooth in juveniles and females; mature males with denticles extending along anterolateral margin of disc, bristle-like denticles on snout tip and well developed alar and malar thorns; alar thorns retractable; malar thorns medium-sized, in a small single patch lateral to eye. Ventral surface entirely smooth in males, females with denticles in very narrow band along disc margin anterior to mouth and on snout tip. Tail with central (and usually with midlateral) rows of sharp widely spaced thorns in males; usually with similar additional lateral rows in females. Pelvic fins medium-sized; anterior lobe short, about 1.25–1.5 in posterior lobe. Dorsal fins small, raked backwards slightly, similar in shape and size, usually separated by up to a first dorsal-fin base length; upper lobe of caudal fin low, base slightly shorter than first dorsal-fin base, usually connected to second dorsal fin; lower lobe ridge-like. Claspers slender, weakly bulbous distally.
- **Colour:** Upper surface pinkish brown to greyish brown with irregular clusters of small, dark, white-edged spots; clusters of spots forming 4–6 bands on tail; rarely with spots on snout. Undersurface mostly pale, sometimes with dark patches around mouth and central disc; pores dark-edged with dusky borders, usually without encircling greyish patches. Dorsal fins uniform pale brown.
- **Size:** Attains about 36 cm; adult males and females similar in size; males mature at 26–30 cm; born at about 17 cm.
- **Distribution:** Tropical eastern Australia on the outer continental shelf and upper slope from Townsville (Queensland) to Byron Bay (New South Wales) in 140–310 m.
- **Remarks:** Common throughout its range. The colour pattern is distinctive, although variable geographically. Northern populations are very heavily spotted, with the intensity of spotting decreasing towards the southern extremity of its range. Some specimens collected off New South Wales have only a few clusters of indistinct spots.
- **Local synonymy:** *Pavoraja polyommata*: Whitley, 1940.
- **Reference:** Ogilby (1910b).

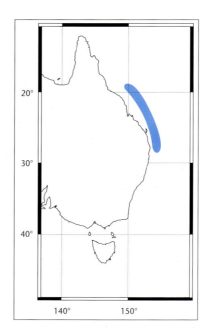

Melbourne Skate — *Raja whitleyi* Iredale, 1938

34.38 *Raja whitleyi* (Plate 53)　　　　　　　　　　　　　　　　**Fish Code:** 00 031006

- **Alternative names:** Great skate, rough skate, wedgenose skate, Whitleys skate.
- **Field characters:** A large skate with a quadrangular disc, a broad snout with a firm rostral cartilage, usually with one or more nuchal thorns, no malar or orbital thorns, a short and broad tail, and the disc entirely covered with fine granular denticles in all but very small juveniles. The dorsal surface is greyish with irregular white flecks.
- **Distinctive features:** Disc quadrangular, broader than long; pectoral-fin apex rather angular; anterior margin of disc almost straight. Snout of moderate length (orbit 5.2–7.3* in preorbital snout), broadly rounded and knob-like at snout tip, supported by a strong medial cartilage; orbit small, 2.2–2.6* in interorbital space. Mouth broad; internasal flap narrow-lobed with dermal fringe. Tail short (about 0.75* times precloacal length), very depressed, broad; lateral skin folds prominent. Usually with 1–5 small thorns in a row on nuchal and shoulder region, no other thorns on midline of disc before cloaca; orbital thorns absent. Dorsal and ventral surfaces of disc and tail uniformly covered with dense, regularly spaced granulations (juveniles under 20 cm disc length less evenly covered on ventral surface, and granulations sometimes absent from ventral surface of pectoral fins in mature specimens); alar thorns retractable; malar thorns absent. Tail with 1–3 rows of thorns, outer rows located near lateral margin. Pelvic fins large, anterior lobe short, about 1.5–1.75 in posterior lobe. Dorsal fins moderate to large, broadly rounded, raked backward, subequal in height, usually connected; caudal fin very small, situated only on dorsal surface; base of upper lobe much shorter than first dorsal-fin base, usually connected to second dorsal fin. Claspers elongate, bulbous distally.
- **Colour:** Dorsal surface greyish or greyish brown, covered with irregularly spaced, white flecks; juveniles with a large dark blotch on each side of disc, sometimes with a few smaller blotches posteriorly. Ventral surface cream or white, sometimes with greyish areas.
- **Size:** Reaches at least 170 cm.
- **Distribution:** Continental shelf between Wollongong (New South Wales) and Albany (Western Australia), including Tasmania. Most abundant in shallow water near the shore but recorded from depths to 170 m.
- **Remarks:** This is the largest Australian skate and may attain more than 50 kg in weight. Care should be taken when removing large adults from fishing nets as their thorny tail is capable of inflicting a minor injury.

- **Local synonymy:** *Raya rostrata* Castelnau, 1873; *Raia scabra* Ogilby, 1888; *Raja ogilbyi* Whitley, 1939.
- **References:** Castelnau (1873a); Whitley (1939b); Last *et al.* (1983).

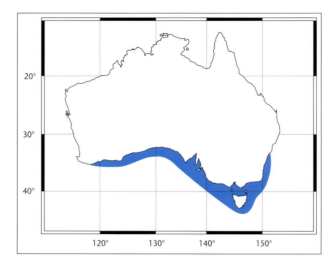

35 Family Anacanthobatidae
LEG SKATES

Leg skates are small, with subcircular to rhomboidal discs, and long snouts that often end in a small filament or a leaf-like appendage. The pelvic fins are so deeply notched that the anterior lobes resemble legs. The tail, which is very slender and mostly short, bears either two dorsal fins or none at all. Few species have thorns (some have thorns around the orbit) and the skin is normally smooth (denticles are sparse when present).

This family contains two genera (about 18 species) which are very unlike in form. Some authors believe that the family should eventually be incorporated, in part at least, with the family Rajidae. At least two species from the most aberrant genus, *Anancanthobatis*, occur in this region. Most of the species live on continental slopes and are rarely caught, even by trawlers.

Key to anacanthobatids

1 Tail bulbous just before apex (fig. 1); more than 22 rows of teeth in each jaw; western Australia .... *Anacanthobatis* sp. A (fig. 3; p. 358)

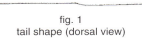
fig. 1
tail shape (dorsal view)

Tail slender, not bulging before apex (fig. 2); less than 22 rows of teeth in each jaw; eastern Australia .... *Anacanthobatis* sp. B (fig. 4; p. 359)

fig. 2
tail shape (dorsal view)

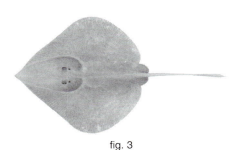

fig. 3

fig. 4

Western Leg Skate — *Anacanthobatis* sp. A

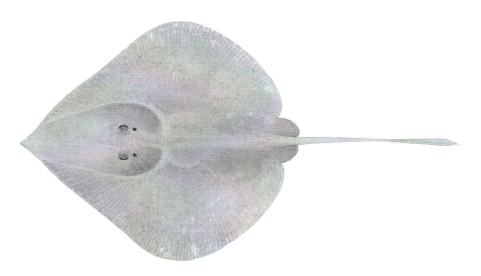

35.1 *Anacanthobatis* sp. A (Plate 64)   **Fish Code:** 00 033001

- **Field characters:** A small, extremely flattened ray with a heart-shaped disc, a thin tail that is broadened and flattened near its tip, no dorsal or caudal fins, 23–25 teeth in each jaw, and leg-like anterior pelvic-fin lobes.

- **Distinctive features:** Disc broadly heart-shaped (more extreme in mature males); its length about equal to breadth in mature males, longer than wide in juveniles and females. Snout very long and triangular, supported by a slender, flexible (sometimes wavy) cartilage; snout tip with a broad fleshy filament (mostly equal to or less than eye diameter in length). Orbit very small, about 8.8–9.9* in preocular snout in females (larger in males); spiracles small, circular. Mouth narrow; internasal flap broadly lobed, with dermal fringe. Pectoral-fin insertion near origin of anterior pelvic-fin lobe in males; more posterior in females (inserted near rear quarter of posterior lobe). Pelvic fins joined posteriorly in females; anterior lobe very slender, leg-like. Tail relatively short (72–93*% of precloacal length), extremely slender, becoming distinctly bulbous and depressed just before apex. Surfaces uniformly smooth, skin rather loose; thorns confined to alar patches in mature males. Dorsal and caudal fins absent. Claspers slender, slightly bulbous distally. Tooth count 23–25/23–25. Pectoral fin radials 68–71. Monospondylous vertebrae 25–28.

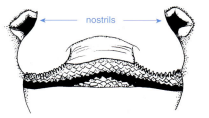

region of nostrils and mouth

tooth of upper jaw

- **Colour:** Upper surface uniformly pale brown to greyish brown. Ventral surface white, semi-translucent.

- **Size:** Attains at least 54 cm; smallest mature male examined was 43 cm; free swimming by 18 cm.

- **Distribution:** Western Australia, from Geraldton to the Ashmore Reef in depths of 420–1120 m (mostly less than 800 m).

- **Remarks:** A few specimens of a third undescribed leg skate were caught recently off the Western Australian coast. This ray, which is dark on all surfaces, is larger than other Australian leg skates. It lives on the mid-continental slope, generally deeper than the western leg skate.

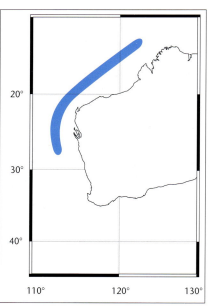

Eastern Leg Skate — *Anacanthobatis* sp. B

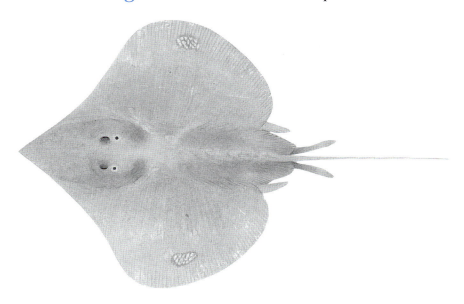

35.2 *Anacanthobatis* sp. **B** (Plate 64)   **Fish Code:** 00 033002

- **Field characters:** A small, extremely flattened ray with a heart-shaped disc, a very thin tail that is uniformly slender to its tip, no dorsal or caudal fins, 18–21 teeth in each jaw, and leg-like anterior pelvic-fin lobes.

- **Distinctive features:** Disc broadly heart-shaped (more extreme in mature males); disc length about equal to breadth in mature males, longer than wide in juveniles and females. Snout very long and triangular, supported by a slender, flexible (sometimes wavy) cartilage; snout tip with a broad fleshy filament (much less than eye diameter in length). Orbit very small, about 8.2–9.7* in preocular snout in females and juveniles (larger in mature males); spiracles small, circular. Mouth narrow; internasal flap broadly lobed, with dermal fringe. Pectoral-fin insertion near origin of anterior pelvic-fin lobe in males; more posterior in females (half to three quarters along posterior pelvic-fin lobe). Pelvic fins joined posteriorly in females; anterior lobe very slender, leg-like. Tail relatively short (78–94*% of precloacal length), extremely slender, uniformly tapering (not becoming distinctly bulbous and depressed just before apex). Surfaces uniformly smooth, skin rather loose; thorns confined to alar patch in mature males. Dorsal and caudal fins absent. Claspers slender, slightly bulbous distally. Tooth count 18–21/19–21*. Pectoral fin radials 72–75*. Monospondylous vertebrae 26–30*.

- **Colour:** Upper surface uniformly pale brown to greyish brown. Ventral surface white, semi-translucent.

- **Size:** Attains at least 57 cm; males mature by 55 cm.

- **Distribution:** Continental slope off Queensland, from Cairns to the Swain Reefs, in 680–880 m.

- **Remarks:** Known from only a few specimens.

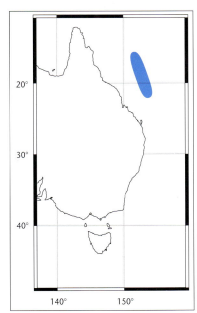

Family Pristidae

SAWFISHES

Sawfishes are large (up to 7 m long), highly modified rays with a blade-like snout armed with enlarged, lateral, tooth-like denticles (rostral teeth). They have elongate, subcylindrical to compressed bodies, rather small pectoral fins that are distinctly detached from the head, two enlarged dorsal fins, and a prominent caudal fin. Their eyes and spiracles are located dorsally. Unlike the superficially similar but smaller saw sharks (which are really sharks rather than rays), sawfishes have all five gill slits situated ventrally on the head (rather than laterally) and there are no barbels on the undersurface of the rostrum.

This family is represented worldwide by two genera and between 4 and 7 species. More taxonomic research is needed to provide definitive identifications of some species. The formidable snout, often referred to as the 'saw', is used to stun and kill prey living on the bottom. It can also cause serious injury to humans when in close contact. All species are ovoviviparous.

Key to pristids

1   Rostral teeth (18–22 on each side) not extending onto basal quarter of saw (fig. 1); lower lobe of caudal fin relatively large, more than half length of upper lobe (fig. 4) ....
    .... **Anoxypristis cuspidata (fig. 4; p. 361)**

    Rostral teeth extending onto basal quarter of saw (almost to base) (fig. 2); lower lobe of caudal fin small, less than half length of upper lobe (fig. 5) ......
    2

2   Rostral teeth on each side near base much further apart than those near apex (fig. 2); first dorsal-fin origin well behind pelvic-fin origin (mostly about over middle of base) (fig. 6) ....
    .... **Pristis zijsron (fig. 5; p. 366)**

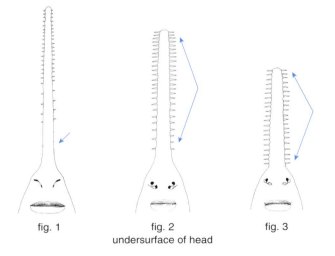

fig. 1   fig. 2   fig. 3
undersurface of head

fig. 4

fig. 5

Rostral teeth equally spaced on each side, basal teeth not further apart than those near apex (fig. 3); first dorsal-fin origin almost above or forward of pelvic-fin origin (figs 7, 8) **3**

3 First dorsal fin originating well forward, pelvic-fin origin closer to level of first dorsal-fin insertion than to its origin (fig. 8); distinct lower lobe on caudal fin (fig. 9) .... ***Pristis microdon*** (**fig. 9; p. 364**)

First dorsal fin originating over or slightly behind pelvic-fin origin (fig. 7); no distinct lower lobe on caudal fin (fig. 10) **4**

4 Pairs of rostral teeth 24–34; a large marine species (up to 5 m) .... ***Pristis pectinata*** (**fig. 10; p. 365**)

Pairs of rostral teeth 18–22 (usually 20); a small coastal marine species (possibly less than 2.5 m) .... ***Pristis clavata*** (**fig. 11; p. 362**)

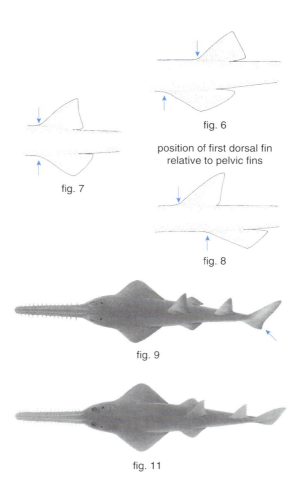

fig. 10

fig. 11

Narrow Sawfish — *Anoxypristis cuspidata* (Latham, 1794)

36.1 *Anoxypristis cuspidata* (Plate 43)   **Fish Code:** 00 025002

- **Field characters:** A moderately large, slender sawfish with gill openings on the undersurface, no teeth on the basal quarter of the saw, the lower lobe of the caudal fin relatively enlarged (more than half length of upper lobe), and with short-based pectoral fins.

- **Distinctive features:** Body shark-like, pectoral fins distinct; head flattened, with a blade-like snout bearing 18–22 pairs of lateral teeth; blade slender, not tapering distally. Nostrils very narrow with small nasal flaps. Rostral teeth short, flattened, broadly triangular, lacking a groove along their posterior margins; absent from basal quarter of blade; not becoming closer to each other toward tip of blade. Skin naked or with denticles widely spaced (in adults). Dorsal fins tall, pointed, first dorsal-fin origin slightly behind pelvic-fin origin. Pectoral fins distinctly triangular with narrow bases and pointed apices; hind margins concave. Caudal-fin lower lobe large (posterior margin of caudal fin deeply concave).

- **Colour:** Greyish above, pale below; fins frequently pale; rostral teeth white.
- **Size:** Locally, attains about 350 cm. Elsewhere reported, rather doubtfully, to attain 600 cm.
- **Distribution:** Indo–Pacific from the Red Sea to Australia, north to Japan. Australian distribution unclear, but moderately common in the Gulf of Carpentaria from inshore areas to 40 m.
- **Remarks:** In other parts of the Indo–Pacific, this species is reported to have more rostral teeth (24–25 compared with 18–22 pairs). Caught for its flesh in parts of Asia; the liver is also rich in oil. Not exploited in Australia.
- **References:** Gloerfelt-Tarp and Kailola (1984); Sainsbury *et al.* (1985).

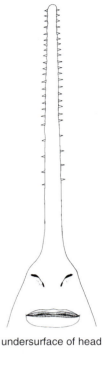

undersurface of head

oral tooth (upper jaw)

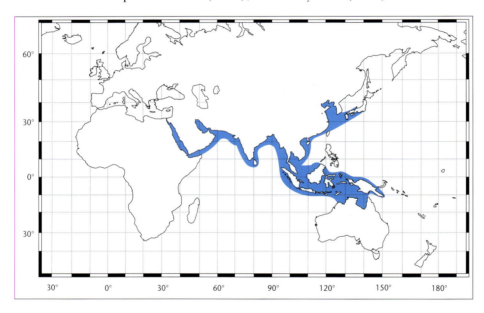

Dwarf Sawfish — *Pristis clavata* Garman, 1906

36.2 *Pristis clavata* (Plate 43) **Fish Code:** 00 025004

- **Alternative name:** Queensland sawfish.
- **Field characters:** A small, robust sawfish with the gill openings located on the undersurface, 18–22 evenly-spaced teeth on the rostrum (teeth start near rostral

base), the dorsal-fin origin over or only slightly forward of the pelvic-fin origin, the lower lobe of the caudal fin feeble (much shorter than half length of upper lobe) and with broad-based pectoral fins.

- **Distinctive features:** Body shark-like, pectoral fins distinct; head flattened, with a blade-like snout bearing 18–22 pairs of lateral teeth; blade broad, not tapering distally. Nostrils broad with large nasal flaps. Rostral teeth slender, with a groove along their posterior margins; present on basal quarter of blade, becoming only slightly closer to each other toward tip of blade. Skin with denticles. Dorsal fins tall and pointed, first dorsal-fin origin over or slightly forward of pelvic-fin origin. Pectoral fins with relatively broad bases, fins broadly triangular; hind margins straight. Caudal-fin lower lobe small (posterior margin of caudal fin almost straight).
- **Colour:** Mostly greenish brown, rarely yellowish; white ventrally; fins paler.
- **Size:** To at least 140 cm.
- **Distribution:** Coastal and estuarine in tropical Australia from Cairns (Queensland) to the Kimberley coast (Western Australia). Common over mudflats in the Gulf of Carpentaria; occurs some distance up rivers, almost into freshwater. Possibly more widely distributed in the Indo–Pacific. A record from the Canary Islands is likely to be erroneous.
- **Remarks:** Little is known of the distribution of this species outside Australia. It is very similar in form to the larger, wide sawfish (36.4), and may have been confused in the past with this species, both in Australia and in other parts of the Indo–Pacific. The dwarf sawfish has fewer rostral teeth than its larger relative (19–22 versus 24–34). Mature males are not represented in collections and are required for research purposes. The flesh is likely to be good eating.
- **Reference:** Grant (1978).

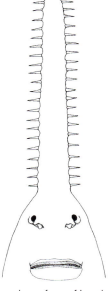
undersurface of head

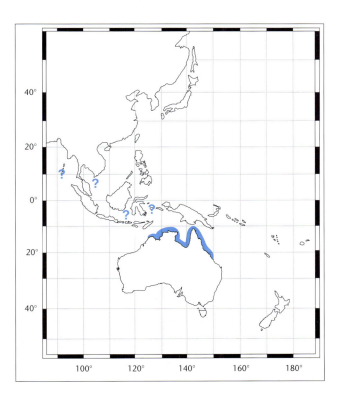

Freshwater Sawfish — *Pristis microdon* Latham, 1794

36.3 *Pristis microdon* (Plate 42) Fish Code: 00 025003

- **Alternative names:** Leichhardts sawfish, smalltooth sawfish.
- **Field characters:** A medium-sized, slender sawfish with the gill openings located on the undersurface, 18–23 evenly-spaced teeth on the rostrum (teeth starting near rostral base), the dorsal fin beginning well forward of the pelvic fin, a short lower caudal-fin lobe (much less than half length of upper lobe), and with rather broad-based pectoral fins.
- **Distinctive features:** Body shark-like, pectoral fins distinct; head flattened, with a blade-like snout bearing 18–23 (mainly 20–22) pairs of lateral teeth; blade broad, not tapering slightly distally. Nostrils broad with large nasal flaps. Rostral teeth slender, with a groove along their posterior margins; present on basal quarter of blade; not noticeably closer to each other toward tip of blade. Skin with denticles. Dorsal fins tall and pointed, first dorsal-fin origin well forward of pelvic-fin origin. Pectoral fins with relatively broad bases, broadly triangular, hind margins straight. Caudal-fin lower lobe small but distinct (posterior margin of caudal fin concave).
- **Colour:** Yellowish to greyish, white ventrally; fin outer margins richer yellowish brown.
- **Size:** Locally attains at least 200 cm; elsewhere reputed to reach 700 cm.
- **Distribution:** Known from several drainages of northern Australia in fresh or weakly saline water: the Fitzroy, Durack and Ord Rivers (Western Australia); the Adelaide, Victoria and Daly rivers (Northern Territory); and the Gilbert, Mitchell, Norman and Leichhardt rivers (Queensland). No confirmed records from the sea off Australia. Elsewhere uncertain; so far confirmed from several major river basins of Indonesia and New Guinea, but possibly west to India. A similar, if not conspecific, form occurs more widely through the tropical and subtropical Atlantic, eastern Pacific and southwestern Indian Oceans.

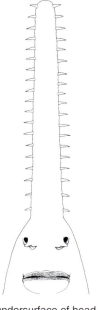

undersurface of head

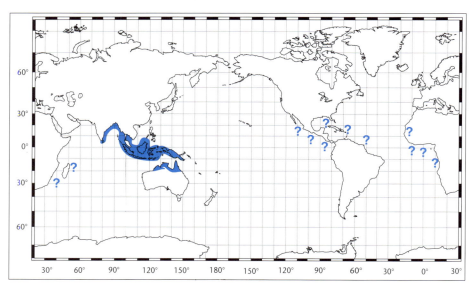

- **Remarks:** This sawfish appears to be confined to freshwater drainages and the upper reaches of estuaries in Australian waters (in some cases more than 100 km from the sea). Small specimens, mostly less than 150 cm, have been caught in remote ponds where they have been isolated for several years between floods. Verifiable records from the sea are misidentifications of other species, the coastal dwarf sawfish (36.2) or the wide sawfish (36.4). Positive sightings, along with a size and photograph, of this species from the sea should be reported to a museum. The relationships between the Australian freshwater sawfish and two other similar species that also enter freshwater, *Pristis pristis* and *P. perotteti*, need to be established. This species is highly vulnerable to gillnet fishing. Populations may be threatened in streams where poaching for barramundi (*Lates calcarifer*) is a common practice.
- **Local synonyms:** *Pristiopsis leichhardti* Whitley, 1945.
- **References:** Whitley (1945b); Merrick and Schmida (1984).

Wide Sawfish — *Pristis pectinata* Latham, 1794

36.4 *Pristis pectinata* (Plate 43) Fish Code: 00 025005

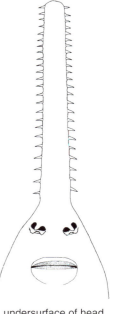

undersurface of head

- **Alternative name:** Smalltooth sawfish.
- **Field characters:** A large, robust sawfish with the gill openings located on the undersurface, 24–34 pairs of evenly-spaced teeth on the rostrum (teeth starting near rostral base), the dorsal-fin origin above the pelvic-fin origin, the lower caudal-fin lobe feeble (much shorter than half length of upper lobe), and with broad-based pectoral fins.
- **Distinctive features:** Body shark-like, pectoral fins distinct; head flattened, with a blade-like snout bearing 24–34 pairs of lateral teeth; blade broad, tapering distally. Nostrils broad with large nasal flaps. Rostral teeth slender, with a groove along their posterior margins; present on basal quarter of blade; not noticeably closer to each other toward tip of blade. Skin with denticles. Dorsal fins tall and pointed, first dorsal-fin origin above pelvic-fin origin. Pectoral fins with broad bases, broadly triangular, hind margins straight. Caudal-fin lower lobe small (posterior margin of caudal fin almost straight).
- **Colour:** Uniform bluish grey to olive green; fins mostly paler; whitish ventrally.
- **Size:** Reported to reach 760 cm, but more commonly to about 550 cm; young born at about 60 cm.

- **Distribution:** Considered to be circumtropical, but its occurrence in the Indo–West Pacific is questionable. Australian records also require validation; so far based on photographs of adult specimens trawled in the Gulf of Carpentaria.
- **Remarks:** This is the largest of the sawfishes. Because of its size, adults are rarely held in museums. In the past, adults have been stuffed or only the highly-prized saw kept. An absence of whole specimens has made comparisons of forms between areas difficult. In southern Africa, this species enters estuaries to give birth to litters of 15–20 young. To avoid damage to the parent, the saws of its young are soft and sheathed before birth. The flesh is of good quality and it is used as a food fish in other parts of the Indo–Pacific. Also, the liver contains high concentrations of oil suitable for medicine, soap making and in leather-tanning.
- **Local synonymy:** *Pristis microdon*: Sainsbury *et al.*, 1985 (misidentification).
- **Reference:** Sainsbury *et al.* (1985).

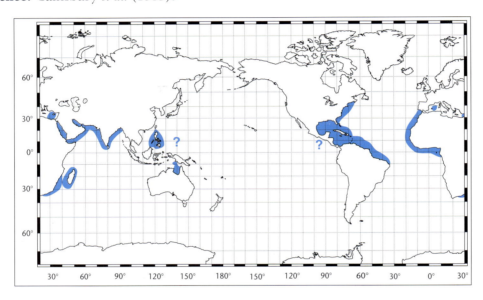

Green Sawfish — *Pristis zijsron* Bleeker, 1851

36.5 *Pristis zijsron* (Plate 43) **Fish Code:** 00 025001

- **Alternative names:** Dindagubba, narrowsnout sawfish, sawfish.
- **Field characters:** A large, robust sawfish with the gill openings on the undersurface, 24–28 pairs of unevenly-spaced teeth on the rostrum (teeth starting near rostral

base), the dorsal-fin origin slightly behind the pelvic-fin origin, the lower lobe of the caudal fin feeble (much shorter than half length of upper lobe), and with broad-based pectoral fins.

- **Distinctive features:** Body shark-like, pectoral fins distinct; head flattened, with a blade-like snout bearing 24–28 [25–32] pairs of lateral teeth; blade slender, not tapering distally. Nostrils broad with large nasal flaps. Rostral teeth slender, with a groove along their posterior margins; present on basal quarter of blade; becoming much closer to each other toward tip of blade (spacing between basal 2–3 teeth more than 3 times wider than spacing between teeth near tip of saw). Skin with denticles. Dorsal fins tall and pointed, first dorsal-fin origin slightly behind pelvic-fin origin. Pectoral fins with broad bases, apices broadly triangular; hind margins straight. Caudal-fin lower lobe small (posterior margin of caudal fin almost straight).
- **Colour:** Greenish brown or olive above; pale whitish below.
- **Size:** Locally to at least 500 cm, although reported to reach 730 cm; males mature by 430 cm.
- **Distribution:** Widely distributed in the northern Indian Ocean (west to South Africa), and off Indonesia and Australia. Locally, most abundant in the tropics but caught occasionally south to Sydney (New South Wales) and Broome (Western Australia). Single record off Glenelg (South Australia).
- **Remarks:** This huge ray is the most commonly encountered sawfish in the Australian region. Large specimens, which are common inshore at certain times of the year, regularly hinder gillnet fishermen target fishing for barramundi (*Lates calcarifer*). A large sawfish thrashing vigorously is difficult to remove from a net and the task must be performed carefully to avoid serious personal injury. The flesh of this ray is good to eat.
- **Reference:** Sainsbury *et al.* (1985).

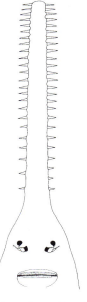

undersurface of head

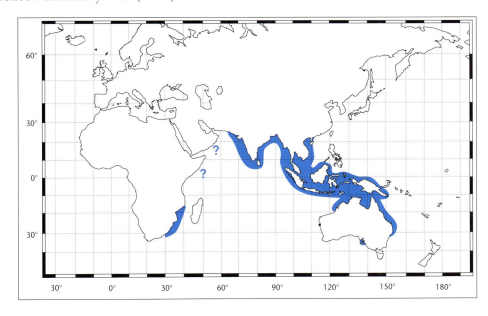

Family Torpedinidae

TORPEDO RAYS

Members of the family Torpedinidae, also known as torpedo rays, are the largest of four families of electric rays. They range from small to quite large in size (up to 1.8 m in length), and all are capable of delivering a substantial electric shock. Most species are similar in appearance, having a broadly oval disc, rather small eyes and a short tail (shorter than the disc). The mouth, which is widely distensible but only slightly protractile, is strongly arched and narrow, and lacks a groove below the lower jaw. The jaws are slender with small unicuspid teeth and lack labial cartilages. The tail has two dorsal fins (the first is partly over the pelvic-fin base), lateral skin folds posteriorly, a large caudal fin with well-developed upper and lower lobes, but no anal fin or stinging spine. Electric charges are produced in two large kidney-shaped organs located on either side of the disc behind the eyes.

The family is represented by a single genus and at least 15 species in temperate and tropical parts of the Atlantic, Indian and Pacific Oceans. Most of the species are found only on the continental shelves within their range, but the two Australian members occur deeper to at least 550 m. Females are viviparous.

Key to torpedinids

1   Tail relatively short, distance from pelvic-fin rear tip to lower lobe of caudal fin less than three-quarters height of caudal fin (fig. 1), less than half width between outer tips of pelvic fins (fig. 1) ....
.... **Torpedo macneilli (fig. 3; p. 370)**

Tail relatively long, distance from pelvic-fin rear tip to lower lobe of caudal fin more than three-quarters height of caudal fin (fig. 2), much more than half width between outer tips of pelvic fins (fig. 2) ....
.... **Torpedo sp. A (fig. 4; p. 369)**

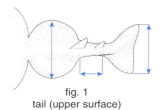

fig. 1
tail (upper surface)

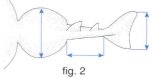

fig. 2
tail (upper surface)

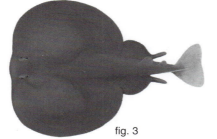

fig. 3

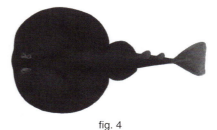

fig. 4

Longtail Torpedo Ray — *Torpedo* sp. A

37.1 *Torpedo* sp. **A** (Plate 65)   **Fish Code:** 00 028006

- **Field characters:** A small electric ray with an almost circular disc that is longer than the tail, and a caudal fin that is much larger than the dorsal fins. The distance from the pelvic-fin rear tip to the lower caudal-fin lobe is greater than three-quarters of the caudal-fin height (often almost equal).

- **Distinctive features:** Disc depressed, usually circular, mostly broader than long, thick centrally; profile of forehead usually straight or slightly convex; lateral margins evenly rounded. Eye very small, orbit diameter more than 4 in snout length, about equal in size to spiracle; no papillae around spiracle. Mouth and internasal flap narrow. Skin surface frequently creased, otherwise smooth, lacking denticles and thorns. Tail relatively long (distance from pelvic-fin rear tip to lower lobe of caudal fin more than three-quarters of caudal-fin height, more than a quarter of disc length); tail tapering gradually. First dorsal fin usually originating forward of pelvic-fin insertion; second dorsal fin behind rear tip of pelvic fin; second dorsal fin usually about half or more size of first dorsal fin, both rather slender. Caudal-fin outer margin truncate to slightly convex. Pelvic fins not joined posteriorly, relatively narrow (overall width less than twice caudal peduncle length).

- **Colour:** Upper surface and fins greyish brown to brownish black, frequently with lighter and darker areas. Ventral surface uniformly white.

- **Size:** Largest specimen examined was an immature male of 34 cm.

- **Distribution:** Continental slope off tropical Australia between Newcastle (New South Wales) and Bowen (Queensland), and Shark Bay to the Rowley Shoals (Western Australia), in 400–560 m.

- **Remarks:** This species appears to be new to science and little is known of its biology.

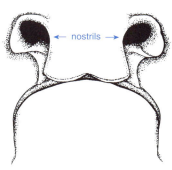

region of nostrils and mouth

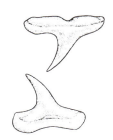

teeth from near symphysis (upper and lower jaw)

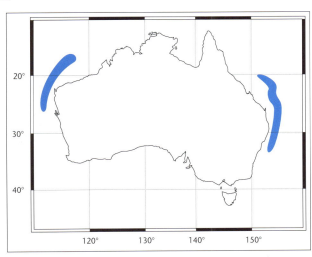

Short-tail Torpedo Ray — *Torpedo macneilli* (Whitley, 1932)

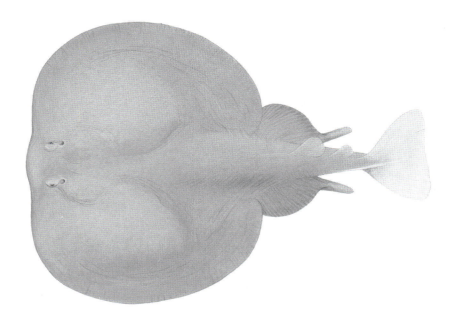

37.2 *Torpedo macneilli* (Plate 65)　　　　　　　　　　**Fish Code:** 00 028003

- **Alternative names:** Electric ray, torpedo.
- **Field characters:** A rather large electric ray with an almost circular disc that is much longer than the tail, and a caudal fin that is much larger than the dorsal fins. The distance from the pelvic-fin rear tip to the lower caudal-fin lobe is less than three-quarters of the caudal-fin height (often only slightly more than a half).
- **Distinctive features:** Disc depressed, almost circular, mostly broader than long, thick centrally; profile of forehead straight or slightly convex; lateral margins evenly rounded. Eye small, orbit diameter about 3–4 in snout length, larger than spiracle; no papillae around spiracle. Mouth and internasal flap narrow. Skin surface frequently creased, otherwise smooth, lacking denticles and thorns. Tail rather short (distance from pelvic-fin rear tip to lower lobe of caudal fin less than three-quarters of caudal-fin height, mostly shorter than a quarter of disc length); tail broad-based, tapering rapidly. First dorsal fin broadly rounded, mostly originating forward of pelvic-fin insertion; second dorsal fin rather slender, located behind rear tip of pelvic fin; second dorsal fin mostly about half size of first dorsal fin. Caudal-fin outer margin truncate to emarginate. Pelvic fins not joined posteriorly, relatively broad (overall width exceeding twice caudal peduncle length).
- **Colour:** Upper surface uniformly greyish, brownish or brownish black; rarely with a few lighter and darker blotches; fins mostly paler. Ventral surface uniformly white.
- **Size:** To at least 100 cm; males mature at about 60 cm.
- **Distribution:** Widespread Australian endemic; southern Australia from Port Hedland (Western Australia) to off the Swain Reefs (Queensland), including Tasmania, in 90–750 m.
- **Remarks:** There are several reports of fishermen being thrown several feet after unwittingly touching large specimens of this ray. The electric organs consist of a battery of hexagonal cells (sometimes visible through the skin in smaller rays) that form a honeycomb structure on each side of the body. Each cell is filled with a jelly-like fluid and is connected to an elaborate nerve network. When fully charged, the organs operate much like a battery, with the upper surface having a positive charge and the lower surface being negative. The wide, tropical–cool temperate distribution of this species is unusual for a ray. Genetic studies may reveal the

presence of more than one species. The New Zealand torpedo ray, *T. fairchildi*, is very similar to this species.

- **Local synonymy:** *Notostrape macneilli* Whitley, 1932.
- **Reference:** Last *et al.* (1983).

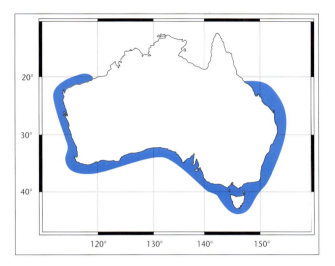

38

Family Hypnidae
COFFIN RAYS

A small family of electric rays with a pear-shaped disc, greatly enlarged pectoral fins, and an extremely short, slender tail. The mouth, which is widely distensible and only slightly protractile, is strongly arched and narrow, and is not surrounded by a deep groove. The jaws are long and slender with tricuspidate teeth and lack labial cartilages. The tail has two dorsal fins (both situated over the pelvic fins), a caudal fin that is barely larger than the dorsal fins, and no lateral skin folds. The eyes are very small and are capable of being raised above the head on short stalks. Large electric organs are situated more or less centrally in each pectoral fin. The decaying bodies of these fishes, which swell in thickness to resemble a coffin, have given rise to their common name, 'coffin ray'.

The family, which is endemic to Australian waters, contains only one recognised species. However, there are minor differences between tropical and temperate forms, which may be distinct species. This group is given subfamily status within the family Torpedinidae by some authors.

Coffin Ray — *Hypnos monopterygium* (Shaw & Nodder, 1795)

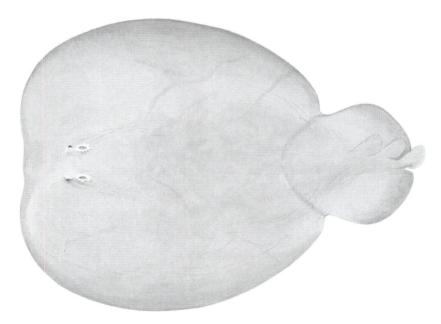

38.1 *Hypnos monopterygium* (Plate 65)　　　　　　　　　　**Fish Code:** 00 028001

- **Alternative names:** Crampfish, electric ray, numbfish, numbie, short-tail elecric ray, torpedo.
- **Field characters:** A small, highly distinctive ray with a very large disc and pelvic fins, a very short tail, and dorsal and caudal fins close together and about the same size.

- **Distinctive features:** Body very depressed, pear-shaped (becoming bloated and more coffin-shaped after death), thick centrally; profile of forehead straight or slightly concave; lateral margins evenly convex. Eye very small, about 5–7.5 in snout length, slightly smaller than spiracle; spiracle and sometimes lower margin of eye bordered with small papillae. Skin surface frequently creased, otherwise smooth, lacking denticles and thorns. Tail very short, protruding only caudal-fin length or less beyond pelvic fins; no stinging spines or anal fin. Caudal and dorsal fins similar in surface area and very close together; dorsal fins thumb-shaped, their sizes variable. Pelvic fins joined together to form a smaller circular second disc.
- **Colour:** Uniform chocolate brown, reddish brown, greyish or pink above; frequently with a few irregular darker and paler spots and blotches; spiracular papillae mostly pale; yellowish or white below.
- **Size:** Reported to reach 60 cm, but specimens rarely exceed 40 cm; males maturing at about 24 cm; young born at about 8–11 cm.
- **Distribution:** Tropical and warm temperate Australia, from St Vincents Gulf (South Australia) to Broome (Western Australia), and from Eden (New South Wales) to at least Caloundra (Queensland). Common inshore but also collected in depths to 220 m.
- **Remarks:** The coffin ray is capable of delivering a strong electric shock that, while not fatal, will generally not be forgotten. The electric organs are so efficient that when water is tipped onto the rays back from a receptacle, a shock can still be received through the water column. In its natural habitat the coffin ray is slow moving and undoubtedly uses its electric organs to stun prey. It feeds on crabs, worms and small fishes. Females are viviparous.
- **Local synonymy:** *Hypnos subnigrum* Duméril, 1852.
- **Reference:** Scott *et al.* (1980).

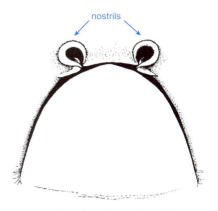

region of nostrils and mouth

teeth from near symphysis (upper and lower jaw)

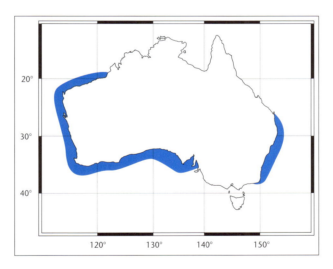

39 Family Narcinidae
NUMBFISHES

Numbfishes are small electric rays with a wedge-shaped to circular disc and a relatively long, broad tail. Their mouth, which is highly protractile but only slightly distensible, is transverse and narrow, and is surrounded by a deep groove. Their jaws are long and stout and have well-developed labial cartilages. The tail has two dorsal fins (the first originating over or slightly behind the pelvic fins), a well-developed caudal fin, and usually has lateral skin folds. The eyes are small in all Australian representatives, but in some members of the family they are minute and non-functional.

The group is represented by four genera and at least 17 species in temperate and tropical parts of the Atlantic, Indian and Pacific Oceans. Five species of a single genus, *Narcine*, occur on soft substrates of the Australian continental shelf.

Key to narcinids

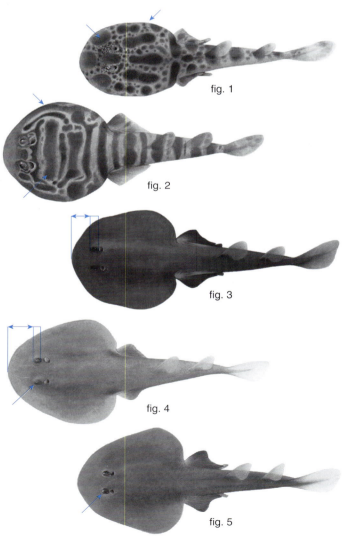

fig. 1

fig. 2

fig. 3

fig. 4

fig. 5

1   Dorsal surface of disc with dark stripes and blotches (figs 1, 2) **2**

    Dorsal surface of disc uniformly yellowish or brownish, lacking darker stripes and blotches **3**

2   Disc distinctly oval-shaped (fig. 1); large, almost circular, spots on upper disc (no broad, dark transverse bars behind eyes) (fig. 1) .... *Narcine* **sp. A (fig. 1; p. 375)**

    Disc almost circular (fig. 2); pattern of broad, dark, transverse bars across upper surface (fig. 2) .... *Narcine westraliensis* **(fig. 2; p. 379)**

3   Dark brown; eye relatively large, snout relatively short (snout length less than 5 times length of exposed part of eye) (fig. 3); temperate Australia .... *Narcine tasmaniensis* **(fig. 3; p. 378)**

    Pale yellow; eye relatively small, snout relatively long (snout length more than 5 times length of exposed part of eye) (fig. 4); tropical Australia **4**

4   Eye usually larger than spiracle; tropical western Australia .... *Narcine* **sp. B (fig. 4; p. 376)**

    Eye usually about same size as spiracle; tropical eastern Australia .... *Narcine* **sp. C (fig. 5; p. 377)**

Ornate Numbfish — *Narcine* sp. A

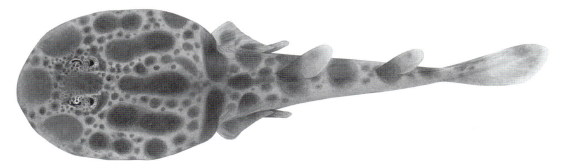

39.1 *Narcine* sp. A (Plate 67) Fish Code: 00 028007

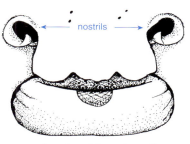

region of nostrils and mouth

tooth of upper jaw

- **Field characters:** A small numbfish with an oval-shaped disc that is shorter than the tail, and a characteristic colour pattern of dark circular spots of varying sizes on the upper surface.
- **Distinctive features:** Body depressed; disc oval, widest near middle of pectoral fin. Eye small, orbit diameter about 3.5–4* in snout length, smaller than spiracle; no papillae around spiracle. Mouth narrow, internasal flap trilobed. Skin surface frequently creased, otherwise smooth, lacking denticles and thorns. Tail elongate (about 1.25 times precloacal length), rather slender, moderately depressed, tapering gradually; lateral skin folds feebly developed. First dorsal fin originating just behind apex of pelvic fin; second dorsal fin about equal in size to first dorsal fin, both fins raked backward. Caudal fin low, broadly rounded posteriorly. Pelvic fins long and narrow, broadest anteriorly.
- **Colour:** Upper surface densely spotted; several large dark brownish pink spots (largest on central disc and tail), smaller spots of similar colour between main spots; spots occasionally joined to form short longitudinal stripes on mid-disc; spiracular membrane with dense array of very fine dark spots; dorsal and caudal fins mostly banded. Ventral surface uniformly white.
- **Size:** Largest specimen a mature male of 17 cm.
- **Distribution:** Arafura Sea, Gulf of Carpentaria and Torres Strait, from the Wessel Islands (Northern Territory) to Cape York (Queensland), in depths of 50–60 m.
- **Remarks:** This small species has been confused with the banded numbfish (39.5), which is found further south off Western Australia. The ornate numbfish, with its rather attractive pattern of different sized spots, is distinctive. Little is known of its biology.
- **Local synonymy:** *Narcine westraliensis*: Gloerfelt-Tarp & Kailola, 1984; Sainsbury *et al.*, 1985(misidentifications).
- **Reference:** Gloerfelt-Tarp and Kailola (1984).

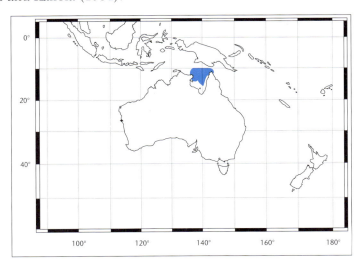

Western Numbfish — Narcine sp. B

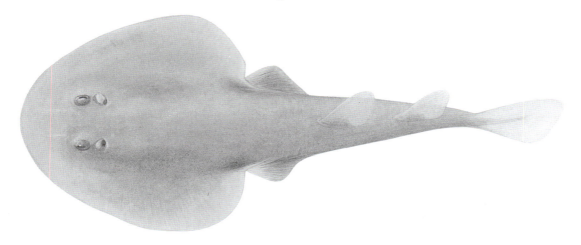

39.2 *Narcine* sp. **B** (Plate 66) **Fish Code:** 00 028004

- **Field characters:** A small, plain yellowish numbfish with a medium-sized eye, and a wedge-shaped disc that is shorter than the tail.

- **Distinctive features:** Body depressed; disc shovel-shaped with a broadly rounded apex (mostly uniformly rounded laterally), widest close to insertion of pectoral fin. Eye medium-sized, orbit diameter normally more than 3* (exposed eye more than 5) in snout length, usually larger than spiracle; no papillae around spiracle. Mouth narrow, internasal flap trilobed. Skin surface frequently creased, otherwise smooth, lacking denticles and thorns. Tail relatively long (about 1.1–1.2* times precloacal length), rather broad-based, somewhat depressed, tapering gradually; lateral skin folds feebly developed. First dorsal fin originating just behind apex of pelvic fin; second dorsal fin about equal in size to first dorsal fin, both raked backward. Caudal fin low, broadly rounded posteriorly. Pelvic fins long and narrow, broadest anteriorly.

- **Colour:** Upper surface uniformly pale yellow; margin of disc frequently white-edged; fins almost translucent. Ventral surface uniformly pale.

- **Size:** Attains at least 33 cm.

- **Distribution:** Outer continental shelf and upper slope off Western Australia, from the Buccaneer Archipelago to Lancelin, in 170–350 m depth.

- **Remarks:** This numbfish appears to have narrow depth and geographical distributions. It is caught occasionally by prawn and scampi fishermen off Western Australia. It is similar to both the eastern (39.3) and Tasmanian (39.4) numbfishes, but the three forms each have unique ventral pore patterns that are too complex to warrant discussion herein. In the absence of genetic studies, we have tentatively regarded them as separate species.

- **Local synonymy:** *Narcine* sp. 1: Gloerfelt-Tarp & Kailola, 1984.

- **Reference:** Gloerfelt-Tarp and Kailola (1984).

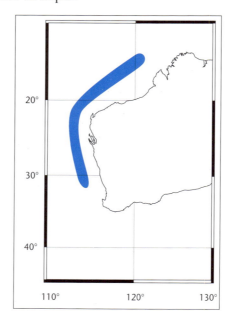

Eastern Numbfish — *Narcine* sp. C

39.3 *Narcine* sp. C (Plate 66) **Fish Code:** 00 028008

- **Field characters:** A small, plain yellowish pink numbfish with a relatively small eye and a shovel-shaped disc that is shorter than the tail. The snout is yellower than the trunk.

- **Distinctive features:** Body depressed; disc shovel-shaped with a broadly rounded apex, widest close to insertion of pectoral fin. Eye relatively small, orbit diameter mostly more than 3.5* (exposed eye more than 5) in snout length, usually about the same size as the spiracle; no papillae around spiracle. Mouth narrow, internasal flap trilobed. Skin surface frequently creased, otherwise smooth, lacking denticles and thorns. Tail relatively long (about 1.1–1.25* times precloacal length), rather broad-based, moderately depressed, tapering gradually; lateral skin folds feebly-developed. First dorsal fin originating over or just behind apex of pelvic fin (further back in females); second dorsal fin about equal in size to first dorsal fin, both raked backward slightly. Caudal fin low, broadly rounded posteriorly (upper lobe slightly more angular in mature males). Pelvic fins long and narrow, broadest anteriorly.

- **Colour:** Upper surface uniformly yellowish pink behind eyes; clearly defined yellowish patch before eyes; fins with pinkish bases, translucent distally; margin of pelvic fin white. Ventral surface uniformly white.

- **Size:** Attains at least 35 cm; males mature by 26 cm.

- **Distribution:** Continental slope off Northern Queensland, from Townsville to Rockhampton, in 325–360 m depth.

- **Remarks:** Only discovered recently, this ray is closely allied to the western numbfish (39.2).

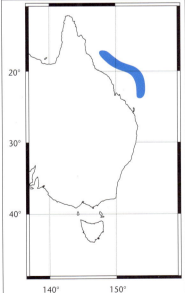

Tasmanian Numbfish — *Narcine tasmaniensis* Richardson, 1841

39.4 *Narcine tasmaniensis* (Plate 66) **Fish Code:** 00 028002

- **Alternative names:** Electric ray, electric torpedo, little numbfish, numbfish.
- **Field characters:** A medium-sized, plain yellowish to brownish numbfish with a medium-sized eye and a shovel-shaped disc (with distinctly concave margins) that is shorter than the tail.
- **Distinctive features:** Body depressed; disc shovel-shaped with a broadly rounded apex (mostly with 2 distinct marginal concavities between snout tip and spiracle level and before maximum width of disc), widest close to insertion of pectoral fin. Eye medium-sized, orbit diameter slightly less than 3* (exposed eye less than 5) in snout length, larger than spiracle; no papillae around spiracle. Mouth narrow, internasal flap trilobed. Skin surface frequently creased, otherwise smooth, lacking denticles and thorns. Tail relatively long (about 1.25* times precloacal length), rather broad-based, moderately depressed, tapering gradually; lateral skin folds well developed. First dorsal fin originating near apex of pelvic fin; second dorsal fin about equal in size to first dorsal fin, both upright. Caudal fin low, broadly rounded posteriorly (upper lobe slightly more angular in mature males). Pelvic fins long and narrow, broadest anteriorly.
- **Colour:** Upper surface mostly uniformly chocolate brown; fins paler brown. Ventral surface uniformly white, occasionally with a few dark blotches. Juveniles frequently with a dark median stripe and dark brown blotches dorsally on disc, and blotches at bases of dorsal fins.
- **Size:** To at least 47 cm; free-swimming by 9 cm.
- **Distribution:** Southeastern Australia from Coffs Harbour (New South Wales) to at least as far west as Beachport (South Australia). Off Tasmania, restricted to the continental shelf (rarely deeper than 100 m and common near the shore). Further north mainly on the continental slope in 200–640 m.
- **Remarks:** This little ray is common throughout its range. The shock from this species is much smaller than that of the larger torpedo and coffin rays. Based on reports from fishermen, it is likely to occur further west into the Great Australian Bight.
- **Reference:** Last *et al.* (1983).

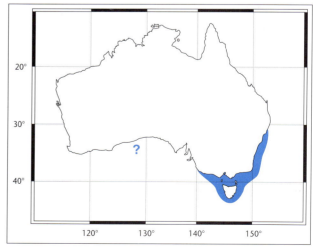

Banded Numbfish — *Narcine westraliensis* McKay, 1966

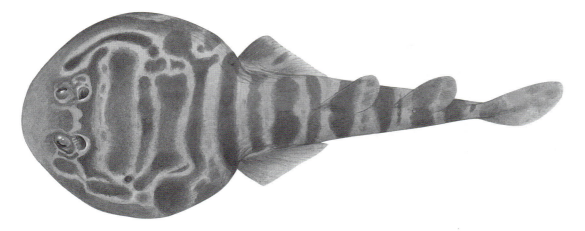

39.5 *Narcine westraliensis* (Plate 67) **Fish Code:** 00 028005

- **Field characters:** A small numbfish with an almost circular disc that is shorter than the tail, and a characteristic pattern of broad dark bands across the upper surface.

- **Distinctive features:** Body depressed; disc subcircular (occasionally oval), widest near middle of pectoral fin. Eye medium-sized, orbit diameter about 2.5–3.5* in snout length, usually about equal in size to spiracle; no papillae around spiracle. Mouth narrow, internasal flap trilobed. Skin surface frequently creased, otherwise smooth, lacking denticles and thorns. Tail rather elongate and broad-based, moderately depressed, tapering gradually; lateral skin folds feebly developed. First dorsal fin originating just behind apex of pelvic fin (further back in mature males); second dorsal fin about equal in size to first dorsal fin, both fins rather broadly rounded (more angular in mature males). Caudal fin low, broadly rounded posteriorly (upper lobe slightly more angular in mature males). Pelvic fins long and narrow, broadest anteriorly.

- **Colour:** Upper surface with prominent pattern of alternating broad dark brown and pale yellowish bars; bars on back sometimes incomplete but never oriented longitudinally; absent from snout. Dorsal fins dark anteriorly, pale posteriorly; caudal fin mostly with two dark bars. Ventral surface uniformly white.

- **Size:** Attains at least 29 cm; females mature at about 18 cm. Young born at about 7 cm.

- **Distribution:** Continental shelf off Western Australia, between Shark Bay and Port Hedland, in 10–70 m depth.

- **Remarks:** This small electric ray is regularly caught by prawn fishermen in Shark Bay. It feeds on the bottom on worms and other small invertebrates. Females are gravid throughout the year, but the main breeding season appears to be September–October. Although its range is restricted, it shares the same latitudes as the western numbfish (39.2). The distributions of the two species do not overlap, however, because the latter lives at greater depths.

- **Reference:** McKay (1966).

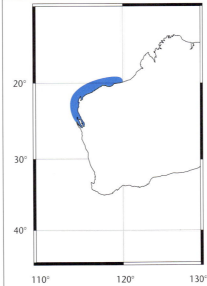

Family Dasyatididae
STINGRAYS

Stingrays are among the largest of cartilaginous fishes, with some species exceeding 2 m in disc width and weighing upwards of 350 kg. The disc is variably depressed and varies in shape from almost circular to rhomboidal. The head is only raised slightly above the pectoral fins and is not demarcated from the the rest of the disc. The mouth, which is armed with numerous small teeth and has several papillae on its floor, is located ventrally. Denticles, thorns and tubercles, if present, are found only on the dorsal surface. The tail, which is mostly longer than the disc, is often whip-like distally, and lacks dorsal, anal and caudal fins. Instead, the tail sometimes has membranous skin folds on the upper and/or lower midlines and most species have one or more serrated stinging spines.

The family is represented by more than 60 living species from 5 or more genera. Because of their large average size, they are poorly represented in museum collections and more research is required to resolve nomenclatural problems. Stingrays are highly adapted fishes and occur in both marine and freshwater habitats. While most species are demersal on continental shelves, some live primarily in the open ocean. They are viviparous with litters of 2–6 that may take up to 12 months to gestate.

Key to dasyatidids

1 No skin folds on tail (fig. 1); base of tail narrow and mostly rounded to slightly compressed in cross-section (slightly depressed in *Himantura granulata*) 2

   Skin fold present on undersurface of tail (may be low) and sometimes on dorsal surface behind sting (fig. 2); base of tail relatively broad, distinctly depressed 10

2 Disc oval to rounded, apex of pectoral fins broadly rounded (figs 3, 6) 3

   Disc more or less quadrangular (fig. 8), apex of pectoral fins narrowly rounded to angular (fig. 8) 5

3 Disc smooth except for a few small denticles on mid-disc; snout tip small, triangular, protruding prominently (fig. 3); freshwater .... *Himantura chaophraya* (**fig. 3; p. 399**)

   Disc granular; snout tip not protruding prominently; marine and estuarine 4

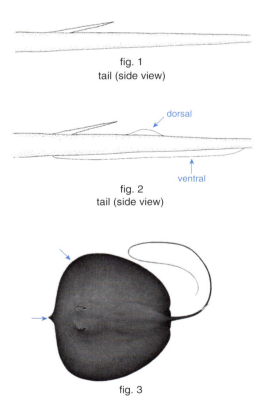

fig. 1
tail (side view)

fig. 2
tail (side view)

fig. 3

4 No stinging spine; tail base rounded (fig. 4), tapering evenly; denticles dense, plate-like over central upper disc; scattered long, sharp thorns present over entire disc in adults ....
.... *Urogymnus asperrimus* (fig. 6; p. 414)

One or more stinging spines (or scar when dislodged) on tail; tail base slightly depressed (fig. 5), tapering abruptly before sting; denticles on upper disc low, comparatively widely spaced, no additional long thorns in adults ....
.... *Himantura granulata* (fig. 7; p. 402)

5 Upper surface of disc with a complex pattern of large dark spots (fig. 8), ocelli (fig. 12), or reticulations (fig. 9) 6

Upper surface of disc uniformly coloured (greyish, pinkish or brownish), lacking a pattern of dark spots, ocelli or reticulations (rarely with a few dark spots confined to disc before pelvic fins) 8

6 Pattern not dense; dark spots on upper surface much more than diameter of largest spot apart (fig. 8); edges of spots normally fuzzy (rarely only with white spots) ..... *Himantura toshi* (fig. 8; p. 405)

Pattern extremely dense; dark spots, ocelli or lines on upper surface less than diameter of largest marking apart (figs 9, 12); edges of markings normally sharply demarcated 7

7 Pattern dominated by fine reticulations in specimens exceeding 80 cm disc width; smaller specimens with relatively small spots and a broadly angular disc (fig. 10) ....
.... *Himantura uarnak* (fig. 9; p. 406)

Pattern dominated by ocelli resembling the spots of a leopard in specimens exceeding 80 cm disc width; smaller specimens with relatively large spots (sometimes connected) with a relatively narrow angular disc (fig. 11) ....
.... *Himantura undulata* (fig. 12; p. 408)

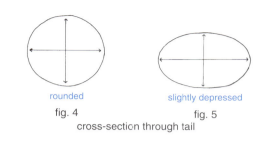

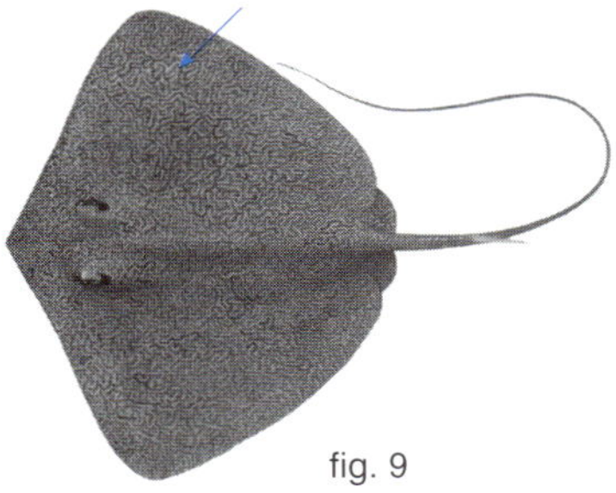

rounded — fig. 4 slightly depressed — fig. 5
cross-section through tail

fig. 6

fig. 7

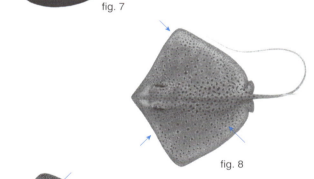

fig. 8

fig. 9

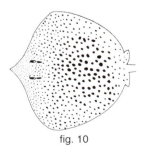

fig. 10

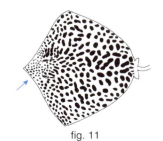

fig. 11

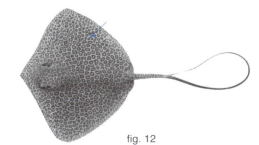

fig. 12

8 Tail with alternating greyish and white bands (fig. 13); eye relatively large, diameter less than 2.3 in preorbital snout length (fig. 13) .... ***Himantura* sp. A (fig. 13; p. 397)**

Tail uniformly dark, not banded; eye smaller, diameter more than 2.3 in preorbital snout length (fig. 16) **9**

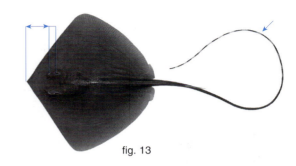
fig. 13

9 Denticles on centre of disc widely spaced (closer together in large adults) (fig. 14); no greatly enlarged thorn-like denticles on disc and tail; greyish pink .... ***Himantura fai* (fig. 16; p. 400)**

Denticles on centre of disc very closely spaced (fig. 15); additional enlarged thorn-like denticles present along midline of disc and tail before stinging spine (fig. 15); yellowish brown .... ***Himantura jenkinsii* (fig. 17; p. 403)**

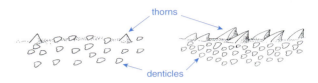
fig. 14   fig. 15
denticles on dorsal midline
above cloaca of adults

10 Disc oval (fig. 20) or subcircular (fig. 21); ventral skin fold tall, extending to tip of tail (fig. 18) **11**

Disc more or less quadrangular (fig. 22); ventral skin fold terminating before tip of tail (sometimes present to tip as a barely detectable ridge) (fig. 19) **12**

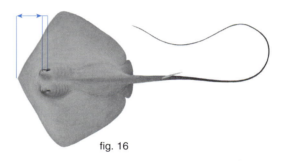
fig. 16

11 Disc oval (fig. 20); upper surface brownish with blue spots and a blue stripe along each side of the tail .... ***Taeniura lymma* (fig. 20; p. 411)**

Disc almost circular (fig. 21); upper surface with fine black and white mottling (no blue spots) .... ***Taeniura meyeni* (fig. 21; p. 412)**

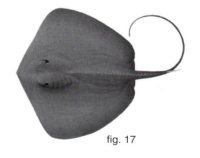
fig. 17

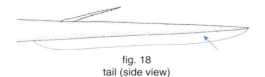
fig. 18
tail (side view)

fig. 19
tail (side view)

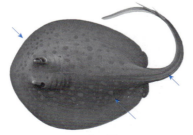
fig. 20

fig. 21

12 Ventral skin fold broad, its height more than twice height of tail above broadest portion (fig. 22); distance from cloaca to stinging spine exceeding half of disc width ....
.... ***Pastinachus sephen*** (**fig. 22; p. 409**)

Ventral skin fold slender, its height less than twice height of tail above broadest portion; distance from cloaca to stinging spine less than half of disc width **13**

13 Anterior margin of disc uniformly convex (fig. 23); ventral surface dark ....
.... ***Dasyatis violacea*** (**fig. 23; p. 396**)

Anterior margin of disc angular (fig. 24); ventral surface pale **14**

14 Tail banded black and white behind sting (fig. 24); dark transverse band through eyes (occasionally faint) (fig. 24); 2 oral papillae on floor of mouth **15**

Tail not banded behind sting; no dark transverse band through eyes ; more than 2 oral papillae on floor of mouth **17**

15 Dorsal surface greyish green, lacking an ornate pattern of spots and reticulations; mostly with up to four short thorns on tail preceding sting; snout tip sharply pointed (fig. 24) ....
.... ***Dasyatis annotata*** (**fig. 24; p. 386**)

Dorsal surface with an ornate pattern of spots and/or reticulations (figs 25, 26); rarely with thorns on tail preceding sting; snout tip comparatively blunt (fig. 25) **16**

16 Dorsal surface with many large bluish spots (fig. 25) .... ***Dasyatis kuhlii*** (**fig. 25; p. 391**)

Dorsal surface without bluish spots, covered in a complex pale honeycomb pattern or with fine black speckles (fig. 26) ....
.... ***Dasyatis leylandi*** (**fig. 26; p. 393**)

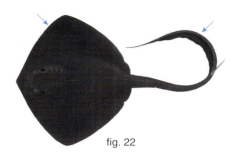

fig. 22

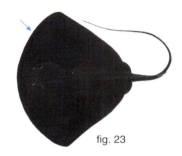

fig. 23

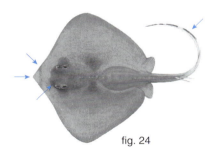

fig. 24

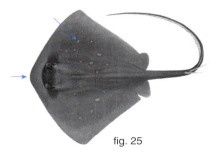

fig. 25

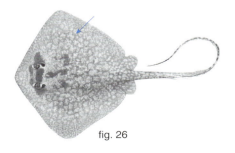

fig. 26

17   Thorn-like denticles rarely present on upper surface of disc; tail shorter than 1.2 times disc length; row of minute white spots located on disc either side of head (fig. 27); transverse furrow present on belly behind gill slits ....
.... ***Dasyatis brevicaudata* (fig. 27; p. 388)**

Thorn-like denticles present on upper surface of disc (except in small juveniles); tail longer than 1.2 times disc length (when undamaged); no minute white spots on either side of head; no furrow on belly behind gill slits **18**

18   Tail with enlarged thorns and/or fine granulations (fig. 28); disc width more than 1.2 times its length .... ***Dasyatis thetidis* (fig. 30; p. 394)**

Tail lacking granulations (thorns absent or confined to midline before sting in adult *D. fluviorum*) (fig. 29); disc width less than 1.2 times its length **19**

19   Nostrils rather long, narrow; ventral skin fold longer than distance from snout to 5th gill opening; thorns along dorsal midline continuous to sting; total vertebrae more than 115 ....
.... ***Dasyatis fluviorum* (fig. 31; p. 390)**

Nostrils short, oval; ventral skin fold equal to or shorter than distance from snout to 5th gill opening; thorns along dorsal midline not continuing on to tail; total vertebrae fewer than 115 .... ***Dasyatis* sp. A (fig. 32; p. 385)**

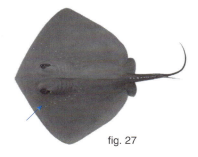
fig. 27

fig. 28
tail (upper surface)

fig. 29
tail (upper surface)

fig. 30

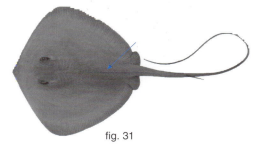
fig. 31

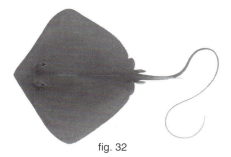
fig. 32

Dwarf Black Stingray — *Dasyatis* sp. A

40.1 *Dasyatis* sp. A (Plate 73)

Fish Code: 00 035021

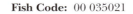

tail (side view)

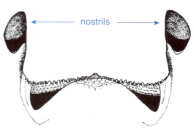

region of nostrils and mouth

teeth from near symphysis
(upper and lower jaw)

- **Field characters:** A small, plainly coloured stingray with a rhomboidal disc and with thorns confined to the mid-disc and to small shoulder patches. The tail is long, whip-like beyond the sting, lacks thorns and denticles, and has low dorsal and ventral cutaneous folds.

- **Distinctive features:** Disc rhomboidal, thickened through trunk, slightly wider than long (width about 1.15–1.2* times length); pectoral-fin apex broadly rounded; snout broadly triangular, tip sharply pointed; anterior margin almost straight to slightly convex. Eyes large, length of eye and spiracle 1.8–1.9* in preorbital snout length; interorbital space relatively narrow. Mouth moderate with 6* papillae on floor (4 centrally and 2 minute laterals); labial furrows and corrugations poorly developed; lower jaw strongly convex. Internasal flap skirt-shaped, rather short and broad, margin fringed; nostrils short, oval. Disc surface smooth except for single row of thorns on dorsal midline (rarely reaching beyond cloaca) and 1–2 thorns on on each shoulder; thorns spear-shaped, flattened, close together; no thorns or denticles on tail. Pelvic fins small. Tail relatively broad-based, depressed slightly; tapering rapidly, becoming nearly cylindrical in cross-section at sting; slender, whip-like beyond sting, slightly depressed posteriorly; length variable (about 1.5 times disc width when undamaged); generally with 1 sting; ventral cutaneous fold very low, elongate (base about equivalent to distance from snout to third or fifth gill slits); dorsal fold very elongate and low (base subequal in length to sting). Total vertebrae 104–106*; monospondylous 35–37*; diplospondylous 69*.

- **Colour:** Dorsal surface yellowish brown to dark greyish brown, disc margin paler; thorns and tail beyond sting pale. Ventral surface white.

- **Size:** To a disc width of 38 cm or more and a total length of at least 85 cm. Males mature at a disc width of 36 cm.

- **Distribution:** Known only from off Port Hedland, northwestern Australia, in 60–125 m.
- **Remarks:** This species superficially resembles the black stingray (40.7) but differs in denticle pattern and maximum size. It is known from only a few individuals collected during trawl surveys; more specimens are required. The genus *Dasyatis*, to which this species belongs, is one of the largest of the stingray genera. Members are presently characterized by their rhomboidal discs, and by having ventral cutaneous folds that are shorter in height than the tail above, and which do not extend to the tail tip. Some species presently ascribed to this group may belong to other genera. The final allocation of these species, however, awaits further research.

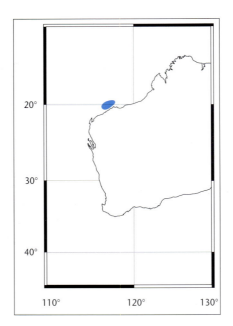

Plain Maskray — *Dasyatis annotata* Last, 1987

40.2 *Dasyatis annotata* (Plate 71)   **Fish Code:** 00 035012

tail (side view)

- **Alternative name:** Brown stingray.
- **Field characters:** A small stingray with a rhomboidal disc, short thorns only on the midline of the disc and tail, and with prominent dorsal and ventral skin folds on the

tail. The dorsal surface is mostly dull greyish green with dark transverse bars about the eyes, and the tail is banded.

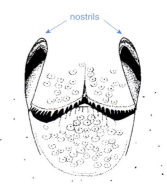

region of nostrils and mouth

- **Distinctive features:** Disc rhomboidal, thin through trunk, slightly wider than long (width about 1.1–1.3* times length); pectoral-fin apex narrowly rounded; snout broadly triangular, tip sharply pointed; anterior margin slightly concave. Eyes small, length of eye and spiracle 2.05–2.4* in preorbital snout length; interorbital space rather narrow. Mouth small, 2 elongate papillae on floor; labial furrows and folds prominent; lower jaw slightly convex. Internasal flap skirt-shaped, rather elongate and narrow, margin fringed, weakly papillate; nostrils long, narrow. Disc surface smooth apart from short thorns in separate rows along dorsal midline; 4–13 small, closely spaced, retrorse nuchal thorns; 0–4 (rarely absent) similar thorns on midline of tail before sting. Pelvic fins moderately large, apices rather pointed. Tail broad-based, depressed, slender beyond sting; its length mostly slightly longer than disc width; generally with 2 stings; ventral cutaneous fold low, elongate (base subequal to distance from snout to about third to fifth gill slit); dorsal fold short, situated behind sting, subequal in height to ventral fold.

- **Colour:** Dorsal surface mostly uniform dull greyish green; sometimes pinkish near disc margin; dark transverse bars before eye, through eye and across snout angle (sometimes indistinct); pair of small dark blotches situated either side of midline behind spiracles. Ventral surface white. Tail with variable black and white bands behind sting (apex usually dark); ventral cutaneous fold pale or dusky.

- **Size:** Reaches a disc width of at least 24 cm and a total length of at least 45 cm.

- **Distribution:** Timor and Arafura Seas, off northern Australia, from Cape Wessels (Northern Territory) to the Bonaparte Archipelago (Western Australia), in depths of 12–62 m.

- **Remarks:** This species, along with two other stingrays (40.5, 40.6), has a distinctive dark bar around the eyes. The group appears to form a subgenus within *Dasyatis* and are referred to here as maskrays. Unlike its relatives, which are highly ornamented with spots, reticulations or blotches, the plain maskray is uniformly coloured (apart from the characteristic mask-like bars around the eye). Little is known of its biology.

- **Local synonymy:** *Amphotistius* sp. 1: Sainsbury *et al.*, 1985.

- **References:** Sainsbury *et al.* (1985); Last (1987).

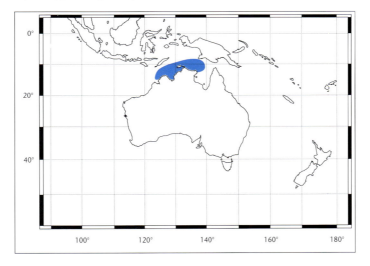

Smooth Stingray — *Dasyatis brevicaudata* (Hutton, 1875)

40.3 *Dasyatis brevicaudata* (Plate 72)   **Fish Code:** 00 035001

tail (side view)

- **Alternative name:** Short-tail stingray.
- **Field characters:** A huge, plainly coloured stingray with a smooth rhomboidal disc, a row of white pores on either side of the head, and with large thorns and tubercles on the tail of adults. The tail is very short with a broad, depressed base that narrows rapidly to the sting and then becomes rather compressed beyond.
- **Distinctive features:** Disc rhomboidal, trunk very thick, slightly wider than long (width about 1.1–1.2* times length); pectoral-fin apex narrowly rounded; snout broadly triangular, tip bluntly pointed; anterior margin almost straight to weakly convex. Eyes small, length of eye and spiracle about 2.0–2.2* in preorbital snout length; interorbital space broad. Mouth of moderate width, 5–7 papillae on floor; with deep labial grooves and corrugations. Internasal flap skirt-shaped, very broad, short, papillate, margin fringed; nostrils rather long, narrow; lower jaw uniformly convex. Disc surface smooth, lacking thorns and denticles. Thorns and tubercles on midline of tail before sting large, either posteriorly directed, spear-like, flat-topped or stellate; juveniles without thorns but at least one thorn usually present by 45 cm disc width; smaller upright thorns (present as granulations on small specimens) on tail behind sting (particularly dense near tail tip). Pelvic fins moderately large, rather truncate posteriorly. Tail short (usually shorter than disc length); broad-based and very depressed, tapering rapidly at sting; slightly compressed posterior to sting; narrowing rapidly to tip; usually with one sting; ventral cutaneous fold well-developed (extending to just beyond sting tip); dorsal skin fold rudimentary (appearing as a low ridge).

- **Colour:** Upper surface greyish brown; dark on tail tip and above eye; inside of spiracles and pores around side of head white. Ventral surface pale; disc margin and undersurface of tail mostly greyish.
- **Size:** Reaches a disc width exceeding 210 cm and a total length of 430 cm or more. Born at about 36 cm disc width.
- **Distribution:** Australia, New Zealand and southern Africa. Most common off southern Australia but occurs north to Maroochydore (southern Queensland) and to Shark Bay (Western Australia) in depths to at least 100 m. Common in 180–480 m off Africa.
- **Remarks:** This species, the largest of all stingrays, may weigh more than 350 kg. Adults are commonly found off beaches and in lower estuaries of southern Australia during summer and autumn. This ray, which will frequently raise its tail above its back in a scorpion-like fashion when approached, is normally inquisitive rather than aggressive. The stinging spine is large and sharp and is capable of inflicting a severe or potentially fatal wound. Through confusion with other large relatives its depth distribution and extension throughout the Southern Hemisphere is unclear. Records from northwestern Australia, however, appear to be misidentifications of the pink whipray (40.11).
- **Local synonymy:** *Trygon thalassia*: Hector, 1872 (misidentification); *Bathytoshia brevicaudata*: Gudger, 1937.
- **References:** Gudger (1937); Garrick (1954a); Ayling and Cox (1982).

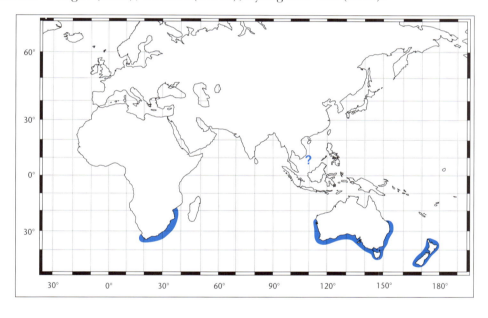

Estuary Stingray — *Dasyatis fluviorum* Ogilby, 1908

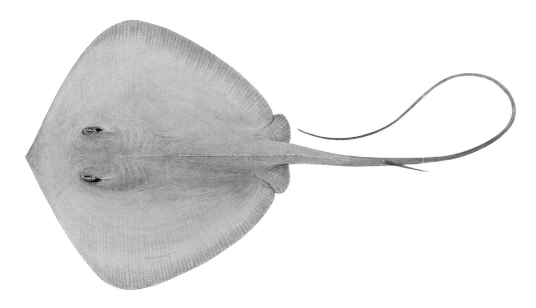

40.4 *Dasyatis fluviorum* (Plate 72)

Fish Code: 00 035008

tail (side view)

- **Alternative names:** Brown stingray, estuary stingaree.
- **Field characters:** A medium-sized, plain coloured stingray with a rhomboidal disc, short thorns on the head and back, a well-defined median row of thorns along the disc and tail, and a long whip-like tail with well developed cutaneous folds.
- **Distinctive features:** Disc rhomboidal, rather broad through trunk, width about 1.0–1.1* times length; pectoral-fin apex broadly rounded; snout broadly triangular, tip pointed; anterior margin convex. Eyes small, length of eye and spiracle about 1.7–2.1* in preorbital snout length; interorbital rather broad. Mouth small with 5 oral papillae on floor (central 3 enlarged, lateral 2 minute); labial furrows and folds prominent. Internasal flap skirt-shaped, rather short and broad, margin weakly fringed; nostrils moderately elongate; lower jaw with concave symphysis. Thorns short, flat-topped and lanceolate, raked backward (those preceding sting often greatly enlarged, spear-shaped); extensive patches above eyes and on mid-disc; single series extending along midline to origin of sting; remainder of disc smooth. Pelvic fins moderately large. Tail broad-based, depressed; tapering rapidly at sting, whip-like beyond; lacking fine granulations; about 2 times disc width when undamaged; frequently with 2 stings; ventral cutaneous fold long, low (base much longer than distance from snout to fifth gill slit); dorsal fold moderately well developed, origin posterior to sting tip. Total vertebrae 121*; monospondylous 39*; diplospondylous 82*.
- **Colour:** Dorsal surface yellowish to olive-brown with paler margins. Ventral surface white. Tail similar to upper surface anteriorly, darker beyond sting.
- **Size:** Reported to reach a disc width of 120 cm; the largest specimen examined was an immature male with a disc width of about 30 cm; born at about 11 cm disc width (35 cm total length).
- **Distribution:** Inshore tropical Australia (including New Guinea) from Forster (New South Wales) to Darwin (Northern Territory). Reported from Port Jackson but its occurrence along the central New South Wales coast requires verification.

- **Remarks:** This species is reported to be common in mangrove swamps and in estuaries of north Queensland and the Gulf of Carpentaria. It has the infamous reputation of being a major predator of shellfish, including farmed oysters. Little is known of its biology and more specimens are required.
- **References:** Ogilby (1908a); McCulloch (1915); Grant (1978).

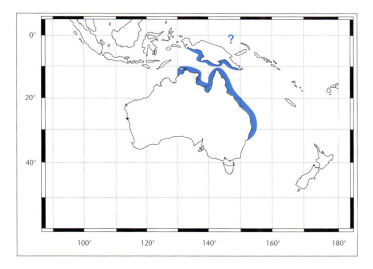

Blue-spotted Maskray — *Dasyatis kuhlii* (Müller & Henle, 1841)

40.5 *Dasyatis kuhlii* (Plate 71)　　　　　　　　　　Fish Code: 00 035004

tail (side view)

- **Alternative names:** Blue-spotted stingaree, blue-spotted stingray.
- **Field characters:** A small to medium-sized stingray with a rhomboidal disc, short thorns confined to the midline of the disc, and with prominent dorsal and ventral

skin folds on the tail. The dorsal surface has prominent bluish spots with dark transverse bars about the eyes, and the tail is banded.

- **Distinctive features:** Disc rhomboidal, trunk thickened, width about 1.15–1.3 times length; pectoral-fin apex angular (narrowly rounded in juveniles); snout broadly rounded, tip seldom pointed; anterior margin almost straight. Eyes large, length of eye and spiracle 1.35–1.8 in preorbital snout length. Mouth small, 2 broad papillae on floor; labial furrows and folds prominent; lower jaw slightly convex. Lower lip and internasal flap papillate; internasal flap skirt-shaped, rather elongate and narrow, margin fringed; nostrils long, narrow. Denticles thorn-like, lanceolate, flat topped, directed posteriorly; confined to single row along disc midline, extending from nuchal region to almost over cloaca in adults (thorns often absent in juveniles); remainder of disc and tail smooth. Pelvic fins moderately large. Tail broad-based, slightly depressed, slender and rather compressed beyond sting; its length mostly longer than disc when undamaged; generally with 2 stings; ventral cutaneous fold low, elongate (mostly longer than distance from snout to fifth gill slit); dorsal fold short, situated beyond sting, subequal in height to ventral fold.

- **Colour:** Dorsal surface greyish to brownish with prominent bluish spots; spots variable in size and number, often bright blue with darker blue borders or whitish with bluish borders; frequently with peppering of fine black spots, concentrated around dark brownish band through orbital region; orbital membrane mostly with fine black spots. Ventral surface mostly pale, slightly darker around disc margin. Tail with alternating black and white bands, tip mostly pale; skin folds pale at base with dark outer margin.

- **Size:** Reaches a disc width of 38 cm and a total length of at least 67 cm; males mature at about 25 cm disc width; born at about 16 cm disc width (33 cm total length).

- **Distribution:** A widespread inshore species of the Indian and western Pacific Oceans, including Melanesia and Micronesia. In Australia, it occurs in inshore tropical areas north of Port Stephens (New South Wales) and Shark Bay (Western Australia), to depths of 90 m.

- **Remarks:** A very common and widely distributed inshore species. Like the blue-spotted fantail ray (40.18), it is covered in bluish spots and is common inshore over coral reefs. It can be readily distinguished from the latter, however, in having a more angular disc, a more slender tail, and narrower skin folds.

- **References:** McCulloch (1921); Grant (1978).

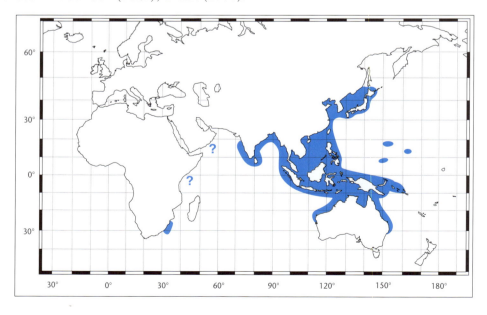

Painted Maskray — *Dasyatis leylandi* Last, 1987

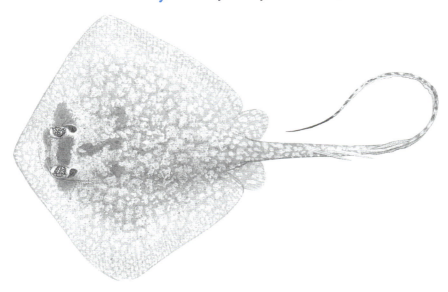

40.6 *Dasyatis leylandi* (Plate 71)

Fish Code: 00 035013

tail (side view)

- **Alternative name:** Brown-reticulate stingray.
- **Field characters:** A small stingray with a rhomboidal disc, short thorns only on the midline of the disc, and with prominent dorsal and ventral skin folds on the tail. The dorsal surface is a mosaic (often with fine black spots but never with blue spots) with dark transverse bars about the eyes, and the tail is banded.
- **Distinctive features:** Disc rhomboidal, trunk relatively thin, slightly wider than long (width about 1.1–1.3* times length); pectoral-fin apex narrowly rounded; snout broadly triangular, tip pointed or slightly rounded; anterior margin slightly concave. Eyes rather small, length of eye and spiracle 1.9–2.3* in preorbital snout length; interorbital space rather narrow. Mouth small, 2 elongate papillae on floor; labial furrows and folds prominent; lower jaw slightly convex. Internasal flap skirt-shaped, rather elongate and narrow, papillate, margin fringed; nostrils long, narrow. Disc surface mostly smooth; 1–9 small, closely-spaced, thorn-like denticles along nuchal area; usually no denticles on tail. Pelvic fins moderately large, apices extended. Tail broad-based, depressed, slender beyond sting; length variable (0.9–1.4 in disc width when undamaged); generally with 2 stings; ventral cutaneous fold low, elongate (much longer than distance from snout to fifth gill slit); dorsal fold short, situated behind sting, subequal in height to ventral fold.
- **Colour:** Dorsal surface pale brownish or yellowish, covered with a network of darker reticulations; western populations with or without light speckling; populations east of the Gulf of Carpentaria with intense black speckling; distinct mask-like bars around eye; tail variably banded black and white (mostly pale near tip); skin folds pale with dark margin. Ventral surface pale.
- **Size:** Reaches a disc width of at least 25 cm (53 cm total length). Born at about 11 cm disc width (24 cm total length).
- **Distribution:** Northern Australia (including New Guinea) between the Monte Bello Islands (Western Australia) and Townsville (Queensland). Western populations in depths of 30–75 m, but shallower (often in less than 5 m) off eastern Australia.
- **Remarks:** Occurs together with two other maskrays, the plain maskray (40.2) and the blue-spotted maskray (40.5) off Western Australia. Until recently, all three species were considered to be variations of the blue-spotted maskray. The two major

colour forms of the reticulated maskray represent either population or species differences (see Plate 71); genetic studies of these forms may be required to solve this problem.

- **Local synonymy:** *Amphotistius* sp. 2: Sainsbury *et al.*, 1985; *Dasyatis* sp.: Allen & Swainston, 1988.
- **References:** Sainsbury *et al.* (1985); Last (1987).

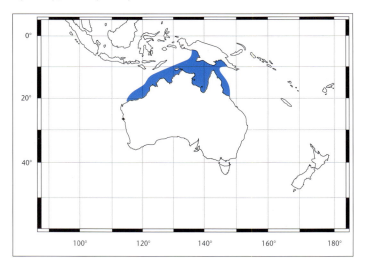

Black Stingray — *Dasyatis thetidis* Ogilby, 1899

40.7 *Dasyatis thetidis* (Plate 72) **Fish Code:** 00 035002

tail (side view)

- **Alternative names:** Thorntail stingray, black stingaree, black skate, longtail stingray.
- **Field characters:** A large, uniformly dark stingray with a rhomboidal disc, sharp thorns over the head and back of adults, and lacking white pores beside the head.

The tail, which has skin folds, is elongate with a broad depressed base, and tapers uniformly posteriorly.

- **Distinctive features:** Disc rhomboidal, trunk deep, width about 1.2–1.25* times length; pectoral-fin apex narrowly rounded; snout broadly triangular, tip pointed; anterior margin almost straight. Eyes small, length of eye and spiracle about 2.1–2.5* in preorbital snout length; interorbital space broad. Mouth small to medium, with 3–5 papillae on floor (3 enlarged centrally, laterals minute); labial furrows poorly developed; lower jaw weakly convex. Internasal flap skirt-shaped, short, very broad, margin with a prominent fringe, lower lip papillate; nostrils short, broad. Juveniles (to 60 cm disc width) smooth or with a few enlarged stellate thorns over head, along midline of back and on snout tip; thorns upright and rather pointed on back, frequently broader based on midline of tail. In adults, disc finely granular with sparse coverage of larger thorns centrally; tail beyond sting uniformly covered with enlarged thorns. Pelvic fins rather small, apices broadly rounded. Tail broad-based, depressed; tapering gradually, not narrowing rapidly at sting origin; slender and almost circular in cross-section beyond sting; about twice disc length when undamaged; mostly with 1 sting; ventral cutaneous fold low, very elongate, extending well beyond sting tip (base almost equivalent to precloacal length); dorsal fold minute, barely detectable.

- **Colour:** Dorsal surface of disc and tail before sting uniform greyish brown to black (often with irregular white flecks where skin is removed); ventral surface mostly white. Tail mostly black beyond sting; ventral skin fold black.

- **Size:** Reported to reach a disc width of 180 cm (about 400 cm total length). Free swimming by 35 cm disc width (about 60 cm total length).

- **Distribution:** Southeastern Africa, New Zealand and southern Australia. Locally from Coffs Harbour (northern New South Wales) to Shark Bay (Western Australia). Common inshore but collected to 360 m depth.

- **Remarks:** One of the largest Australian stingrays, this species may attain a weight of more than 210 kg. It has been regularly confused with the even larger smooth stingray (40.3), which has a similar distribution.

- **References:** McCulloch (1921); Richardson and Garrick (1953); Ayling and Cox (1982).

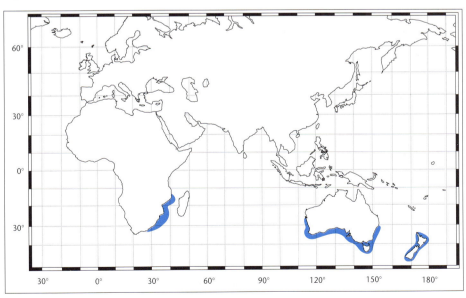

Pelagic Stingray — *Dasyatis violacea* (Bonaparte, 1832)

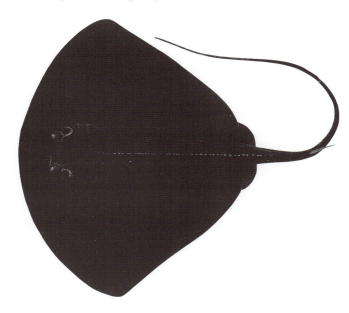

40.8 *Dasyatis violacea* (Plate 72)

Fish Code: 00 035010

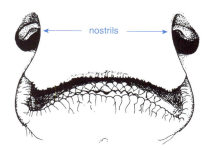

- **Alternative names:** Guilers stingray, violet stingray.
- **Field characters:** A dark, medium-sized stingray with an evenly rounded anterior disc margin, a dark ventral surface, small thorns in a continuous row along the back, and a whip-like tail with a long ventral skin fold.
- **Distinctive features:** Disc with a very thickened trunk; broad, wider than long (width about 1.25–1.4* times length); pectoral-fin apex angular; snout broad, small terminal lobe present; anterior margin evenly convex. Eyes very small, length of eye and spiracle about 1.1–1.3* in preorbital snout length; interorbital space very broad. Mouth small with numerous short, bifurcated papillae in a continuous row across floor; labial furrows and folds prominent; lower jaw weakly convex. Internasal flap skirt-shaped, short, very broad, fringe weak; nostrils short, circular. Disc surface normally without granulations (mid-disc and tail sometimes granular in large females); single row of small, short, sharp thorns commencing near nuchal area and extending along midline to sting. Pelvic fins small, apices broadly rounded. Tail exceeding twice length of disc (when undamaged); broad-based, tapering, slightly depressed in cross-section anteriorly, whip-like beyond sting; 1–2 stings; ventral cutaneous fold low, elongate, extending for about half to three-quarters of tail length beyond sting; dorsal fold rudimentary or absent.
- **Colour:** Dorsal surface, cutaneous folds and whip-like portion of tail uniformly black. Ventral surface of disc and tail dark brownish or black. Cloaca, thorns and sting mostly pale.
- **Size:** Locally, reaches a disc width of at least 59 cm (at least 130 cm total length); females grow larger than males, smallest mature male 35 cm disc width. Elsewhere reported to reach 80 cm disc width.
- **Distribution:** Not adequately defined, but appears to live worldwide in tropical and subtropical seas in the open ocean. Also occurs off temperate Australia south to Tasmania; appears to be rare inshore.
- **Remarks:** This wide-ranging species was originally recorded from Australia on the basis of a beach stranding in the Derwent Estuary near Hobart. It has since been caught regularly by Japanese tuna longliners in the open ocean off eastern and

southern Australia. Possibly the only totally pelagic member of the stingray family, it is immediately recognizable by its characteristic disc shape and dark ventral surface. Elsewhere it is known to feed on jellyfish, squid, crustaceans and fish. The pelagic stingray possesses features unique to the family and has been placed by some authors in a separate genus *Pteroplatytrygon*.

- **Local synonymy:** *Dasyatis guileri* Last, 1979
- **References:** Last (1979); Last *et al.* (1983).

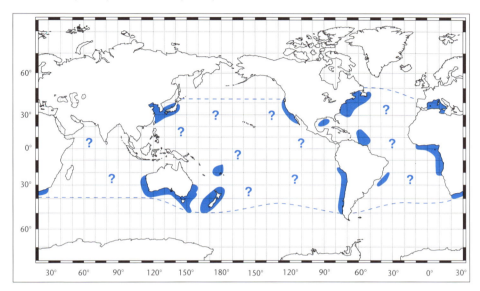

Brown Whipray — *Himantura* sp. A

40.9 *Himantura* sp. A (Plate 67) Fish Code: 00 035022

- **Field characters:** A medium-sized, uniform brownish stingray with a banded, whip-like tail that lacks skin folds.

- **Distinctive features:** Disc quadrangular, trunk not especially thick, width about 1.05–1.15* times length; pectoral-fin apex narrowly rounded; snout broadly triangular, tip distinctly pointed; anterior margin almost straight. Eyes rather large, length of eye and spiracle about 1.95–2.25* in preorbital snout length; interorbital of moderate width. Mouth small, with 4 papillae on floor (2 centrally and 2 laterally); labial furrows weak; lower labial folds and labial papillae prominent; lower jaw concave at symphysis. Internasal flap skirt-shaped, short and broad, margin with fine fringe, not papillate; nostrils long, narrow. Denticles in a band between eyes along central disc onto upper half of tail; 3–4 enlarged spear-like scapular thorns; smaller thorns extending forward to nuchal area in 1–3 rows; other denticles minute, extremely dense, heart-shaped and low (often confined to patch in nuchal area and between eyes in smaller specimens). Pelvic fins small, relatively slender. Tail very slender, elongate, whip-like; almost cylindrical in cross-section at base, slightly flattened posteriorly; very long (about 2.5–3 times disc width when undamaged); mostly with one sting; no cutaneous folds.

- **Colour:** Uniformly dark greenish brown, paler brown around disc margin. Ventral surface uniformly white; disc margin sometimes greyish in smaller specimens. Tail mostly dark with alternating greyish and black bands.

- **Size:** To at least 30 cm in disc width (131 cm total length).

- **Distribution:** Known from tropical eastern and northern Australia from Darwin (Northern Territory) to Moreton Bay (Queensland) in coastal habitats. Likely to be more widely distributed in this region (possibly also in the Indo–West Pacific).

- **Remarks:** Common inshore over muddy bottoms on mangroves flats. Possibly prefers shallower habitats than the closely related black-spotted whipray (40.14). Juveniles appear to move intertidally in search of food. Few specimens are available in museum collections and more are needed to positively identify this species. Similar forms occur off New Guinea and Indonesia. Members of the genus *Himantura*, referred to here as whiprays, have a long slender tail with a sting but no cutaneous folds.

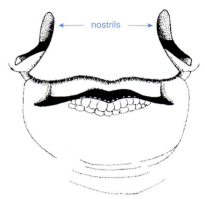

region of nostrils and mouth

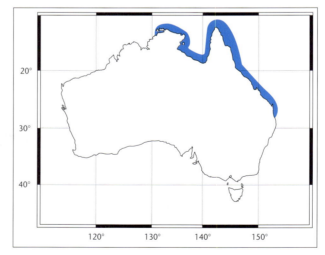

Freshwater Whipray — *Himantura chaophraya* Monkolprasit & Roberts, 1990

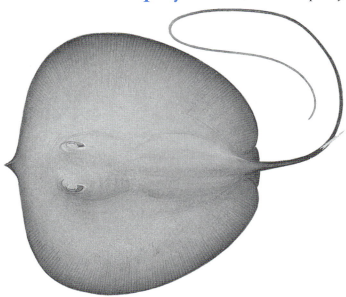

40.10 *Himantura chaophraya* (Plate 69)　　　　　　　　　　**Fish Code:** 00 035023

- **Field characters:** A freshwater stingray with a rounded, plain coloured disc, a prominent snout tip, and a long whip-like tail without cutaneous folds.
- **Distinctive features:** Disc subcircular to oval, trunk thin, slightly longer than wide; pectoral-fin apex broadly rounded; snout extremely broad, almost truncate anteriorly with an enlarged triangular tip. Eyes very small, interorbital space broad; eye and spiracle together 3–4 in preorbital snout length. Mouth small, with 2 enlarged central and up to 4 minute lateral papillae on floor; labial furrows and folds weak or absent; lower jaw arched slightly. Internasal flap skirt-shaped, broad, short, margin with fine fringe, not papillate. Upper surface of disc granular; band of widely spaced, heart-shaped denticles on raised central portion of disc and on upper tail before sting; finer and more dense granular denticles laterally over disc; short prickle-like denticles beyond sting; 2–4 greatly enlarged mid-scapular thorns. Pelvic fins small, narrow. Tail long (about 2–2.5 times disc width), whip-like, almost cylindrical in cross-section; mostly with one sting; no cutaneous folds.
- **Colour:** Upper surface uniform greyish centrally, becoming yellowish and then pinkish laterally; covered in dark brown mucous when alive; prominent dark dendritic pores inward from disc margin. Ventral surface white with broad dark grey margin behind mouth level; mostly with irregular greyish spots centrally. Tail greyish above; pale below forward of sting, blackish behind sting.
- **Size:** Adult size unknown locally, with largest validated specimen slightly exceeding 100 cm disc width (about 270 cm total length). Elsewhere reaches a disc width of almost 200 cm and about 600 kg; males mature by 110 cm disc width; born at about 30 cm disc width.
- **Distribution:** Positive identifications from the Gilbert River (Queensland), the Daly and South Alligator rivers (Northern Territory), and the Ord and Pentecost rivers (Western Australia). Possibly occurs in most large rivers of tropical Australia. Also known from the Fly River basin (New Guinea), the Mahakam basin (Borneo), and several rivers of Thailand.
- **Remarks:** The only Australian stingray to live only in fresh and estuarine waters. This little known species was only discovered in the region a few years ago. All stingrays caught in freshwater habitats of tropical Australia, have been previously misidentified as the estuary stingray (40.4). Local rays appear to be smaller than their relatives occurring in Asian rivers. Specimens from any additional drainages should be reported to museums. *Himantura polylepis* (Bleeker), based on an Indonesian specimen, appears to be an older name for this species.

- **Local synonymy:** *Himantura* sp.: Compagno & Roberts, 1982; *Dasyatis fluviorum*: Merrick & Schmida, 1984 (misidentification).
- **References:** Compagno and Roberts (1982); Monkolprasit and Roberts (1990).

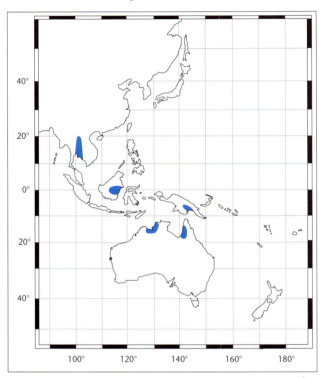

Pink Whipray — *Himantura fai* Jordan & Seale, 1906

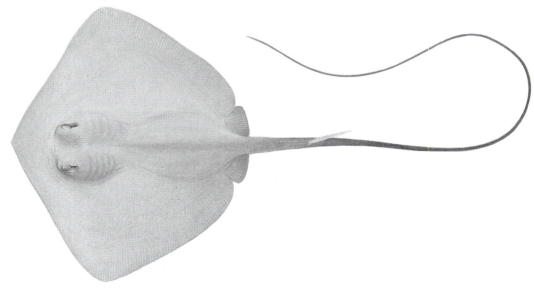

40.11 *Himantura fai* (Plate 69) Fish Code: 00 035024

- **Field characters:** A large, uniformly brownish pink stingray with a very long, whip-like tail that lacks skin folds. The head and trunk are covered with short, widely spaced denticles but there are no enlarged thorns on the upper surface.

- **Distinctive features:** Disc rhomboidal, very robust centrally, quadrangular, width 1.15–1.2* times length; pectoral-fin apex broadly rounded; snout very broad, tip feebly pointed; anterior margin straight to slightly convex. Eyes small, length of eye and spiracle about 2.6–2.7* in preorbital snout length; interorbital space broad. Mouth rather small, with 4 papillae on floor (central pair enlarged, laterals minute); labial furrows and corrugations prominent. Internasal flap skirt-shaped, short and very broad, not papillate, marginal fringe very fine; nostrils long, very narrow; lower jaw slightly concave near symphysis. In juveniles, disc surface and tail mostly smooth, or with low, flat, widely spaced, heart-shaped denticles extending from interorbital posteriorly over centre of disc and onto midline of tail. In adults, denticles more rounded, dense, extending from well before eye, over central disc and over entire tail (sometimes naked on ventral midline near tail base); a few small, sharp, upright thorns on midline (quite dense over tail beyond sting). Pelvic fins small, relatively slender. Tail very long (about twice disc width), rather narrow, almost cylindrical in cross-section; whip-like behind sting; mostly with single sting; no cutaneous folds.
- **Colour:** Upper surface greyish pink. Ventral surface uniformly pale. Tail dark greyish or black beyond sting.
- **Size:** Disc width exceeding 150 cm (greater than 500 cm total length); born at about 55 cm disc width (154 cm total length).
- **Distribution:** Inadequately known. Possibly widespread in the Indian Ocean from South Africa to Micronesia. Locally on the inner continental shelf of tropical Australia from North West Cape (Western Australia) to Stradbroke Island (Queensland).
- **Remarks:** The distribution of this ray in the Indo–Pacific is unclear, but large aggregations occasionally occur on atolls of the Great Barrier Reef and throughout the Caroline Islands. Large writhing specimens are troublesome when caught accidentally by prawn trawlers, as they usually have to be discarded before the catch can be sorted.
- **Reference:** Jordan and Seale (1906).

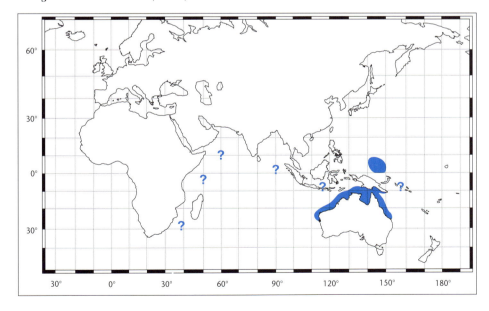

Mangrove Whipray — *Himantura granulata* (Macleay, 1883)

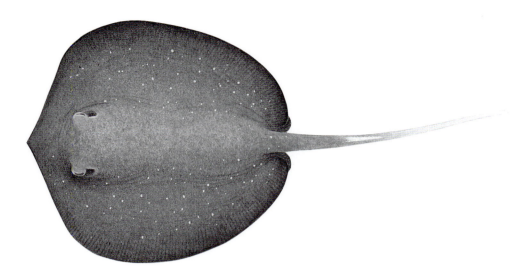

40.12 *Himantura granulata* (Plate 69)   **Fish Code:** 00 035019

- **Alternative names:** Mangrove ray, Macleays coachwhip ray.
- **Field characters:** An oval-shaped ray with a rough upper surface that is usually studded with fine white flecks, one or more stinging spines (when detached a scar is usually visible), and a white, whip-like tail lacking skin folds.
- **Distinctive features:** Disc almost circular to oval, robust though trunk, width about 0.9–1* times length; pectoral-fin apex evenly rounded; snout very broad, tip feebly pointed; anterior margin straight to slightly convex. Eyes moderately large, length of eye and spiracle about 2.1–2.2* in preorbital snout length; interorbital broad. Mouth rather small, with 4–5 papillae on floor (central 2–3 enlarged); labial furrows and corrugations prominent. Internasal flap skirt-shaped, rather short and broad, not papillate, margin with very fine fringe; nostrils long, narrow; lower jaw concave at symphysis. Disc surface granular, in two patches; 1–2 flat, heart-shaped nuchal thorns surrounded by prominent denticles on head, over abdomen and along midline of tail; smaller (often obscure) denticles laterally on disc. Pelvic fins small, relatively slender. Tail of moderate length (about 1.5–2 times disc width), rather broad-based, almost cylindrical in cross-section, whip-like behind sting, slightly flattened posteriorly; with 1–2 stings; no cutaneous folds.
- **Colour:** Usually uniformly dark brownish to greyish brown on central disc; outer area of disc brownish to slate grey; usually covered with fine white spots and flecks (most dense in large specimens); tail uniformly white beyond sting. Ventral surface white; disc margin mostly dark with black blotches.
- **Size:** Attains at least 90 cm disc width (230 cm total length), possibly much larger. Born at about 28 cm disc width (78 cm total length).
- **Distribution:** Possibly widespread in the Indo–West Pacific. Locally, inshore from the Kimberley coast (Western Australia) to Brisbane (Queensland); common off the Northern Territory. Also known from New Guinea, Micronesia, the Santa Cruz and Solomon Islands and the Java Sea.
- **Remarks:** Little is known of this unusual stingray with its distinctive white tail. As the name implies, this species often occurs in mangrove habitats in Australia where it feeds mainly on small crustacea including crabs and prawns. Elsewhere, it has been found over sand flats in shallow water and amongst coral. It resembles the porcupine ray (40.20) in shape but, unlike the latter, has a stinging spine and the disc is less thorny. Adult specimens are required from Australian waters.

- **Reference:** Munro (1967).

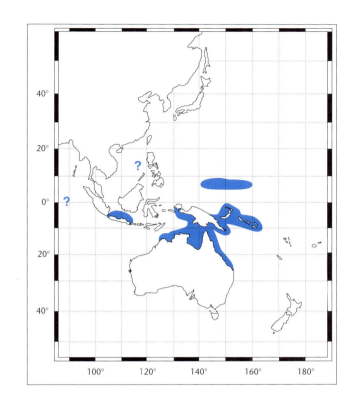

Jenkins Whipray — *Himantura jenkinsii* (Annandale, 1909)

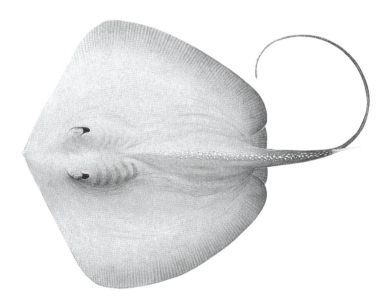

40.13 *Himantura jenkinsii* (Plate 69) **Fish Code:** 00 035025

- **Field characters:** A medium to large, uniformly yellowish brown stingray with a long, whip-like tail that lacks skin folds. A row of enlarged, spear-shaped thorns and a narrow band of closely-spaced denticles extend along the head, back and tail.

- **Distinctive features:** Disc rhomboidal, trunk robust, width about 1.1* times length; pectoral-fin apex broadly rounded; snout broadly triangular, tip feebly pointed; anterior margin almost straight. Eyes moderately large, length of eye and spiracle about 2.4–2.7* in preorbital snout length; interorbital of moderate width. Mouth rather broad, with 4 papillae on floor (central two short); labial furrows and papillae well developed; lower jaw weakly convex, slightly indented at symphysis. Internasal flap skirt-shaped, short and very broad, not papillate, margin with fine fringe; nostrils long, narrow. Disc with a dense band of closely-spaced, flat, heart-shaped denticles extending from between eyes in a broad patch over centre of disc and onto all surfaces of tail; additional irregular row of enlarged upright spear-shaped thorns along midline to sting (forming 2–3 rows before sting in adults); denticle band narrow in small specimens, much wider in scapular region; remainder of upper disc with minute granulations. Pelvic fins small, relatively slender. Tail moderately long (slightly longer than disc width), whip-like, almost round in cross-section; up to 3 stings; no cutaneous folds.
- **Colour:** Upper surface pale brownish; paler over central denticle patch. Ventral surface, including disc margin, white. Tail greyish posteriorly.
- **Size:** Reaches a disc width of at least 104 cm (about 200 cm total length).
- **Distribution:** Poorly defined. Possibly widely distributed but patchy in the Indo–Pacific from Australia to southern Africa. Locally off northern Australia from Port Hedland (Western Australia) to the eastern Gulf of Carpentaria (Queensland) in 25–48 m.
- **Remarks:** Common in other parts of the Indo–Pacific but known from only a few individuals locally. Rays photographed from the Arafura Sea, which resemble *H. jenkinsii* in form but which have dark spots along the posterior margin of the disc, may be *H. draco* (a ray described recently from South Africa). It is also possible that these forms are variants of the same species. More specimens of both forms are needed to resolve this problem.

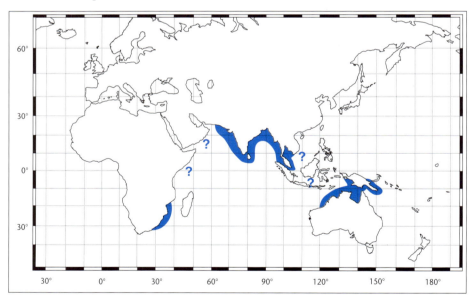

Black-spotted Whipray — *Himantura toshi* Whitley, 1939

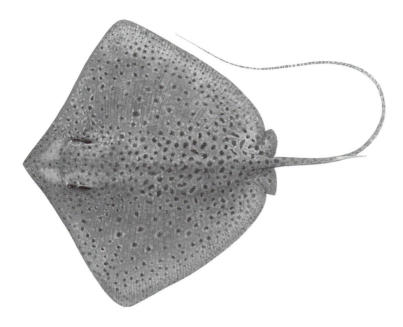

40.14 *Himantura toshi* (Plate 68) **Fish Code:** 00 035020

- **Alternative names:** Coachwhip ray, Toshs longtail ray, wulura.
- **Field characters:** A medium-sized stingray with a banded, whip-like tail that lacks skin folds and with a pattern of small, widely-spaced, dark spots on the dorsal surface.
- **Distinctive features:** Disc rhomboidal, trunk relatively thin, width about 1.1–1.2* times length; pectoral-fin apex rather angular (narrowly rounded in juveniles); snout broadly triangular, tip distinctly pointed; anterior margin almost straight. Eyes small, length of eye and spiracle about 1.8–2.5* in preorbital snout length; interorbital space of moderate size. Mouth moderate, with 4 papillae on floor (2 centrally); labial furrows, folds and papillae poorly developed. Internasal flap skirt-shaped, rather short, broad, not papillate; posterior margin mostly double concave with a fine fringe; lower jaw with a central concavity; nostrils long, narrow. Denticles minute, low, extremely dense, heart-shaped, extending in a broad band from interorbit along central disc and over entire tail in adults (confined to scapular area and head in smaller specimens); narrow band of slightly larger spear-shaped thorns on centre of disc. Pelvic fins small, relatively slender. Tail very slender, elongate (about 2.5–3* times disc width when undamaged), whip-like; base almost cylindrical in cross-section, slightly flattened posteriorly; sting often missing; no cutaneous folds.
- **Colour:** Adults greyish brown with variable black spots; spots sometimes small and very dense, mostly about 0.3 of eye diameter in size and widely spaced, frequently with paler outer borders; white spots also sometimes present; tail variably banded, either only on upper half or on all of tail. Ventral surface uniformly white; upper half of eyeball usually spotted, lower half white. Juveniles variably spotted; neonates with widely spaced, large black spots; spotting usually becoming more dense with age; tail with prominent dark and light bands.
- **Size:** To at least 69 cm disc width (179 cm total length); free-swimming juveniles with a disc width of 20–22 cm.
- **Distribution:** New Guinea and tropical Australia, between Port Hedland (Western Australia) and Mackay (Queensland) in 10–140 m. Possibly also off Indonesia.
- **Remarks:** This species has been misidentified by most ichthyologists as a colour form of the reticulated whipray (40.15). It is very common in the Gulf of Carpentaria, where it is a major predator of prawns.

- **Local synonymy:** *Himantura uarnak*: Whitley, 1940 (misidentification, in part); *H. uarnak* (spotted form): Sainsbury *et al.*, 1985 (misidentification).
- **References:** Whitley (1939b); Sainsbury *et al.* (1985).

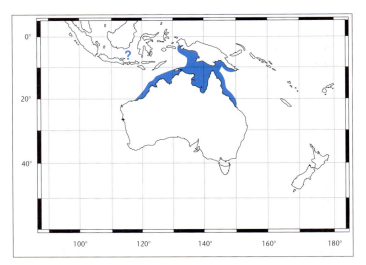

Reticulate Whipray — *Himantura uarnak* (Forsskål, 1775)

40.15 *Himantura uarnak* (Plate 68)   **Fish Code:** 00 035003

- **Alternative names:** Coachwhip ray, longtail ray.
- **Field characters:** A large stingray with a whip-like tail that lacks skin folds and with a dense pattern of fine reticulations on the dorsal surface. Juveniles have a rhomboidal disc densely covered in black spots.
- **Distinctive features:** Disc rhomboidal, trunk thick, width exceeding length (about 1–1.1* times length in juveniles); pectoral-fin apex narrowly rounded (broadly rounded in juveniles); snout broadly triangular (extremely obtuse in juveniles), tip distinctly pointed; anterior margin almost straight. Eyes rather small, length of eye

and spiracle about 1.9–2.2* in preorbital snout length in juveniles; interorbital space broad. Mouth narrow, with 5 papillae on floor (3 centrally); labial furrows weakly developed; lower labial folds and papillae present. Internasal flap skirt-shaped, relatively short and broad, not papillate (almost smooth), margin with fine fringe; nostrils long, narrow; lower jaw deeply concave near symphysis. Denticles low, flat, heart-shaped; in a broad band from interorbit, extending along centre of disc and onto tail (density increasing with size); 2 prominent pearl thorns in centre of disc but enlarged thorns absent from midline of tail; adult tail covered posteriorly with small, sharp thorns. Pelvic fins small, almost triangular. Tail very slender, elongate (about 3–3.5 times disc width in juveniles when undamaged), whip-like beyond sting; slightly depressed near base, tapering rapidly to sting, flattened slightly near tip; mostly with one sting; no cutaneous folds.

- **Colour:** Upper surface of adults highly ornamented; pale to yellowish brown with a dense pattern of dark brown wavy lines or reticulations over disc and tail. Juveniles pale yellowish brown with extremely dense pattern of small, dark brown spots (spaces between spots much smaller than largest spot); more than 5 (usually about 7) spots in a direct line across interspiracular space; tail mostly with 3 distinct rows of spots before sting. Ventral surface uniformly pale. Dorsal and ventral surfaces of tail black behind sting, banded laterally.
- **Size:** To at least 150 cm disc width (about 450 cm total length). Born at about 28 cm disc width (110 cm total length).
- **Distribution:** Widespread in the Indo–Pacific, west to South Africa and the Mediterranean Sea (apparently absent from the western North Pacific). Common on the continental shelf of northern Australia from Shark Bay (Western Australia) to Brisbane (Queensland) from the intertidal zone to at least 45 m.
- **Remarks:** The name, *Himantura uarnak*, has been used for a number of similar species found in the Indo–Pacific region which tend to have dark spotted or reticulated colour patterns. Individual adult rays are not capable of altering their colour pattern to match the substrate as once thought, so colour forms are generally typical of species. Unusually, juveniles of this species complex differ more from each other anatomically and morphologically than do the adults. Major differences include the degree of spotting, the disc shape, birth size and denticle structure.
- **Local synonymy:** *Himantura* sp. 1: Gloerfelt-Tarp & Kailola, 1984.
- **References:** Grant (1978); Coleman (1981).

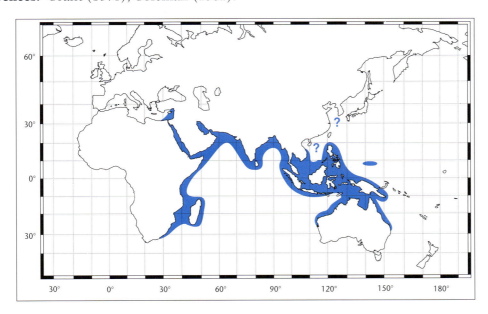

Leopard Whipray — *Himantura undulata* (Bleeker, 1852)

40.16 *Himantura undulata* (Plate 68) Fish Code: 00 035026

- **Field characters:** A large stingray with a whip-like tail that lacks skin folds and with an attractive pattern of leopard-like spots on the dorsal surface. Juveniles have an almost quadrangular disc with a light to medium coverage of black spots.

- **Distinctive features:** Disc quadrangular, trunk deep, slightly longer than wide (width about equal to length in juveniles); pectoral-fin apex narrowly rounded (more angular in juveniles); snout broadly triangular (rather acute in juveniles), tip pointed; anterior margin almost straight. Eyes rather small, length of eye and spiracle about 2.2* in preorbital snout length; interorbital space broad. Mouth narrow; labial furrows weakly developed; lower labial folds and papillae present; lower jaw deeply concave near symphysis. Internasal flap skirt-shaped, smooth, relatively short and broad, margin with fine fringe; nostrils long, narrow. Upper surface granular in adults (smooth in juveniles); one enlarged scapular thorn with an associated thorn patch extending onto nuchal area; no greatly enlarged thorns on midline of tail. Pelvic fins small, subtriangular. Tail very long and slender (over 3.5 times disc width in juveniles when undamaged); slightly depressed or rounded near base, tapering rapidly to sting, whip-like beyond sting; flattened slightly near tip; mostly with one sting; no cutaneous folds.

- **Colour:** Adults sandy brown with dark rings (resembling a leopard's spots) covering most of disc and tail; rings mostly becoming obscure near disc margin. Juveniles greyish to brownish with rather large black spots, the interspaces narrower than the diameter of the largest spots; usually 2–3 spots in interspiracular space. Tail before sting usually with a single row of spots on each side, with black and white bands beyond sting. Ventral surface white.

- **Size:** To at least 140 cm disc width (410 cm total length); born at about 20 cm disc width (92 cm total length).

- **Distribution:** Possibly widespread in the Indo–West Pacific, north to the islands of the Japanese Archipelago (apparently absent from the western Indian Ocean). Locally, known from the inner continental shelf off northern Australia between Ningaloo (Western Australia) and the Torres Strait (Queensland).

- **Remarks:** This species is often confused with the reticulate whipray (40.15). The adults of these species are similar in shape and in denticle structure, but juveniles of the leopard whipray have a different colour pattern, are born at a smaller size and have a more angular disc shape than its relative.

- **Reference:** Masuda *et al.* (1984).

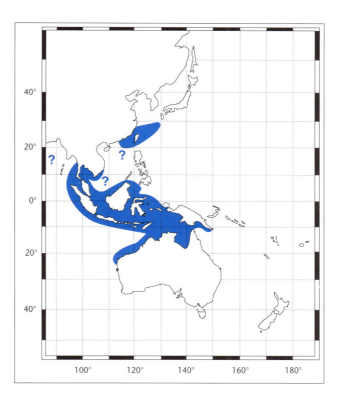

Cowtail Stingray — *Pastinachus sephen* (Forsskål, 1775)

40.17 *Pastinachus sephen* (Plate 70)

Fish Code: 00 035011

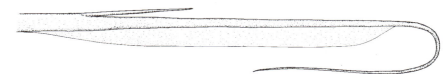

tail (side view)

- **Alternative names:** Banana-tail ray, fantail ray, feathertail stingray, weralli, guergunna.

- **Field characters:** A large, uniformly dark stingray with a dense band of blunt denticles over the central disc. The tail is very broad and flattened anteriorly, has a posteriorly located sting, and the ventral skin fold is very deep but does not reach the tail apex.

- **Distinctive features:** Disc rhomboidal, trunk very thick, width about 1.1–1.3 times length; pectoral-fin apex broadly rounded (more so in juveniles); snout broadly rounded, tip blunt; anterior margin almost straight to slightly convex. Eyes very small, length of eye and spiracle about 1.6 in preorbital snout length; interorbital space very broad. Mouth very narrow, with 5 papillae on floor; labial furrows poorly developed; no lower labial folds, papillae minute; lower jaw truncate. Internasal flap narrow, rather elongate and lobe-like, margin fine, papillae not obvious; nostrils short, narrow. Disc surface with a broad band of small, densely packed, tricuspidate denticles; band extending along mid-disc (beneath eyes) from near snout tip and onto upper surface of tail (only lateral extremity of disc smooth); neonatal juveniles uniformly smooth, denticles developing rapidly after birth; about 4 blunt pearl thorns at disc centre. Pelvic fins of moderate size, broadly rounded. Tail very broad-based and depressed; tapering gradually to sting; slender and almost cylindrical behind sting; moderately elongate (about twice disc width or less when undamaged); mostly with one sting; ventral cutaneous fold very well developed (several times taller than tail height above), terminating abruptly about 2 sting lengths behind sting tip; no dorsal skin fold; tip of tail filamentous.

- **Colour:** Upper surface uniform dark greyish brown to black; tail fold and tail tip black. Ventral surface mostly white.

- **Size:** Adults to at least 180 cm disc width (exceeding 300 cm total length); born at about 18 cm disc width (50 cm total length) or larger.

- **Distribution:** Widespread in the Indo–Pacific from the Red Sea to Australia, including Melanesia and Micronesia. Locally, around northern Australia from Shark Bay (Western Australia) to the Clarence River (New South Wales). Also reported from South Africa. Inshore to 60 m or more.

- **Remarks:** Common inshore in tropical Australian seas. Large adults, immediately recognisable by their characteristic tail fold, are often encountered by divers in and around coral lagoons. They tend to be inquisitive and will approach closely if fish have been speared. Adults are sometimes accompanied by remoras or members of the trevally family. This ray also ventures into estuaries and freshwater.

- **Local Synonymy:** *Pastinachus sephen ater* Macleay, 1883.

- **Reference:** Grant (1978).

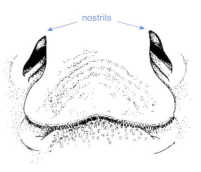

region of nostrils and mouth

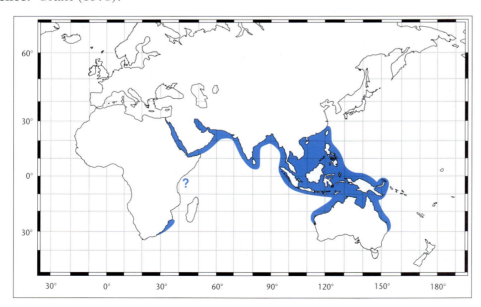

Blue-spotted Fantail Ray — *Taeniura lymma* (Forsskål, 1775)

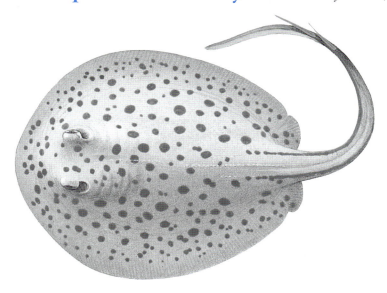

40.18 *Taeniura lymma* (Plate 70)

Fish Code: 00 035009

tail (side view)

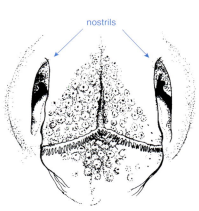

region of nostrils and mouth

- **Alternative names:** Blue-spotted lagoon ray, blue-spotted ribbontail ray, lagoon ray, lesser fantail ray, reef ray.
- **Field characters:** A small stingray with a smooth, oval disc with blue spots on the dorsal surface and blue stripes along the tail. The ventral skin cutaneous fold is deep and extends to the tail tip.
- **Distinctive features:** Disc oval, trunk robust, width about 0.8* times length; snout bluntly rounded, tip not pointed; anterior margin slightly convex. Eyes moderately large, length of eye and spiracle about 2* in preorbital snout length; interorbital space narrow. Mouth small, with 2 large central papillae on floor; labial furrows deep, lower lip and internasal flap with prominent papillae. Internasal flap narrow, elongate; posterior margin deeply concave with prominent fringe; lower jaw slightly concave near symphysis; nostrils slender, narrow. Disc smooth except for small patch of widely spaced thorns in nuchal area; sometimes with a short row posteriorly. Pelvic fins moderate, slender. Tail rather broad-based, depressed, relatively short (about 1.5 times disc width when undamaged), tapering rapidly to sting, slightly compressed beyond sting; sting located well back on tail; mostly with 2 stings; ventral cutaneous fold deep, long, extending to tail tip; dorsal midline with very low fleshy ridge.
- **Colour:** Dorsal surface brownish or yellowish brown; disc and pelvic fins covered in blue spots; largest spots about 0.5 of eye diameter. Ventral surface mostly uniformly white. Tail lacking spots, with long blue stripe midlaterally before sting; blue posteriorly.
- **Size:** To a disc width of at least 30 cm (70 cm total length). Reports of specimens reaching 240 cm total length are probably inaccurate.
- **Distribution:** Widespread in the Indo–West Pacific from southern Africa to at least the Solomon Islands (including the Red Sea). Not recorded from the Japanese Archipelago. Locally, throughout tropical Australia from Ningaloo Reef (Western Australia) to Bundaberg (Queensland) in depths to at least 20 m.
- **Remarks:** Possibly the most abundant stingray in coral reef habitats of tropical Australia. It migrates *en masse* into shallow sandy areas during the rising tide to feed

on molluscs, and then disperses on the falling tide to seek shelter in caves and under ledges. Members of the genus *Taeniura* are referred to here as fantail rays.

- **Reference:** Grant (1978).

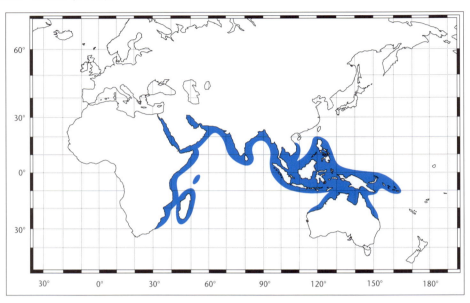

Blotched Fantail Ray — *Taeniura meyeni* Müller & Henle, 1841

40.19 *Taeniura meyeni* (Plate 70)   **Fish Code:** 00 035017

tail (side view)

- **Alternative names:** Black-blotched stingray, black-spotted stingray, giant reef ray, round ribbontail ray.

- **Field characters:** A large stingray with a circular disc, no thorns, a black and white mottled upper surface, and a deep and prominent ventral skin fold that extends to the tail tip.
- **Distinctive features:** Disc subcircular, trunk robust, slightly wider than long; margin of disc uniformly convex, tip not pointed. Eyes moderate, length of eye and spiracle about half preorbital snout length; interorbital space broad. Mouth rather broad, with 7 short papillae on floor (5 centrally); labial furrows and folds weakly developed. Internasal flap short, very broad, papillae absent, margin with short fringe; lower jaw uniformly convex; nostrils large, oval. Disc surface covered uniformly with short widely spaced granulations; no thorns. Pelvic fins small, elongate. Tail relatively broad-based, depressed, tapering rapidly at sting, slender and compressed beyond sting; short (slightly longer than disc width when undamaged); generally with one sting; ventral cutaneous fold very well developed and extending to tail tip; several times deeper than tail above; dorsal fold absent.
- **Colour:** Dorsal surface mottled black and white, sometimes brownish; tail uniformly black behind sting. Ventral surface of disc uniformly pale; margins and undersurface of tail greyish brown to black.
- **Size:** Maximum disc width at least 180 cm (about 330 cm total length); free-swimming by 35 cm disc width (67 cm total length).
- **Distribution:** Indo–West Pacific, from the Red Sea to tropical Australia (including the Japanese Archipelago, Lord Howe Island and Micronesia). Also occurs along the eastern African coast to Natal (including Madagascar). Not well known locally but records include northern Australia from Ningaloo (Western Australia) to Townsville (Queensland).
- **Remarks:** A large stingray with a distinctive disc shape and colour pattern, and a long, broad ventral skin fold. Possibly widespread along the Great Barrier Reef, although it was only recently recorded from the Australian region. More widely known as *T. melanospila*, the scientific name used here was based on the description of a juvenile specimen. Although not normally aggressive, it has been responsible for at least one human fatality.
- **Local synonymy:** *Taeniura melanospilos* Bleeker, 1853; *Taeniura mortoni* Macleay, 1883.
- **References:** Bleeker (1853a); Coleman (1981).

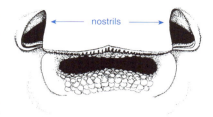

region of nostrils and mouth

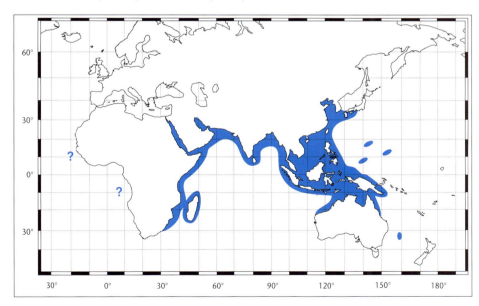

Porcupine Ray — *Urogymnus asperrimus* (Bloch & Schneider, 1801)

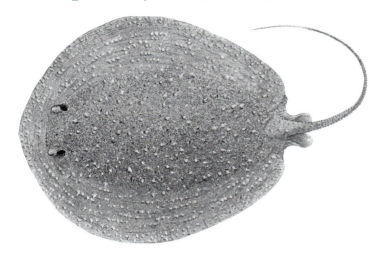

40.20 *Urogymnus asperrimus* (Plate 70)   **Fish Code:** 00 035027

- **Alternative names:** Solanders ray, roughskin stingaree, thorny ray.
- **Field characters:** An oval-shaped ray with an extremely rough upper surface consisting of plate-like denticles and sharp thorns (also present on outer disc). The dark tail lacks stinging spines and skin folds.
- **Distinctive features:** Disc oval, trunk very robust; slightly longer than wide; snout tip blunt. Eyes small, length of eye and spiracle 1.5–2 in preorbital snout length. Mouth narrow, with 3–5 papillae on floor; labial furrows and papillae prominent. Internasal flap skirt-shaped, papillate, margin with prominent fringe; nostrils elongate, narrow. Upper disc extremely prickly; denticles flat, plate-like, in dense patch on raised central part of disc and on tail; interspersed with taller and sharper, more widely spaced thorns that extend laterally toward disc margin (no enlarged thorns in small specimens); no outsized scapular thorns. Pelvic fins small, slender. Tail about equal to disc length, almost cylindrical in cross-section, tapering rapidly; no stinging spine or cutaneous folds.
- **Colour:** Dark brown to greyish above, white below; tail tip dark.
- **Size:** To at least 100 cm disc width.
- **Distribution:** Widespread in the western Pacific and Indian Oceans from Natal (South Africa) to Fiji (apparently not found in the Japanese Archipelago). Also in

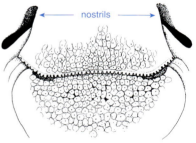

nostrils and mouth

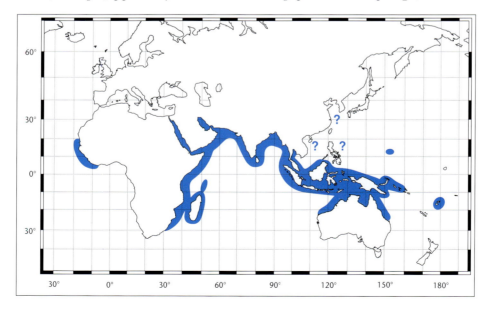

the tropical eastern Atlantic, off central Africa. Australian records are based on anecdotal information, photographs (from the Great Barrier Reef) and live sightings (off Western Australia). Probably more widespread locally.

- **Remarks:** Little is known of the biology of this highly distinctive ray and its occurrence locally is based on very few verified accounts. Rather surprisingly, a ray captured by James Cook off the Endeavour River (Queensland) in 1770 appears to have been this species. Recent records from Western Australia are based on two adults seen on seagrass inside the main reef off Ningaloo. Specimens and distributional records are needed because more than one member of this genus may exist. If caught, adults should be handled carefully because the thorns are particularly sharp.
- **Local synonymy:** *Raja africana* Bloch & Schneider, 1801; *Urogymnus asperrimus solanderi* Whitley, 1939.
- **Reference:** Whitley (1939b).

41 Family Urolophidae
STINGAREES

Stingarees resemble stingrays in body shape, but have a shorter tail with a well-developed caudal fin. They are small to medium rays (0.3–2 m in total length) with oval, almost circular or rhomboidal discs, and the head is not prominently elevated above the pectoral fins. The mouth, which is located ventrally, has numerous small teeth and usually has several papillae on its floor. All Australian species, apart from one, lack denticles and thorns. The tail is slender (not whip-like), about equal in length to the disc, may have a small dorsal fin, and normally has one or more serrated stinging spines. Skin folds may be present laterally but are absent from the dorsal and ventral midlines.

The generic relationships of this family, which contains about 40 species, are not fully understood. Three of the currently recognised genera are represented in the Australian region by 22 species, including several new or recently described rays. Some of these are difficult to identify on external features alone. Most occur demersally on the continental shelf but some species occur down the continental slope to at least 700 m in this region. Many live inshore and some venture into estuaries. Stingarees are viviparous, with litters of 2–4 that take about 3 months to gestate.

Key to urolophids

1. Upper surface covered with small, close-set denticles; snout long, more than 6 times orbit diameter (fig. 1) .... *Plesiobatis daviesi* (fig. 1; p. 420)

   Upper surface uniformly smooth, without denticles; snout rather short, much less than 6 times orbit diameter (fig. 6) **2**

2. A broad flattened fleshy lobe on mid-lateral margin of nostril (fig. 2) **3**

   No broad flattened fleshy lobe on mid-lateral margin of nostril (fig. 3) **8**

3. Upper surface of disc with distinct dark markings centrally and around eyes (figs 6, 7); dorsal fin present (fig. 4) **4**

   Upper surface of disc with a uniform colour or pattern, lacking distinct dark markings centrally and around eyes; dorsal fin absent (or if present its height is less than the fin's base length) (fig. 5) **5**

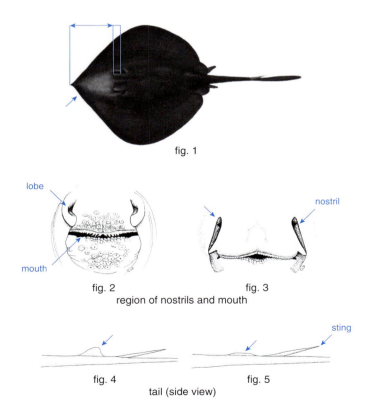

fig. 1

region of nostrils and mouth
fig. 2   fig. 3

tail (side view)
fig. 4   fig. 5

4   Disc oval (fig. 6); central disc with a pair of dark blotches which extend as dark stripes posteriorly along disc and tail (fig. 6) ....
    .... *Trygonoptera ovalis* **(fig. 6; p. 425)**

    Disc almost circular (fig. 7); central disc with a broad dark blotch which does not extend posteriorly along disc and tail (fig. 7) ....
    .... *Trygonoptera personata* **(fig. 7; p. 426)**

5   A low dorsal fin or narrow ridge of skin present before the stinging spine(s) (fig. 5); central eastern Australia .... *Trygonoptera testacea* **(fig. 8; p. 427)**

    No evidence of a dorsal fin before the stinging spine(s); southern or western Australia **6**

6   Disc angular at snout tip (fig. 9); pale to medium yellowish above; usually with 9 or more oral papillae; central western Australia ....
    .... *Trygonoptera* sp. A **(fig. 9; p. 422)**

    Disc not especially angular at snout tip (figs 13, 14); greyish, dark brown or black (sometimes yellowish) above; usually with less than 9 oral papillae **7**

7   Distance from snout tip to cranium several times more than diameter of nasal capsules (fig. 10); small to medium-sized, shorter than 50 cm; southern and southwestern Australia ....
    .... *Trygonoptera mucosa* **(fig. 13; p. 424)**

    Distance from snout tip to cranium about equal to diameter of nasal capsules (fig. 11); large, exceeds 50 cm; southeastern Australia ....
    .... *Trygonoptera* sp. B **(fig. 14; p. 423)**

8   Prominent lobes on anterior borders of internasal flap (fig. 12) .... *Urolophus lobatus* **(fig. 15; p. 436)**

    No prominent lobes on anterior borders of internasal flap (fig. 3) **9**

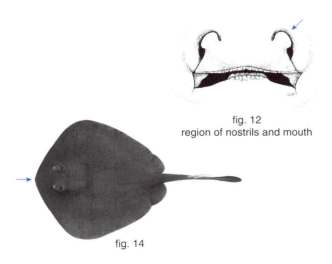

fig. 12
region of nostrils and mouth

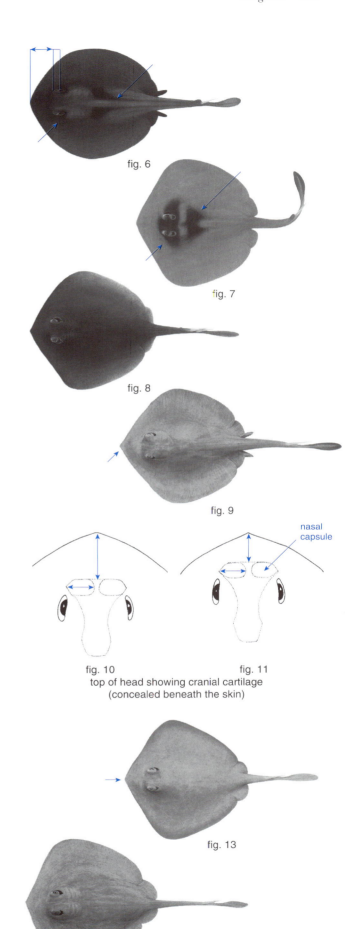

fig. 6

fig. 7

fig. 8

fig. 9

fig. 10   fig. 11
top of head showing cranial cartilage
(concealed beneath the skin)

fig. 13

fig. 14   fig. 15

9 Tail oval to almost circular in cross-section near its base (fig. 16); no lateral skin folds **10**

   Tail not as above, flattened slightly above and below (fig. 17); lateral skin folds present (sometimes weak) (fig. 17) **14**

10 Dorsal fin present (more than just a low skin fold) (fig. 4); outer margin of upper surface of disc spotted (sometimes faint) (figs 18, 19) **11**

   Dorsal fin absent (rarely there may be a low skin fold) (fig. 5); outer margin of upper surface of disc never spotted **12**

11 Central disc with an indistinct ring of large, white-edged, black spots (fig. 18) .... ***Urolophus circularis*** (**fig. 18; p. 431**)

   Central disc with aggregations of small, white spots (sometimes faint) (fig. 19) .... ***Urolophus gigas*** (**fig. 19; p. 435**)

12 No dark patches around eyes and on central disc (upper surface mostly plain yellow) .... ***Urolophus sufflavus*** (**fig. 20; p. 440**)

   Dark markings on upper surface, prominent around eyes and on central disc (figs 23, 26) **13**

13 Dark stripe along middle of disc and tail (fig. 23); dark markings intricate and usually connected; caudal fin very short (fig. 21) .... ***Urolophus cruciatus*** (**fig. 23; p. 432**)

   No dark stripe along middle of disc and tail; dark markings simple and unconnected (fig. 26); caudal fin rather elongate (fig. 22) .... ***Urolophus orarius*** (**fig. 26; p. 438**)

14 Internasal flap bell-shaped, margin bulging near nostrils (fig. 24) **15**

   Internasal flap skirt-shaped (fig. 25) **16**

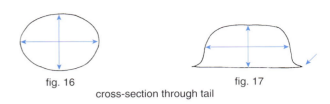

fig. 16   fig. 17
cross-section through tail

fig. 18

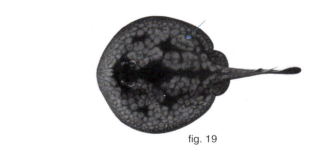

fig. 19

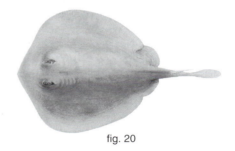

fig. 20

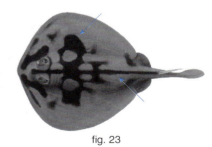

fig. 23

fig. 26

fig. 21   fig. 22
caudal fins

fig. 24   fig. 25
region of nostrils and mouth

15 Dorsal fin absent; no prominent dark marking beside eye (sometimes with a V-shaped bar between eyes, fig. 27); Victorian and Tasmanian specimens usually with two or more white spots on disc (fig. 27) .... ***Urolophus paucimaculatus*** (**fig. 27; p. 439**)

Dorsal fin present (fig. 29); prominent dark blotch usually present beside eye (fig. 28); no white spots on disc .... ***Urolophus* sp. A (fig. 28; p. 428)**

16 Prominent dorsal fin present (fig. 29) **17**

Dorsal fin absent or reduced to a low skin fold (fig. 30) **19**

17 Upper surface plain coloured or with a peppering of fine brownish spots (fig. 31); tail moderately elongate (usually longer than 76% of disc length) (fig. 31) .... ***Urolophus* sp. B (fig. 31; p. 429)**

Upper surface with an ornate pattern of white spots or reticulations (rarely faint) (figs 32, 33); tail relatively short (usually shorter than 76% of disc length) (fig. 32) **18**

18 Upper surface with fine reticulations and spots (fig. 32); spots on centre of disc not appreciably larger and not more widely spaced than those near edge of disc; southeastern Australia .... ***Urolophus bucculentus* (fig. 32; p. 430)**

Upper surface with a honeycomb pattern of large, white spots with dark borders (fig. 33); spots on centre of disc appreciably larger and more widely spaced than those near edge of disc; north-eastern and northwestern Australia .... ***Urolophus flavomosaicus* (fig. 33; p. 434)**

19 Upper surface with a very distinctive pattern of large, widely spaced blotches and lines each composed of fine spots (fig. 34); central disc mostly greenish, reddish brown near margin, markings bluish ....***Urolophus mitosis* (fig. 34; p. 437)**

Not as above; plain or with faint and simple markings **20**

20 Pale brownish or yellowish; usually with a low skin fold before stinging spine (fig. 30) .... ***Urolophus westraliensis* (fig. 35; p. 442)**

Greenish or greyish; no skin fold before stinging spine **21**

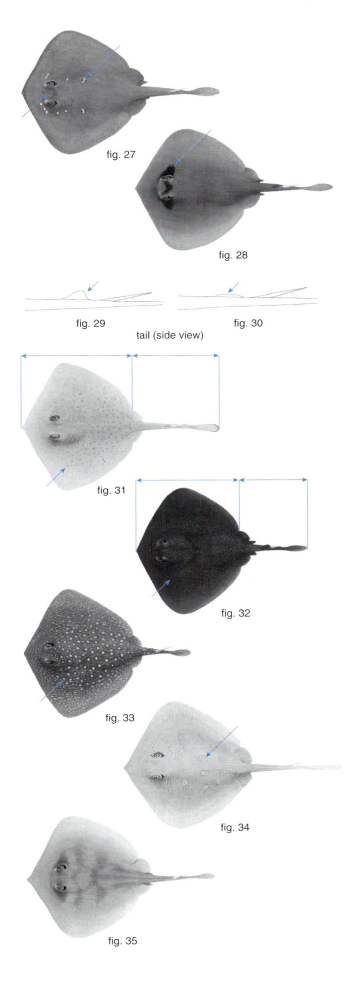

fig. 27

fig. 28

fig. 29   fig. 30
tail (side view)

fig. 31

fig. 32

fig. 33

fig. 34

fig. 35

21 Disc relatively broad (length usually less than 90% width); 2–3 greyish transverse bands on disc (sometimes faint) (fig. 36); southwestern Australia and Great Australian Bight ....
.... ***Urolophus expansus* (fig. 36; p. 433)**

Disc slightly narrower (length mostly more than 90% width); no transverse bands on disc; southeastern Australia .... ***Urolophus viridis* (fig. 37; p. 441)**

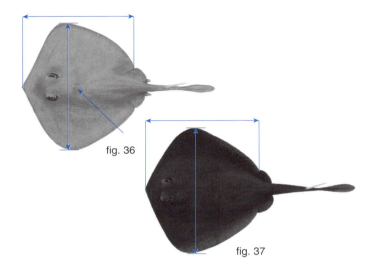

Giant Stingaree — *Plesiobatis daviesi* (Wallace, 1967)

41.1 *Plesiobatis daviesi* (Plate 73) **Fish Code:** 00 038023

- **Alternative name:** Deepwater stingray.
- **Field characters:** A very large, greyish brown to black deepwater stingaree with a very long snout, a skirt-shaped internasal flap, granular skin on the tail and dorsal surface of the disc, no oral papillae, and lacking both tail folds and a dorsal fin.
- **Distinctive features:** Disc not especially broad, oval, usually longer than wide; broadest part 2–4 eye diameters behind level of spiracles; anterior profile obtuse.

Snout long, thin, tip extended into lobe. Eye small (10–14*% preocular snout length). Posterior margin of spiracle usually angular. Mouth very large; papillae absent from floor. Internasal flap very broad, rectangular, posterior angle not extended into distinct lobe; fringe well developed. Nostrils almost circular, posterolateral border not forming a broad, flattened, fleshy lobe; no papillae below mouth. Disc covered in fine denticles (less pronounced in juveniles). Tail slightly depressed to oval in cross-section; covered in fine denticles; long (93–102*% disc length); no lateral cutaneous tail folds or dorsal fin; caudal fin extremely long, slender.

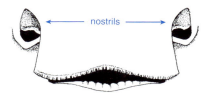

region of nostrils and mouth

- **Colour:** Dorsal surface and caudal fin mostly greyish brown to black. Ventral surface mainly white, with narrow dark grey or black lateral margin. Tail greyish or black, darker distally. Caudal fin black.

- **Size:** Reaches at least 200 cm in Australian waters; elsewhere reported to attain 270 cm. Smallest mature male examined was 130 cm.

- **Distribution:** Southeastern Africa and western Indian and west–central Pacific Oceans. Collected mainly from tropical Australia, between Wooli (New South Wales) and Townsville (Queensland) in the east, and Shark Bay and Rowley Shoals (Western Australia), in depths of 350–680 m.

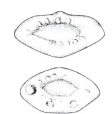

teeth from near symphysis (upper and lower jaw)

- **Remarks:** This is the largest and most widespread member of the family and also lives in deeper water than any of its relatives. It is distinct in shape from all other Australian stingarees. Some ichthyologists have chosen to place this ray alone in a separate family.

- **Local synonymy:** *Urotrygon* sp. 1: Gloerfelt-Tarp & Kailola, 1984.

- **Reference:** Wallace (1967a).

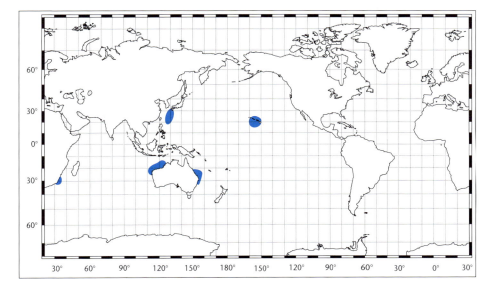

Yellow Shovelnose Stingaree — *Trygonoptera* sp. A

41.2 *Trygonoptera* sp. **A** (Plate 78)   **Fish Code:** 00 038013

- **Field characters:** A small to medium-sized, yellowish brown stingaree with a relatively acute snout, broad lobes on the posterolateral border of the nostrils, a skirt-shaped internasal flap, usually more than eight papillae on the floor of the mouth, and no dorsal fin or tail folds.

- **Distinctive features:** Disc not especially broad, subcircular to rhomboidal, wider than long; broadest about an eye diameter behind level of spiracles; anterior profile obtuse. Snout fleshy, tip not extended. Eye of moderate size (24–26% preocular snout length). Posterior margin of spiracle usually angular. Mouth small; 9–10 papillae on floor; lower jaw papillate. Internasal flap skirt-shaped, posterior angle not extended into distinct lobe; fringe prominent. Posterolateral border of nostril forming a broad, flattened, fleshy lobe. Disc upper surface smooth. Tail slightly depressed to rounded in cross-section; of moderate length (71–87% disc length); no lateral cutaneous tail folds or dorsal fin; caudal fin lanceolate.

- **Colour:** Dorsal surface uniform yellowish brown. Caudal fin greyish to greyish brown in adults. Ventral surface white or pale yellowish brown; sometimes with darker greyish brown lateral margins on posterior half of disc.

- **Size:** Reaches about 44 cm. Smallest mature male examined was 32 cm.

- **Distribution:** From the Abrolhos Islands to Shark Bay (Western Australia) in about 200 m.

- **Remarks:** This species is closely related to two other rays referred to here as shovelnose stingarees. The western shovel-nose stingaree (41.4) also occurs off Western Australia. The yellow shovelnose stingaree differs from the latter in upper surface colour, disc shape, and possibly in meristics. Further research is required to find more reliable characters for identifying each of the three shovelnose stingarees.

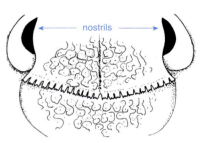

region of nostrils and mouth

teeth from near symphysis (upper and lower jaw)

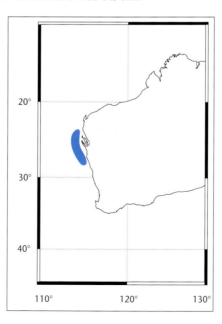

Eastern Shovelnose Stingaree — *Trygonoptera* sp. **B**

41.3 *Trygonoptera* sp. **B** (Plate 78)   Fish Code: 00 038014

- **Field characters:** A large, medium to dark brown stingaree with a thick body, broad lobes on the posterolateral border of the nostrils, a skirt-shaped internasal flap, fewer than eight papillae on the floor of the mouth, and lacking both a dorsal fin and tail folds.

- **Distinctive features:** Disc not especially broad, subcircular, wider than long; broadest part 1.5–2 eye diameters behind level of spiracles; anterior profile obtuse. Snout fleshy, tip not extended. Eye of moderate size (20–23*% preocular snout length). Posterior margin of spiracle angular. Mouth small; about 6 papillae on floor; lower jaw papillate. Internasal flap skirt-shaped, posterior angle not extended into distinct lobe; fringe prominent. Posterolateral border of nostril forming a broad, flattened, fleshy lobe. Disc upper surface smooth. Tail depressed and distinctly oval in cross-section; of moderate length (81–89*% disc length); no lateral cutaneous tail folds or dorsal fin; caudal fin broadly lanceolate.

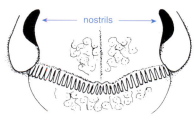

region of nostrils and mouth

- **Colour:** Dorsal surface medium to dark brown, often darker on central disc and tail; sometimes with a few irregularly distributed small, brownish or yellowish spots. Caudal-fin distal margin sometimes dark. Ventral surface white with broad, dark lateral margins; paler in larger adults.

- **Size:** Reaches at least 80 cm; males mature at about 56 cm.

- **Distribution:** Southeastern Australia from Beachport (South Australia) to Bermagui (New South Wales), excluding Tasmania, in coastal areas to at least 120 m in depth.

- **Remarks:** This large coastal species has been confused in the past with the smaller common stingaree (41.7), from which it differs externally in lacking a dorsal fin. It also superficially resembles the much smaller western shovel-nosed stingaree (41.4), which also lacks a dorsal fin, but appears to be different internally. This stingaree is common in Port Phillip and Western Port bays (Victoria). Like most other Australian stingarees, little is known of its biology. The flesh is edible but tends to be chewy unless prepared properly.

- **Local synonymy:** *Urolophus testaceus* : Scott *et al.*, 1980 (misidentification).

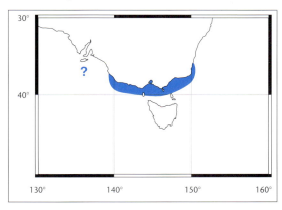

Western Shovelnose Stingaree — *Trygonoptera mucosa* Whitley, 1939

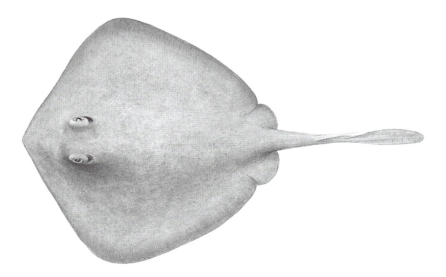

41.4 *Trygonoptera mucosa* (Plate 78) Fish Code: 00 038015

- **Alternative names:** Kejetuck, bebil, western stingaree.

- **Field characters:** A small to medium-sized, uniformly dark stingaree with 7–9 oral papillae, a skirt-shaped internasal flap, broad lobes on the posterolateral borders of the nostrils, and lacking both a dorsal fin and tail folds.

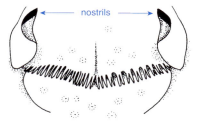

region of nostrils and mouth

- **Distinctive features:** Disc not especially broad, subcircular, slightly wider than long; broadest part about 1.5–2 eye diameters behind level of spiracles; anterior profile obtuse. Snout fleshy, tip rounded. Eye of moderate size (20–27% preocular snout length). Posterior margin of spiracle angular. Mouth small; about 7–9 small unbranched papillae on floor; lower jaw papillate. Internasal flap skirt-shaped, posterior angle not extended into distinct lobe; fringe prominent; posterolateral border of nostril forming a broad flattened, fleshy lobe. Disc upper surface smooth. Tail almost oval in cross-section; rather elongate (71–91% disc length); no lateral cutaneous tail folds or dorsal fin; caudal fin lanceolate.

- **Colour:** Upper surface greyish, yellowish brown or brownish black, sometimes with irregular scattered yellow and dusky spots. Caudal fin mostly dusky or black. Ventral surface white or yellowish, sometimes with very broad, dark brown lateral margins and blotches (more pronounced in juveniles) on disc and tail.

- **Size:** Reaches at least 44 cm; males are mature from about 30 cm.

- **Distribution:** Southwestern Australia from Glenelg (South Australia) to at least Dongara (Western Australia) in depths to 20 m.

- **Remarks:** This ray is quite common in shallow water on sandy substrates or near seagrasses in the Great Australian Bight. Similar in form to the other undescribed shovelnose stingarees (41.2 and 41.3), specimens from a wide range of geographic areas need to be collected and examined carefully to determine the relationships between these forms.

- **References:** Whitley (1939b); McKay (1966); Coleman (1974).

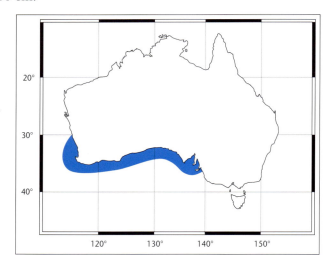

Striped Stingaree — *Trygonoptera ovalis* Last & Gomon, 1987

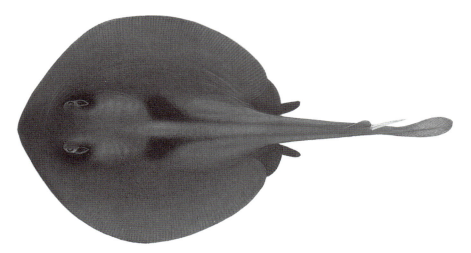

41.5 *Trygonoptera ovalis* (Plate 77) Fish Code: 00 038016

- **Field characters:** A small to medium-sized, greyish to greyish brown stingaree with an oval-shaped disc and a pair of dark longitudinal stripes extending along the central disc and tail. It has a skirt-shaped internasal flap, broad lobes on the posterolateral borders of the nostrils, a dorsal fin, but lacks tail folds.

- **Distinctive features:** Disc almost oval, slightly longer than wide; broadest part 1–3 eye diameters behind level of spiracles; anterior profile obtuse. Snout fleshy, tip rounded, not extended. Eye of moderate size (17–22% preocular snout length). Posterior margin of spiracle usually angular. Mouth small; about 4 minute papillae on floor; lower jaw strongly papillate. Internasal flap skirt-shaped, posterior angle not usually extended into distinct lobe; fringe very long. Posterolateral border of nostril forming a broad, flattened and fleshy lobe. Disc upper surface smooth. Tail oval in cross-section, more depressed anteriorly; of moderate length (75–100% disc length); no lateral cutaneous tail folds; dorsal fin small; caudal fin relatively large, short and deep.

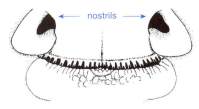

region of nostrils and mouth

- **Colour:** Dorsal surface greyish to greyish brown; dark mask-like markings around eyes, also often with a dark stripe extending to snout tip (more obvious on juveniles); paired dark blotches near centre of disc extend as narrow stripes posteriorly along central disc and tail (pale along midline between stripes); sometimes dark markings obscure. Caudal fin greyish or black, with darker margin. Ventral surface whitish; tail, margins of disc and ventral fins mostly dark.

- **Size:** Reaches at least 61 cm; males mature at about 35 cm.

- **Distribution:** Occurs off the southern coast of Western Australia from the Abrolhos Islands to Eucla in depths to 43 m.

- **Remarks:** This attractive little ray differs from other members of the genus with a dorsal fin, the common (41.7) and masked (41.6) stingarees, in having a more oval-shaped disc and a distinctive pair of dark longitudinal stripes on the dorsal midline. Trawled occasionally but little is known of its biology.

- **Local synonymy:** *Urolophus testaceus*: Hutchins, 1979 (misidentification); *Urolophus* sp. 1: Hutchins & Thompson, 1983; *Urolophus* sp.: Hutchins & Swainston, 1986.

- **References:** Hutchins and Swainston (1986); Last and Gomon (1987).

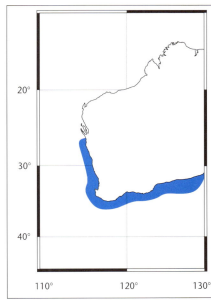

Masked Stingaree — *Trygonoptera personata* Last & Gomon, 1987

41.6 *Trygonoptera personata* (Plate 77)   Fish Code: 00 038017

- **Field characters:** A small stingaree with a yellowish to greyish upper surface, dark markings around the eyes, a central dark blotch on the disc, a skirt-shaped internasal flap, broad lobes on the posterolateral border of the nostrils, a dorsal fin, but lacks tail folds.

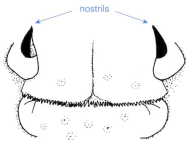

region of nostrils and mouth

- **Distinctive features:** Disc not especially broad, subcircular, length equal to or slightly longer than width; broadest part slightly more than eye diameter behind level of spiracles; anterior profile obtuse. Snout fleshy, tip not extended. Eye of moderate size (21–28% preocular snout length). Posterior margin of spiracle usually angular. Mouth very small; 3–4* papillae on floor; lower jaw weakly papillate. Internasal flap skirt-shaped, posterior angle not extended into distinct lobe; fringe prominent. Posterolateral border of nostril forming a broad, flattened, fleshy lobe. Disc upper surface smooth. Tail slightly depressed to oval in cross-section; of moderate length (67–86% disc length); no lateral cutaneous tail folds; dorsal fin rather large; caudal fin moderately elongate (lanceolate in young).

- **Colour:** Dorsal surface yellowish brown to greyish with continuous dark mask-like markings around and between eyes, and a large dark blotch on central disc; these patches frequently connected medially and laterally by thin lines. Caudal-fin margin and dorsal fin black in young, greyish in adults. Ventral surface white; dusky along lateral margins.

- **Size:** Attains at least 47 cm. Smallest mature male examined was 36 cm.

- **Distribution:** Known to occur between Bunbury and Shark Bay (Western Australia), inshore to depths of at least 70 m.

- **Remarks:** Readily distinguished from close Western Australian relatives, the striped (41.5) and western shovelnose stingarees (41.4), by its colour pattern, disc shape and dorsal-fin size. Little is known of its biology.

- **Local synonymy:** *Urolophus* sp. 2: Hutchins & Thompson, 1983; *Urolophus* sp.: Hutchins & Swainston, 1986.

- **References:** Hutchins and Swainston (1986); Last and Gomon (1987).

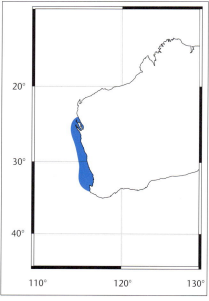

Common Stingaree — *Trygonoptera testacea* Müller & Henle, 1841

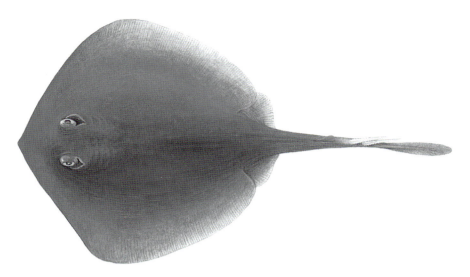

41.7 *Trygonoptera testacea* (Plate 78)   Fish Code: 00 038006

- **Alternative name:** Stingaree.
- **Field characters:** A medium-sized, dark brown to grey, coastal stingaree with a skirt-shaped internasal flap, broad lobes on the posterolateral border of the nostrils, a dorsal fin (sometimes as a thin skin fold), and no tail folds.
- **Distinctive features:** Disc not especially broad, subcircular, slightly wider than long; broadest about 1–1.5 eye diameters behind level of spiracles; anterior profile obtuse. Snout fleshy, tip sometimes extended slightly. Eye of moderate size (21–27% preocular snout length). Posterior margin of spiracle angular. Mouth small; 3–5 papillae on floor; lower jaw papillate. Internasal flap skirt-shaped, posterior angle not extended into distinct lobe; fringe prominent. Posterolateral border of nostril forming a broad, flattened, fleshy lobe. Disc upper surface smooth. Tail slightly depressed (anteriorly) to oval in cross-section; moderately elongate (83–96% disc length); lateral cutaneous tail folds absent; dorsal fin small (rarely reduced to a low fold); caudal fin lanceolate, elongate.

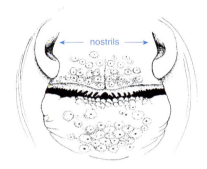

region of nostrils and mouth

- **Colour:** Upper surface uniform brownish to greyish, paler near edges. Caudal fin black in juveniles, paler greyish or brownish in adults; extreme edge darkest. Dorsal fin greyish. Ventral surface white, sometimes with broad, dusky lateral margins.
- **Size:** Reaches at least 47 cm; males mature at about 31 cm.
- **Distribution:** Eastern Australia from Cape Howe (New South Wales) to Caloundra (southern Queensland). Common in estuaries and shallow coastal waters in depths less than 60 m (rarely to 135 m).
- **Remarks:** This stingaree is common around the coast of eastern Australia where it may venture well upstream in estuaries. It appears to favour sand and reef habitats near the coast but is often caught by trawlers further offshore. It can be distinguished from the other eastern member of the genus, the eastern shovelnose stingaree (41.3), by the presence of a dorsal fin, which is sometimes reduced to a very thin fold of skin. Earlier reports of this species attaining 76 cm may have been misidentifications of the latter.
- **Local synonymy:** *Trygonoptera muelleri* Steindachner, 1866; *Trygonoptera henlei* Steindachner, 1866; *Trygonoptera australis* Steindachner, 1866.
- **Reference:** McCulloch (1916).

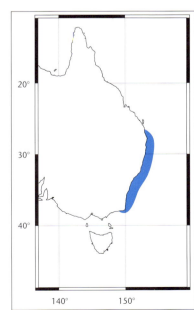

Kapala Stingaree — *Urolophus* sp. A

41.8 *Urolophus* sp. A (Plate 74) Fish Code: 00 038018

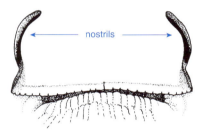

region of nostrils and mouth

- **Field characters:** A medium-sized, greenish stingaree with dark markings around the eyes, a bell-shaped internasal flap, a dorsal fin, tail folds, but no broad lobes on the posterolateral border of the nostrils.

- **Distinctive features:** Disc not especially broad, subcircular to rhomboidal, wider than long; broadest about 1–2 eye diameters behind level of spiracles; anterior profile obtuse. Snout fleshy, tip slightly extended. Eye small to moderate in size (19–28% preocular snout length). Posterior margin of spiracle rounded to slightly angular. Mouth small; 5–6 papillae on floor; short patch of papillae beneath mouth. Internasal flap bell-shaped, posterior angle extended into distinct lobe; fringe weak. Posterolateral border of nostril forming a weak knob, not forming a broad flattened, fleshy lobe. Disc upper surface smooth. Tail very depressed in cross-section; rather elongate (75–100% disc length); lateral cutaneous folds rather well developed; dorsal fin long based and low; caudal fin usually broadly lanceolate.

- **Colour:** Dorsal surface uniform greenish with a pinkish outer disc; dark blotches beneath each eye and a broad, V-shaped interorbital bar; mostly with additional, large dark blotches above pelvic-fin bases. Caudal fin dark in juveniles, paler with a dark margin in adults. Ventral surface whitish with dusky lateral margins.

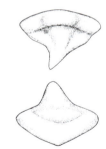

teeth from near symphysis (upper and lower jaw)

- **Size:** Reaches at least 45 cm; males mature at about 28 cm.

- **Distribution:** Eastern Australia off the New South Wales coast between Yamba and Jervis Bay, inshore in depths of 18–85 m.

- **Remarks:** Resembles the sparsely-spotted stingaree (41.19), except that it always has a dorsal fin, lacks white spots on the dorsal surface, and has a distinctive purplish pink disc margin. It also has a narrower distributional range and appears to be confined to New South Wales.

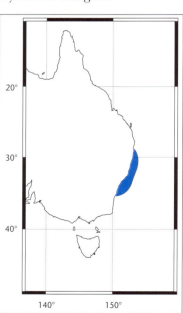

Coral Sea Stingaree — *Urolophus* sp. **B**

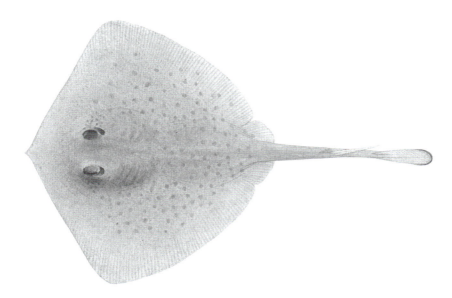

41.9 *Urolophus* sp. **B** (Plate 77)　　　　　　　　　　　　　　　　**Fish Code:** 00 038019

- **Field characters:** A medium-sized, pale brown to grey stingaree with a skirt-shaped internasal flap, a dorsal fin, sometimes with weak tail folds, but no broad lobes on the posterolateral border of the nostrils.

- **Distinctive features:** Disc somewhat broad, rhomboidal, wider than long; broadest about an eye diameter or less behind level of spiracles; anterior profile obtuse. Snout fleshy, tip angular and distinctly extended. Eye moderate to large (22–26% preocular snout length). Posterior margin of spiracle slightly angular to rounded. Mouth of moderate width; about 7–9* papillae on floor; lower jaw very weakly papillate. Internasal flap skirt-shaped, posterior angle extended into weak lobe; fringe very short. Posterolateral border of nostril straight (not knob-like or lobed). Disc upper surface smooth. Tail depressed (anteriorly) to oval in cross-section; rather short (76–85% disc length); lateral cutaneous tail folds weak or absent; dorsal fin low, extending over origin of spine; caudal fin short and deep.

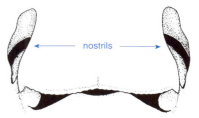

region of nostrils and mouth

- **Colour:** Dorsal surface pale brownish or greyish; either uniform or with a few minute, dark brown spots (most dense in juveniles). Tail usually with a dark brown median stripe in juveniles. Caudal and dorsal fins brownish, margins usually dark. Ventral surface of disc white; lateral margins sometimes distinctly dark.

- **Size:** Reaches at least 48 cm; smallest mature male examined was 23 cm; born at about 10 cm.

- **Distribution:** Northern Queensland from Mackay to Cairns (including the Saumarez and Marion Reefs) in 280–350 m.

- **Remarks:** Closely related to the sandyback (41.10) and patchwork (41.14) stingarees, but differs in size, colour pattern and disc shape. Some specimens of the Coral Sea stingaree appear to have much smaller eyes and narrower mouths, and more than one species may be present. Further research is required after more specimens have been collected. A similar, if not identical, species occurs off New Caledonia.

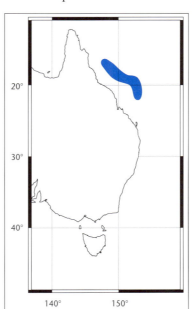

Sandyback Stingaree — *Urolophus bucculentus* Macleay, 1884

41.10 *Urolophus bucculentus*   (Plate 75)

Fish Code: 00 038001

- **Alternative name:** Great stingaree.
- **Field characters:** A large stingaree with a skirt-shaped internasal flap, tail folds, a dorsal fin, but no broad lobes on the posterolateral border of the nostrils. Yellowish or brownish above and usually with a complex pattern of light spots and reticulations.
- **Distinctive features:** Disc broad, rhomboidal, much wider than long; broadest part usually almost an eye diameter behind level of spiracles; anterior profile angular, obtuse. Snout fleshy, tip slightly extended. Eye small (20–27% preocular snout length). Posterior margin of spiracle rounded or slightly angular. Mouth moderate to large; 14–16 papillae on floor; narrow area of papillae below mouth. Internasal flap skirt-shaped, posterior angle not extended; fringe fine. Posterolateral border of nostril sometimes forming a ridge of skin but not forming a broad flattened, fleshy lobe. Disc upper surface smooth. Tail strongly depressed in cross-section; rather short (62–73% disc length); lateral cutaneous tail folds narrow; dorsal fin rather large; caudal fin very short and broad.

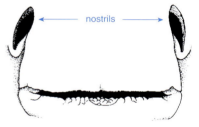

region of nostrils and mouth

- **Colour:** Dorsal surface yellowish to brownish, often with lighter spots and fine reticulations. Caudal and dorsal fins brownish to black in young, greyish or brown (sometimes mottled) in adults. Ventral surface white, sometimes with darker blotches on tail.
- **Size:** Reaches at least 80 cm; males mature at about 40 cm.
- **Distribution:** Continental shelf and upper slope off southeastern Australia from Beachport (South Australia) to Stradbroke Island (Queensland), south to the Hippolyte Rocks (Tasmania), in depths of 100 to 230 m.
- **Remarks:** This ray, which is the largest stingaree in southern Australia, is most common on the outer half of the continental shelf and rarely occurs inshore. It is the temperate relative of the tropical patchwork stingaree (41.14). Little is known of its biology. Edible, but not used commercially.
- **References:** Macleay (1884b); McCulloch (1916); Last *et al.* (1983).

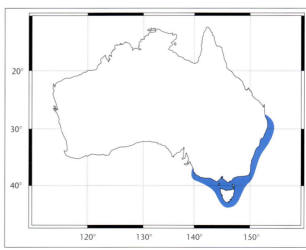

Circular Stingaree — *Urolophus circularis* McKay, 1966

41.11 *Urolophus circularis* (Plate 73) Fish Code: 00 038020

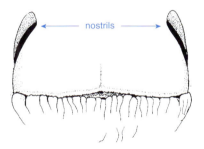

region of nostrils and mouth

- **Field characters:** A rather large, strikingly coloured stingaree with a skirt-shaped internasal flap, a dorsal fin, no tail folds, and no broad lobes on the posterolateral border of the nostrils. A ring of mostly white-centred, bluish grey spots on the mid-disc is distinctive.

- **Distinctive features:** Disc almost circular, length subequal to width; broadest part about twice eye diameter behind level of spiracles; anterior profile obtuse to rounded. Snout fleshy, tip slightly extended. Eye large (about 28*% preocular snout length). Posterior margin of spiracle mostly rounded. Mouth large; 10* papillae on floor. Internasal flap skirt-shaped, posterior angle extended into long lobe; posterior margin very short. Posterolateral border of nostril smooth (not knob-like or lobed). Disc upper surface smooth. Tail oval to rounded in cross-section; short (about 66*% disc length); lateral cutaneous tail folds absent; dorsal fin rather large; caudal fin very short and deep.

- **Colour:** Dorsal surface bluish (light sandy grey in preservative) with light spots, blotches and rings, and a diagnostic ring of mostly white-centred blue or grey spots on the disc centre. Caudal-fin margin and dorsal fin bluish or brownish. Ventral surface pale; tail light dusky brown.

- **Size:** Attains at least 60 cm. Smallest mature male examined was 53 cm.

- **Distribution:** Southwestern Australia from Esperance to Rottnest Island (Western Australia) inshore to 120 m.

- **Remarks:** This highly attractive species does not seem to be common but is sometimes observed by divers on offshore reefs of southern Western Australia. Unlike other stingarees, it is frequently found on rocky bottoms or amongst kelp. The spotted pattern makes it rather difficult to distinguish on some substrates. Few specimens are held in collections.

- **References:** McKay (1966); Hutchins and Swainston (1986).

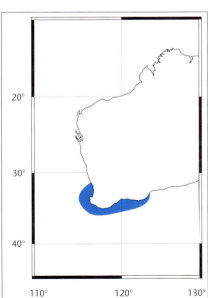

Banded Stingaree — *Urolophus cruciatus* (Lacépède, 1804)

41.12 *Urolophus cruciatus* (Plate 76) Fish Code: 00 038002

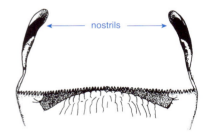

region of nostrils and mouth

- **Alternative name:** Crossback stingaree.
- **Field characters:** A medium-sized stingaree with a distinctive pattern of dark stripes and crossbars, a skirt-shaped internasal flap, no tail folds, no lobes on the posterolateral border of the nostrils, and usually lacking a dorsal fin.
- **Distinctive features:** Disc not especially broad, oval or subcircular, slightly wider than long; broadest part 1–2.5 eye diameters behind level of spiracles (further back in juveniles); anterior profile obtuse. Snout fleshy, tip rarely extended. Eye small (25–28% preocular snout length). Posterior margin of spiracle rounded or angular. Mouth small; 3–6 papillae on floor; lower jaw moderately papillate. Internasal flap skirt-shaped, posterior angle not extended into distinct lobe; fringe prominent. Posterolateral border of nostril sometimes forming a small knob, not forming a broad, flattened, fleshy lobe. Disc upper surface smooth. Tail rather depressed and oval in cross-section; usually short (63–84% disc length); lateral cutaneous tail folds absent; dorsal fin usually absent (sometimes present as a low skin fold in juveniles); caudal fin very short, deep, broadly rounded posteriorly.
- **Colour:** Dorsal surface greyish to yellowish brown (rarely dark brown or black) with a pattern of dark bars and stripes over disc; central median dark stripe with transverse bars radiating from eye, mid-gill and mid-disc (pattern strongest in southern populations). Caudal fin greyish brown. Ventral surface whitish; disc margin sometimes pale grey; tail sometimes with greyish brown blotches.
- **Size:** Reaches about 50 cm. Smallest mature male examined was 25 cm.
- **Distribution:** Continental shelf off southeastern Australia from Tathra (New South Wales) to Beachport (South Australia), including Tasmania, to a depth of 160 m.
- **Remarks:** The banded stingaree is an inshore species, preferring muddy bottoms in the mouths of estuaries and in bays. It is rather inactive, lying concealed beneath the substrate where it is accidentally disturbed by divers and swimmers. It should be handled with care as it is capable of inflicting a very painful wound with its stinging spine. The tail base is more flexible than most other rays and thrusts can reach to any point on the disc if the animal is touched. If the tip of the sting breaks off in the wound, its serrations make it difficult to remove without surgery.
- **Local synonymy:** *Urolophus ephippiatus* Richardson, 1845.
- **References:** McCulloch (1916); Last *et al.* (1983).

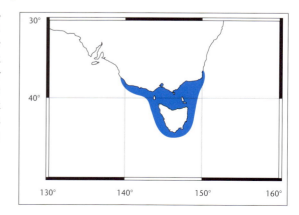

Wide Stingaree — *Urolophus expansus* McCulloch, 1916

41.13 *Urolophus expansus* (Plate 74)   Fish Code: 00 038008

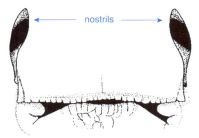

region of nostrils and mouth

- **Field characters:** A broad, medium-sized, deepwater stingaree with faint bluish cross bars over the disc, a skirt-shaped internasal flap, tail folds, no dorsal fin, and no lobes on the posterolateral border of the nostrils.

- **Distinctive features:** Disc broad, subcircular to rhomboidal, much wider than long; broadest part about 1–1.5 eye diameters behind level of spiracles; anterior profile obtuse. Snout fleshy, tip slightly extended. Eye rather large (25–30% preocular snout length). Posterior margin of spiracle mostly rounded. Mouth medium-sized; 6–9 simple or irregular papillae on floor; narrow area of papillae below mouth. Internasal flap skirt-shaped, posterior angle usually extended; fringe feeble. Posterolateral border of nostril straight (not knob-like or lobed). Disc upper surface smooth. Tail very depressed anteriorly; moderately elongate (71–93% disc length); lateral cutaneous tail folds prominent; dorsal fin absent; caudal fin lanceolate to moderately elongate.

- **Colour:** Dorsal surface greyish green; mostly with two rather faint, greyish blue crossbars posterior to eyes and similar oblique bars extending laterally from the front of each eye. Caudal fin dark in juveniles, paler in adults. Ventral surface white or yellowish with darker lateral margin; tail with dark blotches.

- **Size:** Attains at least 47 cm. Smallest mature male examined was 41 cm.

- **Distribution:** Outer continental shelf and upper continental slope of southwestern Australia and the Great Australian Bight, from Perth (Western Australia) to Port Lincoln (South Australia), in depths of 140–420 m (mainly 200–300 m).

- **Remarks:** Little is known of this rather deep-water stingaree which is caught occasionally by trawlers. Edible, but not used commercially.

- **References:** McCulloch (1916); Scott *et al.* (1980).

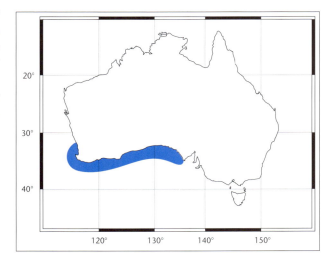

Patchwork Stingaree — *Urolophus flavomosaicus* Last & Gomon, 1987

41.14 *Urolophus flavomosaicus* (Plate 75)

Fish Code: 00 038010

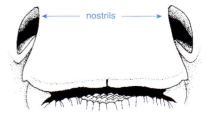

region of nostrils and mouth

- **Field characters:** A rather large, yellowish stingaree with a complex pattern of large, whitish spots encircled by rings, a skirt-shaped internasal flap, weak tail folds, a dorsal fin, but no lobes on the posterolateral border of the nostrils.

- **Distinctive features:** Disc rhomboidal, much wider than long; broadest part about an eye diameter behind spiracles; anterior margin obtuse. Snout tip prominent, extended. Eye small (19–28% preocular snout length). Posterior margin of spiracle usually angular. Mouth large; about 8–14 very short papillae on floor; narrow area of large papillae below lower jaw. Internasal flap skirt-shaped, posterior angle not lobed; fringe fine. Posterolateral border of nostril flat or forming weak knob, not forming broad, flattened lobe. Disc upper surface smooth. Tail very depressed in cross-section; short (67–79% disc length); lateral cutaneous tail folds present, sometimes obscure but prominent in juveniles; dorsal fin prominent; caudal fin short and broad.

- **Colour:** Dorsal surface yellowish with numerous paler spots encircled by darker yellowish brown rings; spots almost regularly spaced, sometimes interspersed with pale, narrow reticulations (largest on central disc). Caudal and dorsal fins usually pale; dark with very dark distal margins in newborn.

- **Size:** Reaches at least 59 cm. Smallest mature male examined was 38 cm.

- **Distribution:** Occurs off tropical Australia, from the Abrolhos Islands to Port Hedland (Western Australia) and from Caloundra to Townsville (Queensland), in depths from 60–300 m.

- **Remarks:** One of the largest tropical Australian stingarees, this species closely resembles the more temperate, sandyback stingaree (41.10), with which it overlaps in distribution in southern Queensland. These species differ from each other subtly in shape and have quite different colour patterns. The sandy-backed stingaree is either uniformly yellow or brown and, although it may be covered in fine reticulations and spots, it never has a broad mosaic pattern.

- **Local synonymy:** *Trygonoptera* sp. 1: Gloerfelt-Tarp & Kailola, 1984; *Urolophus* sp. 2: Sainsbury *et al.*, 1985.

- **References:** Sainsbury *et al.* (1985); Last and Gomon (1987).

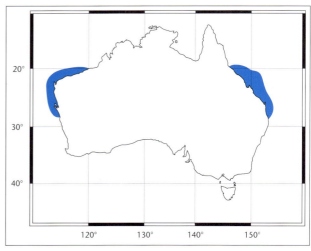

Spotted Stingaree — *Urolophus gigas* Scott, 1954

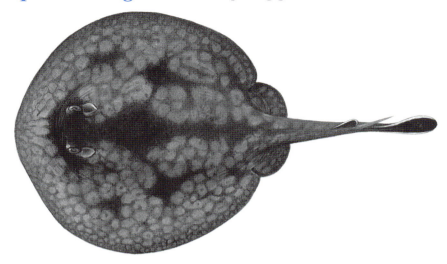

41.15 *Urolophus gigas* (Plate 73)　　　　　　　　　　**Fish Code:** 00 038003

- **Alternative name:** Sinclairs stingaree.
- **Field characters:** A large, dark coloured stingaree with a complex pattern of pale spots on the upper surface, a skirt-shaped internasal flap, a dorsal fin, no tail folds, and no lobes on the posterolateral border of the nostrils.

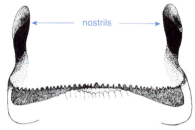

region of nostrils and mouth

- **Distinctive features:** Disc not especially broad, oval, slightly longer than wide; broadest part 1.5–4 eye diameters behind level of spiracles (most posterior in small specimens); anterior profile rounded. Snout fleshy, tip rarely extended. Eye very small (16–25*% preocular snout length). Posterior margin of spiracle rounded. Mouth small; 9–12 papillae on floor; narrow area of papillae on lower jaw. Internasal flap skirt-shaped, posterior angle not extended into distinct lobe; fringe fine. Posterolateral border of nostril straight (not knob-like or lobed). Disc upper surface smooth. Tail rounded to oval in cross-section, deep, moderately short (76–80*% disc length); no lateral cutaneous tail folds; dorsal fin rather large; caudal fin very short, deep.
- **Colour:** Upper surface dark brown or black, merging to paler brown to white toward the disc margin; border with 2 or 3 rows of small cream spots which may extend to tail; central and hind parts of disc with patches of much larger creamish or white spots; disc sometimes also with irregularly distributed, dark spots. Caudal and dorsal fins dark brown or black; margins pale. Ventral surface white or cream; often with blotches and broad, greyish or brownish, lateral margins; tail with pale median stripe in small individuals, sometimes with cream spots in adults.
- **Size:** Reaches about 70 cm. Smallest mature male examined was 55 cm.
- **Distribution:** Continental shelf off southern Australia from Albany (Western Australia) to Lakes Entrance (eastern Victoria), including the northern coast of Tasmania, in depths to 50 m.
- **Remarks:** This species is occasionally observed by divers, often partly concealed, in sand near seagrass beds. The spotted pattern is sometimes indistinct on live fish but becomes more pronounced after death. The western form of this stingaree differs slightly in colour from the eastern form and is considered by some to be a different species. Further work is required to determine the validity of the western form which recent authors have called Sinclairs stingaree.
- **Local synonymy:** *Urolophus* sp.: Hutchins & Swainston, 1986.
- **References:** Scott *et al.* (1980); Hutchins and Swainston (1986).

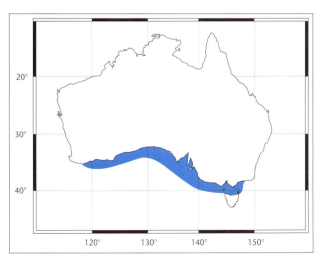

Lobed Stingaree — *Urolophus lobatus* McKay, 1966

41.16 *Urolophus lobatus* (Plate 76)   **Fish Code:** 00 038021

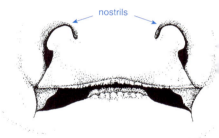

region of nostrils and mouth

- **Field characters:** A rather small, pale brown stingaree with unique lobes on the anterior border of the internasal flap, a skirt-shaped internasal flap, tail folds, no dorsal fin, and no lobes on the posterolateral border of the nostrils.

- **Distinctive features:** Disc broad, subcircular, much wider than long; broadest part an eye diameter or less behind level of spiracles; anterior profile obtuse. Snout fleshy, tip sometimes slightly extended. Eye of moderate size (22–26% and 32–35*% preocular snout length in mature specimens and juveniles, respectively). Posterior margin of spiracle mostly rounded. Mouth small; 9–10 papillae on floor; few papillae on lower jaw. Internasal flap with the anterior angle extended into a very distinct semi-circular lobe; fringe moderate. Posterolateral border of nostril forming an angular fleshy lobe. Disc upper surface smooth. Tail very depressed in cross-section, rather elongate (87–100% disc length); lateral cutaneous tail folds well developed; dorsal fin absent; caudal fin elongate, narrow, lanceolate.

- **Colour:** Dorsal surface sandy brown with slightly paler lateral margins; sometimes with irregular blotches; tail occasionally with a narrow, dark median stripe. Caudal fin dark posteriorly, paler anteriorly. Ventral surface whitish, sometimes with darker blotches.

- **Size:** Reaches at least 38 cm. Males mature at about 28 cm.

- **Distribution:** Continental shelf off southern Western Australia, between Esperance and Rottnest Island, in depths to 30 m.

- **Remarks:** This stingaree is easily distinguished from related species in the same area by the prominent lobes on the front borders of the internasal flap. The function of this modification is unknown.

- **Reference:** McKay (1966).

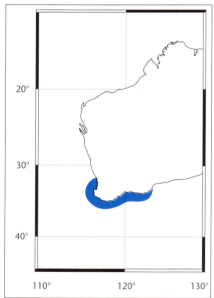

Mitotic Stingaree — *Urolophus mitosis* Last & Gomon, 1987

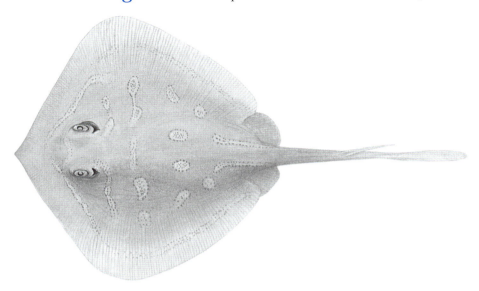

41.17 *Urolophus mitosis* (Plate 75) Fish Code: 00 038011

- **Alternative name:** Blotched Stingaree.
- **Field characters:** A small stingaree with a skirt-shaped internasal flap, weak tail folds, no dorsal fin, and no broad lobes on the posterolateral border of the nostrils. The upper surface has an attractive pattern of blotches and stripes each comprised of minute spots and streaks.

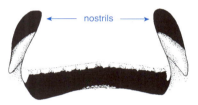

region of nostrils and mouth

- **Distinctive features:** Disc subcircular to rhomboidal, slightly wider than long; broadest part about an eye diameter behind spiracles; anterior margin obtuse. Snout tip slightly extended. Eye large (33–43% preocular snout length). Posterior margin of spiracle rounded or acute. Mouth moderately large; 3–4 papillae on floor; few papillae below mouth. Internasal flap skirt-shaped, posterior angle extended into distinct lobe; fringe very short. Posterolateral border of nostril flat or forming weak knob, not forming broad, flattened lobe. Disc upper surface smooth. Tail very depressed in cross-section; elongate (85–104% disc length); lateral cutaneous folds weak; dorsal fin absent; caudal fin long, lanceolate.
- **Colour:** Dorsal surface of disc with very ornate markings; pale green centrally, reddish brown near margin; sparsely covered with several large, widely spaced, granular blotches and stripes (granulations resembling living cells in the process of mitotic division); markings pale blue, mostly variable in shape but regular in position. In preserved specimens, dorsal surface pale yellow, granulations remaining obvious. Caudal fin and ventral surface pale.
- **Size:** Reaches at least 29 cm; males mature at about 25 cm.
- **Distribution:** Known only from off Port Headland, northwestern Australia, in depths of about 200 m.
- **Remarks:** This attractive little stingaree is immediately recognizable by its characteristic colour pattern. Little is known of its biology and more specimens are required.
- **Local synonymy:** *Urolophus* sp. 2: Gloerfelt-Tarp & Kailola, 1984; *Urolophus* sp. 3: Sainsbury *et al.*, 1985.
- **References:** Sainsbury *et al.* (1985); Last and Gomon (1987).

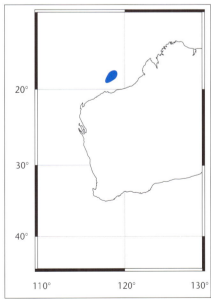

Coastal Stingaree — *Urolophus orarius* Last & Gomon, 1987

41.18 *Urolophus orarius* (Plate 76)　　　　　　　　　　**Fish Code:** 00 038022

- **Field characters:** A rather small, greyish brown stingaree with dark blotches (but no stripes), a skirt-shaped internasal flap, no tail folds, no dorsal fin, and no broad lobes on the posterolateral border of the nostrils.

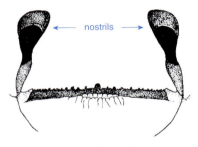

region of nostrils and mouth

- **Distinctive features:** Disc broadly subcircular, usually slightly wider than long; broadest part about 1.5 eye diameters behind spiracles; anterior margin broadly rounded. Snout tip rarely extended. Eye moderate (28–34% preocular snout length). Posterior margin of spiracle mostly rounded. Mouth moderately large; 4–5 papillae on floor; lower jaw with a few papillae. Internasal flap skirt-shaped or slightly bell-shaped, posterior angle not extended into distinct lobe; fringe weak. Posterolateral border of nostril a weak knob, not forming broad flattened lobe. Disc upper surface smooth. Tail deep, round to oval in cross-section; moderately short (72–80% disc length); lateral cutaneous folds and dorsal fin absent; caudal fin rather short in adults, moderately elongate in juveniles.

- **Colour:** Dorsal surface greyish brown with dark markings (pattern more pronounced in small specimens); dark patches around eyes extend onto back; enlarged blotches on middle of pectoral fins and above bases of pelvic fins (sometimes indistinct in large specimens); caudal and pelvic fins of juveniles dark. Ventral surface pale, margins greyish brown to black; lower surface of tail pigmented, mostly dark.

- **Size:** Reaches at least 31 cm; males mature at about 23 cm.

- **Distribution:** Known from the eastern Great Australian Bight, between Ceduna and Beachport (South Australia), in depths of 20–50 m.

- **Remarks:** This ray is similar in general form to the closely related banded (41.12) and yellow-backed stingarees (41.20), but differs in colour pattern and in the relative lengths of the tail and caudal fin lobe. The coastal stingaree lacks a dark median stripe, which is always present in the banded stingaree and sometimes present in the yellow-backed stingaree, but possesses a large dark blotch near each pelvic fin base that is absent in the yellow-backed stingaree. The coastal stingaree has a relatively longer tail and longer caudal fin than its relatives. Possibly more widely distributed in the Great Australian Bight.

- **Local synonymy:** *Urolophus cruciatus*: Scott *et al.*, 1980 (misidentification).

- **Reference:** Last and Gomon (1987).

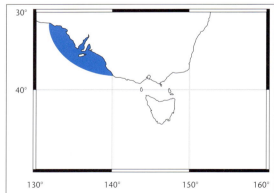

Sparsely-spotted Stingaree — *Urolophus paucimaculatus* Dixon, 1969

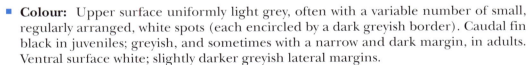

41.19 *Urolophus paucimaculatus* (Plate 74)   Fish Code: 00 038004

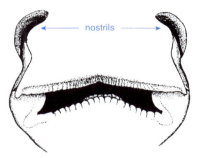

region of nostrils and mouth

- **Alternative names:** Dixons stingaree, white-spotted stingaree.
- **Field characters:** A medium-sized stingaree with a distinctive bell-shaped internasal flap, prominent tail folds, no dorsal fin, and no broad lobes on the posterolateral border of the nostrils. The upper surface is greyish with a dark V-shaped interorbital bar, and sometimes with a few small white spots on the pectoral fins.
- **Distinctive features:** Disc variable in shape, not especially broad, rhomboidal, wider than long; broadest part less than an eye diameter behind level of spiracles; anterior profile obtuse. Snout rather fleshy, tip barely extended. Eye small (22–28% preocular snout length). Posterior margin of spiracle angular or rounded. Mouth small; 5–6 mostly bifurcated papillae on floor; papillae weak below jaw. Internasal flap bell-shaped, posterior angle rarely produced; fringe weak. Posterolateral border of nostril with prominent knob but not forming a broad, flattened, fleshy lobe. Disc upper surface smooth. Tail very depressed anteriorly; rather elongate (77–98% disc length); lateral cutaneous tail folds prominent; dorsal fin absent; caudal fin rather deep, lanceolate.
- **Colour:** Upper surface uniformly light grey, often with a variable number of small, regularly arranged, white spots (each encircled by a dark greyish border). Caudal fin black in juveniles; greyish, and sometimes with a narrow and dark margin, in adults. Ventral surface white; slightly darker greyish lateral margins.
- **Size:** Reaches at least 44 cm; smallest mature male examined was 25 cm.
- **Distribution:** Widely distributed on the continental shelf off southern Australian from Crowdy Head (northern New South Wales) to Lancelin (Western Australia), including Tasmania. Inshore to a depth of 150 m or more.
- **Remarks:** This ray is seldom marketed despite large catches frequently being taken by trawlers and the flesh being of reasonable quality. Populations from cool temperate Australia (Victoria and Tasmania) usually have distinctive white spots on the disc and are most abundant in shallow water (generally less than 30 m). Further north, the species occurs mainly in deeper water (usually deeper than 100 m) and generally lacks spots.
- **References:** Dixon (1969); Edwards (1980).

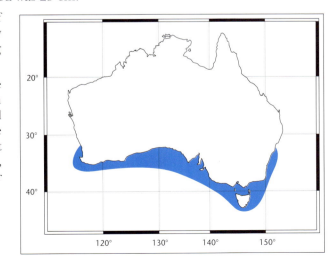

Yellowback Stingaree — *Urolophus sufflavus* Whitley, 1929

41.20 *Urolophus sufflavus* (Plate 76)   **Fish Code:** 00 038005

- **Field characters:** A medium-sized, plain yellowish brown stingaree with a skirt-shaped internasal flap, no tail folds, no dorsal fin, and no broad lobes on the posterolateral border of the nostrils.

- **Distinctive features:** Disc not broad, subcircular, width about equal to or slightly longer than length, rather deep centrally; broadest part 1.5–2 eye diameters behind level of spiracles; anterior profile obtuse. Snout fleshy, tip not extended. Eye moderate to large (24–30% preocular snout length). Posterior margin of spiracle mostly angular. Mouth small; 3–4 papillae on floor; papillae weak below mouth. Internasal flap skirt-shaped, posterior angle not extended into distinct lobe; fringe weak. Posterolateral border of nostril not forming a broad, flattened, fleshy lobe. Disc upper surface smooth. Tail deep, oval to rounded in cross-section; length short (64–76% disc length); no lateral cutaneous tail folds or dorsal fin; caudal fin short, deep.

- **Colour:** Dorsal surface uniform yellowish brown, sometimes with pinkish margin and sometimes with a faint, brown median stripe extending onto tail. Caudal fin similar to dorsal coloration. Ventral surface white with pinkish margin; tail sometimes dark or with dark blotches.

- **Size:** Reaches at least 42 cm. Males mature at about 23 cm.

- **Distribution:** Eastern Australia between Stradbroke Island (Queensland) and Green Cape (New South Wales) in depths of 45–300 m (mainly on the outer continental shelf in 100–160 m).

- **Remarks:** The yellowback and banded stingarees (41.12) may hybridize as some specimens from southern New South Wales display faint markings intermediate between the two forms. The relationship between the two species needs to be investigated further. Little is known of its biology.

- **Local synonymy:** *Urolophus aurantiacus*: McCulloch, 1916 (misidentification).

- **References:** McCulloch (1916); Whitley (1929b).

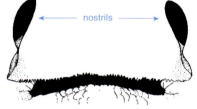

region of nostrils and mouth

Greenback Stingaree — *Urolophus viridis* McCulloch, 1916

41.21 *Urolophus viridis* (Plate 74)   Fish Code: 00 038007

- **Field characters:** A medium-sized, plain greenish stingaree with a skirt-shaped internasal flap, tail folds (often narrow), no dorsal fin, and no broad lobes on the posterolateral border of the nostrils.

- **Distinctive features:** Disc broad, rhomboidal, wider than long; broadest part about one eye diameter behind level of spiracles; anterior profile obtuse. Snout angular, fleshy, tip slightly extended. Eye rather large (27–33% preocular snout length). Posterior margin of spiracle rounded or slightly projecting. Mouth of moderate size; 4–7 irregular papillae on floor; narrow band of papillae beneath mouth. Internasal flap skirt-shaped, posterior angle usually extended, fringe weak. Posterolateral border of nostril straight (not knob-like or lobed). Disc upper surface smooth. Tail very depressed anteriorly; moderately elongate (75–91% disc length); lateral cutaneous tail folds well developed (rarely narrow); dorsal fin absent; caudal fin broadly lanceolate.

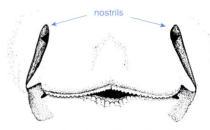

region of nostrils and mouth

- **Colour:** Upper surface uniform light green with paler edges. Caudal-fin posterior margin dark brown in young, greenish brown in adults. Ventral surface usually whitish with a purplish or pinkish lateral margin (sometimes with dark brown lateral margin and/or a few dark brown blotches); snout tip and undersurface of tail sometimes dark.

- **Size:** Reaches at least 44 cm. Smallest mature male examined was 27 cm.

- **Distribution:** Southeastern Australia, from Portland (Victoria) to Stradbroke Island (Queensland), including Tasmania. Demersal on the outer continental shelf in depths of 20–200 m (mainly 80–180 m).

- **Remarks:** This common ray lives in deeper water than most of the other stingarees caught off southeastern Australia. Another similar, if not identical, species has been collected recently from deepwater off southern Western Australia. The latter has a higher pectoral-fin ray count (106–107) than the greenback stingaree (maximum of 100).

- **References:** McCulloch (1916); Last *et al.* (1983).

Brown Stingaree — *Urolophus westraliensis* Last & Gomon, 1987

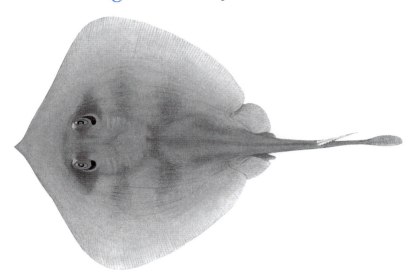

41.22 *Urolophus westraliensis* (Plate 75) Fish Code: 00 038009

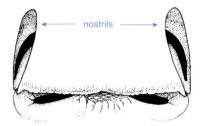

region of nostrils and mouth

- **Field characters:** A small, pale yellowish or brownish stingaree with indistinct dusky bars across the disc, a skirt-shaped internasal flap, a tail with or without a dorsal fin, poorly developed tail folds, and no lobes on the posterolateral border of the nostrils.

- **Distinctive features:** Disc rhomboidal, slightly wider than long; broadest part about an eye diameter behind spiracles; anterior margin obtuse. Snout tip slightly extended. Eye medium-sized (22–28% preocular snout length). Posterior margin of spiracle mostly rounded. Mouth moderately large; 5–6 small papillae on floor; few papillae below mouth. Internasal flap skirt-shaped, posterior angle extended slightly into distinct lobe, fringe very short. Posterolateral border of nostril flat or forming weak knob, not forming broad flattened lobe. Disc upper surface smooth. Tail moderately depressed in cross-section; rather short (66–80*% disc length); lateral cutaneous folds barely noticeable; dorsal fin either absent or reduced to a low fold; caudal fin short, usually broad.

- **Colour:** Dorsal surface and caudal fin uniformly pale yellow to light brown; lacking spots but sometimes with 3 indistinct dusky bars (across eyes, middle of gills and mid-disc). Ventral surface uniformly pale. Juveniles pale yellow; margin of caudal fin black, remainder of fin yellow.

- **Size:** Reaches at least 36 cm. Smallest mature male examined was 24 cm. Born at about 10 cm.

- **Distribution:** Outer continental shelf off northwestern Australia, between Dampier and the Buccaneer Archipelago, in depths of 60–210 m.

- **Remarks:** This species is closely allied to the sympatric mitotic stingaree (41.17), but has smaller eyes, a relatively shorter tail, and the colour pattern is plain (rather than ornate). Little is known of this species.

- **Local synonymy:** *Urolophus* sp. 1: Sainsbury *et al.*, 1985.

- **References:** Sainsbury *et al.* (1985); Last and Gomon (1987).

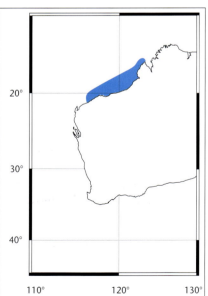

42

Family Gymnuridae
BUTTERFLY RAYS

Butterfly rays are medium to large, very flat fishes with a characteristic lozenge-shaped disc that is much broader than long, and which resembles a butterfly in shape. The head, with its very obtuse snout, is not elevated above the disc and cannot be distinguished from the pectoral fins. The mouth, which is located ventrally, has numerous small teeth and lacks papillae on its floor. The very slender tail is much shorter than the disc. These rays sometimes have a small dorsal fin and/or a stinging spine, but never have caudal or anal fins.

The group is represented worldwide in warm temperate and tropical seas by two genera and at least 12 living species. Most species live on the bottom on the upper continental shelf. Embryonic development is viviparous with litters of 2–6 young. Only one colour-variable species has been recorded from Australian waters, although other relatives occur off Indonesia and may occur here rarely.

Australian Butterfly Ray — *Gymnura australis* (Ramsay & Ogilby, 1886)

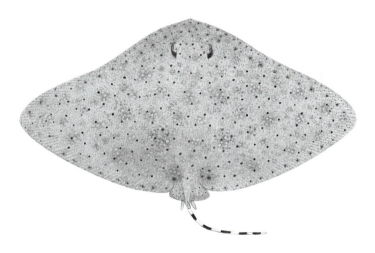

42.1 *Gymnura australis* (Plate 79)　　　　　　　　　　　　　　Fish Code: 00 037001

- **Alternative names:** Rat-tail ray, skate.
- **Field characters:** A distinctive, medium-sized ray with a very broad and flattened butterfly-shaped disc, and an extremely short, filamentous tail without a stinging spine.
- **Distinctive features:** Disc lozenge-shaped, extremely flattened, much broader than long (length 51–54% disc width); anterior margin slightly concave beside spiracle. Snout very broadly angular, with a short, broad fleshy lobe at tip; pectoral-fin apex narrowly rounded. Body firm, without denticles or thorns. Eyes very small, widely separated. Spiracles large, about equal to eye diameter; long tentacle near posterior margin. Mouth of moderate width; internasal flap broad, short. Tail filamentous, very flexible and short (rarely longer than half disc length); single dorsal fin originating near tip of pelvic fin or well beyond; no stinging spine or caudal fin. Pelvic fins very small, barely extending beyond disc.

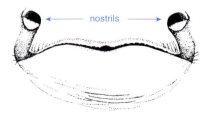

region of nostrils and mouth

- **Colour:** Mostly greenish grey above (sometimes greyish or yellowish) with dense peppering of fine black spots over a delicate mosaic pattern (spots sometimes very faint and mostly in groups); irregular blotch near each pectoral-fin insertion. Tail with alternating black and white bands. Ventral surface white.
- **Size:** To at least 73 cm disc width; males maturing at about 35–40 cm.
- **Distribution:** New Guinea and tropical and warm temperate parts of Australia, between Broken Bay (New South Wales) and Dampier (Western Australia). Inshore to at least 50 m depth. Possibly also Indonesia.
- **Remarks:** These unusual rays are caught regularly by prawn trawlers. The flesh is of reasonable quality and is occasionally sold under the name of skate.
- **Reference:** Grant (1978).

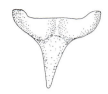

tooth from near symphysis (upper jaw)

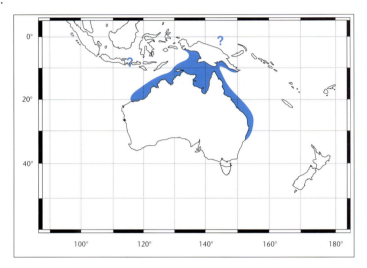

43

Family Hexatrygonidae
SIXGILL STINGRAYS

These very unusual deepwater rays were only discovered about a decade ago. Unlike other rays, they have six pairs of gill openings and six gill arches with well-developed gill filaments. The snout is long and pointed, but its shape is variable. The nostrils are wide apart, not connected to the mouth by a groove, and the internasal flap does not overlap the mouth. The skin is flabby and lacks thorns and denticles. Apart from the characters above, they most closely resemble stingarees in having a caudal fin and 1–2 stinging spines, but never have dorsal fins.

The family is represented in the Indo–Pacific by a single genus and five species. Most of the species are known from single specimens and some, if not most, are probably invalid. Five specimens have been collected in Australian waters, the first in 1985 off Queensland. More specimens are required to solve problems of nomenclature and to find the correct name for the Australian species.

Sixgill Stingray — *Hexatrygon* sp. **A**

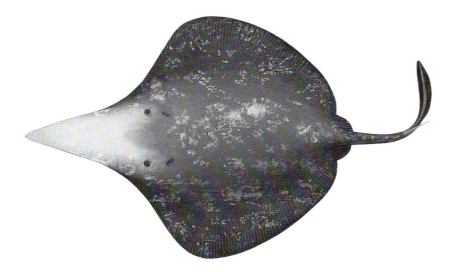

43.1 *Hexatrygon* sp. **A** (Plate 64) **Fish Code:** 00 037002

- **Field characters:** An unusually flabby ray with 6 gill openings, a long pointed snout, a short tail with a sting, and a long, low caudal fin.
- **Distinctive features:** Disc rather thick, almost circular behind nostrils; snout very long, narrow, triangular (length 36–38*% disc length); anterior disc margin very deeply concave beside eye; disc longer than broad. Body and snout flabby, surface lacking denticles and thorns. Eyes very small; interorbital space very broad, about a tenth of total length. Spiracles large, about 2 or more times eye diameter. Mouth broad; internasal flap very broad, short. Teeth numerous, small, blunt. Tail short (about half disc length); stinging spine prominent; caudal fin very long and low. Pelvic fins very small.

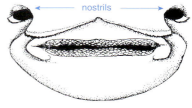

region of nostrils and mouth

- **Colour:** Upper surface of postorbital disc uniformly brownish pink (white patches where skin removed); slightly darker brown toward disc margin; snout almost uniformly pale; caudal fin brownish black. Ventral surface mostly pale, darker brownish black around posterior margin of disc and on tail.
- **Size:** To at least 80 cm.
- **Distribution:** Continental slope off Flinders Reef (Queensland), and from the Exmouth Plateau to Shark Bay (Western Australia), in 900–1120 m.
- **Remarks:** The shape of the snout is considered to be important in distinguishing between the species of *Hexatrygon*. Four specimens, taken recently from the continental slope off Western Australia, have variable snout shapes and lengths. The snout tip, which is highly flexible and capable of lateral and vertical movement, is presumably used for probing around within the substrate for food. Any specimens caught should be kept and sent to a museum.

tooth from near symphysis (upper jaw)

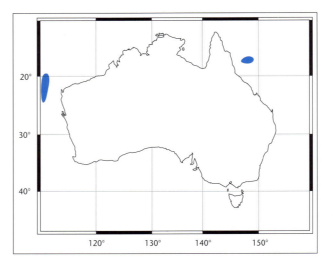

44 Family Myliobatididae
EAGLE RAYS

Eagle rays are medium to large, heavy-bodied fishes with a lozenge-shaped disc that is much broader than long. The head protrudes slightly beyond the disc and the snout is rounded (rather than indented anteriorly) and lacks paired fleshy lobes. The pectoral fins meet the head laterally beside or below the eyes. The mouth is broad, located ventrally, and has plate-like teeth arranged in 1–7 rows (teeth of the middle series are always broadest). Denticles and thorns are present around the eye and along the midline of the disc in some species. The tail is longer than the disc, is filamentous distally, lacks a caudal fin, and has a small dorsal fin near its base. Stinging spines are either present or absent.

Eagle rays live mainly inshore in temperate and tropical seas. They are viviparous with 2–6 young. Of the four genera and 22 or so species, five species from three genera are known to occur in the Australian region. Some of the species are rare in collections and, hence, are not well known.

Key to myliobatidids

1   Fleshy lobe around snout almost continuous with pectoral fins (fig. 1) (sometimes less obvious in *Myliobatis hamlyni*); lobe relatively short, broadly rounded (fig. 5); internasal distance usually equal to or longer than direct distance from nostrils to snout tip (fig. 3) **2**

   Fleshy lobe around snout not continuous with pectoral fins, situated well below level of pectoral-fin origin (fig. 2); lobe rather elongate, narrowly rounded (fig. 9); internasal distance shorter than direct distance from nostrils to snout tip (fig. 4) **3**

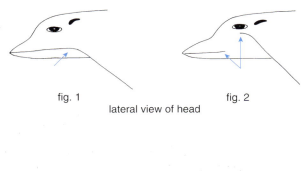

fig. 1   fig. 2
lateral view of head

2   Dorsal-fin origin near rear tips of pelvic fins (fig. 5); pale bluish or grey bands on upper disc (fig. 5) .... ***Myliobatis australis* (fig. 5; p. 453)**

   Dorsal-fin origin well behind rear tips of pelvic fins, by more than half base of dorsal fin (fig. 6); no pale bluish or grey bands on upper disc (fig. 6) .... ***Myliobatis hamlyni* (fig. 6; p. 454)**

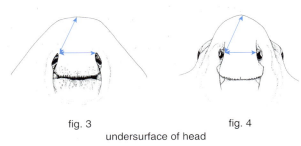

fig. 3   fig. 4
undersurface of head

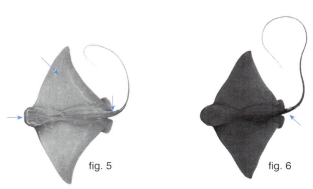

fig. 5   fig. 6

3 Upper surface uniformly covered in fine white spots; teeth in a single row in each jaw; hind margin of internasal flap V-shaped (fig. 7) ....
.... *Aetobatus narinari* (fig. 9; p. 448)

Upper surface not covered in fine white spots; teeth in more than one row in each jaw; internasal flap skirt-shaped (fig. 8) 4

4 Upper surface covered in a network of fine black lines (fig. 10); mouth more than 3 eye diameters behind snout tip ....
.... *Aetomylaeus vespertilio* (fig. 10; p. 451)

Upper surface not covered in a network of fine black lines, plain or with faint transverse bands (fig. 11); mouth less than 3 eye diameters behind snout tip .... *Aetomylaeus nichofii* (fig. 11; p. 450)

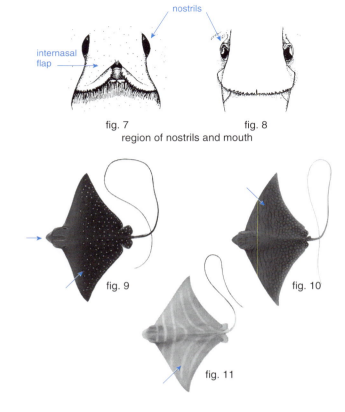

White-spotted Eagle Ray — *Aetobatus narinari* (Euphrasen, 1790)

44.1 *Aetobatus narinari* (Plate 79)   Fish Code: 00 039003

- **Alternative names:** Bonnet skate, duckbill ray, spotted eagle ray.
- **Field characters:** A distinctive white-spotted ray with a very angular disc, a V-shaped internasal flap, a stinging spine, and a long fleshy lobe around the snout that is not connected to the pectoral fins.

- **Distinctive features:** Disc very broad and short (length about 40*% of width); anterior margin deeply notched beside eye. Head and body very thick, former almost circular in cross-section; rostral lobe long, duck-bill shaped, fleshy, distinct from upper snout, much lower posteriorly than origin of pectoral fin. Spiracles large, originating near pectoral-fin origin, broadly oval, distinctly visible dorsally. Mouth located ventrally, about 2 or more orbit diameters behind snout tip; mouth width much less than preoral length; lower jaw teeth convex distally, protruding well beyond mouth; internasal flap long, almost an inverted V-shape, with a broad flap-like fringe. Teeth plate-like, single row in each jaw (those in lower jaw V-shaped). Skin surface uniformly smooth, without thorns and denticles. Pectoral-fin posterior margin deeply concave; apices angular. Pelvic fins slender, narrowly rounded distally. Tail extremely long, whip-like, 2.5–3 times disc width when undamaged; dorsal fin small, posterior margin almost upright, origin just behind pelvic-fin insertion; stinging spine rather short (distinctly shorter than preoral snout), located just behind dorsal fin.

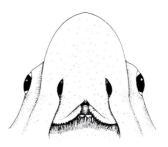

undersurface of head

tooth plates
(upper and lower jaws)

- **Colour:** Upper surface uniformly greenish to pinkish with numerous small white or pale blue spots with faint margins; disc and pelvic-fin margins dark. Ventral surface uniformly white. Tail dark.
- **Size:** Locally, attains at least 300 cm disc width (total length almost 880 cm when tail undamaged) but rarely exceeding 180 cm disc width; elsewhere reported to reach a disc width of 330 cm; born at about 26 cm disc width.
- **Distribution:** Worldwide in tropical and warm temperate seas. Locally known in northern waters from Shark Bay (Western Australia) to Sydney (New South Wales) in inshore habitats to 60 m. Occasionally enters estuaries.
- **Remarks:** The highly attractive white-spotted eagle ray appears to spend much of its time swimming actively in open water, often near the surface in groups. It is capable of leaping well clear of the water. Like other members of the family, the plate-like teeth are used to crush shellfish such as clams, oysters, whelks and other large molluscs. It reaches sexual maturity after 4–6 years and bears up to 4 young. The flesh is edible, but the ray is seldom eaten in this country. A powerful fighter when caught on a line.
- **Reference:** Grant (1978).

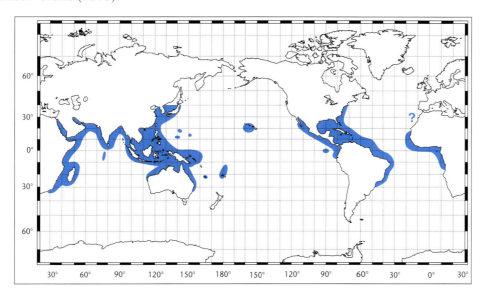

Banded Eagle Ray — *Aetomylaeus nichofii* (Schneider, 1801)

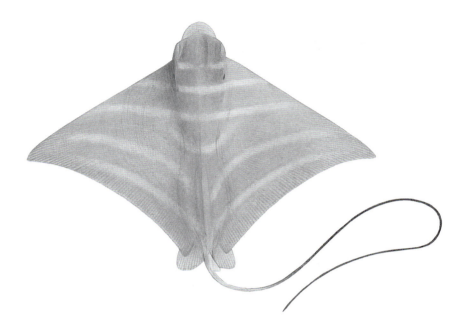

44.2 *Aetomylaeus nichofii* (Plate 79)

Fish Code: 00 039002

- **Field characters:** An eagle ray with a plain or faintly banded upper disc, no stinging spine, a skirt-shaped internasal flap, and a single fleshy lobe around the snout that is not connected to the pectoral fins.

- **Distinctive features:** Disc very broad and short (length about 58–68*% of width); anterior margin deeply notched slightly beneath hind margin of eye. Head rather thick (but flattened above), not especially broad; rostral lobe moderately long, duck-bill shaped, very fleshy, distinct from upper snout, much lower posteriorly than origin of pectoral fin. Spiracles large, originating much less than an eye diameter from pectoral-fin origin, slit-like and usually not visible dorsally. Mouth located ventrally, 2 orbit diameters or less behind snout tip; mouth width less than preoral length; lower jaw teeth not protruding; internasal flap relatively long, skirt-shaped, with a fine fringe. Teeth plate-like, mostly with 7 rows in each jaw (outer rows sometimes concealed or reduced); central row very broad, each tooth more than 5 times wider than those adjacent. Pectoral fin originating just below and behind eye; posterior margin of fin deeply concave; apices angular. Pelvic fins slender, narrowly rounded distally. Skin surface uniformly smooth, without thorns and denticles (large adults with minute, very widely spaced denticles on upper surface). Tail extremely long, whip-like, about 1.5–2 times disc width when undamaged; dorsal fin relatively large, triangular, origin over or just in front of pelvic-fin insertion; stinging spine absent.

- **Colour:** Adults greyish brown above; fin margins and area below spiracle white; tail grey with faint evidence of banding posteriorly. Juveniles with about 8 broad, dark bands (dark-edged in some specimens) across a paler disc; posterior pectoral-fin margin dark; prominent dark band across interorbital; tail greyish above, black beneath, with alternating dark and light bands posteriorly. Ventral surface white.

- **Size:** Attains at least 58 cm disc width (about 100 cm total length); born at about 17 cm disc width (42 cm total length).

- **Distribution:** Indo–West Pacific from southern Japan to Australia, west to India. Locally, demersal on the tropical continental shelf between the Bonaparte Archipelago (Western Australia) and Cairns (Queensland) to depths of at least 70 m.

undersurface of head

tooth plates
(upper and lower jaws)

- **Remarks:** A widely distributed but little known ray. Viviparous with litters of up to 4 young. There is a small-scale commercial fishery for this species in parts of Asia but it is not eaten in Australia.
- **Reference:** Gloerfelt-Tarp and Kailola (1984).

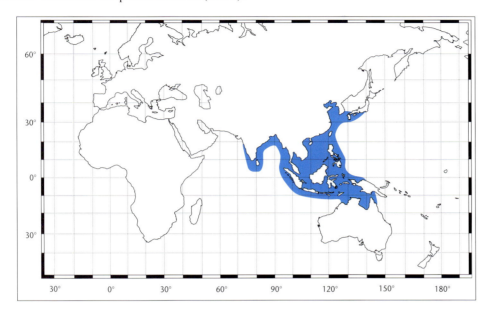

Ornate Eagle Ray — *Aetomylaeus vespertilio* (Bleeker, 1852)

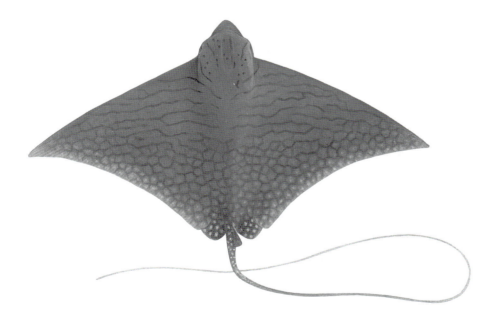

44.3 *Aetomylaeus vespertilio* (Plate 79)　　　　　　　　　　**Fish Code:** 00 039005

- **Field characters:** An eagle ray with a network of dark lines over the upper disc, a skirt-shaped internasal flap, a single fleshy lobe around the snout that is not connected to the pectoral fins and no stinging spine.

- **Distinctive features:** Disc extremely broad and short (length about half of width); anterior margin deeply notched slightly behind eye. Head very thick (but flattened above), broad; rostral lobe very long, duck-bill shaped, fleshy, distinct from upper snout, much lower posteriorly than origin of pectoral fin. Spiracles medium-sized, originating slightly less than an eye diameter from pectoral-fin origin, oval, barely visible dorsally. Mouth located ventrally, 3–4 orbit diameters behind snout tip; mouth width much less than preoral length; outer teeth of lower jaw almost straight, protruding only slightly; internasal flap skirt-shaped, with an indistinct fringe. Teeth plate-like, mostly with 7 rows in each jaw; central row very broad. Pectoral fin originating behind eye; posterior margin deeply concave; apices angular. Pelvic fins slender, narrowly rounded distally. Skin surface without thorns; adults with a narrow band of tiny, flat denticles along midline of disc and predorsal tail; tail beyond fin covered in fine widely spaced granulations. Tail extremely long, whip-like, nearly 3 times disc width when undamaged; dorsal fin relatively large, triangular, origin behind pelvic-fin insertion; stinging spine absent.

- **Colour:** Dorsal surface dark to light brown with black lines arranged transversely on front half of disc, forming an open network posteriorly; head with a few black spots and stripes; posterior margin of disc pale brown with white spots; white spots on pelvic fins, extending onto anterior tail. Ventral surface white. Tail black, brownish with faint blackish rings near base.

- **Size:** Attains 160 cm disc width (about 385 cm total length).

- **Distribution:** Not well known; so far from Taiwan, Indonesia and the Arafura Sea. Possibly more widely distributed in the Indo–West Pacific. Caught locally in about 50 m depth.

- **Remarks:** This very attractive ray was described almost 150 years ago but few specimens have been recorded since. Specimens collected in recent months from off northern Australia are the first from this region. The lack of specimens appears to have caused nomenclatural problems which are not fully resolved in this book. A ray, *Aetomylaeus reticulatus*, described recently from Taiwan as a new species, is likely to be the ornate eagle ray. Also, scientific names older than the one used here may apply to earlier growth stages of this species.

undersurface of head

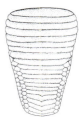
tooth plates
(upper and lower jaws)

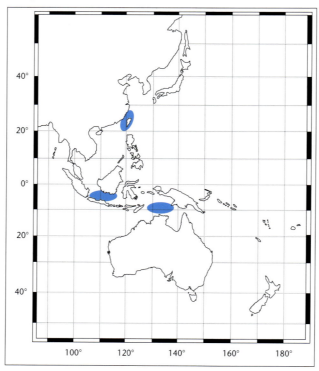

Southern Eagle Ray — *Myliobatis australis* Macleay, 1881

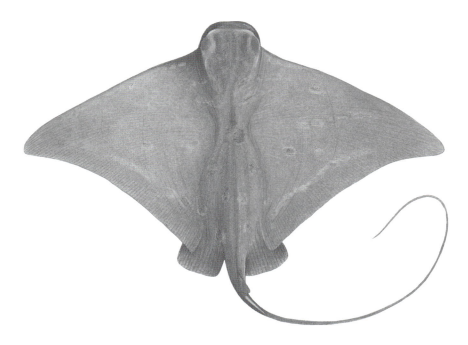

44.4 *Myliobatis australis* (Plate 80) Fish Code: 00 039001

- **Alternative names:** Bull ray, cowfish, mill ray, whip ray, whiptail ray.
- **Field characters:** An eagle ray with a faintly banded upper disc, a skirt-shaped internasal flap, a single fleshy lobe around the snout that is almost connected to the pectoral fin, a stinging spine, and a dorsal-fin origin near the rear tips of the pelvic fins.
- **Distinctive features:** Disc very broad and short (length about 55–60*% of width); anterior margin deeply notched beside eye. Head rather thick, broad; rostral margin broadly rounded, fleshy (forming a lobe), distinct from upper snout. Eye of medium size. Spiracles large, originating about an orbit diameter from pectoral-fin margin, oblique, not visible dorsally. Mouth located ventrally, about 2 orbit diameters behind snout tip; mouth width narrower than preoral length; fleshy lobe of snout almost continuous with pectoral fins; internasal flap skirt-shaped, relatively short and broad, with a long fringe. Teeth plate-like, mostly with 7 rows in each jaw (outer rows sometimes concealed or reduced); central row very broad, each tooth more than 5 times wider than those adjacent. Skin surface uniformly smooth, without thorns and denticles. Pectoral-fin origin below eye; posterior margin of fin deeply concave; apices very angular. Pelvic fins broad, rather angular. Tail elongate, whip-like, about 1.3–1.7 times disc width when undamaged; dorsal fin small, apex and posterior margin broadly rounded, origin over or just behind pelvic-fin rear tips; stinging spine length about equal to preoral length, located just behind dorsal fin.
- **Colour:** Olive green to yellowish above (paler near disc margin) with bluish spots and crescentic bars. Pale ventrally, sometimes greyish along disc margin.
- **Size:** To at least 120 cm disc width (almost 190 cm total length). Free swimming by 32 cm disc width.
- **Distribution:** Southern Australia from Jurien Bay (Western Australia) to Moreton Bay (Queensland), including Tasmania, inshore to 85 m. May also occur off New Zealand.
- **Remarks:** This ray is common off beaches and over sandflats in shallow water. It feeds mainly on crabs and shellfish. Adults appear to migrate, at least off Tasmania where it occurs in the southernmost bays only in the warmest months. The New

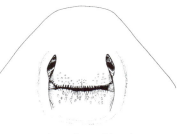
undersurface of head

tooth plates
(upper and lower jaws)

Zealand eagle ray, *Myliobatis tenuicaudatus*, may be the same species. If so, the latter will become the new scientific name for the Australian form. A powerful fighter when caught on a line.

- **References:** Macleay (1881a); Grant (1978); Last *et al.* (1983).

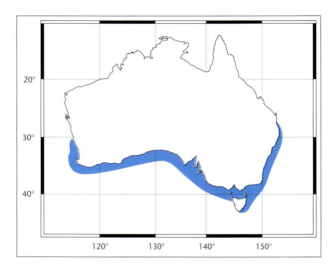

Purple Eagle Ray — *Myliobatis hamlyni* Ogilby, 1911

44.5 *Myliobatis hamlyni* (Plate 80)   Fish Code: 00 039004

- **Field characters:** An eagle ray with a plain upper disc, a skirt-shaped internasal flap, a single fleshy lobe around the snout that is separated slightly from the pectoral fin, a stinging spine, and a dorsal-fin origin well behind the pelvic fins. Adult males have a small thorn over each eye.

- **Distinctive features:** Disc very broad and short (length about 54*% of width); anterior margin deeply notched beside eye. Head relatively broad; rostral margin broadly rounded, fleshy, distinct from upper snout. Eye relatively large. Spiracles large, originating slightly less than an orbit diameter from pectoral-fin margin, slightly oblique, not visible dorsally. Mouth broad, located ventrally, about 1.5–2 orbit diameters behind snout tip; mouth width equal to or less than preoral length; fleshy lobe of snout continuous with or slightly below pectoral-fin origin; internasal flap skirt-shaped, relatively short and broad, with a coarse fringe. Teeth plate-like, with 7 rows in each jaw; central row very broad. Skin surface uniformly smooth, without thorns and denticles, except for a short, blunt thorn above each eye in mature males. Pectoral-fin origin below eye; posterior margin of fin concave; apices broadly pointed. Pelvic fins not especially broad, rather angular. Tail elongate, whip-like, about 1.5 times disc width when undamaged; dorsal fin small, apex broadly rounded, origin half its base length or more behind pelvic-fin rear tips; stinging spine well developed, located just behind dorsal fin.

- **Colour:** Dorsal surface purplish to greenish grey centrally, grading to olive brown laterally on disc. Ventral surface cream. Tail purple or white beyond sting.

- **Size:** Attains at least 48 cm disc width (total length about 90 cm).

- **Distribution:** Poorly known; so far only from single specimens off Cape Moreton (Queensland) and off Forestier Island (Western Australia) in about 200 m. Possibly more widely distributed in the Indo–Pacific to South Africa.

- **Remarks:** Extremely rare in collections and more specimens are required. The type, collected from Queensland, differs slightly from the Western Australian specimen and different growth stages of the two forms need to be compared. Apart from having a different colour pattern, the purple eagle ray has a wider mouth and internasal flap, larger eyes, smaller pelvic fins, and a more posteriorly placed dorsal fin than the closely related southern eagle ray (44.4). It also possibly lives in deeper water than other members of the family. Two other Indo–Pacific eagle rays, *Myliobatis aquila* and *M. tobijei*, are very similar and may even be older names for the Australian species.

- **Reference:** Ogilby (1911).

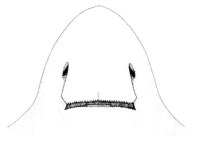

undersurface of head

tooth plates
(upper and lower jaws)

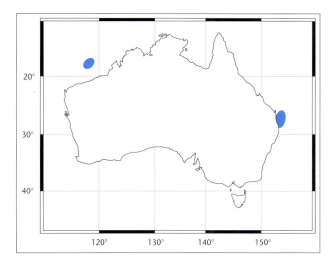

45

Family Rhinopteridae

COWNOSE RAYS

Cownose rays are medium to large, heavy-bodied fishes with a lozenge-shaped disc that is much broader than long. The shape of the head, which protrudes anteriorly beyond the disc, is characteristic of the family. The snout is indented anteriorly and has two large, fleshy lobes in front of a fringed internasal flap. The pectoral fins meet the head laterally just behind the eyes. The mouth is broad and ventral with plate-like teeth arranged in seven or more rows. The tail, which is normally longer than the disc, is filamentous distally, lacks a caudal fin, but has a small dorsal fin and one or more serrated spines near its base.

Members of the family are represented in warm temperate and tropical seas worldwide, except for the islands of the western Pacific. They are most abundant in coastal waters and frequently enter mouths of estuaries. Embryonic development is viviparous, but there is no placental connection between the mother and her young. Only one genus, *Rhinoptera*, represented by about 10 species, is currently recognized. A single species has been recorded from Australia but other widespread Indo–Pacific species may also occur here.

Australian Cownose Ray — *Rhinoptera neglecta* Ogilby, 1912

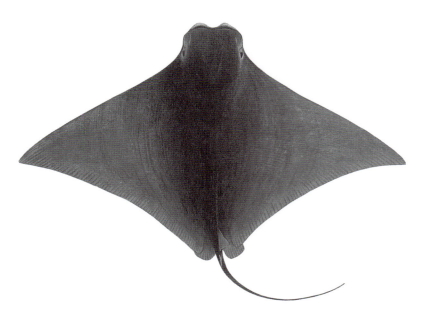

45.1 *Rhinoptera neglecta* (Plate 80)　　　　　　　　　　　**Fish Code:** 00 040001

- **Alternative name:** Cow ray, flapray.
- **Field characters:** A stocky ray with a lozenge-shaped disc, a whip-like tail and a distinctive bilobed forehead with a pair of lobe-like flaps located underneath.
- **Distinctive features:** Disc broad, short, length about 60% of disc width; anterior margin deeply notched behind eye. Head thick and broad; rostral margin deeply notched; lobe-like rostral flaps partly depressible into shallow grooves. Spiracles large, about equal to eye diameter, situated above pectoral-fin origin. Mouth located ventrally, about 2 eye diameters behind snout tip; width subequal to preoral length.

undersurface of head

Teeth plate-like, 9 rows in each jaw; 3 central rows much broader than those laterally; middle row about 1.4–1.5 times wider than those adjacent. Pectoral-fin tips angular. Tail elongate, subequal in length to disc width; dorsal fin prominent, origin just in front of pelvic-fin insertion; serrated spine about equal in length to dorsal-fin base.

- **Colour:** Uniform dark greyish brown above; white below.
- **Size:** To at least 86 cm disc width.
- **Distribution:** Known only off eastern Australia, from Newcastle (New South Wales) to Cairns (Queensland), but certainly likely to be more widely distributed to the north.
- **Remarks:** This ray is considered by some authors to be identical to the wide-ranging Javanese cownose ray, *Rhinoptera javanica* (see Plate 80). In the absence of more thorough research, however, the two forms are provisionally regarded herein as valid species. While *R. javanica* is also likely to occur in Australian waters, early reports need to be validated. It differs from *R. neglecta* in having fewer rows of teeth (7 in each jaw) and may reach a larger size (up to 1.5 m). Cownose rays are sometimes netted on tidal mudflats of northern Queensland, but little is known of their distribution through the Gulf of Carpentaria and into Western Australia. Specimens are required for taxonomic study.

tooth plates
(upper and lower jaws)

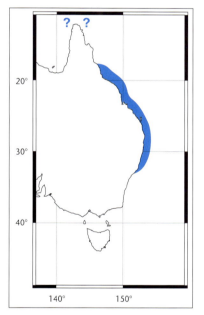

46

Family Mobulidae

DEVILRAYS

Devilrays are the largest of all rays and may exceed 9 m in disc width. The lozenge-shaped disc is much broader than long and the head, which is broad, protrudes anteriorly beyond eye level. Prominent fleshy extensions of the pectoral fins, known as cephalic lobes, project forward on each side of the head. Eyes and spiracles are located laterally and the mouth is very broad and either terminal or almost terminal. Minute teeth form bands in either one or both jaws. The tail is mostly filamentous with a small dorsal fin near its base, and a serrated stinging spine is sometimes present.

Members of the group are represented worldwide in warm temperate and tropical seas. They are mostly pelagic over continental shelves and near offshore islands. Often occurring in groups, they spend much of their time near the surface. Ten species are currently recognised as valid, but their distribution in this part of the Indo–Pacific is not well known. Specimens are rarely caught and, because of their large size, are not accurately identified or deposited in museum collections. From literature records and photographs available to the authors, it appears that at least three and possibly four species may occur in Australian waters.

Key to mobulids

1. Mouth at end of snout tip (fig. 1) .... **Manta birostris (fig. 4; p. 459)**

   Mouth located on undersurface of head (fig. 2) **2**

2. Head relatively long (fig. 2); distance from corner of mouth to tip of cephalic fin more than 16% of disc width (more than 15% in juveniles) .... **Mobula eregoodootenkee (fig. 5; p. 460)**

   Head relatively short (fig. 3); distance from corner of mouth to tip of cephalic fin less than 16% of disc width **3**

3. Anterior margin of disc concave initially then becoming convex near apex (fig. 6); no stinging spine .... **Mobula thurstoni (fig. 6; p. 463)**

   Anterior margin of disc straight or slightly convex (fig. 7); stinging spine mostly present .... **Mobula japanica (fig. 7; p. 461)**

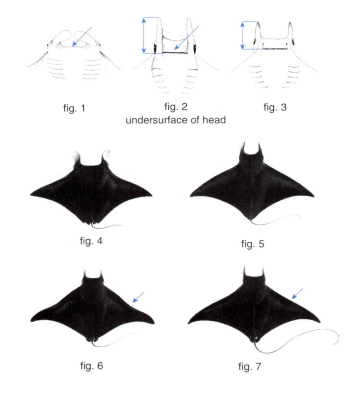

Manta Ray — *Manta birostris* (Donndorff, 1798)

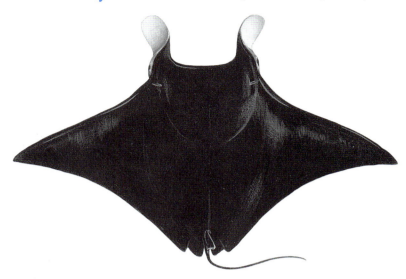

46.1 *Manta birostris* (Plate 81) Fish Code: 00 041004

undersurface of head

tooth (upper jaw)

- **Alternative names:** Australian devilray, devilfish, manta, munguna, Prince Alfreds ray.
- **Field characters:** A gigantic devilray with a lozenge-shaped disc, prominent cephalic lobes, and a very broad, terminal mouth with no teeth in the upper jaw.
- **Distinctive features:** Disc broad, short, length (measured from mid-snout) about 43*% of disc width; anterior margin straight to slightly convex. Head very short, distance from mid-snout to fifth gill slit more than 5 in disc width; cephalic lobes long, tip of cephalic lobe to corner of mouth about 14*% of disc width; rostral margin straight. Spiracles small, furrow-like. Mouth located at tip of snout; teeth small in lower jaw (often concealed by skin); no teeth in upper jaw. Body surface rough. Tail shorter than disc width, lacking a stinging spine.
- **Colour:** Greyish blue to greenish brown above; frequently with irregular, paler shoulder patches; dorsal fin similar to upper disc. White below with greyish margin, sometimes with irregular dark patches on disc.
- **Size:** Reaches a disc width of at least 670 cm and reported to 910 cm; specimens of 400 cm disc width or more are common.
- **Distribution:** Circumtropical and pelagic. Locally, mainly over the continental shelf off northern Australia. Occasionally observed in temperate areas, south to at least Montague Island (New South Wales) and Rottnest Island (Western Australia).

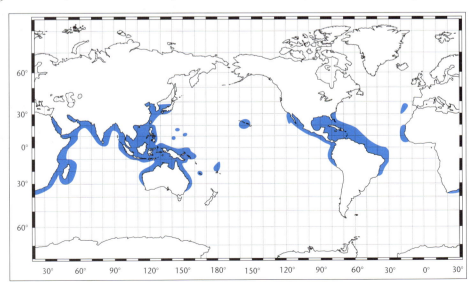

- **Remarks:** The manta is the largest ray and one of the largest living fishes. This graceful giant is quite common off tropical Australia and occasionally migrates into temperate waters. It is capable of rapid speed and will sometimes leap well clear of the water, landing with a loud slap. It is occasionally encountered by divers, as it tends to be rather inquisitive. Reports of divers riding on large adults are not uncommon. Like other devilfishes it feeds on small planktonic organisms which are ingested through its gaping mouth and sieved through gills which are modified into complex filtering plates. Viviparous.
- **Local synonymy:** *Ceratoptera alfredi* Krefft, 1868; *Manta alfredi*: Munro, 1956.
- **References:** Whitley (1936a); Notarbartolo-di-Sciara (1987).

Pygmy Devilray — *Mobula eregoodootenkee* (Cuvier, 1829)

46.2 Mobula eregoodootenkee (Plate 81) **Fish Code:** 00 041001

- **Alternative names:** Ox ray, smaller devilray, diamond fish, eregoodoo.
- **Field characters:** A small devilray with a long head and prominent cephalic lobes (distance between lobe tip and corner of mouth less than 6.3 in disc width), a ventral mouth, a quadrangular tail base, and no stinging spine or white tip on the dorsal fin.
- **Distinctive features:** Disc broad, short, length (measured from mid-snout) 57–59% of disc width; anterior margin straight to slightly convex. Head long, distance from mid-snout to fifth gill slit about 3.6–3.8 in disc width; cephalic lobes long, tip of cephalic lobe to corner of mouth about 17–18% of disc width; rostral margin deeply concave. Spiracles very small, almost circular, situated below level of pectoral fins. Mouth on undersurface of head; teeth small, crown width exceeding tooth height. Gill filter-plates reduced, with 4–6 short lateral lobes and a leaf-shaped terminal lobe without ridges. Surface mostly smooth, some denticles aggregated around dorsal-fin base. Tail shorter than disc width, lacking a stinging spine; base quadrangular in section.
- **Colour:** Grey to greyish brown above; dorsal fin uniformly dark. White below with dark anterior pectoral-fin margins.
- **Size:** To about 100 cm disc width.

undersurface of head

tooth (upper jaw)

- **Distribution:** Northern Indian Ocean from the Red Sea to Australia. Locally, known from off northern Australia from Townsville (Queensland) to Port Hedland (Western Australia); pelagic, probably more widely distributed within the tropics.
- **Remarks:** This small devil ray is fairly common off the Queensland coast. Misidentified previously as *Mobula diabolus*, it is also likely to have been confused regularly with other *Mobula* species. The rather unusual scientific name was coined from the common name for the ray in the Coromandel region of India. Specimens are required for further study.
- **Local synonymy:** *Mobula diabolus*: Whitley, 1936 (misidentification).
- **References:** Whitley (1936a); Notarbartolo-di-Sciara (1987).

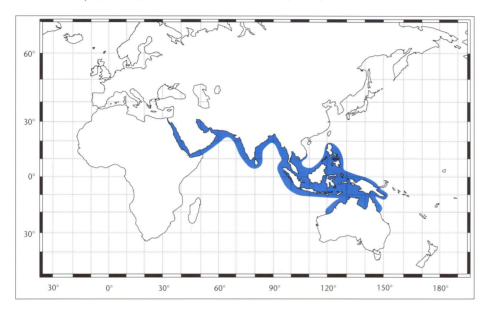

Japanese Devilray — *Mobula japanica* (Müller & Henle, 1841)

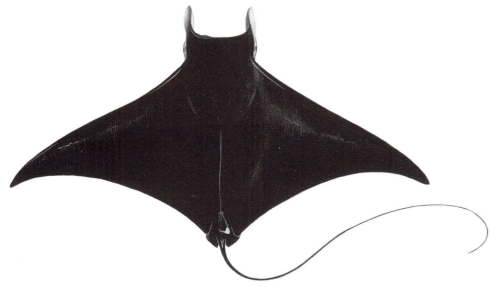

46.3 *Mobula japanica* (Plate 81) Fish Code: 00 041002

- **Field characters:** A medium to large devilray with a short head and cephalic lobes (distance between lobe tip and corner of mouth more than 6.3 in disc width), a ventral mouth, a slightly compressed tail base, a white tip on the dorsal fin, and mostly with a stinging spine. The tail is long and wiry, with prominent, white lateral denticles.

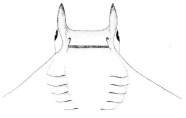
undersurface of head

- **Distinctive features:** Disc broad, short, length (measured from mid-snout) 44–53% of disc width; anterior margin uniformly straight to slightly convex. Head very short, distance from mid-snout to fifth gill slit about 4.5–5.2 in disc width; cephalic lobes short, tip of cephalic lobe to corner of mouth about 11–13% of disc width; rostral margin straight. Spiracles elliptical, transverse, slit-like, situated above level of pectoral fins. Mouth on undersurface of head; teeth small, tooth height greatly exceeding crown width. Gill filter-plates with 18–25 lateral lobes; terminal lobe leaf-shaped with longitudinal ridges. Denticles dense. Tail long, when undamaged subequal to or longer than disc width; mostly with a stinging spine; base oval, slightly compressed.

- **Colour:** Upper surface bluish black; two white crescentic patches on shoulders in juveniles, sometimes fading in adults; dorsal-fin tip white; inside of mouth dark. Disc white ventrally, often with dark patches in adults; no broad dark margin anteriorly.

- **Size:** Attains at least 310 cm disc width; usually smaller than 250 cm in the eastern Pacific; born at about 85 cm disc width.

- **Distribution:** Possibly circumtropical and pelagic, but presently known from many localities in the Atlantic, Pacific and Indian Oceans. Recorded off New Zealand, but its occurrence off Australia needs to be validated.

- **Remarks:** Little is known of the biology and distribution of this species in the western Pacific. Specimens are required.

- **References:** Whitley (1936a); Paulin *et al.* (1982); Notarbartolo-di-Sciara (1987).

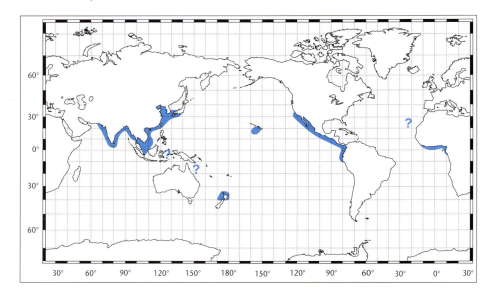

Bentfin Devilray — *Mobula thurstoni* (Lloyd, 1908)

46.4 *Mobula thurstoni* (Plate 81)   Fish Code: 00 041003

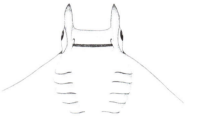
undersurface of head

- **Field characters:** A medium-sized devilray with a short head and cephalic lobes (distance between lobe tip and corner of mouth more than 6.3 in disc width), a ventral mouth, a depressed tail base, a white tip on the dorsal fin, and lacking a stinging spine. The anterior margin of the disc is wavy.

- **Distinctive features:** Disc broad, short, length (measured from mid-snout) about 47–56% of disc width; anterior margin concave then distinctly convex before tip. Head relatively short, distance from mid-snout to fifth gill slit about 3.9–4.6 in disc width; cephalic lobes short, tip of cephalic lobe to corner of mouth about 11–14% of disc width; rostral margin straight. Spiracles small, almost circular, situated below level of pectoral fins. Mouth on undersurface of head; teeth small, crown width exceeding tooth height. Gill filter-plates with 15–20 lateral lobes; terminal lobe leaf-shaped or elliptical, sometimes with median longitudinal ridges. Denticles sparse. Tail shorter than disc width, lacking a caudal spine; base dorsally flattened.

- **Colour:** Upper surface dark bluish; silvery grey around eye; dorsal-fin tip white; inside of mouth pale. Disc white ventrally, often with silvery patches near pectoral-fin apices; anterior margin of disc dark.

- **Size:** Attains at least 180 cm disc width; born at 65–85 cm.

- **Distribution:** Possibly circumglobal and pelagic in tropical seas. Locally, only from off Mackay (Queensland).

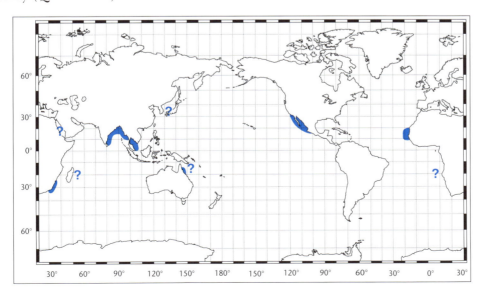

- **Remarks:** This species is called the bentfin devilray because of its distinctively curved anterior disc margin. A provisional record for Australia is based on colour transparencies of an unidentified *Mobula* caught recently off Queensland. More specimens are required from this region.
- **References:** Whitley (1936a); Notarbartolo-di-Sciara (1987).

47

Family Callorhinchidae
ELEPHANT FISHES

Elephant fishes, also known as plownose chimaeras, have a smooth, rather elongate body, a heterocercal tail, a single gill opening, and a long snout terminating in a flexible hoe-shaped structure. The head has a well-developed complex of mucous canals and sensory pores. The dentition consists of pairs of plate-like teeth in both jaws, and an additional pair of tooth plates in the upper jaw. The two dorsal fins are widely separated; the first is erectile, triangular in shape, and preceded by a serrated spine; the second is short-based (compared with other chimaeras) and tallest anteriorly. The caudal fin is preceded by a tall, narrow-based anal fin and lacks a caudal filament. Females are oviparous; the egg cases are large, flask-shaped and have broad horizontal flanges. Males have simple rod-like intromittent claspers without dilated tips. Like their relatives, the longnose and shortnose chimaeras, elephant fishes have a pair of retractable, prepelvic claspers and a thumb-shaped head clasper that are used to hold the female during copulation. The complex, spoon-shaped prepelvic claspers are made up of frill-like lobes, a globular gland, and multicuspid denticles along the inner face of the blade.

These highly distinctive fishes are confined to temperate areas of Australasia, South America and southern Africa. The family contains a single genus, *Callorhinchus*, and one of the three species lives in the Australian region.

Elephant Fish — *Callorhinchus milii* (Bory de Saint-Vincent, 1823)

47.1 *Callorhinchus milii* (Plate 82)   **Fish Code:** 00 043001

- **Alternative names:** Elephant shark, ghost shark, reperepe, whitefish.
- **Field characters:** A chimaera with a bizarre, fleshy, plough-shaped snout, an anal fin, an arched caudal fin, and a rather short-based and tall second dorsal fin.

- **Distinctive features:** Body rather elongate, slightly compressed; profile of forehead highly convex; eye small, diameter about a half to a third of distance between eye and dorsal spine; mouth directed anteroventrally. Snout tapering rapidly to a plough-shaped process; outer limb of process depressed, directed posteriorly. Teeth plate-like, not forming a cutting edge laterally; anterior upper plates small, forming a weak beak anteriorly; posterior upper plates large, with V-shaped ridges; lower plates with broad longitudinal ridges; gums of upper jaw with enlarged papillae. First dorsal fin and spine tall; tip of depressed spine well short of origin of second dorsal fin. Second dorsal-fin base relatively short, less than twice length of pelvic fin; anterior third of fin very tall. Base of upper lobe of caudal fin only slightly longer than base of lower lobe; caudal fin broadest at origin of lower lobe; no caudal filament; anal fin taller than caudal fin. Claspers tapering gradually toward apex; outer portion indented slightly. Head clasper spiny near apex with additional spines near front of clasper groove.
- **Colour:** Silvery white; often with irregular, brownish black blotches behind eye and on sides. Fins similar, usually with brownish markings.
- **Size:** Reaches about 120 cm; males mature at about 65 cm. Egg cases large, about 25 cm long and 10 cm wide.
- **Distribution:** Continental shelves of cool temperate areas of Australia and New Zealand in depths to at least 200 m. Locally, southern Australia from Esperance (Western Australia) to Sydney (New South Wales), including Tasmania.
- **Remarks:** In Australian waters, most abundant south of Bass Strait where it is caught commercially and sold as whitefish fillets. Migrates into large estuaries and inshore bays (usually shallower than 40 m) in spring to breed. The egg cases, which are sometimes washed ashore on nearby beaches during storms, are laid on sandy or muddy substrates and take up to 8 months to hatch. In New Zealand, the annual catch sometimes approaches 1000 tonnes with a substantial quantity exported to Australia. Small quanities are taken by commercial vessels off Victoria and Tasmania.
- **Local synonymy:** *Callorhynchus tasmanius* Richardson, 1840; *Callorhynchus australis* Owen, 1852; *Callorhynchus antarcticus*: McCoy, 1886 (misidentification).
- **Reference:** McCoy (1886).

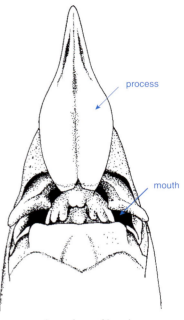

undersurface of head

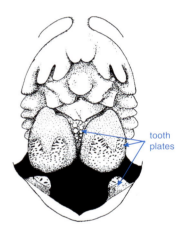

mouth

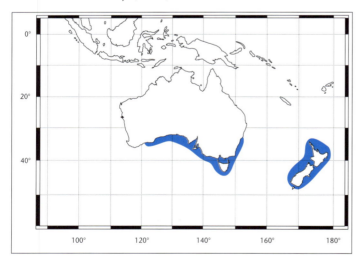

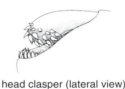

head clasper (lateral view)

pre-pelvic clasper

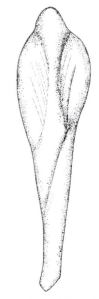

upper surface of right clasper

Family Chimaeridae
SHORTNOSE CHIMAERAS

Shortnose chimaeras, also known as ratfishes, have long, smooth bodies, a short rounded or feebly pointed snout, a diphycercal tail and a single gill opening. The head has a well-developed network of mucous canals and sensory pores. The mouth is small and located ventrally. The dentition consists of pairs of beak-like cutting teeth in both jaws and an additional pair of tooth plates in the upper jaw. The first dorsal fin, triangular in shape and preceded by a smooth or serrate spine, is erectile. The second dorsal fin, which is almost connected to the upper caudal-fin lobe, is very long and low with a straight or slightly convex outer margin. A caudal filament of variable length is usually present. A small anal fin is present or absent. Females are oviparous; the large egg cases are tadpole-shaped with a terminal filament. Males have bifid or trifid claspers with expanded tips; accessory appendages are present on the head and in the prepelvic area of mature males. Head clasper greatly expanded at tip with fine spines on anterior surface. Prepelvic clasper more or less disc-shaped with short thorns on ventral margin.

Members of this family are found worldwide in both temperate and tropical seas. Some species occur near the shore at high latitudes, but most are benthopelagic on continental slopes and abyssal plains. Like other chimaeroid fishes they are slow swimmers. They appear to feed mainly on small fishes and invertebrates, particularly cephalopods. The two genera, *Chimaera* and *Hydrolagus,* are both represented in the Australian region but the systematics of these groups needs further attention. Although some of the species described below appear to be new to science, their distributions are not well known. Unidentified species from New Zealand are similar. Some species could be of minor commercial importance.

Key to chimaerids

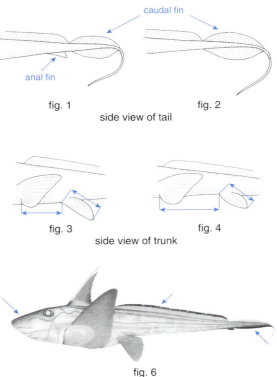

1 Anal fin present (fig. 1) **2**
   Anal fin absent (fig. 2) **6**

2 Pelvic-fin length less than 1.5 in interspace between pectoral and pelvic-fin bases (fig. 3) **3**
   Pelvic-fin length more than 1.5 in interspace between pectoral and pelvic-fin bases (fig. 4) **4**

3 Body purplish, snout rather short (fig. 5); second dorsal fin and lower lobe of caudal fin uniform in colour .... ***Chimaera* sp. D (fig. 5; p. 473)**
   Body silvery, snout rather long (fig. 6); margins of second dorsal fin and lower lobe of caudal fin darker than remainder of fish (fig. 6) ....
   .... ***Chimaera* sp. E (fig. 6; p. 474)**

4 Dorsal spine long (when undamaged longer than first dorsal fin) (fig. 7); margin of first dorsal fin black (fig. 7); mature clasper short, less than 10% of distance from snout tip to insertion of second dorsal fin .... **Chimaera sp. C (fig. 7; p. 471)**

Dorsal spine short (when undamaged mostly shorter than first dorsal fin) (fig. 8); first dorsal fin uniformly dark or with pale posterior margin; mature clasper long, more than 10% of distance from snout tip to insertion of second dorsal fin **5**

5 Body pale to medium brown; first dorsal, pectoral and pelvic fins dusky with pale outer margins .... **Chimaera sp. A (fig. 8; p. 469)**

Body blackish; first dorsal, pectoral and pelvic fins uniformly black .... **Chimaera sp. B (fig. 9; p. 470)**

6 Body and fins uniformly black (except where skin has been removed); claspers divided for about a third of their length (fig. 10) .... **Hydrolagus sp. A (fig. 12; p. 475)**

Body and fins not uniformly black; claspers divided for a half or more of their length (fig. 11) **7**

7 Sides marbled with a pattern of light and dark reticulations (rarely faint) (fig. 13) .... **Hydrolagus sp. B (fig. 13; p. 476)**

Sides without a network of light and dark reticulations (sometimes with a few large blotches or stripes) **8**

8 First dorsal fin uniformly brownish or bluish black; dark margin of second dorsal fin broader than a third of fin height (fig. 14) .... **Hydrolagus ogilbyi (fig. 14; p. 478)**

First dorsal fin with a black posterior margin (fig. 15); dark margin of second dorsal fin narrower than a third of fin height (fig. 15) .... **Hydrolagus lemures (fig. 15; p. 477)**

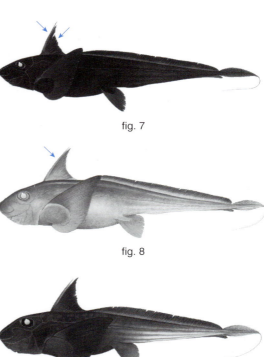

fig. 7

fig. 8

fig. 9

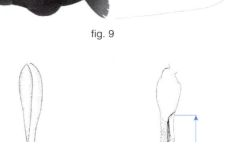

fig. 10   fig. 11

claspers

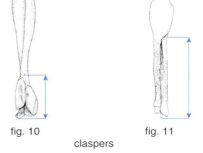

fig. 12

fig. 13

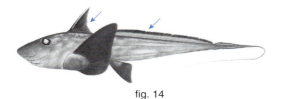

fig. 14

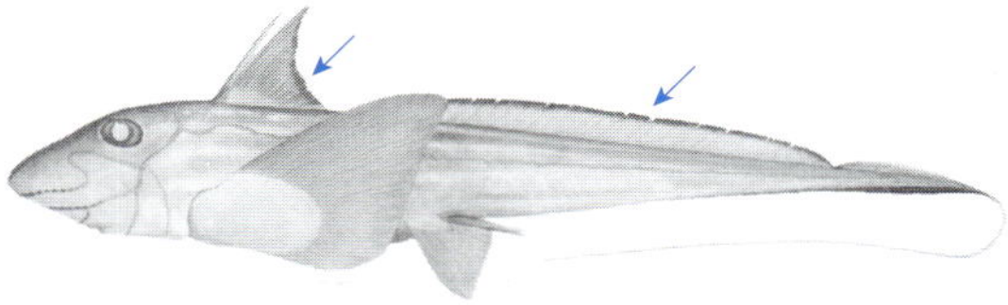

fig. 15

Southern Chimaera — *Chimaera* sp. **A**

48.1 *Chimaera* sp. **A** (Plate 83)　　　　　　　　　　**Fish Code:** 00 042005

- **Alternative name:** Deepwater Ghostshark.
- **Field characters:** A large, pale brownish, shortnose chimaera with deciduous skin, an anal fin, large claspers with bulbous tips, and usually a pale posterior margin on the first dorsal fin. The first dorsal fin is mostly slightly taller than the dorsal spine and the pelvic fin longer than 1.5 in the pectoral–pelvic interspace.
- **Distinctive features:** Body moderately robust, tapering rapidly from massive head; eye large, about 1.4–1.9* in snout length of adults. Snout short, rather blunt. Anterior teeth beak-like; upper anteriors rather small, posteriors plate-like with broad longitudinal ridges; lowers with a weak, inner longitudinal ridge. Tail relatively short, distance from cloaca to end of upper caudal-fin lobe 1.0–1.35* times precloacal length. Lateral line only slightly wavy. First dorsal fin equal to or slightly taller than dorsal spine; its tip extending beyond origin of second dorsal fin; undamaged caudal filament mostly subequal to pectoral-fin length in adults (longer in juveniles). Anal fin small; its insertion under, or just posterior to, insertion of second dorsal fin. Pectoral and pelvic fins relatively short; pectoral-fin tip short of, or just extending to, pelvic-fin origin. Claspers large (about 14.8–16.5*% of distance from snout tip to insertion of second dorsal fin when mature), trifurcate, divided for slightly less than half of their length, bulbous distally with large patches of bristles.
- **Colour:** Silvery pink to pale brown; mostly paler on belly, lower surface of head, and around mouth; sides of tail usually with 3–4 faint longitudinal stripes; teeth pale; eye silvery white with green pupil; caudal filament white; claspers pale. In southern populations, first dorsal fin dusky near base, often whitish along posterior margin; other fins dusky, except for slightly pale, translucent or white areas along

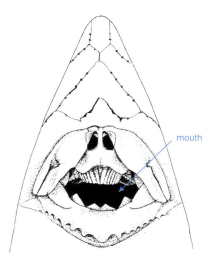

undersurface of head

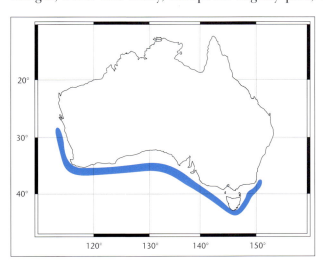

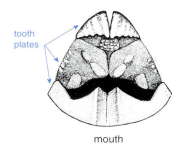

mouth

dorsal view of right clasper

basal half to third of second dorsal fin, and along margins of pectoral, pelvic and caudal fins. In western populations, all fins distinctly bluish; only posterior margin of pelvic fin with broad white margin.

- **Size:** Attains at least 90 cm (without the caudal filament). Smallest mature males examined were 68 cm. Egg case about 3.5* cm wide, 22* cm long.
- **Distribution:** Continental slope off southern Australia (including Tasmania) between the Abrolhos Islands (Western Australia) and Eden (New South Wales). Lives on or near the bottom in 300–850 m.
- **Remarks:** The southern chimaera is frequently caught by trawlers fishing for blue grenadier. The flesh is of good quality and small quantities find their way to fish markets of southern Australia. Two colour forms exist and their relationship to the closely related shortspine chimaera (48.2) and to each other needs to be investigated further. Little is known of its biology.
- **Local synonymy:** *Hydrolagus* sp.: Last *et al.*, 1983; *Chimaera* sp. 1: May & Maxwell, 1986.
- **Reference:** Last *et al.* (1983).

head clasper (lateral view)

pre-pelvic clasper

Shortspine Chimaera — *Chimaera* sp. B

48.2 *Chimaera* sp. B (Plate 83) **Fish Code:** 00 042006

- **Field characters:** A dark, moderately robust, shortnose chimaera with deciduous skin, a relatively short tail, an anal fin, large claspers with bulbous tips, and a uniformly dark first dorsal fin. The first dorsal fin is usually taller than the dorsal spine.
- **Distinctive features:** Body moderately robust, tapering rapidly from massive head; eye large, about 1.2–1.6* in snout length of adults. Snout short, blunt. Anterior teeth beak-like; upper anteriors rather small, posteriors plate-like with small knobs; lowers sharp edged with a deeply concave inner margin. Tail relatively short, distance from cloaca to end of upper caudal-fin lobe 1.1–1.25* times precloacal length. Lateral line slightly wavy. First dorsal fin taller than dorsal spine; its tip extending well beyond origin of second dorsal fin; pectoral fin extending at least to origin of pelvic fin in large specimens; caudal filament mostly longer than pectoral-fin length. Anal fin small; its insertion mostly just posterior to insertion of second dorsal fin. Claspers large (when mature 13.8–15.8*% of distance from snout tip to insertion of second dorsal fin), trifurcate, divided for less than half their length, bulbous distally with large patches of bristles.

- **Colour:** Uniformly dark brownish to black, pale where skin deciduous; belly, teeth, undersurface of head and snout dark; narrow, indistinct stripes on lower half of tail; dorsal spine mostly dark; first dorsal, pectoral and pelvic fins black; second dorsal fin brownish black with narrow translucent basal portion; caudal filament and claspers mostly pale.
- **Size:** Attains at least 69 cm (without the caudal filament). Smallest mature males examined were 54 cm.
- **Distribution:** Demersal on the continental slope of eastern Australia, between Townsville (Queensland) and Ulladulla (New South Wales), in 450–1000 m.
- **Remarks:** This species is rather common throughout its range but nothing is known of its biology. It differs from the southern chimaera (48.1), to which it is closely related, in being smaller and darker in colour. It could be a variant of the latter.

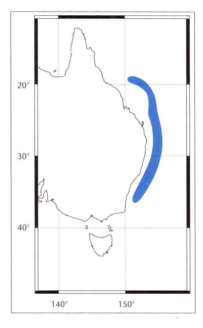

dorsal view of right clasper

Longspine Chimaera — *Chimaera* sp. C

48.3 *Chimaera* sp. C (Plate 83)　　　　　　　　　　　　Fish Code: 00 042007

- **Field characters:** A rather large, dark brownish, shortnose chimaera with deciduous skin, a relatively long tail, an anal fin, relatively short claspers, and a black posterior margin on the first dorsal fin. The first dorsal fin is shorter than the dorsal spine.

- **Distinctive features:** Body relatively slender, tapering rapidly from rather large head; eye large, about 1.2–1.4* in snout length. Snout short, rather blunt. Anterior teeth beak-like; upper anteriors rather small, posteriors plate-like with narrow longitudinal ridges; lowers with a weakly concave margin. Tail relatively long, distance from cloaca to end of upper caudal-fin lobe 1.3–1.45* times precloacal length. Lateral line slightly wavy. First dorsal fin much shorter than height of dorsal spine; its tip extending well behind origin of second dorsal fin; caudal filament often short. Anal fin small; its insertion just posterior to insertion of second dorsal fin. Claspers relatively short (8.4–8.8 *% of distance from snout tip to insertion of second dorsal fin when mature), trifurcate, divided for slightly less than half of their length, slightly enlarged distally with bristle patches.
- **Colour:** Chocolate brown, whitish on head and abdomen where skin has been removed; belly and pectoral region dark; snout slightly paler; teeth dark greyish or brownish; first dorsal fin dark brown with black posterior margin; pectoral and pelvic fins bluish brown; dorsal spine moderately dark; caudal filament pale; faint stripes present on lower half of tail below lateral line.
- **Size:** Attains at least 77 cm (without the caudal filament). Smallest mature males examined were 72 cm.
- **Distribution:** Continental slope of warm temperate and tropical Australia; between Perth and North West Cape (Western Australia), from Byron Bay to Wollongong (New South Wales), and on the Queensland Plateau, in 440–1300 m (mainly deeper than 800 m). Possibly also in New Zealand waters.
- **Remarks:** Little information is available on this chimaera. Its geographic range partly overlaps with the closely related shortspine (48.2) and southern chimaeras (48.1), but it normally occurs in deeper water than its relatives.

dorsal view of right clasper

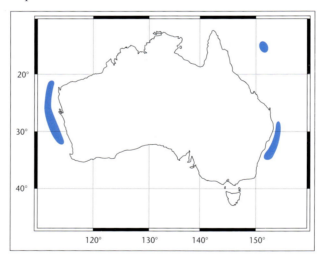

Giant Chimaera — *Chimaera* sp. **D**

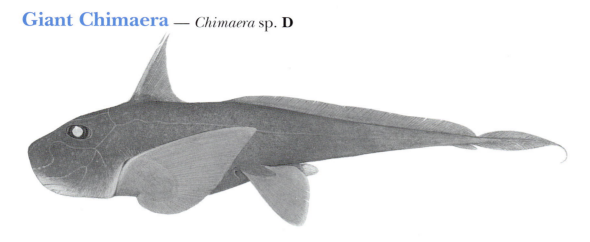

48.4 *Chimaera* sp. **D** (Plate 83) Fish Code: 00 042008

- **Field characters:** A very large, purplish blue, shortnose chimaera with a long tail, an anal fin, a uniformly bluish first dorsal fin, and short, very robust claspers. The first dorsal fin is usually taller than the dorsal spine and the pelvic-fin length is less than 1.5 in the pectoral–pelvic interspace.

- **Distinctive features:** Body very robust, tapering gradually; head massive in adults; eye large, about 1.5* in snout length. Snout short, blunt. Anterior teeth beak-like; upper anteriors rather small, posteriors plate-like with knobs; lowers with a weakly concave margin. Tail rather short, distance from cloaca to end of upper caudal-fin lobe 1.4–1.45* times precloacal length. Lateral line only slightly wavy on tail. First dorsal fin taller than dorsal spine, when depressed its tip just reaching origin of second dorsal fin; pectoral and pelvic fins huge, pectoral fin extending beyond pelvic-fin insertion in large specimens; caudal filament mostly short. Anal fin small; its insertion just anterior to insertion of second dorsal fin. Claspers massive, rather short (when mature about 12.5–14*% of distance from snout tip to insertion of second dorsal fin), trifurcate, divided for slightly less than half of their length, bulbous distally with large patches of bristles.

dorsal view of right clasper

- **Colour:** Uniformly purplish blue; pale around mouth and undersurface of snout; teeth whitish; faint stripes along lower half of tail. Fins uniformly bluish, without distinct light or dark margins.

- **Size:** Attains at least 110 cm (without the caudal filament).

- **Distribution:** Mid-continental slopes of southern Australia (so far only off Tasmania and seamounts to the south) and New Zealand in depths exceeding 900 m.

- **Remarks:** This species is rarely caught and nothing is known of its biology. More specimens are required.

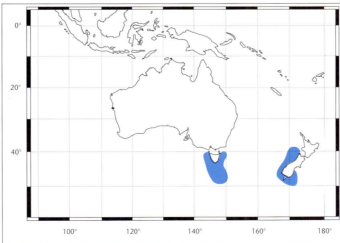

Whitefin Chimaera — *Chimaera* sp. E

48.5 *Chimaera* sp. E (Plate 84)   Fish Code: 00 042009

- **Field characters:** A slender, medium to large, shortnose chimaera with a long snout, a very tall dorsal spine, a black anal fin, a wavy lateral line, and a pale posterior margin on the first dorsal fin.

- **Distinctive features:** Body slender, tapering gradually from massive head; eye large, about 2–2.5 in snout length, about 3.6–4.5 in dorsal spine length of adults. Snout relatively long, broadly pointed. Anterior teeth beak-like; upper anteriors rather small, posteriors plate-like with an enlarged posterior ridge; lower teeth with a similar, enlarged posterior ridge and a weakly concave margin. Tail relatively long, distance from cloaca to end of upper caudal-fin lobe 1.3–1.4 times precloacal length. Lateral line very wavy. Dorsal spine very tall, first dorsal fin distinctly shorter than dorsal spine, its tip extending well beyond origin of second dorsal fin; caudal-fin lobes relatively tall; caudal filament mostly longer than pectoral-fin length. Anal fin small. Claspers massive, very elongate (when mature about 18% of distance from snout to insertion of second dorsal fin), bifurcate, divided for about half of their length, extremely bulbous distally.

- **Colour:** Uniformly pale, sometimes with a slightly darker stripe above the lateral line (often faint); undersurface of head and snout pale. First dorsal fin dusky with a white outer margin; pectoral and pelvic fins pale grey; basal half or so of second dorsal fin pale, margin black. Lower caudal lobe usually with a broad, dark edge; upper lobe with a dark base and pale margin. Dorsal spine pale; anal fin black; caudal filament white.

- **Size:** Attains at least 78 cm (without the caudal filament).

- **Distribution:** Known from the continental slope off northern Western Australia, from Shark Bay to the Rowley Shoals, in 440–520 m.

- **Remarks:** This species more closely resembles a *Hydrolagus* than a *Chimaera*. It is abundant throughout its range and could be of commercial value.

dorsal view of right clasper

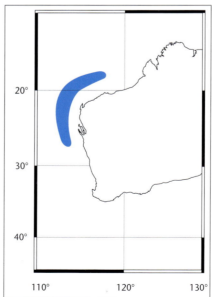

Black Ghostshark — *Hydrolagus* sp. A

48.6 *Hydrolagus* sp. A (Plate 84)
Fish Code: 00 042010

- **Field characters:** A moderately large, uniformly dark brownish black, shortnose chimaera with highly deciduous skin, uniformly black fins, slender claspers that are divided for only about a third of their length, and lacking an anal fin.

- **Distinctive features:** Body moderately robust, tapering rapidly from a rather large head; eye large, about 1.4–1.9* in snout length, about 2.8–4* in dorsal-spine length of adults. Snout of medium length, pointed in adults, much longer in juveniles. Anterior teeth long and beak-like; lowers and upper posteriors with several longitudinal ridges. Tail relatively long, distance from cloaca to end of upper caudal-fin lobe 1.25–1.55* times precloacal length. Lateral line straight or only slightly wavy. Dorsal spine relatively tall; first dorsal fin usually much longer than dorsal spine, its tip extending beyond origin of second dorsal fin; caudal filament mostly shorter than pectoral-fin length. Anal fin absent. Claspers relatively short (when mature about 7.7–9*% of distance from snout tip to insertion of second dorsal fin), trifurcate, divided for about a third of their length, slender distally with small bristle patches.

- **Colour:** Black or brownish black, undersurface of head and fins black; dorsal spine greyish to black; teeth dark; eye bluish black; caudal filament and claspers dark.

- **Size:** Attains at least 93 cm (without the caudal filament). Smallest mature males examined were 85 cm.

- **Distribution:** Continental slope of southeastern Australia between Ulladulla (New South Wales) and Portland (Victoria), including Tasmania and the seamounts to the south (Cascade Plateau and South Tasman Rise), in 900–1400 m. Also occurs off New Zealand.

- **Remarks:** This chimaera is commonly taken by trawlers operating in the orange roughy (*Hoplostethus atlanticus*) grounds off Tasmania. The flesh is of good edible quality. Little is known of its biology.

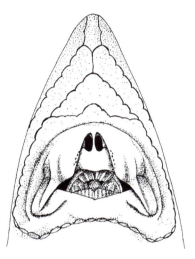

undersurface of head

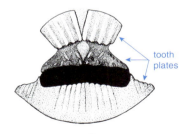

mouth

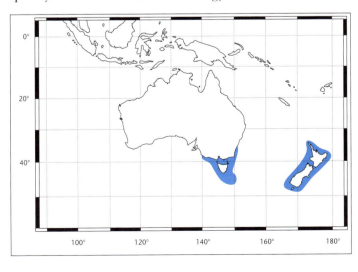

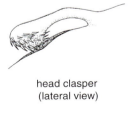

head clasper (lateral view)

pre-pelvic clasper

right clasper (dorsal view)

Marbled Ghostshark — *Hydrolagus* sp. **B**

48.7 *Hydrolagus* sp. **B** (Plate 84) Fish Code: 00 042011

- **Field characters:** A rather small shortnose chimaera with deciduous skin, slender claspers, a uniformly dusky first dorsal fin, a distinctive network of light and dark reticulations on the flanks, and lacking an anal fin.

- **Distinctive features:** Body rather robust, tapering gradually from a moderately large head; eye large, about 1.1–1.8* in snout length, about 2.4–2.6 in dorsal-spine length of adults. Snout rather short, blunt. Anterior teeth long and beak-like; posteriors plate-like, thickened, with pronounced knobs. Tail relatively short, distance from cloaca to end of upper caudal-fin lobe 1.15–1.3* times precloacal length. Dorsal spine relatively short; first dorsal fin slightly taller than spine, its tip extending beyond origin of second dorsal fin; caudal filament mostly longer than pectoral fin. No anal fin. Lateral line only slightly wavy. Claspers slender, very long (when mature about 18*% of distance from snout tip to insertion of second dorsal fin), trifurcate, divided for more than half of their length, bristle patches over much of divided portions.

- **Colour:** Pale silvery white or cream with prominent, broad, dark brown to greyish brown reticulations on sides; head and belly pale brown; pale around mouth and on undersurface of snout; first dorsal fin, pectoral and pelvic fins, and lower lobe of the caudal fin brownish (mostly uniform); basal two-thirds of the second dorsal fin pale, dark brown beyond; dorsal spine rather dark; caudal filament white.

- **Size:** Attains at least 53 cm (without the caudal filament), males are mature at this size.

- **Distribution:** Continental slope off eastern Australia, between Townsville (Queensland) and Sydney (New South Wales), in 450–850 m.

- **Remarks:** Little is known of the biology of this rather diminutive chimaera.

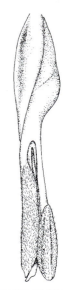
dorsal view of right clasper

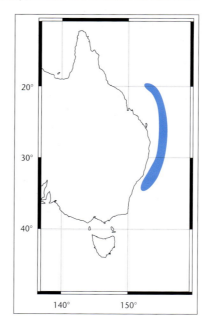

Blackfin Ghostshark — *Hydrolagus lemures* (Whitley, 1939)

48.8 *Hydrolagus lemures* (Plate 84)

Fish Code: 00 042003

- **Alternative name:** Bight ghostshark.
- **Field characters:** A slender, pale chimaera with a rather long snout, the dorsal spine shorter than the first dorsal fin, bulbous claspers and no anal fin. The posterior margin of the first dorsal fin is dark, the body colour is not reticulated, and the second dorsal fin has a narrow, dark margin (its height mostly less than a quarter of fin height).
- **Distinctive features:** Body slender, tapering gradually from rather large head; eye large, about 1.6–2.3 in snout length, about 2.3–3.3 in dorsal spine length in adults. Snout relatively long, broadly pointed. Anterior teeth beak-like, with very deep concavities on lowers; posterior upper teeth plate-like, thickened, with broad ridges. Tail relatively long, distance from cloaca to end of upper caudal-fin lobe 1.1–1.4 times precloacal length. Lateral line very wavy. First dorsal spine rather short; first dorsal fin distinctly taller than dorsal spine, its tip extending well beyond origin of second dorsal fin; caudal-fin lobes relatively low; caudal filament usually longer than pectoral fin. No anal fin. Claspers robust, bifurcate, about 18–20% of distance from snout tip to insertion of second dorsal fin; divided for about two-thirds of their length; tips bulbous.
- **Colour:** Pale silvery white, with pale brownish areas, but without distinct reticulations; undersurface of head and snout uniformly pale; upper snout dark; first dorsal fin dusky with a broad, black posterior margin (more extreme in juveniles); pectoral and pelvic fins dark greyish, their anterior margins mostly black (margins silvery white in juveniles); basal half to three-quarters of second dorsal fin pale, outer portion black; lower caudal-fin lobe black-edged, upper lobe uniformly white or translucent, with a dark margin; dorsal spine mostly dark; caudal filament silvery white.

dorsal view of right clasper

- **Size:** Attains at least 58 cm (without the caudal filament). Males mature at about 50 cm.
- **Distribution:** Appears to be widespread on the Australian continental slope, from Portland (Victoria) to Exmouth Gulf (Western Australia), from Cairns (Queensland) to eastern Bass Strait (excluding Tasmania) and off Darwin. Lives on or near the bottom in 200–510 m.
- **Remarks:** This species was originally described from two immature specimens from the Great Australian Bight off Eucla, Western Australia. As for other members of this family, little is known of its biology. The flesh is of high quality.
- **Reference:** Whitley (1939b).

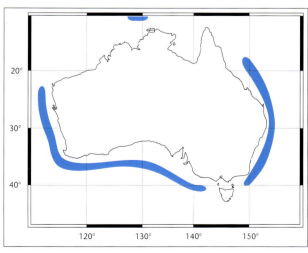

Ogilbys Ghostshark — *Hydrolagus ogilbyi* (Waite, 1898)

48.9 *Hydrolagus ogilbyi* (Plate 84)

Fish Code: 00 042001

- **Alternative names:** Ghostshark, spookfish, whitefish.
- **Field characters:** A slender, silvery brown chimaera with a rather long snout, relatively short dorsal spine, slender claspers, and no anal fin. The first dorsal fin is uniformly dark, the body colour is not reticulated, and the second dorsal fin has a broad dark margin (equal to or more than half fin height).
- **Distinctive features:** Body rather elongate, tapering behind abdomen; eye large, 1.7–2.1* in snout length, 2.9–3.8* in dorsal-spine length of adults. Snout relatively long, narrow, pointed. Anterior teeth beak-like, with deep concavities on lowers; posterior upper teeth plate-like, thickened, with broad ridges. Tail rather long, distance from cloaca to end of upper caudal-fin lobe 1.25–1.5* times precloacal length. Lateral line very wavy. Dorsal spine relatively short; first dorsal fin much taller than spine, its tip extending beyond origin of second dorsal fin; caudal filament mostly longer than pectoral fin. Anal fin absent. Claspers slender (when mature about 17*% of distance from snout tip to insertion of second dorsal fin), trifurcate, divided for much more than half of their length, distal portion not bulbous.
- **Colour:** Pale silvery white, without reticulations; sometimes with faint brownish blotches on sides (becoming darker after death); juveniles with dark brown stripes across lateral line, the lower stripe ending above the pelvic fins; undersurface of snout and head pale; first dorsal fin, pectoral and pelvic fins, and lower lobe of the caudal fin brownish or bluish black (mostly uniform); basal half or less of the second dorsal fin pale, margin dark brown; dorsal spine rather dark; caudal filament whitish.
- **Size:** Attains at least 85 cm (without the caudal filament). Males mature at about 70 cm.
- **Distribution:** Continental shelf and upper slope off southeastern Australia, between Coffs Harbour (New South Wales) and Beachport (South Australia), including Tasmania, in depths of 120–350 m.
- **Remarks:** This is the common shallow water chimaera of southern Australia. It is abundant in Tasmanian waters where it is caught by vessels fishing for flathead (*Platycephalus* spp), latchet (*Pterygotrigla polyommata*) and morwong (*Nemadactylus macropterus*). The flesh is of good quality and small quantities find their way to fish markets. Little is known of its biology.
- **Local synonymy:** *Hydrolagus (Psychichthys) waitei* Fowler, 1908.
- **References:** Fowler (1908a); Last *et al.* (1983).

dorsal view of right clasper

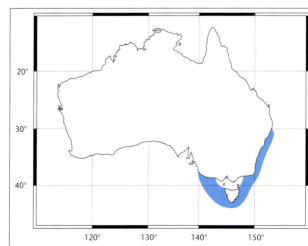

49 Family Rhinochimaeridae
SPOOKFISHES

Spookfishes, also known as longnose chimaeras or longnose rabbitfishes, have a smooth, rather elongate body, a long pointed snout, a diphycercal tail, and a single gill opening. The head has a well-developed network of mucous canals and sensory pores. The dentition consists of pairs of beak-like cutting teeth in both jaws and an additional pair of tooth plates in the upper jaw. The erectile first dorsal fin is triangular in shape and preceded by a smooth or serrated spine. The second dorsal fin is long and low, with a straight or only slightly convex outer margin, but has a shorter base than in the closely related shortnose chimaeras. A caudal filament of variable length is present (unless damaged), but Australian representatives lack an anal fin. Females are oviparous; the large egg cases are tadpole-shaped with wide lateral membranes and many transverse ribs. Males have simple rod-like claspers with expanded, spiny tips, as well as the typical chimaeroid accessory appendages on the head and prepelvic area.

The family is represented worldwide on the continental slopes and abyssal plains of tropical and temperate regions. Of the three recognized genera and at least seven species, two genera and three species occur in Australian waters.

Key to rhinochimaerids

1 Teeth forming a narrow beak, tooth plates with smooth surfaces (fig. 1); mouth located forward of eye (fig. 3); snout about 3 times pelvic-fin length (fig. 3) .... *Rhinochimaera pacifica* (**fig. 3; p. 482**)

Teeth forming a broad beak, tooth plates with ridges and knobs (fig. 2); mouth located forward of or below eye (fig. 4); snout about twice pelvic-fin length (fig. 4) **2**

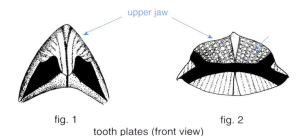

fig. 1 tooth plates (front view) fig. 2

2 Upper caudal-fin lobe longer than lower lobe (fig. 4); middle of snout distinctly conical; first dorsal fin barely taller than second dorsal fin (fig. 4) .... *Harriotta haeckeli* (**fig. 4; p. 480**)

Upper caudal-fin lobe shorter than lower lobe (fig. 5); middle of snout depressed; first dorsal fin more than twice height of second dorsal fin (fig. 5) ....*Harriotta raleighana* (**fig. 5; p. 481**)

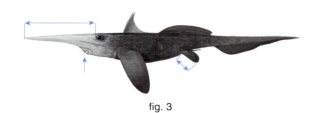

fig. 3

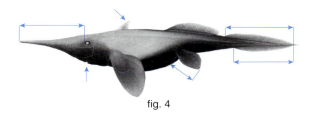

fig. 4

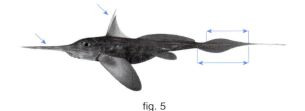

fig. 5

Smallspine Spookfish — *Harriotta haeckeli* Karrer, 1972

49.1 *Harriotta haeckeli* (Plate 82) **Fish Code:** 00 044003

- **Field characters:** A spookfish with a relatively broad and short, conical snout tapering rapidly to form a long rostral process distally, very small eyes situated above the mouth, a very short first dorsal fin and spine, knobbly tooth plates, and the upper caudal-fin lobe longer than the lower lobe.

- **Distinctive features:** Body rather elongate; head robust; eye very small, diameter more than 4 in distance between eye and dorsal spine; mouth directed almost ventrally. Snout conical near base; apex firm, slender, upturned slightly; ventral mucous canals not diverging abruptly halfway along snout. Anterior teeth beak-like, edges sharp; inner surfaces of lower and posterior upper teeth with large knobs. First dorsal fin only slightly taller than second dorsal fin; dorsal spine very short, strongly recurved, its tip well short of origin of second dorsal fin. Second dorsal-fin base relatively short, less than twice length of pelvic fin. Base of upper lobe of caudal fin longer than base of lower lobe, almost joined to second dorsal fin, lacking tubercles; caudal filament short. Pectoral-fin apex broadly rounded. No anal fin.

- **Colour:** Uniform pale brown above, darker ventrally; dorsal spine pale.

- **Size:** Reaches at least 65 cm (without the caudal filament).

- **Distribution:** Until recently, known only from a few North Atlantic specimens taken in 1800–2600 m. Specimens collected by Russian vessels from submarine seamounts of the Indian Ocean in 1400–1730 m are the first records of the species outside the Atlantic. Two Australian specimens from off St Helens (Tasmania) in 1480–1700 m represent additional records for the Southern Hemisphere.

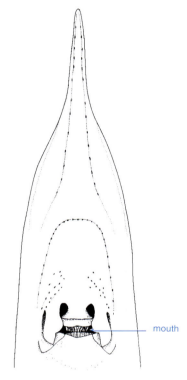

undersurface of head

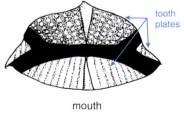

mouth

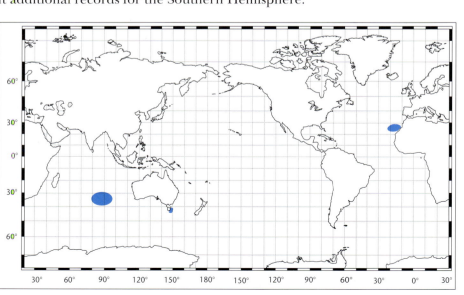

- **Remarks:** This species is rare in collections and almost nothing is known of its biology. Juvenile specimens appear to have deeper bodies and relatively longer first dorsal fins than adults. This spookfish is thought to live at greater depths than the bigspine spookfish (49.2). The Australian specimens were taken in the deepest demersal trawl sample ever taken by an Australian vessel and it may be more common beyond the depths fished by the commercial fleet.

Bigspine Spookfish — *Harriotta raleighana* Goode & Bean, 1895

49.2 *Harriotta raleighana* (Plate 82)   Fish Code: 00 044001

- **Alternative name:** Narrownose chimaera.
- **Field characters:** A spookfish with a rather long, narrow, depressed snout, a small eye situated above or behind the mouth, a rather long first dorsal fin and spine, knobbly tooth plates, and with the lower caudal-fin lobe longer than the upper lobe.
- **Distinctive features:** Body moderately elongate; forehead mostly convex in profile; eye small, diameter about half of distance between eye and dorsal spine; mouth directed anteroventrally. Snout slightly depressed, greatest width near midlength; apex firm, tapering, often upturned slightly; ventral mucous canals diverging abruptly halfway along snout. Anterior tooth plates relatively broad and beak-like; lower and upper posterior teeth with broad, elevated ridges. First dorsal fin more than twice height of second dorsal fin; dorsal spine long, straight or only slightly recurved, its tip (when undamaged) reaching to or beyond origin of second dorsal fin. Second dorsal-fin base more than twice length of pelvic fin. Base of upper lobe of caudal fin shorter than base of lower lobe, lacking tubercles; origin of lower lobe below insertion of second dorsal fin; caudal filament longer than head in some undamaged specimens. Pectoral-fin apex rather angular. Anal fin absent. Claspers simple, enlarged distally with fine spines. Prepelvic clasper with large thorns along inner margin, and an enlarged granular apex. Head clasper with dense spines over anterior surface.
- **Colour:** Uniform chocolate brown; edges of fins darker; pelvic fins brownish black; often pale where the skin has been abraded.
- **Size:** To at least 94 cm without the caudal filament off Australia, but elsewhere attains 120 cm; smallest free-swimming specimens about 13 cm without filament; egg cases about 16 cm long.
- **Distribution:** Widely distributed, but patchy, in the Atlantic, Indian and Pacific Oceans; reported to 2600 m depth. Caught occasionally on the continental slopes of New

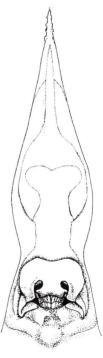

undersurface of head

head clasper (lateral view)

Zealand and southern Australia, from Brush Island (New South Wales) to Geraldton (Western Australia), including Tasmania, in 350–1000 m (mainly 700–900 m).

- **Remarks:** Very little is known of the biology of this spookfish. It appears to feed mainly on shellfish and crustaceans that live on the bottom of the continental slope.
- **References:** Garrick (1971); Garrick and Inada (1975).

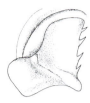
pre-pelvic clasper

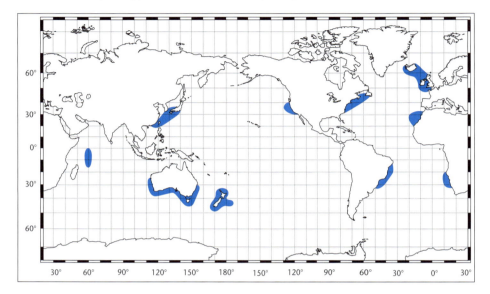

dorsal surface of right clasper

Pacific Spookfish — *Rhinochimaera pacifica* (Mitsukuri, 1895)

49.3 *Rhinochimaera pacifica* (Plate 82)   **Fish Code:** 00 044002

- **Alternative name:** Narrownose chimaera.
- **Field characters:** A spookfish with a very long, narrow, conical snout, a small eye situated behind the mouth, a rather long first dorsal fin and spine, relatively smooth tooth plates, and with the lower caudal-fin lobe longer than the upper lobe.
- **Distinctive features:** Body elongate; head narrow; eye small, diameter 2–3 in distance between eye and dorsal spine; mouth directed anteroventrally. Snout very long, flexible, conical near base; often more depressed near apex, not upturned; ventral mucous canals not diverging abruptly halfway along snout. Tooth plates smooth, sharp-edged. First dorsal fin much taller than second dorsal fin; dorsal spine moderately elongate, almost straight, its tip short of second dorsal fin.

Second dorsal-fin base more than twice length of pelvic fin. Upper lobe of caudal fin with a row of fleshy tubercles along fin margin in adults (more pronounced in mature males); base shorter than base of lower lobe; caudal filament mostly short. Pectoral-fin apex rather angular. No anal fin. Claspers long, simple, slightly enlarged distally with fine spines. Prepelvic clasper with large thorns along inner margin, no glandular tissue at apex. Head clasper with dense spines over anterior surface.

- **Colour:** Pale brown on upper surface, greyish brown on abdomen and around gill cover; snout and area around mouth mostly white; teeth black; eye black with silvery overtones; fins bluish grey.
- **Size:** Attaining more than 120 cm (without caudal filament) and maturing at almost 100 cm.
- **Distribution:** Irregularly distributed in the Pacific and eastern Indian Oceans; from the Japanese Archipelago to the South China Sea, and off Australasia and Peru. Common along the continental slopes off New Zealand and southern Australia, between Broken Bay (New South Wales) and North West Cape (Western Australia), including the South Tasman Rise, Tasmania and the Norfolk Ridge, in depths of 760–1290 m.
- **Remarks:** This spookfish is very similar to *Rhinochimaera atlantica* from the Atlantic Ocean. A second species, *R. africana*, was collected recently from seamounts south of Tasmania. It differs from the Pacific Spookfish in having a longer, somewhat depressed, paddle-shaped snout. Little is known of the biology of these spookfishes, however, *R. atlantica* is reported to aggregate in schools of the same sex and size.
- **Reference:** Inada and Garrick (1979).

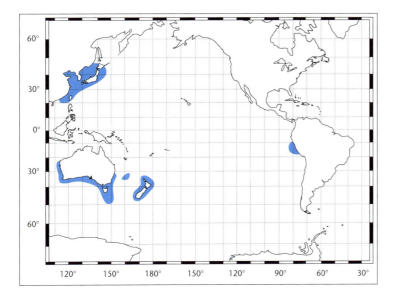

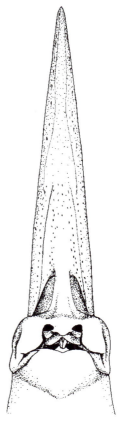
undersurface of head

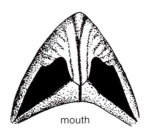
mouth

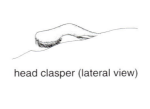
head clasper (lateral view)

pre-pelvic clasper

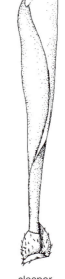
clasper

REFERENCES

Abe, T. (1966). Description of a new squaloid shark, *Centroscyllium kamoharai*, from Japan. *Japanese Journal of Ichthyology* **13**: 190–194.

Allen, G.R. and Swainston, R. (1988). *The marine fishes of north–western Australia. A guide for anglers and divers.* Western Australian Museum, Perth, 201 pp.

Annandale, N. (1909). Report on the fishes taken by the Bengal fisheries steamer 'Golden Crown', Part 1: Batoidei. *Memoirs of the Indian Museum* **2** (1): 1–58.

Anon. (1989). Status report on southern shark fishery. *Australian Fisheries* **48** (7): 14–19.

Ayling, T. and Cox, G.J. (1982). *Collins guide to the sea fishes of New Zealand.* Collins, Auckland, 343 pp.

Barrett, C. (1933). *Water Life.* Sun Nature Book 4, Sun News Pictorial: Melbourne, 44 pp.

Bass, A.J. (1979). Records of little-known sharks from Australian waters. *Proceedings of the Linnean Society of New South Wales* **103**: 247–254.

Bass, A.J., D'Aubrey, J.D. and Kistnasamy, N. (1973). Sharks of the east coast of southern Africa. I. The genus *Carcharhinus* (Carcharhinidae). *South African Association for Marine Biological Research. Oceanographic Research Institute Investigational Report* **33**: 1–168.

Bass, A.J., D'Aubrey, J.D. and Kistnasamy, N. (1975a). Sharks of the east coast of Southern Africa. II. The families Scyliorhinidae and Pseudotriakidae. *South African Association for Marine Biological Research. Oceanographic Research Institute Investigational Report* **37**: 1–64.

Bass, A.J., D'Aubrey, J.D. and Kistnasamy, N. (1975b). Sharks of the east coast of southern Africa. III. The families Carcharhinidae (excluding *Mustelus* and *Carcharhinus*) and Sphyrnidae. *South African Association for Marine Biological Research. Oceanographic Research Institute Investigational Report* **38**: 1–100.

Bass, A.J., D'Aubrey, J.D. and Kistnasamy, N. (1975c). Sharks of the east coast of southern Africa. IV. The families Odontaspididae, Scapanorhynchidae, Isuridae, Cetorhinidae, Alopiidae, Orectolobidae and Rhiniodontidae. *South African Association for Marine Biological Research. Oceanographic Research Institute Investigational Report* **39**: 1–102.

Bass, A.J., D'Aubrey, J.D. and Kistnasamy, N. (1975d). Sharks of the east coast of southern Africa. V. The families Hexanchidae, Chlamydoselachidae, Heterodontidae, Pristiophoridae, and Squatinidae. *South African Association for Marine Biological Research. Oceanographic Research Institute Investigational Report* **38**: 1–50.

Bass, A.J., D'Aubrey, J.D. and Kistnasamy, N. (1976). Sharks of the east coast of southern Africa. VI. The families Oxynotidae, Squalidae, Dalatiidae and Echinorhinidae. *South African Association for Marine Biological Research. Oceanographic Research Institute Investigational Report* **45**: 1–103.

Bennett, E.T. (1830). Fishes. In: *Memoir of the life and public services of Sir Stamford Raffles.* Lady Raffles (Ed.). London, 694 pp.

Berra, T.M. and Hutchins, J.B. (1990). A specimen of megamouth shark, *Megachasma pelagios* (Megachasmidae) from Western Australia. *Records of the Western Australian Museum* **14** (4): 651–656.

Bibron, G. (1839). In: *Systematische beschreibung der plagiostomen.* J. Müller, and F.G.J. Henle (Eds). Veit & Co., Berlin, pp. 29–102.

Bigelow, H.B. and Schroeder, W.C. (1944). New sharks from the western North Atlantic. *Proceedings of the New England Zoology Club* **23**: 21–36.

Bigelow, H.B. and Schroeder, W.C. (1948). Fishes of the western North Atlantic. 1. Lancelets, cyclostomes and sharks. *Memoir Sears Foundation for Marine Research* **1** (1): 56–576.

Bigelow, H.B. and Schroeder, W.C. (1953). Fishes of the western North Atlantic. 2. Sawfishes, guitarfishes, skates, rays and chimaeroids. *Memoir Sears Foundation for Marine Research* **1** (2), 588 pp.

Bigelow, H.B. and Schroeder, W.C. (1957). A study of the sharks of the suborder Squaloidea. *Bulletin of the Museum of Comparative Zoology Harvard* **117**: 1–150.

Bleeker, P. (1851). Zesde bijdrage tot de kennis der ichthyologische fauna van Borneo. Visschen van Pamangkat, Bandjermassing, Praboekarta end Sampit. *Natuurkundig Tijdschrift voor Nederlandsch–Indië*, **3**: 407–442.

Bleeker, P. (1852). Bijdrage tot de kennis der plagiostomen van den Indischen Archipel. *Verhandelingen van het Bataviaasch Genootschap van Kunsten en Wetenschappen* **24**: 1–92.

Bleeker, P. (1853a). Diagnostische beschrijvingen van nieuwe of weinig bekende vischsoorten van Batavia. Tiental I–VI. *Natuurkundig Tijdschrift voor Nederlandsch–Indië* **4**: 451–516.

Bleeker, P. (1853b). Zevende bijdrage tot de kennis der ichthyologische fauna van Borneo. Zoetwatervisschen van Sambas, Pontianak en Pangaron. *Natuurkundig Tijdschrift voor Nederlandsch–Indië* **5**: 414–442.

Bleeker, P. (1856). *Carcharias (Prionodon) amblyrhynchus*, eene nieuwe haaisoort, gevangen nabij het eiland Solombo. *Natuurkundig Tijdschrift voor Nederlandsch–Indië* **10**: 467–468.

Bleeker, P. (1860). Elfde bijdrage tot de kennis der vischfauna van Amboina. *Acta Societatis Scientiarum Indo–Neerlandicae* **8**: 1–14.

Bleeker, P. (1867). Description et figure d'une espece inedite de *Crossorhinus* de l'archipel des Moluques. *Archives Neerlandaises des Sciences Naturelles* **2**: 400–402.

Bloch, M.E. and Schneider, J.G. (1801). M.E. Blochii Systema ichthyologiae iconibus ex illustratum. Post orbitum auctoris opus inchoatum absolvit, correxit, interpolavit J.G. Schneider, Saxo. Berolini, 584 pp.

Bocage, J.V.B. du and Capello, F. de B. (1864). Sur quelques espèces inédites de Squalidae de la tribu Acanthiana, Gray, qui fréquentent les côtes du Portugal. *Proceedings of the Zoological Society of London* **24**: 260–263.

Bonaparte, C.L. (1832–41). *Iconografia della fauna italica per le quattro classe degli animali vertebrati. Tomo III, Pesci.* Roma. 556 pp.

Bonnaterre, J. P. (1788). *Tableau encyclopédique et méthodique des trois règnes de la nature. Ichthyologie.* Paris, 215 pp.

Bory de Saint-Vincent, J.B.G.M. (1823). *Dictionnaire classique d'histoire naturelle.* Rey et Gravier, Paris, Vol. 3, 592 pp.

Cadenat, J. and. Blache, J. (1981). Requins de Méditerranée et d'Atlantique (plus particulièrement de la Côte Occidentale d'Afrique). *Faune Tropicale* (ORSTOM: Paris) **21**: 1–330.

Castelnau, F.L. (1873a). Contribution to the ichthyology of Australia. 8. Fishes of Western Australia. *Proceedings of the Zoological and Acclimatisation Society of Victoria* **2**: 123–149.

Castelnau, F.L. (1873b). Intercolonial exhibition essays, 1872–3. No. 5. Notes on the edible fishes of Victoria. In: *London international exhibition of 1873: Official Record.* Melbourne, pp. 1–17.

Castro, J.I. (1983). *The sharks of North American waters.* Texas A & M University Press, College Station, 180 pp.

Chan, W.L. (1966). New sharks from the south China Sea. *Journal of Zoology (London)* **148**: 218–237.

Chen, J.T.F. (1963). A review of the sharks of Taiwan. *Biological Bulletin Department of Biology College of Science Tunghai University (Ichthyological Series)* **19**: 1–102.

Chen, C., Taniuchi, T. and Nose, Y. (1979). Blainville's dogfish, *Squalus blainville,* from Japan, with notes on *S. mitsukurii* and *S. japonicus. Japanese Journal of Ichthyology* **26** (1): 26–42.

Coleman, N. (1974). *Australian marine fishes in colour.* A.H. and A.W. Reed Pty Ltd, Auckland, 108 pp.

Coleman, N. (1980). *Australian sea fishes south of 30° S.* Doubleday Australia, Sydney, 302 pp.

Coleman, N. (1981). *Australian sea fishes north of 30° S.* Doubleday Australia, Sydney, 297 pp.

Collett, R. (1879). Fiske fra Nordhavs-expeditionens sidste togt, sommeren 1878. *Forhandlinger i Videnskabsselskabet i Kristiania* 1878 (1879), (14):1–106.

Compagno, L.J.V. (1984). FAO species catalogue. Vol. 4, Sharks of the world. An annotated and illustrated catalogue of shark species known to date. Part 1 – Hexanchiformes to Lamniformes: viii, 1–250. Part 2 – Carcharhiniformes: x, 251–655. *FAO Fisheries Synopsis* 125: 1-655.

Compagno, L.J.V. (1986). Sharks of the eastern Cape coast. J.L.B. Smith Institute of Ichthyology, *Ichthos Field Guide* **1**: 1–27.

Compagno, L.J.V. (1988). *Sharks of the order Carcharhiniformes.* Princetown University Press, New Jersey, 486 pp.

Compagno, L.J.V., Ebert, D.A. and Smale, M.J. (1989). *Guide to the sharks and rays of southern Africa.* Struik Publishers, Cape Town, 160 pp.

Compagno, L.J.V. and Roberts, T.R. (1982). Freshwater stingrays (Dasyatidae) of southeast Asia and New Guinea, with description of a new species of *Himantura* and reports of unidentified species. *Environmental Biology of Fishes* **7** (4): 321–339.

Cuvier. G.L. (1816). *La règne animal distribué d'après son organisation, pour servir de base a l'histoire naturelle des animaux et d'introduction à l'anatomie comparée.* Edn 1, tome II. Les reptiles, les poissons, les mollusques et les annélides. Deterville, Paris, 532 pp. [Not 1817: see Roux, 1976, *Journal of the Society for the Bibliography of Natural History* **8** (1): 31].

Cuvier. G.L. (1829). *La règne animal distribué d'après son organisation, pour servir de base a l'histoire naturelle des animaux et d'introduction à l'anatomie comparée.* Edn 2, tome II. Deterville, Paris, 406 pp.

Davenport, S.R. and Deprez, P.P. (1989). Market opportunities for shark liver oil. *Australian Fisheries* **48** (11): 8–10.

Davenport, S.R. and Stevens, J.D. (1988). Age and growth of two commercially important sharks (*Carcharhinus tilstoni* and *C. sorrah*) from northern Australia. *Australian Journal of Marine and Freshwater Research* **39**: 417–33.

De Buen, F. (1960a). Tiburones, rayas y quimeras en la estacion de biologia marina de Montemar, Chile. *Revista de Biologia Marina* **10**: 1–50.

De Buen, F. (1960b). Los peces de la Isla de Pasqua. Catalogo descriptivo e ilustrado. *Boletin de la Sociedad de Biologia de Concepcion* 1960 (35–36): 3–80.

Deprez, P.P., Volkman, J.K. and Davenport, S.R. (1990). Squalene content and neutral lipid composition of livers from deep–sea sharks caught in Tasmanian waters. *Australian Journal of Marine and Freshwater Research* **41**: 375–87.

de Vis, C.W. (1883). Descriptions of new genera and species of Australian fishes. *Proceedings of the Linnean Society of New South Wales* (1) **8** (2): 283–289.

Dingerkus, G. and DeFino, T. (1983). A revision of the Orectolobiform shark family Hemiscillidae (Chondrichthyes, Selachii). *Bulletin of the American Museum of Natural History* **176** (1): 1–94.

Dixon, J.M. (1969). A new species of ray of the genus *Urolophus* (Elasmobranchii: Urolophidae) from Victoria. *Victorian Naturalist* **86** (1): 11–18.

Donndorf, J.A. (1798). *Zoologische beitrage zur izten ausgabe des linneschen natursystems.* 3 vols. Leipzig.

Duméril, A.H.A. (1852). Monographie de la famille des torpédiniens, ou poissons plagiostomes électriques, comprenant la description d'un genre nouveau, de 3 espèces nouvelles, et de 2 espèces nommées dans le Musée de Paris, mais non encore décrites. *Revue Magasin Zoologie* Series 2 (**4**): 176–189, 227–244, 270–285.

Duméril, A.H.A. (1853). Monographie de la tribu des sylliens ou roussettes (poissons plagiostomes) comprenant deux espèces nouvelles. *Revue Magasin Zoologie* Series 2 (**5**): 8–25, 73–87, 119–130.

Edgar, G.J., Last, P.R. and Wells, M.W. (1982). *Coastal fishes of Tasmania and Bass Strait.* Cat and Fiddle Press, Hobart, 176 pp.

Edwards, R.R. (1980). Aspects of the population dynamics and ecology of the white spotted stingaree, *Urolophus paucimaculatus* Dixon, in Port Phillip Bay, Victoria. *Australian Journal of Marine and Freshwater Research* **31**: 459–467.

Euphrasen, B.A. (1790). Raja (*Aetobatis narinari*) beskrifven. *Kungliga Svenska Vetenskapsakademiens Handlingar* **11**: 217–219.

Forsskål, P. (1775). Descriptiones animalium, avium, amphibiorum, iscium, insectorum, vermium; quae in itirene orientali observavit. Molleri, Hauniae, 164 pp.

Fourmanoir, P. (1961). Requins de la côte ouest de Madagascar. *Memoires de l'Institut de Recherche Scientifique de Madagascar* (Serie F) **4**: 1–81.

Fourmanoir, P. (1975–1979). Requins de Nouvelle-Caledonie. *Nature Caledonienne* 1975, **11–12**: 23–29; 1976, **13**: 5–9; 1977, **14**: 13–17; 1979, **16**: 11–14.

Fourmanoir, P. and Rivaton, J. (1979). Poissons de la pente recifale externe de Nouvelle-Caledonie et des Nouvelles-Hebrides. *Cahiers du Indo–Pacifique* **4**: 405–443.

Fowler, H.W. (1908a). A collection of fishes from Victoria, Australia. *Proceedings of the Academy of Natural Sciences of Philadelphia* **59** (3): 419–444.

Fowler, H.W. (1908b). Notes on sharks. *Proceedings of the Academy of Natural Sciences of Philadelphia* **60**: 52–70.

Fowler, H.W. (1941a). The fishes of the groups Elasmobranchii, Holocephali, Isospondyli, and Ostariophysi obtained by United States bureau of fisheries steamer 'Albatross' in 1907 to 1910, chiefly in the Philippine Islands and adjacent seas. *Bulletin United States National Museum* (100) **13**: 1–879.

Fowler, H.W. (1941b). Notes on Florida fishes with descriptions of seven new species. *Proceedings of the Academy of Natural Sciences of Philadelphia* **93**: 81–106.

Francis, M.P., Stevens, J.D. and Last, P.R. (1988). New records of *Somniosus* (Elasmobranchii: Squalidae) from Australasia, with comments on the taxonomy of the genus. *New Zealand Journal of Marine and Freshwater Research* **22**: 401–409.

Garman, S. (1884). An extraordinary shark (*Chlamydoselachus anguineus*). *Bulletin of the Essex Institute* **16**: 47–55.

Garman, S. (1906). New Plagiostomia. *Bulletin of the Museum of Comparative Zoology Harvard* **46** (11): 203–208.

Garman, S. (1913). The Plagiostomia (sharks, skates, and rays). *Memoirs of the Museum of Comparative Zoology of Harvard College* **36**: 1–515.

Garrick, J.A.F. (1954a). Studies on New Zealand Elasmobranchii. Part II. A description of *Dasyatis brevicaudatus* (Hutton), Batoidei, with a review of records of the species outside New Zealand. *Transactions of the Royal Society of New Zealand* **82** (1): 189–198.

Garrick, J.A.F. (1954b). Studies on New Zealand Elasmobranchii. Part III. A new species of *Triakis* (Selachii) from New Zealand. *Transactions of the Royal Society of New Zealand* **82**: 695–702.

Garrick, J.A.F. (1955). Studies on New Zealand Elasmobranchii. Part IV. The systematic position of *Centroscymnus waitei* (Thompson, 1930), Selachii. *Transactions of the Royal Society of New Zealand* **83** (1): 227–239.

Garrick, J.A.F. (1956). Studies on New Zealand Elasmobranchii. Part V. *Scymnodalatias* n.g. Based on *Scymnodon sherwoodi* Archey, 1921 (Selachii). *Transactions of the Royal Society of New Zealand* **83** (3): 555–571.

Garrick, J.A.F. (1957). Studies on New Zealand Elasmobranchii. Part VI. Two new species of *Etmopterus* from New Zealand. *Bulletin of the Museum of Comparative Zoology Harvard* **116** (3): 169–190.

Garrick, J.A.F. (1959a). Studies on New Zealand Elasmobranchii. Part VII. The identity of specimens of *Centrophorus* from New Zealand. *Transactions of the Royal Society of New Zealand* **86** (1–2): 127–141.

Garrick, J.A.F. (1959b). Studies on New Zealand Elasmobranchii. Part VIII. Two Northern Hemisphere species of *Centroscymnus* in New Zealand waters. *Transactions of the Royal Society of New Zealand* **87** (1–2): 75–89.

Garrick, J.A.F. (1959c). Studies on New Zealand Elasmobranchii. Part IX. *Scymnodon plunketi* (Waite, 1910). An abundant deep-water shark of New Zealand waters. *Transactions of the Royal Society of New Zealand* **87** (3–4): 271–282.

Garrick, J.A.F. (1960a). Studies on New Zealand Elasmobranchii. Part X. The genus *Echinorhinus*, with an account of a second species, *E. cookei* Pietschmann, 1928, from New Zealand waters. *Transactions of the Royal Society of New Zealand* **88** (1): 105–117.

Garrick, J.A.F. (1960b). Studies on New Zealand Elasmobranchii. Part XI. Squaloids of the genera *Deania*, *Etmopterus*, *Oxynotus* and *Dalatias* in New Zealand waters. *Transactions of the Royal Society of New Zealand* **88** (3): 489–517.

Garrick, J.A.F. (1960c). Studies on New Zealand Elasmobranchii. Part XII. The species of *Squalus* from New Zealand and Australia; and a general account and key to the New Zealand Squaloidea. *Transactions of the Royal Society of New Zealand* **88**: 519–557.

Garrick, J.A.F. (1967). Revision of sharks of genus *Isurus* with description of a new species (Galeoidea, Lamnidae). *Proceedings of the U.S. National Museum* **118** (3537): 663–690.

Garrick, J.A.F. (1971). *Harriotta raleighana*, a long-nosed chimaera (family Rhinochimaeridae), in New Zealand waters. *Journal of the Royal Society of New Zealand* **1** (3–4): 203–213.

Garrick, J.A.F. (1974). First record of an odontaspidid shark in New Zealand waters. *New Zealand Journal of Marine and Freshwater Research* **8** (4): 621–630.

Garrick, J.A.F. (1982). Sharks of the genus *Carcharhinus*. *National Oceanic and Atmospheric Administration Technical Report, National Marine Fisheries Service Circular* 445, 194 pp.

Garrick, J.A.F. (1985). Additions to a revision of the shark genus *Carcharhinus*: synonymy of *Aprionodon* and *Hypoprion*, and description of a new species of *Carcharhinus* (Carcharhinidae). *National Oceanic and Atmospheric Administration Technical Report, National Marine Fisheries Service* 34, 26 pp.

Garrick, J.A.F. and Inada, T. (1975). Dimensions of long-nosed chimaera *Harriotta raleighana* from New Zealand. *New Zealand Journal of Marine and Freshwater Research* **9** (12): 159–167.

Garrick, J.A.F. and Moreland, J.M. (1968). Notes on a bramble shark, *Echinorhinus cookei*, from Cook Strait, New Zealand. *Records of the Dominion Museum Wellington* **6** (10): 133–139.

Garrick, J.A.F. and Paul, L.J. (1971a). *Heptranchias dakini* Whitley, 1931, a synonym of *H. perlo* (Bonnaterre, 1788), the sharpsnouted sevengill or perlon shark, with notes on sexual dimorphism in this species. *Zoology Publications from Victoria University of Wellington* **54**: 1–14.

Garrick, J.A.F. and Paul, L.J. (1971b). *Cirrigaleus barbifer* (Fam. Squalidae), a little-known Japanese shark from New Zealand waters. *Zoology Publications from Victoria University of Wellington* **55**: 1–13.

Gilbert, C.R. (1967). A revision of the hammerhead sharks (Family Sphyrnidae). *Proceedings of the United States National Museum* **119** (3539): 1–88.

Gloerfelt-Tarp, T. and Kailola, P.J. (1984). *Trawled fishes of southern Indonesia and northwestern Australia*. Australian Development Assistance Bureau; Directorate General of Fisheries, Indonesia; German Agency for Technical Cooperation, 406 pp.

Glover, C.J.M. (1974). The whitetip oceanic shark *Pterolamiops longimanus* (Poey) 1861 — a first record from off the southern Australian coast. *The South Australian Naturalist* **49** (1): 12–13.

Gmelin, J.F. (1789). Amphibia, Pisces. In: *Systema Naturae*. 13th edn, Caroli a Linné (Ed.). Leiden, Delamolliere, **1** (3): 1126–1516.

Gohar, H.A.F. and Mazhar, F. M. (1964). The elasmobranchs of the north–western Red Sea. *Publications of the Marine Biological Station Al-Ghardaqa (Red Sea)* 13: 1–144.

Goode, G.B. and Bean, T.H. (1895). Scientific results of explorations by the U.S. fish commission steamer 'Albatross'. No. 30. On *Harriotta*, a new type of chimaeroid fish from the deeper waters of the northwestern Atlantic. *Proceedings of the United States National Museum* **17** (1014): 471–473.

Goode, G.B. and Bean, T.H. (1896). Oceanic ichthyology. *United States National Museum Special Bulletin* **1**: 1–553.

Gordon, I. (1992). A new record extending the southerly distribution of the shark ray (*Rhina ancylostoma*), and notes on its behaviour in captivity. *Australian Journal of Marine and Freshwater Research* **43** (1): 319–323.

Grant, E.M. (1978). *Guide to fishes*. Department of Harbours and Marine: Brisbane, 768 pp.

Gray, J.E. (1827). Pisces. In: *Narrative of a survey of the intertropical and western coasts of Australia performed between the years 1818 and 1822*. P.P. King (Ed.). John Murray, London, **2** (436), pp. 435–437.

Gray, J.E. (1831). Description of three new species of fish, including two undescribed genera, discovered by John Reeves, Esq., in China. *Zoological Miscellany* **1**: 4–5.

Gray, J.E. (1830–35). *Illustrations of Indian zoology; chiefly selected from the collection of Major-General Hardwicke*. 20 parts in 2 vols. London.

Griffith, E. and Smith, C.H. (1834). *The class Pisces, arranged by the Baron Cuvier, with supplementary additions by Edward Griffith, F.R.S., &c. and Lieut.-Col. Charles Hamilton Smith, F.R., L.S.S., &c*. London, 680pp.

Gudger, E.W. (1937). *Bathytoshia*, the giant stingaree of Australia. The largest of the stingray tribe in the seven seas. *Australian Museum Magazine* April–June, 1937: 205–210.

Gunnerus, J.E. (1765). Brugden (*Squalus maximus*). *Trondhjemsk Selskab Skrifter* **3**: 33–49.

Günther, A. (1870). *Catalogue of the fishes in the British Museum*. vol. 8. Catalogue of the Physostomi containing the families Gymnotidae, Symbranchidae, Muraenidae, Pegasidae and of the Lophobranchii, Plectognathi, Dipnoi, Ganoidei, Chondropterygii, Cyclostomata, Leptocardii in the collection of the British Museum, London, 549 pp.

Günther, A. (1871). Report on several collections of fishes recently obtained for the British Museum. *Proceedings of the Zoology Society London* (**1871**): 652–675.

Günther, A. (1877). Preliminary notes on new fishes collected in Japan during the expedition of H.M.S. 'Challenger'. *Annals and Magazine of Natural History* (4) **20** (119): 433–446.

Günther, A. (1880a). Report on the shore fishes. In: Zoology of the voyage of 'Challenger'. *Challenger Reports, Zoology*, **1** (6): 1–82.

Günther, A. (1880b). A contribution to the knowledge of the fish-fauna of the Rio de la Plata. *Annals and Magazine of Natural History* (5) **6** (31): 7–13.

Haacke, J.W. (1885). Diagnosen zweier bemerkenswerther südaustralischer Fische. *Zoologischer Anzeiger* **8** (203–4): 508–509.

Hamilton-Buchanan, F. (1822). *An account of the fishes found in the river Ganges and its branches*. Archibald Constable & Co., Edinburgh; Hurst, Robinson & Co., Cheapside, London, 5 pp.

Heald, D.I. (1987). The commercial shark fishery in temperate waters of Western Australia. *Fisheries Department of Western Australia Report Number* 75: 1–71.

Hector, J. (1872). Notes on the edible fishes. In: *Fishes of New Zealand*. F.W. Hutton and J. Hector (Eds). Colonial Museum and Geological Survey Department, Wellington, pp. 97–133.

Heemstra, P.C. (1973). A revision of the shark genus *Mustelus* (Squaliformes Carcharhinidae). M.A. thesis, University of Miami, 41 pp.

Hermann, J. (1783). *Tabula affinitatum animalium*. Argensorati. (Treuttel) 370 pp.

Hutchins, B. (1979). *A guide to the marine fishes of Rottnest Island*. Creative Research, Perth, 103 pp.

Hutchins, B. and Swainston, R. (1986). *Sea fishes of southern Australia. Complete field guide for anglers and divers*. Swainston Publishing, Perth, 180 pp.

Hutchins, B. and Thompson, M. (1983). *The marine and estuarine fishes of south-western Australia. A field guide for anglers and divers*. Western Australian Museum, Perth, 103 pp.

Hutton, F.W. (1875). Descriptions of new species of New Zealand fish. *Annals and Magazine of Natural History* (4) **16** (41): 313–317.

Inada, T. and Garrick, J.A.F. (1979). *Rhinochimaera pacifica*, a long-snouted chimaera (Rhinochimaeridae), in New Zealand waters. *Japanese Journal of Ichthyology* **25** (4): 235–243.

Iredale, T. (1938). *Raja whitleyi*, the great skate. *Australian Zoology* **9** (2): 169.

Jahn, A.E. and Haedrich, R.L. (1987). Notes on the pelagic squaloid shark *Isistius brasiliensis*. *Biological Oceanography* **5**: 297–309.

Johnson, R.H. (1978). *Sharks of tropical and temperate seas*. Les editions du Pacifique, Papeete, 170 pp.

Jordan, D.S. (1898). Description of a species of fish (*Mitsukurina owstoni*) from Japan, the type of a distinct family of lamnoid sharks. *Proceedings of the California Academy of Sciences, 3rd Series* **1**: 199–204.

Jordan, D.S. and Seale, A. (1906). The fishes of Samoa. Description of the species found in the archipelago, with a provisional check-list of the fishes of oceania. *U.S. Department of Commerce and Labor. Bulletin of the Bureau of Fisheries* **25** (1905): 173–455.

Jordan, D.S. and Snyder, J.O. (1902). Descriptions of two new species of squaloid sharks from Japan. *Proceedings of the United States National Museum* **25** (1279): 79–87.

Jordan, D.S. and Snyder, J.O. (1903). In: A review of the elasmobranchiate fishes of Japan. D.S. Jordan and H.W. Fowler (Eds). *Proceedings of the United States National Museum* **26** (1324): 593–674.

Karrer, C. (1972). Die gattung *Harriotta* Goode and Bean, 1895 (Chondrichthyes, Chimaeriformes, Rhinochimaeridae) mit beschreibung einer neuen art aus dem Nordatlantik. *Mitteilungen aus dem zoologischen Museum in Berlin* **48** (1): 203–221.

Kato, S., Springer, S. and Wagner, M.H. (1967). Field guide to eastern Pacific and Hawaiin sharks. *United States Fish and Wildlife Service Circular* **271**: 1–47.

Klunzinger, C.B. (1871). Synopsis der fische des Rothen Meeres. II Theil. *Verhandlungen der K. K. Zoologisch–botanischen Gesellschaft in Wien* **21**: 441–688.

Klunzinger, C.B. (1872). Zur fische-fauna von süd Australien. *Archiv fur Naturgeschichte* **38** (1): 17–47.

Kner, R. (1865). Fische. In: *Reise der österreichischen fregatte 'Novara' um die Erde in der Jahren 1857–59, unter den Befehlen des Commodore B. von Wullerstorf-Urbain.* Zoologischer Theil, Erster Band. Wien. 433 pp.

Krefft, G. (1868). *Deratoptera alfredi* (Prince Alfred's ray). Illustrated Sydney News (11 July) **5** (50): 1–16.

Lacépède, B.G. 1798–1803. *Histoire naturelle des poissons.* 5 vols, Paris (1802 vol. 4, 728 pp.).

Lacépède, B.G. (1804). Mémoire sur plusieurs animaux de la Nouvelle Hollande dont la description n'a pas encore été publiée. *Annals and Magazine of Natural History* **4**: 184–211.

Last, P.R. (1979). A new species of stingray (F. Dasyatidae) with a key to the Australian species. *Papers and Proceedings of the Royal Society of Tasmania* **113**: 169–175.

Last, P.R. (1987). New Australian fishes. Part 14. Two new species of *Dasyatis* (Dasyatididae). *Memoirs of the Museum of Victoria* **48** (1): 57–61.

Last, P.R. and Gomon, M.F. (1987). New Australian fishes. Part 15. New species of *Trygonoptera* and *Urolophus* (Urolophidae). *Memoirs of the Museum of Victoria* **48** (1): 63–72.

Last, P.R., Scott, E.O.G. and Talbot, F.H. (1983). *Fishes of Tasmania.* Tasmanian Fisheries Development Authority, Hobart, 563 pp.

Latham, J. (1794). An essay on the various species of sawfish. *Transactions of the Linnean Society of London* **2** (25): 273–282.

Lavery, S. and Shaklee, J.B. (1989). Population genetics of two tropical sharks, *Carcharhinus tilstoni* and *C. sorrah*, in northern Australia. *Australian Journal of Marine and Freshwater Research* **40**: 541–557.

Lenanton, R.J.C., Heald, D.I., Platell, M., Cliff, M. and Shaw, J. (1990). Aspects of the reproductive biology of the gummy shark, *Mustelus antarcticus* Günther, from waters off the south coast of Western Australia. *Australian Journal of Marine and Freshwater Research* **41**: 807–822.

Lesson, R.P. (1830–31). Poissons. In: L.I. Duperrey. Voyage autour du monde, execute par ordre du roi, sur la corvette de la Majesté, 'La Coquille', pendant les années 1822, 1823, 1824 et 1825…Zoologie. Paris, Zool. **2** (1): 66–238.

Lesueur, C.A. (1818). Description of several new species of North American fishes. *Journal of the Academy of Natural Sciences of Philadelphia* **1** (2): 222–235, 359–368.

Lesueur, C.A. (1822). Description of a *Squalus*, of a very large size, which was taken on the coast of New Jersey. *Journal of the Academy of Natural Sciences of Philadelphia* **2**: 343–352.

Linnaeus, C. (1758). *Systema Naturae.* 10th edn, Tome 1, 824 pp. Nantes & Pisces: 230–338.

Lloyd, R.E. (1908). On two species of eagle-rays with notes on the skull of the genus *Ceratoptera. Records of the Indian Museum (Calcutta)* **2** (2): 175–180.

Lowe, R.T. (1839). A supplement to a synopsis of the fishes of Madeira. *Proceedings of the Zoological Society of London* **7**: 76–92 (also published in *Transactions of the Zoological Society of London,* 1842, **3** (1): 1–20.

Lyle, J.M. (1986). Mercury and selenium concentrations in sharks from northern Australia. *Australian Journal of Marine and Freshwater Research* **37**: 309–321.

Lyle, J.M. (1987a). Observations on the biology of *Carcharhinus cautus* (Whitley), *C. melanopterus* (Quoy & Gaimard) and *C. fitzroyensis* (Whitley) from northern Australia. *Australian Journal of Marine and Freshwater Research* **38**: 701–710.

Lyle, J.M. (1987b). Northern pelagic fish stock research programme: summary of catch and effort data. Department of Industries and Development, Darwin. Fishery Report No. 16, 55 pp.

Lyle, J.M. and Timms, G.J. (1984). Survey shows potential for NT shark fishery. *Australian Fisheries* **43** (11): 2–5.

Lyle, J.M. and Timms, G.J. (1987). Predation on aquatic snakes by sharks from northern Australia. *Copeia* 1987(3): 802–803.

McCoy, F. (1874). On a new *Parascyllium* from Hobson's Bay. *Annals and Magazine of Natural History* (4) **13** (73): 15.

McCoy, F. (1878–1890). *Prodromous of the zoology of Victoria.* Government Printer, Melbourne, 375 pp.

McCulloch, A.R. (1911–1926). Report on some fishes obtained by the F.I.S. 'Endeavour' on the coasts of Queensland, New South Wales, Victoria, Tasmania, South and south–western Australia. Parts 1–4. Zoological (biological) results of the fishing experiments carried out by F.I.S. 'Endeavour', 1909–1914. 1911, Part 1, **1**: 1–87; 1914, Part 2, **2**: 77–199; 1915, Part 3, **3**: 97–170; 1916, Part 4, **4**: 169–199.

McCulloch, A.R. (1921). Notes on, and descriptions of Australian fishes. *Proceedings of the Linnean Society of New South Wales* **46** (1): 458–472.

McCulloch, A.R. (1929–1930). A check-list of the fishes recorded from Australia. *Memoirs of the Australian Museum* 1–534.

McDonald, J.D. and Barron, C. (1868). On a supposed new species of *Galeocerdo* from the southern seas. *Proceedings of the Zoological Society of London* 1868: 368–371.

McEachran, J.D. (1982). Chondrichthyes. In: *Synopsis and classification of living organisms.* S.P. Parker (Ed.). McGraw Hill, New York, 2 volumes, pp. 831–844.

McEachran, J.D. and Fechhelm, J.D. (1982). A new species of skate from Western Australia with comments on the status of *Pavoraja* Whitley, 1939 (Chondrichthyes: Rajiformes). *Proceedings of the Biological Society of Washington* **95** (1): 1–12.

McKay, R.J. (1966). Studies on western Australian sharks and rays of the families Scyliorhinidae, Urolophidae, and Torpedinidae. *Journal of the Royal Society of Western Australia* **49** (3): 65–82.

McLaughlin, R.H. and O'Gower, A.K. (1970). Underwater tagging of the Port Jackson shark, *Heterodontus portusjacksoni* (Meyer). *Bulletin of the Institute of Oceanography Monaco* **69** (1410): 1–11.

McLaughlin, R.H. and O'Gower, A.K. (1971). Life history and underwater studies of a heterodont shark. *Ecological Monographs* **41** (4): 271–289.

Macleay, W.J. (1881a). Descriptive catalogue of the fishes of Australia. Parts 1, 2 & 4. *Proceedings of the Linnean Society of New South Wales* (Series 1). Part 1, **5**: 302–444; Part 2, **5**: 510–629; Part 4, **6**: 202–387.

Macleay, W.J. (1881b). *Descriptive catalogue of Australian fishes*. F. W. White, Sydney, 323 pp.

Macleay, W.J. (1882). Notes on the Pleuronectidae of Port Jackson with descriptions of two hitherto unobserved species. *Proceedings of the Linnean Society of New South Wales* (1) **7** (1): 11–15.

Macleay, W.J. (1883). Notes on a collection of fishes from the Burdekin and Mary rivers, Queensland. *Proceedings of the Linnean Society of New South Wales* **8** (2): 199–213.

Macleay, W.J. (1884a). Some results of trawl fishing outside Port Jackson. *Proceedings of the Linnean Society of New South Wales* (1) **8** (4): 457–462.

Macleay, W.J. (1884b). Notices of new fishes. *Proceedings of the Linnean Society of New South Wales* (1) **9** (1): 170–172.

Masuda, H., Amaoka, K., Araga, C., Uyeno, T. and Yoshino, T. (Eds). (1984). *The fishes of the Japanese Archipelago*. Tokai University Press, Tokyo, 437 pp.

Matsubara, K. (1936a). A new carcharoid shark found in Japan. *Zoological Magazine* (Tokyo) **48** (7): 380–382.

Matsubara, K. (1936b). Order Plagiostomi I (sharks). *Fauna Nipponica* **15**, 2(1): 1–160.

Matsubara, K. (1936c). Order Plagiostomi II (rays). *Fauna Nipponica* **15**, 2(2): 1–70.

May, J.L. and Maxwell, J.G.H. (1986). *Field guide to trawl fish from temperate waters of Australia*. Revised edition. CSIRO, Melbourne, 492 pp.

Maxwell, J.G.H. (1980). A field guide to trawl fish from the temperate waters of Australia. *CSIRO Division of Fisheries and Oceanography Circular* **8** (1069): 1–201.

Merrick, J.R. and Schmida, G.E. (1984). *Australian freshwater fishes. Biology and management*. Griffin Press, Netley, South Australia, 409 pp.

Meyer, F.A. (1793). *Systematisch–summarische Uebersicht der neuesten zoologischen entdeckungen in Neuholland und Africa*. Dykirchen, Leipzig, 178 pp.

Mitsukuri, K. (1895). On a new genus of the chimaeroid group *Harriotta*. *Zoological Magazine (Tokyo)* **7** (80): 97–98.

Miyosi, Y. (1939). Description of three new species of elasmobranchiate fishes collected at Hyuga Nada, Japan. *Bulletin of the Biogeographical Society of Japan* **9** (5): 91–97.

Monkolprasit, S. (1984). *The cartilagenous fishes (Class Elasmobranchii) found in Thai waters and adjacent areas*. Department of Fishery Biology, Kasetsart University, Bangkok, 175 pp.

Monkolprasit, S. and Roberts, T.R. (1990). *Himantura chaophraya*, a new giant freshwater stingray from Thailand. *Japanese Journal of Ichthyology* **37** (3): 203–208.

Moulton, P.L., Saddlier, S.R. and. Knuckey, I.A. (1989). New time-at-liberty record set by tagged school shark *Galeorhinus galeus* caught off southern Australia. *North American Journal of Fisheries Management* **9**: 254–255.

Müller, J. and Henle, F.G.J. (1838–1841). *Systematische beschreibung der plagiostomen*. 1838a, pp. 1–28 ; 1839, pp. 27–28 (reset), pp. 29–102 ; 1841, pp. 103–200, Veit & Co, Berlin.

Müller, J. and Henle, F.G.J. (1838b). On the generic characters of cartilaginous fishes, with descriptions of new genera. *Magazine of Natural History* (new series) **2**: 33–37, 88–91.

Müller, J. and Henle, F.G.J. (1838c). Poissons cartilagineux. *L'Institut* **6**: 63–65.

Müller, J. and Henle, F.G.J. (1838d). Ueber die gattungen der plagiostomen. *Archiv fur Naturgeschichte* **4**: 83–85.

Munro, I.S.R. (1956–1961). Handbook of Australian fishes. Nos 1–42. *Fisheries Newsletter*. Elasmobranchs 1956, **15** (7): 1–8; **15** (8): 9–12; **15** (9): 13–16; **15** (10): 17–20.

Munro, I.S.R. (1967). *The fishes of New Guinea*. Department of Agriculture, Stock and Fisheries, Port Moresby, 650 pp.

Nakamura, H. (1935). On the two species of the thresher shark from Formosan waters. *Memorial Faculty of Science and Agriculture Taihoku Imperial University* **14** (1): 1– 6.

Nakaya, K. (1975). Taxonomy, comparative anatomy and phylogeny of Japanese catsharks, Scyliorhinidae. *Memoirs of the Faculty of Fisheries Hokkaido University* **23**: 1–94.

Nakaya, K. (1988a). Morphology and taxonomy of *Apristurus longicephalus* (Lamniformes, Scyliorhinidae). *Japanese Journal of Ichthyology* **34** (4): 431–442.

Nakaya, K. (1988b). Records of *Apristurus herklotsi* (Lamniformes, Scyliorhinidae) and discussion of its taxonomic relationships. *Japanese Journal of Ichthyology* **35** (2): 133–141.

Nakaya, K. and Bass, A.J. (1978). The frill shark *Chlamydoselachus anguineus* in New Zealand seas. *New Zealand Journal of Marine and Freshwater Research* **12** (4): 397– 398.

Nardo, G.D. (1827). Prodromus observationum et disquisitionum ichthyologiae Adriaticae. *Oken's Isis* **20** (6): 472–631.

Notarbartolo-Di-Sciara, G. (1987). A revisionary study of the genus *Mobula* Rafinesque, 1810 (Chondrichthyes: Mobulidae) with the description of a new species. *Zoological Journal of the Linnean Society of London* **91**: 1–91.

Ogilby, J.D. (1885). Descriptions of three new fishes from Port Jackson. *Proceedings of the Linnean Society of New South Wales* **10**: 445–447.

Ogilby, J.D. (1886). Notes on the distribution of some Australian sharks and rays, with a description of *Rhinobatus bougainvillei*, Müller & Henle. *Proceedings of the Linnean Society of New South Wales* **10** (3): 463–466.

Ogilby, J.D. (1888). Catalogue of the fishes in the collection of the Australian Museum. Part 1. Recent Palaeichthyan fishes. *Australian Museum Catalogue* **14**: 1–26.

Ogilby, J.D. (1893). Description of a new shark from the Tasmanian Coast. *Records of the Australian Museum* **2** (5): 62–63.

Ogilby, J.D. (1898). In: *Handbook of Sydney and the country of Cumberland*. W.M. Hamlet (Ed.). George Robertson, Sydney, 198 pp.

Ogilby, J.D. (1899). In: Scientific results of the trawling expedition of H.M.C.S. 'Thetis' off the coast of New South Wales in February and March 1898, Part 1. Fishes. E.R. Waite (Ed.). *Memoirs of the Australian Museum* **4**: 1–132.

Ogilby, J.D. (1908a). On new genera and species of fishes. *Proceedings of the Royal Society of Queensland* **21**: 1–26.

Ogilby, J.D. (1908b). Descriptions of new Queensland fishes. *Proceedings of the Royal Society of Queensland* **21**: 87–98.

Ogilby, J.D. (1908c). New or little known fishes in the Queensland Museum. *Annals of the Queensland Museum* **9** (1):1–41.

Ogilby, J.D. (1910a). On new or insufficiently described fishes. *Proceedings of the Royal Society of Queensland* **23** (1): 1–55.

Ogilby, J.D. (1910b). On some new fishes from the Queensland coast. Endeavour Series No. 1 (originally intended for publication in *Proceedings of the Royal Society of Queensland* **23**: 85–139).

Ogilby, J.D. (1911). Descriptions of new or insufficiently described fishes from Queensland waters. *Annals of the Queensland Museum* **10**: 36–58.

Ogilby, J.D. (1912). On some Queensland fishes. *Memoirs of the Queensland Museum* **1**: 26–65.

Ogilby, J.D. (1915a). On some new or little–known Australian fishes. *Memoirs of the Queensland Museum* **3**: 117–129.

Ogilby, J.D. (1915b). Ichthyological notes (no. 2). *Memoirs of the Queensland Museum* **3**: 130–136.

Ogilby, J.D. (1916). Checklist of the cephalochordates, selachians, and fishes of Queensland. 1. Cephalochordata, Selachii, Isospondyli (part). *Memoirs of the Queensland Museum* **5**: 70–98.

O'Gower, A.K. and Nash, A.R. (1978). Dispersion of the Port Jackson shark in Australian waters. In: *Sensory biology of sharks, skates and rays.* E.S. Hodgson and R.F. Mathewson (Eds). Office of Naval Research, Department of the Navy, Arlington, pp. 529–544.

Olsen, A.M. (1954). The biology, migration, and growth rate of the school shark, *Galeorhinus australis* (Macleay) (Carcharhinidae) in south–eastern Australian waters. *Australian Journal of Marine and Freshwater Research* **5**: 353–410.

Olsen, A.M. (1959). The status of the school shark fishery in south–eastern Australian waters. *Australian Journal of Marine and Freshwater Research* **10**: 150–176.

Olsen, A.M. (1984). Synopsis of biological data on the school shark, *Galeorhinus australis* (Macleay 1881). *FAO Fisheries Synopsis* 139: 1–42.

Owen, R. (1852). *Descriptive catalogue of the fossil organic remains of Reptilia and Pisces contained in the museum of the Royal College of Surgeons of England.* London, 184 pp.

Owen, R. (1853). *Descriptive catalogue of the osteological series contained in the museum of the Royal College of Surgeons of England.* Vol.1. Taylor & Francis, London, 350 pp.

Parker, T.J. and Haswell, W.A. (1897). *Textbook of zoology.* Macmillan, London, 2 volumes, 779 pp. and 683 pp.

Paulin, C.D., Habib, G., Carey, C.L., Swanson, P.M. and Vos, G.J. (1982). New records of *Mobula japanica* and *Masturus lanceolatus*, and further records of *Luvaris imperialis* (Pisces: Mobulidae, Louvaridae) from New Zealand. *New Zealand Journal of Marine and Freshwater Research* **16**: 11–17.

Paulin, C., Stewart, A., Roberts, C. and McMillan, P. (1989). *New Zealand fish: a complete guide.* National Museum of New Zealand, Wellington, 279 pp.

Paxton, J.R., Hoese, D.F., Allen, G.R. and Hanley, J.E. (1989). *Zoological catalogue of Australia. Vol. 7. Pisces, Petromyzontidae to Carangidae.* Australian Government Publishing Service, Canberra, 665 pp.

Péron, F. (1807). *Squalus cepedianus* (p. 337). In: *Voyage de découvertes aux Terres Australes...1800–1802.* F. Péron and L.C. Freycinet, 1807–1816, Paris, 2 volumes and atlas.

Péron, F. and Lesueur, C.A. (1822). Description of a *Squalus*, of a very large size, which was taken on the coast of New Jersey. *Journal of the Academy of Natural Sciences of Philadelphia* **2**: 343–352.

Peters, W.C.H. (1864). Uber eine neue percoiden-gattung *Plectroperca*, aus Japan und eine neue art von haifischen, *Crossorhinus tentaculatus* aus Neu Holland. *Monatsberichte der Koniglich Preussichen Akademie der Wissenschaften zu Berlin* 1864: 121–126.

Phillipps, W.J. (1931). New species of piked dogfish. *New Zealand Journal of Science and Technology* **12** (6): 360–361.

Phillipps, W.J. (1932). Notes on new fishes from New Zealand. *New Zealand Journal of Science and Technology* **13**: 226–234.

Phillipps, W.J. (1935). Sharks of New Zealand: No. 4. *New Zealand Journal of Science and Technology* **16**: 236–241.

Phillipps, W.J. (1936). Notes on new fishes from New Zealand. *New Zealand Journal of Science and Technology* **17** (4): 226–234.

Pietschmann, V. (1928). Neue fish-arten aus dem Pacifischen Ozean. *Anzeiger der Akademie der Wissenschaften, Wien* **65**: 297–298.

Poey, F. (1858–1861). *Memorias sobre la historia natural de la Isla de Cuba, acompañadas de sumarios Latinos y extractos Francés. La habana.* Vol. 2. Viuda de Barcina, Havana, 442 pp.

Quoy, J.R.C. and Gaimard, P. (1824-25). Description de poissons. Chapt 9. In: *Voyage autour de monde...exécuté sur les corvettes de L. M. "L'Uranie" et "La Physicienne", pendant les années 1817, 1818, 1819, et 1820.* L. de Freycinet (Ed). Pillet Aine, Paris, 192–401.

Rafinesque (Schmaltz), C.S. (1810a). *Caratteri di alcuni nuovi generi e nuove specie di animali (principalmente di pesca) e piante della Sicilia, con varie osservazioni sopra i medisimi.* Palermo, 105 pp.

Rafinesque (Schmaltz), C.S. (1810b). *Indice d'ittiologia Siciliana; ossia, catalogo metodico...pesci Sicilian.* Giovanni del Nobolo, Messina, 70 pp.

Ramsay, E. P. (1880). Notes on *Galeocerdo rayneri*, with a list of other sharks taken in Port Jackson. *Proceedings of the Linnean Society of New South Wales* (1) **5**: 95–97.

Ramsay, E.P. and Ogilby, J.D. (1886). Descriptions of new or rare Australian fishes. *Proceedings of the Linnean Society of New South Wales* **10** (4): 575–579.

Ramsay, E.P. and Ogilby, J.D. (1887). On an undescribed shark from Port Jackson. *Proceedings of the Linnean Society of New South Wales* (2) **2**: 163–164.

Ramsay, E.P. and Ogilby, J.D. (1888). Descriptions of two new Australian fishes. *Proceedings of the Linnean Society of New South Wales* (2) **3** (3): 1310–1312.

Randall, J.E. (1986). *Sharks of Arabia.* IMMEL Publishing, London, 148 pp.

Regan, C.T. (1906a). Descriptions of some new sharks in the British Museum collection. *Annals and Magazine of Natural History* (7) **18** (65): 435–440.

Regan, C.T. (1906b). A classification of the selachian fishes. *Proceedings of the Zoological Society of London* 1906: 722–758.

Regan, C.T. (1906c). Descriptions of new or little-known fishes from the coast of Natal. *Annals of the Natal Museum* **1** (1): 1–6.

Regan, C.T. (1909). A new specific name for an orectolobid shark. *Annals and Magazine of Natural History* (8) **3**: 529.

Richardson, J. (1840). On some new species of fishes from Australia. *Proceedings of the Zoological Society of London* **8**: 25–30.

Richardson, J. (1841). On some new or little known fishes from Australian seas. *Proceedings of the Zoological Society of London* **9**: 21–22.

Richardson, J. (1843). *Icones piscium, or plates of rare fishes*. Part 1. R. and J.E. Taylor, London, 8 pp.

Richardson, J. (1845). Ichthyology (17–52). In: *The zoology of the voyage of H.M.S. 'Erebus' and 'Terror' under the command of Captain Sir James Clark Ross, R.N., F.R.S., during the years 1839–43*. Vol. 2. Part 2. J. Richardson and J.E. Gray (Eds). E.W. Janson, London.

Richardson, L.R. and Garrick, J.A.F. (1953). *Dasyatis thetidis* Waite, a second species of giant stingray in New Zealand waters. *Transactions of the Royal society of New Zealand* **81**: 319–320.

Risso, A. (1810). *Ichthyologie de Nice, ou histoire naturelle des poissons du departement des Alpes maritimes*. F. Schaell, Paris, 388 pp.

Risso, A. (1826). *Histoire naturelle des principales productions de l'Europe méridionale, et particulièrement de celles des environs de Nice et des Alpes maritimes*. Paris et Strasbourg **3**, 486 pp.

Rivero, L.H. (1936). A new shark from Tasmania. *Occasional Papers of the Boston Society of Natural History* **8**: 267–268.

Rüppell, W.P.E. (1837). Fische des Rothen Meeres. In: *Neue wirbelthiere zu der fauna von Abyssinien gehörig*. Frankfurt am Main, pp. 53–80.

Sainsbury, K.J., Kailola, P.J. and Leyland, G.G. (1985). *Continental shelf fishes of northern and north–western Australia. An illustrated guide*. CSIRO Division of Fisheries Research; Clouston & Hall and Peter Pownall Fisheries Information Service, Canberra, 375 pp.

Sasaki, K. and Uyeno, T. (1987). *Squaliolus aliae*, a dalatiid shark distinct from *S. laticaudus*. *Japanese Journal of Ichthyology* **34** (3): 373–376.

Schneider, J.G. (1801). In: *M. E. Blochii, Systema ichthyologiae iconibus ex illustratum. Post orbitum auctoris opus inchoatum absolvit, correxit, interpolavit J.G. Schneider, Saxo*. Berolini, 584 pp.

Scott, E.O.G. (1935). Observations on some Tasmanian fishes. Part 2. *Proceedings of the Royal Society of Tasmania* 1934: 63–73.

Scott, E.O.G. (1963). Observations on some Tasmanian fishes. Part XI. *Papers and Proceedings of the Royal Society of Tasmania* **97**: 1–31.

Scott, E.O.G. (1976). Observations on some Tasmanian fishes. Part 22. *Papers and Proceedings of the Royal Society of Tasmania* **110**: 157–217.

Scott, E.O.G. (1980). Observations on some Tasmanian fishes. Part 26. *Papers and Proceedings of the Royal Society of Tasmania* **114**: 85–144.

Scott, T.D. (1954). Four new fishes from South Australia. *Records of the South Australian Museum* **11** (2):105–112.

Scott, T.D., Glover, C.J.M. and Southcott, R. V. (1980). *The marine and freshwater fishes of South Australia* (2nd edn). Government Printer, South Australia, 392 pp.

Seret, B. (1990). *Aulohalaelurus kanakorum* n.sp., a new species of catshark (Carcharhiniformes, Scyliorhinidae, Atelomycterinae) from New Caledonia. *Records of the Australian Museum* **42**:127–136.

Shaw, G. (1803–1804). General zoology or systematic natural history…with plates from the first authorities and most select specimens. London, 14 volumes, fishes in vol. 4. 1804 (2): 251–463.

Shaw, G. and Nodder, F.B. (1794). *The naturalist's miscellany, or coloured figures of natural objects, drawn and described from nature*. London, vol. 5, pls 162–182, unnumbered pages.

Shaw, G. and Nodder, F.B. (1795). *The naturalist's miscellany, or coloured figures of natural objects, drawn and described from nature*. London, vol. 6, unnumbered pages.

Shaw, G. and Nodder, F.B. (1806). *The naturalist's miscellany, or coloured figures of natural objects, drawn and described from nature*. London vol, 17, unnumbered pages.

Shcherbachev, Y.N. (1987). Preliminary list of thalassobathyal fishes of the tropical and subtropical waters of the Indian Ocean. *Journal of Ichthyology* **27** (2): 37–46.

Sherrard, J.E. (1893–1896). *Illustrated official handbook to the aquarium*. Government Printer, Melbourne, 8 volumes, 120 pp.

Sibley, G. 1985. (Ed.). Biology of the white shark. A symposium. *Memoirs of the Southern California Academy of Sciences* **9**: 1–150.

Smith, A. (1828). Descriptions of new, or imperfectly known objects of the animal kingdom found in the south of Africa. *The South African Commercial Advertiser* **3** (145): 2.

Smith, A. (1829). Additions a la zoologie du Sud de l'Afrique. *Bulletin des Sciences naturelles* **18** (166): 272–278.

Smith, H.M. and Radcliffe, L. (1912). The squaloid sharks of the Philippine archipelago with descriptions of new genera and species. *Proceedings of the United States National Museum* **41** (1877): 677–685.

Smith, J.L.B. and Smith, M.M. (1963). *The fishes of Seychelles*. Department of Ichthyology, Rhodes University, Grahamstown, 215 pp.

Snodgrass, R.E. and Heller, E. (1905). Papers from the Hopkins-Stanford Galapagos expedition, 1898–1899. xvii. Shore fishes of the Revillagigedo, Clipperton, Cocos and Galapagos Islands. *Proceedings of the Washington Academy of Science* **6**: 333–427.

Springer, S. (1950a). A revision of North American sharks allied to the genus *Carcharhinus*. *American Museum Novitates* **1451**: 1–13.

Springer, S. (1950b). Natural history notes on the lemon shark, *Negaprion brevirostris*. *Texas Journal of Science* 1950 (3): 349–359.

Springer, S. (1979). A revision of the catsharks, family Scyliorhinidae. *National Oceanic and Atmospheric Administration Technical Report, National Marine Fisheries Circular* **422**: 1–152.

Springer, S. and D'Aubrey, J.D. (1972). Two new scyliorhinid sharks from the east coast of Africa, with notes on related species. *South African Association for Marine Biological Research. Oceanographic Research Institute Investigational Report* (29): 1–19.

Springer, S. and Waller, R.A. (1969). *Hexanchus vitulus*, a new sixgill shark from the Bahamas. *Bulletin of Marine Science* **19** (1): 159–174.

Springer, V.G. (1964). A revision of the carcharhinid shark genera *Scoliodon, Loxodon, and Rhizoprionodon. Proceedings of the United States National Museum* **115**: 559–632.

Stead, D.G. (1938). The 'bronze whaler'. An undescribed Australian shark. *Australian Naturalist* **10** (3): 98–105.

Stead, D.G. (1963). *Sharks and rays of Australian seas.* Angus and Robertson, Sydney, 211 pp.

Stehmann, M. and Burkel, D.L. (1984). In: *Fishes of the northeastern Atlantic and Mediterranean. Rajidae.* P.J.P. Whitehead, M.-L. Bauchot, J.C. Hureau and E. Tortonese (Eds). UNESCO, Paris, pp. 163–196.

Steindachner, F. (1866). Zur fischfauna von Port Jackson in Australien. *Sitzungsberichte der Akademie der Wissenschaften in Wien* **53** (1): 424–480.

Stevens, J.D. (1983). Observations on reproduction in the shortfin mako *Isurus oxyrinchus. Copeia* (1): 126–130.

Stevens, J.D. (1984). Biological observations on sharks caught by sport fishermen off New South Wales. *Australian Journal of Marine and Freshwater Research* **35**: 573–590.

Stevens, J.D. and Church, A.G. (1984). Northern tagging project yields interesting results. *Australian Fisheries* **43** (11): 6–10.

Stevens, J.D. and Cuthbert, G.J. (1983). Observations on the identification and biology of *Hemigaleus* (Selachii: Carcharhinidae) from Australian waters. *Copeia* (2): 487–497.

Stevens, J.D., Dunning, M.C. and Machida, S. (1983). Occurrence of the porbeagle shark, *Lamna nasus*, in the Tasman Sea. *Japanese Journal of Ichthyology* **30** (3): 301–307.

Stevens, J.D., and Lyle, J.M. (1989). Biology of three hammerhead sharks (*Eusphyra blochii, Sphyrna mokarran* and *S. lewini*) from northern Australia. *Australian Journal of Marine and Freshwater Research* **40** (2): 129–46.

Stevens, J.D. and McLoughlin, K.J. (1991). Distribution, size and sex composition, reproductive biology and diet of sharks from northern Australia. *Australian Journal of Marine and Freshwater Research* **42** (2): 151–199.

Stevens, J.D. and Paxton, J.R. (1985). A new record of the goblin shark, *Mitsukurina owstoni* (Family Mitsukurinidae), from eastern Australia. *Proceedings of the Linnean Society of New South Wales* **108** (1): 37–45.

Stevens, J.D. and Wiley, P.D. (1986). Biology of two commercially important carcharhinid sharks from northern Australia. *Australian Journal of Marine and Freshwater Research* **37**: 671–688.

Suckow, G.A. (1799). *Anfungsgrunde der Theoretischen und angewandten Naturgeschichte der Thiere. Fishes.* Leipzig, vol. 4.

Tachikawa, H., Taniuchi, T. and Arai, R. (1989). *Etmopterus baxteri*, a junior synonym of *E. granulosus* (Elasmobranchii, Squalidae). *Bulletin of the National Science Museum* Series A (Zoology) **15** (4): 235–241.

Tanaka, S. (1911–1921). *Figures and descriptions of the fishes of Japan, including Riukiu Islands, Bonin Islands, Formosa, Kurile Islands, Korea, and Southern Sakhalin.* Vols 1–36, pp. 1–692. Tokyo.

Taniuchi, T. and Garrick, J.A.F. (1986). A new species of *Scymnodalatias* from the Southern Oceans, and comments on other squaliform sharks. *Japanese Journal of Ichthyology* **33** (2): 119–134.

Taniuchi, T., Shimizu, M., Sano, M., Baba, O. and Last, P.R. (1991). Descriptions of freshwater elasmobranchs collected from three rivers in northern Australia. In: Studies on elasmobranchs collected from seven river systems in northern Australia and Papua New Guinea. M. Shimizu and T. Taniuchi (Eds). *The University Museum, The University of Tokyo, Nature and Culture* 3: 11–26.

Taylor, L.R., Compagno, L.J.V. and Struhsaker, P.J. (1983). Megamouth — a new species, genus, and family of lamnoid shark (*Megachasma pelagios*, Family Megachasmidae) from the Hawaiian Islands. *Proceedings of the California Academy of Sciences* **43** (8): 87–110.

Teng, H.T. (1959). Studies on the elasmobranch fishes from Formosa. Part 4. *Squaliolus alii*, a new species of deep sea squaloid shark from Tung-Kang, Formosa. *Taiwan Fisheries Research Institute, Laboratory of Fishery Biology, Report No. 8,* pp. 1–6.

Teng, H.T. (1962). Classification and distribution of the Chondrichthys of Taiwan. Ph.D. thesis, Kyoto University, 304 pp. (in Japanese).

Tester, A.L. (1969). *Co-operative shark research and control program. Final report 1967–69.* University of Hawaii, Hawaii, 47 pp.

Thompson, E.F. (1930). New records of the genera *Centrophorus* and *Hoplichthys* in New Zealand. *Records of the Canterbury Museum* **3**: 275–279.

Tinker, S.W. and DeLuca, C.J. (1973). *Sharks and rays: A handbook of the sharks and rays of Hawaii and the central Pacific Ocean.* C.E. Tuttle Company, Vermont, 80 pp.

Valenciennes, A. (1839). In: *Systematische beschreibung der plagiostomen.* J. Müller and F.G.J. Henle (Eds). Veit & Co., Berlin, pp. 29–102.

Waite, E.R. (1898). *Report upon trawling operations off the coast of New South Wales between the Manning River and Jervis Bay, carried on by H.M.C.S. 'Thetis'.* Scientific report on the fishes. Government Printer, Sydney, pp. 17–45.

Waite, E.R. (1899). Scientific results of the trawling expedition of H.M.C.S. 'Thetis' off the coast of New South Wales in February and March 1898, Part 1. Fishes. *Memoirs of the Australian Museum* **4**: 20–22, 27–51.

Waite, E.R. (1900). Additions to the fish fauna of Lord Howe Island. *Records of the Australian Museum* **3** (7): 194–197.

Waite, E.R. (1905). Notes on fishes from Western Australia, No. 3. *Records of the Australian Museum* **6** (2): 55–82.

Waite, E.R. (1906). Studies in Australian sharks. No. 3. *Records of the Australian Museum* **6**: 226–229.

Waite, E.R. (1910). Notes on New Zealand fishes. *Transactions and Proceedings of the New Zealand Institute* **42**: 384–391.

Walker, T.I. (1988). The southern shark fishery. In: *Proceedings of the workshop on scientific advice for fisheries management: getting the message across.* M. Williams (Ed.). Australian Government Publishing Service, Canberra, pp. 24–31.

Wallace, J.H. (1967a). The batoid fishes of the east coast of southern Africa. Part II: Manta, eagle, duckbill, cownose, butterfly and sting rays. *South African Association for Marine Biological Research. Oceanographic Research Institute Investigational Report* (16): 1–56.

Wallace, J.H. (1967b). The batoid fishes of the east coast of southern Africa. Part I: Sawfishes and guitarfishes. *South African Association for Marine Biological Research. Oceanographic Research Institute Investigational Report* (15): 1–32.

Wallace, J.H. (1967c). The batoid fishes of the east coast of southern Africa. Part III: Skates and electric rays. *South African Association for Marine Biological Research. Oceanographic Research Institute Investigational Report* (17): 1–62.

West, J.G. and Carter, S. (1990). Observations on the development and growth of the epaulette shark *Hemiscyllium ocellatum* (Bonnaterre) in captivity. *Journal of Aquariculture and Aquatic Sciences* **5** (4): 111–117.

Wheeler, J.F.G. (1953). Report on the Mauritius–Seychelles fisheries survey 1948–49. Part 1. The bottom fishes of economic importance. *Fishery Publication. Colonial Office Great Britain* **3**:1– 57.

Wheeler, J.F.G. (1959a). Sharks of the western Indian Ocean. I. *Loxodon macrorhinus* M. & H. *East African Agricultural Journal* **25** (2): 106–109.

Wheeler, J.F.G. (1959b). Sharks of the western Indian Ocean. II. *Triaenodon obesus* (Rüppell). *East African Agricultural Journal* **25** (3): 202–204.

Wheeler, J.F.G. (1960). Sharks of the western Indian Ocean. III. *Carcharhinus menisorrah* (Müller and Henle). *East African Agricultural Journal* **25** (4): 271–273.

Wheeler, J.F.G. (1962). Notes on the three common species of sharks in the Mauritius–Seychelles area. *Proceedings of the Royal Society of Arts and Science Mauritius* **2** (2): 146–160.

Whitley, G.P. (1928). Studies in ichthyology. No. 2. *Records of the Australian Museum* **16** (4): 211–239.

Whitley, G.P. (1929a). Studies in ichthyology No. 3. *Records of the Australian Museum* **17** (3): 101–143.

Whitley, G.P. (1929b). Additions to the check-list of the fishes of New South Wales. No. 2. *Australian Zoologist* **5** (4): 353–357.

Whitley, G.P. (1930). The teeth of fishes. *Australian Museum Magazine* **4** (3): 92–99.

Whitley, G.P. (1931a). New names for Australian fishes. *Australian Zoologist* **6**: 310– 334.

Whitley, G.P. (1931b). Studies in ichthyology. No. 5. *Records of the Australian Museum* **18** (4): 138–160.

Whitley, G.P. (1931c). Studies in ichthyology. No. 4. *Records of the Australian Museum* **18** (3): 96–133.

Whitley, G.P. (1932). Studies in ichthyology. No. 6. *Records of the Australian Museum* **18** (6): 321–348.

Whitley, G.P. (1934a). Notes on some Australian sharks. *Memoirs of the Queensland Museum* **10** (4): 180–200.

Whitley, G.P. (1934b). Studies in ichthyology. No. 8. *Records of the Australian Museum* **19** (2): 155–163.

Whitley, G.P. (1936a). The Australian devil ray *Daemomanta alfredi* (Krefft), with remarks on the superfamily Mobuloidea (Order Batoidei). *Australian Zoologist* **8** (3): 164–188.

Whitley, G.P. (1936b). More ichthyological miscellania. *Memoirs of the Queensland Museum* **11** (1): 23–51.

Whitley, G.P. (1937a). Studies in ichthyology No. 10. *Records of the Australian Museum* **20** (1): 3–24.

Whitley, G.P. (1937b). The Middleton and Elizabeth Reefs, South Pacific Ocean. *Australian Zoologist* **8** (4): 199–273.

Whitley, G.P. (1939a). Studies in ichthyology No. 12. *Records of the Australian Museum* **20** (4): 264–267.

Whitley, G.P. (1939b). Taxonomic notes on sharks and rays. *Australian Zoologist* **9** (3): 227–262.

Whitley, G.P. (1940). The fishes of Australia. Part I. The sharks, rays, devilfish, and other primitive fishes of Australia and New Zealand. *Australian Zoological Handbook*. Royal Zoological Society of New South Wales, Sydney, 280 pp.

Whitley, G.P. (1943a). Ichthyological notes and illustrations (Part 2). *Australian Zoologist* **10** (2): 167–187.

Whitley, G.P. (1943b). A new Australian shark. *Records of the South Australian Museum* **7** (4): 397–399.

Whitley, G.P. (1943c). Ichthyological descriptions and notes. *Proceedings of the Linnean Society of New South Wales* **68** (3, 4): 114–144.

Whitley, G.P. (1944a). New sharks and fishes from Western Australia. *Australian Zoologist* **10** (3): 252–273.

Whitley, G.P. (1944b). Illustrations of some Western Australian fishes. *Proceedings of the Royal Zoological Society of New South Wales* 1943–1944: 25–29.

Whitley, G.P. (1945a). New sharks and fishes from Western Australia. Part 2. *Australian Zoologist* **11** (1): 1–41.

Whitley, G.P. (1945b). Leichhardt's sawfish. *Australian Zoologist* **11** (1): 43–45.

Whitley, G.P. (1947). New sharks and fishes from Western Australia. Part 3. *Australian Zoologist* **11** (2): 129–150.

Whitley, G.P. (1949). A new shark from Papua. *Proceedings of the Royal Zoological Society of New South Wales* 1947–1948, 24 pp.

Whitley, G.P. (1950a). Studies in ichthyology No. 14. *Records of the Australian Museum* **22** (3): 234–245.

Whitley, G.P. (1950b). A new shark from north–western Australia. *The Western Australian Naturalist* **2** (5): 100–105.

Whitley, G.P. (1951a). New fish names and records. *Proceedings of the Royal Zoological Society of New South Wales* 1949–1950: 61–68.

Whitley, G.P. (1951b). Shark attacks in Western Australia. *The Western Australian Naturalist* **2** (8): 185–194.

Whitley, G.P. (1951c). Studies in ichthyology No. 15. *Records of the Australian Museum* **22** (4): 389–408.

Whitley, G.P. (1964a). Fishes from the Coral Sea and the Swain Reefs. *Records of the Australian Museum* **26** (5): 145–195.

Whitley, G.P. (1964b). Sharks. *Australian Natural History* 14: 287–290.

Whitley, G.P. (1964c). Some fish genera scrutinized. *Proceedings of the Royal Zoological Society of New South Wales* 1964–65: 25–26.

Whitley, G.P. (1964d). Presidential address. A survey of Australian ichthyology. *Proceedings of the Linnean Society of New South Wales* **89**: 11–127.

Whitley, G.P. (1981). *Sharks of Australia*. M. Lincoln Smith (Ed.). Jack Pollard, Sydney, 135 pp.

Wolfson, F.H. (1986). Occurrences of the whale shark, *Rhincodon typus* Smith. In: *Indo–Pacific Fish Biology: Proceedings of the Second International Conference on Indo–Pacific Fishes*. T. Uyeno, R. Arai, T. Taniuchi and K. Matsuura (Eds). Ichthyological Society of Japan, Tokyo, pp. 208–226.

Yamakawa T., Taniuchi, T. and Nose, Y. (1986). Review of the *Etmopterus lucifer* group (Squalidae) in Japan. In: *Indo–Pacific Fish Biology: Proceedings of the Second International Conference on Indo–Pacific Fishes*. T. Uyeno, R. Arai, T. Taniuchi and K. Matsuura (Eds). Ichthyological Society of Japan, Tokyo, pp. 197–207.

Yano, K. and Tanaka, S. (1984a). Some biological aspects of the deep sea squaloid shark *Centroscymnus* from Suruga Bay, Japan. *Bulletin of the Japanese Society of Scientific Fisheries* **50** (2): 249–256.

Yano, K. and Tanaka, S. (1984b). Review of the deep sea squaloid shark genus *Scymnodon* of Japan, with a description of a new species. *Japanese Journal of Ichthyology* **30** (4): 341–360.

Zeitz, A.H.C. (1908a). Description of a hitherto undescribed species of shark from Investigator Strait. *Transactions of the Royal Society of South Australia* **32**: 287.

Zeitz, A.H.C. (1908b). A synopsis of the fishes of South Australia. Part I. *Transactions of the Royal Society of South Australia* **32**: 288–293.

CHECKLIST OF AUSTRALIAN SHARKS, RAYS AND CHIMAERAS

5	**CHLAMYDOSELACHIDAE**	
5:1	*Chlamydoselachus anguineus* Garman, 1884	Frilled Shark
6	**HEXANCHIDAE**	
6:1	*Heptranchias perlo* (Bonnaterre, 1788)	Sharpnose Sevengill Shark
6:2	*Hexanchus griseus* (Bonnaterre, 1788)	Bluntnose Sixgill Shark
6:3	*Hexanchus nakamurai* Teng, 1962	Bigeye Sixgill Shark
6:4	*Notorynchus cepedianus* (Péron, 1807)	Broadnose Sevengill Shark
7	**ECHINORHINIDAE**	
7:1	*Echinorhinus brucus* (Bonnaterre, 1788)	Bramble Shark
7:2	*Echinorhinus cookei* Pietschmann, 1928	Prickly Shark
8	**SQUALIDAE**	
8:1	*Centrophorus granulosus* (Bloch & Schneider, 1801)	Gulper Shark
8:2	*Centrophorus harrissoni* McCulloch, 1915	Harrisons Dogfish
8:3	*Centrophorus moluccensis* Bleeker, 1860	Endeavour Dogfish
8:4	*Centrophorus squamosus* (Bonnaterre, 1788)	Leafscale Gulper Shark
8:5	*Centrophorus uyato* (Rafinesque, 1810)	Southern Dogfish
8:6	*Centroscyllium kamoharai* Abe, 1966	Bareskin Dogfish
8:7	*Centroscymnus coelolepis* Bocage & Capello, 1864	Portuguese Dogfish
8:8	*Centroscymnus crepidater* (Bocage & Capello, 1864)	Golden Dogfish
8:9	*Centroscymnus owstoni* Garman, 1906	Owstons Dogfish
8:10	*Centroscymnus plunketi* (Waite, 1910)	Plunkets Dogfish
8:11	*Cirrhigaleus barbifer* Tanaka, 1912	Mandarin Shark
8:12	*Dalatias licha* (Bonnaterre, 1788)	Black Shark
8:13	*Deania calcea* (Lowe, 1839)	Brier Shark
8:14	*Deania quadrispinosa* (McCulloch, 1915)	Longsnout Dogfish
8:15	*Etmopterus* sp. A	Smooth Lantern Shark
8:16	*Etmopterus* sp. B	Bristled Lantern Shark
8:17	*Etmopterus* sp. C	Pygmy Lantern Shark
8:18	*Etmopterus* sp. D	Pink Lantern Shark
8:19	*Etmopterus* sp. E	Blackmouth Lantern Shark
8:20	*Etmopterus* sp. F	Lined Lantern Shark
8:21	*Etmopterus brachyurus* Smith & Radcliffe, 1912	Short-tail Lantern Shark
8:22	*Etmopterus granulosus* (Günther, 1880)	Southern Lantern Shark
8:23	*Etmopterus lucifer* Jordan & Snyder, 1902	Blackbelly Lantern Shark
8:24	*Etmopterus molleri* Whitley, 1939	Mollers Lantern Shark
8:25	*Etmopterus pusillus* (Lowe, 1839)	Slender Lantern Shark
8:26	*Euprotomicrus bispinatus* (Quoy & Gaimard, 1824)	Pygmy Shark
8:27	*Isistius brasiliensis* (Quoy & Gaimard, 1824)	Cookie-cutter Shark
8:28	*Scymnodalatias albicauda* Taniuchi & Garrick, 1986	Whitetail Dogfish
8:29	*Somniosus pacificus* Bigelow & Schroeder, 1944	Pacific Sleeper Shark
8:30	*Squaliolus aliae* Teng, 1959	Smalleye Pygmy Shark
8:31	*Squalus* sp. A	Bartail Spurdog
8:32	*Squalus* sp. B	Eastern Highfin Spurdog
8:33	*Squalus* sp. C	Western Highfin Spurdog
8:34	*Squalus* sp. D	Fatspine Spurdog
8:35	*Squalus* sp. E	Western Longnose Spurdog
8:36	*Squalus* sp. F	Eastern Longnose Spurdog
8:37	*Squalus acanthias* Linnaeus, 1758	White-spotted Spurdog
8:38	*Squalus megalops* (Macleay, 1881)	Piked Spurdog
8:39	*Squalus mitsukurii* Jordan & Snyder, 1903	Greeneye Spurdog
8:40	*Zameus squamulosus* (Günther, 1877)	Velvet Dogfish
9	**OXYNOTIDAE**	
9:1	*Oxynotus bruniensis* (Ogilby, 1893)	Prickly Dogfish

10	**PRISTIOPHORIDAE**	
10:1	*Pristiophorus* sp. A	Eastern Sawshark
10:2	*Pristiophorus* sp. B	Tropical Sawshark
10:3	*Pristiophorus cirratus* (Latham, 1794)	Common Sawshark
10:4	*Pristiophorus nudipinnis* Günther, 1870	Southern Sawshark
11	**HETERODONTIDAE**	
11:1	*Heterodontus galeatus* (Günther, 1870)	Crested Horn Shark
11:2	*Heterodontus portusjacksoni* (Meyer, 1793)	Port Jackson Shark
11:3	*Heterodontus zebra* (Gray, 1831)	Zebra Horn Shark
12	**PARASCYLLIIDAE**	
12:1	*Parascyllium* sp. A	Ginger Carpet Shark
12:2	*Parascyllium collare* Ramsay & Ogilby, 1888	Collared Carpet Shark
12:3	*Parascyllium ferrugineum* McCulloch, 1911	Rusty Carpet Shark
12:4	*Parascyllium variolatum* (Duméril, 1853)	Varied Carpet Shark
13	**BRACHAELURIDAE**	
13:1	*Brachaelurus colcloughi* Ogilby, 1908	Colcloughs Shark
13:2	*Brachaelurus waddi* (Bloch & Schneider, 1801)	Blind Shark
14	**ORECTOLOBIDAE**	
14:1	*Eucrossorhinus dasypogon* (Bleeker, 1867)	Tasselled Wobbegong
14:2	*Orectolobus* sp. A	Western Wobbegong
14:3	*Orectolobus maculatus* (Bonnaterre, 1788)	Spotted Wobbegong
14:4	*Orectolobus ornatus* (de Vis, 1883)	Banded Wobbegong
14:5	*Orectolobus wardi* Whitley, 1939	Northern Wobbegong
14:6	*Sutorectus tentaculatus* (Peters, 1865)	Cobbler Wobbegong
15	**HEMISCYLLIIDAE**	
15:1	*Chiloscyllium punctatum* Müller & Henle, 1838	Grey Carpet Shark
15:2	*Hemiscyllium ocellatum* (Bonnaterre, 1788)	Epaulette Shark
15:3	*Hemiscyllium trispeculare* Richardson, 1843	Speckled Carpet Shark
16	**STEGASTOMATIDAE**	
16:1	*Stegastoma fasciatum* (Hermann, 1783)	Zebra Shark
17	**GINGLYMOSTOMATIDAE**	
17:1	*Nebrius ferrugineus* (Lesson, 1830)	Tawny Shark
18	**RHINCODONTIDAE**	
18:1	*Rhincodon typus* (Smith, 1828)	Whale Shark
19	**ODONTASPIDIDAE**	
19:1	*Carcharias taurus* Rafinesque, 1810	Grey Nurse Shark
19:2	*Odontaspis ferox* (Risso, 1810)	Sand Tiger Shark
20	**MITSUKURINIDAE**	
20:1	*Mitsukurina owstoni* Jordan, 1898	Goblin Shark
21	**PSEUDOCARCHARIIDAE**	
21:1	*Pseudocarcharias kamoharai* (Matsubara, 1936)	Crocodile Shark
22	**MEGACHASMIDAE**	
22:1	*Megachasma pelagios* Taylor, Compagno & Struhsaker, 1983	Megamouth Shark
23	**ALOPIIDAE**	
23:1	*Alopias pelagicus* Nakamura, 1935	Pelagic Thresher
23:2	*Alopias superciliosus* (Lowe, 1840)	Bigeye Thresher
23:3	*Alopias vulpinus* (Bonnaterre, 1788)	Thresher Shark
24	**CETORHINIDAE**	
24:1	*Cetorhinus maximus* (Gunnerus, 1765)	Basking Shark
25	**LAMNIDAE**	
25:1	*Carcharodon carcharias* (Linnaeus, 1758)	White Shark
25:2	*Isurus oxyrinchus* Rafinesque, 1810	Shortfin Mako
25:3	*Lamna nasus* (Bonnaterre, 1788)	Porbeagle

26	**SCYLIORHINIDAE**	
26:1	*Apristurus* sp. A	Freckled Catshark
26:2	*Apristurus* sp. B	Bigfin Catshark
26:3	*Apristurus* sp. C	Fleshynose Catshark
26:4	*Apristurus* sp. D	Roughskin Catshark
26:5	*Apristurus* sp. E	Bulldog Catshark
26:6	*Apristurus* sp. F	Bighead Catshark
26:7	*Apristurus* sp. G	Pinocchio Catshark
26:8	*Apristurus longicephalus* Nakaya, 1975	Smoothbelly Catshark
26:9	*Asymbolus* sp. A	Dwarf Catshark
26:10	*Asymbolus* sp. B	Blotched Catshark
26:11	*Asymbolus* sp. C	Variegated Catshark
26:12	*Asymbolus* sp. D	Orange Spotted Catshark
26:13	*Asymbolus* sp. E	Pale Spotted Catshark
26:14	*Asymbolus* sp. F	Western Spotted Catshark
26:15	*Asymbolus analis* (Ogilby, 1885)	Grey Spotted Catshark
26:16	*Asymbolus vincenti* (Zietz, 1908)	Gulf Catshark
26:17	*Atelomycterus* sp. A	Banded Catshark
26:18	*Atelomycterus macleayi* Whitley, 1939	Marbled Catshark
26:19	*Aulohalaelurus labiosus* (Waite, 1905)	Black-spotted Catshark
26:20	*Cephaloscyllium* sp. A	Whitefin Swell Shark
26:21	*Cephaloscyllium* sp. B	Saddled Swell Shark
26:22	*Cephaloscyllium* sp. C	Northern Draughtboard Shark
26:23	*Cephaloscyllium* sp. D	Narrowbar Swell Shark
26:24	*Cephaloscyllium* sp. E	Speckled Swell Shark
26:25	*Cephaloscyllium fasciatum* Chan, 1966	Reticulate Swell Shark
26:26	*Cephaloscyllium laticeps* (Duméril, 1853)	Draughtboard Shark
26:27	*Galeus* sp. A	Slender Sawtail Shark
26:28	*Galeus* sp. B	Northern Sawtail Shark
26:29	*Galeus boardmani* (Whitley, 1928)	Sawtail Shark
26:30	*Halaelurus* sp. A	Dusky Catshark
26:31	*Halaelurus boesemani* Springer & D'Aubrey, 1972	Speckled Catshark
26:32	*Parmaturus* sp. A	Short-tail Catshark
27	**TRIAKIDAE**	
27:1	*Furgaleus macki* (Whitley, 1943)	Whiskery Shark
27:2	*Galeorhinus galeus* (Linneaus, 1758)	School Shark
27:3	*Hemitriakis* sp. A	Sicklefin Hound Shark
27:4	*Hemitriakis* sp. B	Darksnout Hound Shark
27:5	*Hypogaleus hyugaensis* (Miyosi, 1939)	Pencil Shark
27:6	*Iago garricki* (Fourmanoir & Rivaton, 1979)	Longnose Hound Shark
27:7	*Mustelus* sp. A	Grey Gummy Shark
27:8	*Mustelus* sp. B	White-spotted Gummy Shark
27:9	*Mustelus antarcticus* Günther, 1870	Gummy Shark
28	**HEMIGALEIDAE**	
28:1	*Hemigaleus microstoma* Bleeker, 1852	Weasel Shark
28:2	*Hemipristis elongata* (Klunzinger, 1871)	Fossil Shark
29	**CARCHARHINIDAE**	
29:1	*Carcharhinus albimarginatus* (Rüppell, 1837)	Silvertip Shark
29:2	*Carcharhinus altimus* (Springer, 1950)	Bignose Shark
29:3	*Carcharhinus amblyrhynchoides* (Whitley, 1934)	Graceful Shark
29:4	*Carcharhinus amblyrhynchos* (Bleeker, 1856)	Grey Reef Shark
29:5	*Carcharhinus amboinensis* (Müller & Henle, 1839)	Pigeye Shark
29:6	*Carcharhinus brachyurus* (Günther, 1870)	Bronze Whaler
29:7	*Carcharhinus brevipinna* (Müller & Henle., 1839)	Spinner Shark
29:8	*Carcharhinus cautus* (Whitley, 1945)	Nervous Shark
29:9	*Carcharhinus dussumieri* (Valenciennes, 1839)	Whitecheek Shark
29:10	*Carcharhinus falciformis* (Bibron, 1839)	Silky Shark
29:11	*Carcharhinus fitzroyensis* (Whitley, 1943)	Creek Whaler
29:12	*Carcharhinus galapagensis* (Snodgrass & Heller, 1905)	Galapagos Shark
29:13	*Carcharhinus leucas* (Valenciennes, 1839)	Bull Shark
29:14	*Carcharhinus limbatus* (Valenciennes, 1839)	Common Blacktip Shark
29:15	*Carcharhinus longimanus* (Poey, 1861)	Oceanic Whitetip Shark
29:16	*Carcharhinus macloti* (Müller & Henle, 1839)	Hardnose Shark
29:17	*Carcharhinus melanopterus* (Quoy & Gaimard, 1824)	Blacktip Reef Shark

29:18	*Carcharhinus obscurus* (Lesueur, 1818)	Dusky Shark
29:19	*Carcharhinus plumbeus* (Nardo, 1827)	Sandbar Shark
29:20	*Carcharhinus sorrah* (Valenciennes, 1839)	Spot-tail Shark
29:21	*Carcharhinus tilstoni* (Whitley, 1950)	Australian Blacktip Shark
29:22	*Galeocerdo cuvier* (Péron & Lesueur, 1822)	Tiger Shark
29:23	*Glyphis* sp. A	Speartooth Shark
29:24	*Loxodon macrorhinus* Müller & Henle, 1839	Sliteye Shark
29:25	*Negaprion acutidens* (Rüppell, 1837)	Lemon Shark
29:26	*Prionace glauca* (Linnaeus, 1758)	Blue Shark
29:27	*Rhizoprionodon acutus* (Rüppell, 1837)	Milk Shark
29:28	*Rhizoprionodon oligolinx* Springer, 1964	Grey Sharpnose Shark
29:29	*Rhizoprionodon taylori* (Ogilby, 1915)	Australian Sharpnose Shark
29:30	*Triaenodon obesus* (Rüppell, 1837)	Whitetip Reef Shark
30	**SPHYRNIDAE**	
30:1	*Eusphyra blochii* (Cuvier, 1816)	Winghead Shark
30:2	*Sphyrna lewini* (Griffith & Smith, 1834)	Scalloped Hammerhead
30:3	*Sphyrna mokarran* (Rüppell, 1837)	Great Hammerhead
30:4	*Sphyrna zygaena* (Linnaeus, 1837)	Smooth Hammerhead
31	**SQUATINIDAE**	
31:1	*Squatina* sp. A	Eastern Angel Shark
31:2	*Squatina* sp. B	Western Angel Shark
31:3	*Squatina australis* Regan, 1906	Australian Angel Shark
31:4	*Squatina tergocellata* McCulloch, 1914	Ornate Angel Shark
32	**RHINOBATIDAE**	
32:1	*Aptychotrema* sp. A	Spotted Shovelnose Ray
32:2	*Aptychotrema rostrata* (Shaw & Nodder, 1794)	Eastern Shovelnose Ray
32:3	*Aptychotrema vincentiana* (Haacke, 1885)	Western Shovelnose Ray
32:4	*Rhinobatos* sp. A	Goldeneye Shovelnose Ray
32:5	*Rhinobatos typus* Bennett, 1830	Giant Shovelnose Ray
32:6	*Trygonorrhina* sp. A	Eastern Fiddler Ray
32:7	*Trygonorrhina fasciata* Müller & Henle, 1841	Southern Fiddler Ray
32:8	*Trygonorrhina melaleuca* Scott, 1954	Magpie Fiddler Ray
33	**RHYNCHOBATIDAE**	
33:1	*Rhina ancylostoma* Bloch & Schneider, 1801	Shark Ray
33:2	*Rhynchobatus djiddensis* (Forsskål, 1775)	White-spotted Guitarfish
34	**RAJIDAE**	
34:1	*Bathyraja* sp. A	Abyssal Skate
34:2	*Irolita* sp. A	Western Round Skate
34:3	*Irolita waitii* (McCulloch, 1911)	Southern Round Skate
34:4	*Notoraja* sp. A	Blue Skate
34:5	*Notoraja* sp. B	Pale Skate
34:6	*Notoraja* sp. C	Ghost Skate
34:7	*Notoraja* sp. D	Blotched Skate
34:8	*Pavoraja* sp. A	Eastern Looseskin Skate
34:9	*Pavoraja* sp. B	Western Looseskin Skate
34:10	*Pavoraja* sp. C	Sandy Skate
34:11	*Pavoraja* sp. D	Mosaic Skate
34:12	*Pavoraja* sp. E	False Peacock Skate
34:13	*Pavoraja* sp. F	Dusky Skate
34:14	*Pavoraja alleni* McEachran & Fechelm, 1982	Allens Skate
34:15	*Pavoraja nitida* (Günther, 1880)	Peacock Skate
34:16	*Raja* sp. A	Longnose Skate
34:17	*Raja* sp. B	Grey Skate
34:18	*Raja* sp. C	Grahams Skate
34:19	*Raja* sp. D	False Argus Skate
34:20	*Raja* sp. E	Oscellate Skate
34:21	*Raja* sp. F	Leylands Skate
34:22	*Raja* sp. G	Pale Tropical Skate
34:23	*Raja* sp. H	Blacktip Skate
34:24	*Raja* sp. I	Wengs Skate
34:25	*Raja* sp. J	Deepwater Skate
34:26	*Raja* sp. K	Queensland Deepwater Skate
34:27	*Raja* sp. L	Maugean Skate

34:28	*Raja* sp. M	Pygmy Thornback Skate
34:29	*Raja* sp. N	Thintail Skate
34:30	*Raja* sp. O	Sawback Skate
34:31	*Raja* sp. P	Challenger Skate
34:32	*Raja australis* Macleay, 1884	Sydney Skate
34:33	*Raja cerva* Whitley, 1939	White-spotted Skate
34:34	*Raja gudgeri* (Whitley, 1940)	Bight Skate
34:35	*Raja hyperborea* Collett, 1879	Boreal Skate
34:36	*Raja lemprieri* Richardson, 1845	Thornback Skate
34:37	*Raja polyommata* Ogilby, 1910	Argus Skate
34:38	*Raja whitleyi* Iredale, 1938	Melbourne Skate
35	**ANACANTHOBATIDAE**	
35:1	*Anacanthobatis* sp. A	Western Leg Skate
35:2	*Anacanthobatis* sp. B	Eastern Leg Skate
36	**PRISTIDAE**	
36:1	*Anoxypristis cuspidata* (Latham, 1794)	Narrow Sawfish
36:2	*Pristis clavata* Garman, 1906	Dwarf Sawfish
36:3	*Pristis microdon* Latham, 1794	Freshwater Sawfish
36:4	*Pristis pectinata* Latham, 1794	Wide Sawfish
36:5	*Pristis zijsron* Bleeker, 1851	Green Sawfish
37	**TORPEDINIDAE**	
37:1	*Torpedo* sp. A	Longtail Torpedo Ray
37:2	*Torpedo macneilli* (Whitley, 1932)	Short-tail Torpedo Ray
38	**HYPNIDAE**	
38:1	*Hypnos monopterygium* (Shaw & Nodder, 1795)	Coffin Ray
39	**NARCINIDAE**	
39:1	*Narcine* sp. A	Ornate Numbfish
39:2	*Narcine* sp. B	Western Numbfish
39:3	*Narcine* sp. C	Eastern Numbfish
39:4	*Narcine tasmaniensis* Richardson, 1841	Tasmanian Numbfish
39:5	*Narcine westraliensis* McKay, 1966	Banded Numbfish
40	**DASYATIDIDAE**	
40:1	*Dasyatis* sp. A	Dwarf Black Stingray
40:2	*Dasyatis annotata* Last, 1987	Plain Maskray
40:3	*Dasyatis brevicaudata* (Hutton, 1875)	Smooth Stingray
40:4	*Dasyatis fluviorum* Ogilby, 1908	Estuary Stingray
40:5	*Dasyatis kuhlii* (Müller & Henle, 1841)	Blue-spotted Maskray
40:6	*Dasyatis leylandi* Last, 1987	Painted Maskray
40:7	*Dasyatis thetidis* Ogilby, 1899	Black Stingray
40:8	*Dasyatis violacea* (Bonaparte, 1832)	Pelagic Stingray
40:9	*Himantura* sp. A	Brown Whipray
40:10	*Himantura chaophraya* Monkolprasit & Roberts, 1990	Freshwater Whipray
40:11	*Himantura fai* Jordan & Seale, 1906	Pink Whipray
40:12	*Himantura granulata* (Macleay, 1883)	Mangrove Whipray
40:13	*Himantura jenkinsii* (Annandale, 1909)	Jenkins Whipray
40:14	*Himantura toshi* Whitley, 1939	Black-spotted Whipray
40:15	*Himantura uarnak* (Forsskål, 1775)	Reticulate Whipray
40:16	*Himantura undulata* (Bleeker, 1852)	Leopard Whipray
40:17	*Pastinachus sephen* (Forsskål, 1775)	Cowtail Stingray
40:18	*Taeniura lymma* (Forsskål, 1775)	Blue-spotted Fantail Ray
40:19	*Taeniura meyeni* Müller & Henle, 1841	Blotched Fantail Ray
40:20	*Urogymnus asperrimus* (Bloch & Schneider, 1801)	Porcupine Ray
41	**UROLOPHIDAE**	
41:1	*Plesiobatis daviesi* (Wallace, 1967)	Giant Stingaree
41:2	*Trygonoptera* sp. A	Yellow Shovelnose Stingaree
41:3	*Trygonoptera* sp. B	Eastern Shovelnose Stingaree
41:4	*Trygonoptera mucosa* Whitley, 1939	Western Shovelnose Stingaree
41:5	*Trygonoptera ovalis* Last & Gomon, 1987	Striped Stingaree
41:6	*Trygonoptera personata* Last & Gomon, 1987	Masked Stingaree
41:7	*Trygonoptera testacea* (Müller & Henle, 1841)	Common Stingare
41:8	*Urolophus* sp. A	Kapala Stingaree

41:9	*Urolophus* sp. B	Coral Sea Stingaree
41:10	*Urolophus bucculentus* Macleay, 1884	Sandyback Stingaree
41:11	*Urolophus circularis* McKay, 1966	Circular Stingaree
41:12	*Urolophus cruciatus* (Lacépède, 1804)	Banded Stingaree
41:13	*Urolophus expansus* McCulloch, 1916	Wide Stingaree
41:14	*Urolophus flavomosaicus* Last & Gomon, 1987	Patchwork Stingaree
41:15	*Urolophus gigas* Scott, 1954	Spotted Stingaree
41:16	*Urolophus lobatus* McKay, 1966	Lobed Stingaree
41:17	*Urolophus mitosis* Last & Gomon, 1987	Mitotic Stingaree
41:18	*Urolophus orarius* Last & Gomon, 1987	Coastal Stingaree
41:19	*Urolophus paucimaculatus* Dixon, 1969	Sparsely-spotted Stingaree
41:20	*Urolophus sufflavus* Whitley, 1929	Yellowback Stingaree
41:21	*Urolophus viridis* McCulloch, 1916	Greenback Stingaree
41:22	*Urolophus westraliensis* Last & Gomon, 1987	Brown Stingaree
42	**GYMNURIDAE**	
42:1	*Gymnura australis* (Ramsay & Ogilby, 1886)	Australian Butterfly Ray
43	**HEXATRYGONIDAE**	
43:1	*Hexatrygon* sp. A	Sixgill Stingray
44	**MYLIOBATIDIDAE**	
44:1	*Aetobatus narinari* (Euphrasen, 1790)	White-spotted Eagle Ray
44:2	*Aetomylaeus nichofii* (Schneider, 1801)	Banded Eagle Ray
44:3	*Aetomylaeus vespertilio* (Bleeker, 1852)	Ornate Eagle Ray
44:4	*Myliobatis australis* Macleay, 1881	Southern Eagle Ray
44:5	*Myliobatis hamlyni* Ogilby, 1911	Purple Eagle Ray
45	**RHINOPTERIDAE**	
45:1	*Rhinoptera neglecta* Ogilby, 1912	Australian Cownose Ray
46	**MOBULIDAE**	
46:1	*Manta birostris* (Donndorff, 1798)	Manta Ray
46:2	*Mobula eregoodootenkee* (Cuvier, 1829)	Pygmy Devilray
46:3	*Mobula japanica* (Müller & Henle, 1841)	Japanese Devilray
46:4	*Mobula thurstoni* (Lloyd, 1908)	Bentfin Devilray
47	**CALLORHINCHIDAE**	
47:1	*Callorhinchus milii* (Bory de Saint-Vincent, 1823)	Elephant Fish
48	**CHIMAERIDAE**	
48:1	*Chimaera* sp. A	Southern Chimaera
48:2	*Chimaera* sp. B	Shortspine Chimaera
48:3	*Chimaera* sp. C	Longspine Chimaera
48:4	*Chimaera* sp. D	Giant Chimaera
48:5	*Chimaera* sp. E	Whitefin Chimaera
48:6	*Hydrolagus* sp. A	Black Ghostshark
48:7	*Hydrolagus* sp. B	Marbled Ghostshark
48:8	*Hydrolagus lemures* (Whitley, 1939)	Blackfin Ghostshark
48:9	*Hydrolagus ogilbyi* (Waite, 1898)	Ogilbys Ghostshark
49	**RHINOCHIMAERIDAE**	
49.1	*Harriotta haeckeli* Karrer, 1972	Smallspine Spookfish
49:2	*Harriotta raleighana* Goode & Bean, 1895	Bigspine Spookfish
49:3	*Rhinochimaera pacifica* (Mitsukuri, 1895)	Pacific Spookfish

INDEX
SCIENTIFIC NAMES

Note: Family names are in bold, those in italics are the accepted scientific names. Page numbers in bold refer to major descriptions of species. Plates refer to the colour illustrations.

A

acanthias, Squalus 48, **98–9**, Plate 5
Acanthias vulgaris see *Squalus acanthias*
Acanthidium quadrispinosum see *Deania quadrispinosa*
acanutus, Apristurus 174
acutidens, Negaprion 222, **262–3**, Plate 31
acutus, Rhizoprionodon 223, **265–6**, Plate 30
Aetobatus narinari **448–9**, Plate 79
Aetomylaeus nichofii 448, **450–1**, Plate 79
Aetomylaeus reticulatus 452
Aetomylaeus vespertilio 448, **451–2**, Plate 79
africana, Rhinochimaera 483
albicauda, Scymnodalatias 54, **88–9**, Plate 9
albimarginatus, Carcharhinus 224, **228–9**, Plate 31
aliae, Squaliolus 54, **90–1**, Plate 9
alleni, Pavoraja 305, **320–1**, Plate 45
Alopias caudatus see *Alopias vulpinus*
Alopias greyi see *Alopias vulpinus*
Alopias pelagicus 154, **155**, Plate 12
Alopias superciliosus 154, **156**, Plate 12
Alopias vulpinus 154, **157–8**, Plate 12
Alopiidae 31, **154–8**
altimus, Carcharhinus 224, **230–1**, Plate 34
amblyrhynchoides, Carcharhinus 227, **231–2**, Plate 33
amblyrhynchos, Carcharhinus 224, **232–3**, Plate 32
amboinensis, Carcharhinus 226, **234–5**, Plate 34
Amphotistius sp. 1 see *Dasyatis annotata*
Amphotistius sp. 2 see *Dasyatis leylandi*
Anacanthobatidae 29, **357–9**
Anacanthobatis sp. A 357, **358**, Plate 64
Anacanthobatis sp. B 357, **359**, Plate 64
analis, Asymbolus 172, **187**, Plate 21
ancylostoma, Rhina 295, **296–7**, Plate 42
anguineus, Chlamydoselachus **36–7**, Plate 1
annotata, Dasyatis 383, **386–7**, Plate 71
Anoxypristis cuspidata 360, **361–2**, Plate 43
antarcticus, Mustelus 206, **215–16**, Plate 28
Aprionodon acutidens queenslandicus see *Negaprion acutidens*
Apristurus acanutus 174
Apristurus brunneus 175
Apristurus herklotsi 179
Apristurus longicephalus 168, **180**, Plate 19
Apristurus platyrhynchus 174
Apristurus sinensis 173
Apristurus sp. A 169, **173**, Plate 19
Apristurus sp. B 169, **174**, Plate 19
Apristurus sp. C 169, **175**, Plate 20
Apristurus sp. D 168, **176**, Plate 20
Apristurus sp. E 168, **177**, Plate 20
Apristurus sp. F 169, **178**, Plate 20
Apristurus sp. G 168, **179**, Plate 19
Aptychotrema rostrata 284, **286–7**, Plate 40

Aptychotrema sp. 1 see *Aptychotrema vincentiana*
Aptychotrema sp. 2 see *Aptychotrema* sp. A
Aptychotrema sp. A 284, **285–6**, Plate 40
Aptychotrema sp.: Allen & Swainston see *Aptychotrema* sp. A
Aptychotrema sp.: Gloerfelt-Tarp & Kailola see *Aptychotrema* sp. A
Aptychotrema vincentiana 284, **287–8**, Plate 40
aquila, Myliobatis 455
asperrimus, Urogymnus 381, **414–15**, Plate 70
Asymbolus analis 172, **187**, Plate 21
Asymbolus sp. A 172, **181**, Plate 22
Asymbolus sp. B 171, **182**, Plate 22
Asymbolus sp. C 172, **183**, Plate 21
Asymbolus sp. D 172, **184**, Plate 21
Asymbolus sp. E 172, **185**, Plate 21
Asymbolus sp. F 172, **186**, Plate 21
Asymbolus sp.: Hutchins & Swainston see *Asymbolus* sp. C
Asymbolus vincenti 172, **188**, Plate 22
Atelomycterus macleayi 167, **190**, Plate 22
Atelomycterus sp. A 167, **189**, Plate 22
atlantica, Rhinochimaera 483
Aulohalaelurus labiosus 168, **191**, Plate 24
australis, Gymnura **443–4**, Plate 79
australis, Myliobatis 447, **453–4**, Plate 80
australis, Raja 303, **346–7**, Plate 61
australis, Squatina 277, **280–1**, Plate 39

B

barbifer, Cirrhigaleus 48, **68**, Plate 6
Bathyraja sp. A 303, **306**, Plate 53
Bathytoshia brevicaudata see *Dasyatis brevicaudata*
birostris, Manta 458, **459–60**, Plate 81
bispinatus, Euprotomicrus 54, **85–6**, Plate 9
blochii, Eusphyra 270, **271–2**, Plate 37
boardmani, Galeus 170, **201**, Plate 18
boesemani, Halaelurus 171, **203–4**, Plate 18
bogimba, Galeolamna see *Carcharhinus leucas*
Brachaeluridae 33, **122–4**
Brachaelurus colcloughi 122, **123**, Plate 17
Brachaelurus waddi 122, **124**, Plate 17
brachyurus, Carcharhinus 8, 225, **235–6**, Plate 35
brachyurus, Etmopterus 51, **79–80**, Plate 7
brasiliensis, Isistius 54, **86–7**, Plate 3
brevicaudata, Dasyatis 384, **388–9**, Plate 72
brevipinna, Carcharhinus 227, **237–8**, Plate 33
brucus, Echinorhinus 44, **45**, Plate 11
bruniensis, Oxynotus **104–5**, Plate 11
brunneus, Apristurus 175
bucculentus, Urolophus 419, **430**, Plate 75

C

calcea, Deania 51, **70–1**, Plate 3
Callorhinchidae 35, **465–6**
Callorhinchus milii 10, **465–6**, Plate 82
Callorhynchus australis see *Callorhinchus milii*
Callorhynchus tasmanius see *Callorhinchus milii*
Carcharhinidae 35, **221–69**
Carcharhinus albimarginatus 224, **228–9**, Plate 31

Carcharhinus altimus 225, **230–1**, Plate 34
Carcharhinus amblyrhynchoides 227, **231–2**, Plate 33
Carcharhinus amblyrhynchos 224, **232–3**, Plate 32
Carcharhinus amboinensis 226, **234–5**, Plate 34
Carcharhinus brachyurus 8, 225, **235–6**, Plate 35
Carcharhinus brevipinna 227, **237–8**, Plate 33
Carcharhinus cautus 227, **238–9**, Plate 32
Carcharhinus cyrano see *Galeorhinus galeus*
Carcharhinus dussumieri 224, **239–40**, Plate 36
Carcharhinus falciformis 225, **241–2**, Plate 35
Carcharhinus fitzroyensis 227, **242–3**, Plate 36
Carcharhinus galapagensis 226, **243–4**, Plate 35
Carcharhinus leucas 7, 226, **244–6**, Plate 34
Carcharhinus limbatus 228, **246–7**, Plate 33
Carcharhinus longimanus 8, 224, **247–8**, Plate 31
Carcharhinus macki see *Prionace glauca*
Carcharhinus macloti 223, **249–50**, Plate 36
Carcharhinus melanopterus 227, **250–1**, Plate 32
Carcharhinus obscurus 8, 226, **252–3**, Plate 35
Carcharhinus plumbeus 226, **253–4**, Plate 34
Carcharhinus sorrah 10, 225, **255–6**, Plate 32
Carcharhinus tilstoni 10, 228, **256–7**, Plate 33
Carcharhinus wheeleri 233
Carcharias arenarius see *Carcharias taurus*
carcharias, Carcharodon 161, **162–3**, Plate 14
Carcharias dussumieri see *Carcharhinus dussumieri*
Carcharias falciformis see *Carcharhinus falciformis*
Carcharias leucas see *Carcharhinus leucas*
Carcharias limbatus see *Carcharhinus limbatus*
Carcharias sorrah see *Carcharhinus sorrah*
Carcharias spenceri see *Carcharhinus leucas*
Carcharias stevensi see *Carcharhinus plumbeus*
Carcharias taurus 144, **145–6**, Plate 15
Carcharias tricuspidatus see *Carcharias taurus*
Carcharodon albimors see *Carcharodon carcharias*
Carcharodon carcharias 161, **162–3**, Plate 14
Catulus labiosus see *Auiohalaelurus labiosus*
cautus, Carcharhinus 227, **238–9**, Plate 32
Centrophorus foliaceus see *Centrophorus squamosus*
Centrophorus granulosus 52, **55–6**, Plate 4
Centrophorus harrissoni 52, **56–7**, Plate 4
Centrophorus kaikourae see *Deania calcea*
Centrophorus moluccensis 52, **57–8**, Plate 4
Centrophorus nilsoni see *Centrophorus squamosus*
Centrophorus scalpratus see *Centrophorus moluccensis*
Centrophorus squamosus 52, **58–9**, Plate 4
Centrophorus uyato 52, **60–1**, Plate 4
Centrophorus waitei see *Centroscymnus plunketi*
Centroscyllium kamoharai 47, **61–2**, Plate 9
Centroscymnus coelolepis 53, **62–3**, Plate 10
Centroscymnus crepidater 53, **64–5**, Plate 10
Centroscymnus owstoni 53, **65–6**, Plate 10
Centroscymnus plunketi 53, **66–7**, Plate 10
cepedianus, Notorynchus 38, **42–3**, Plate 1
Cephaloscyllium fasciatum 170, **197**, Plate 23
Cephaloscyllium isabella 192
Cephaloscyllium isabella laticeps see *Cephaloscyllium laticeps*
Cephaloscyllium isabella nascione see *Cephaloscyllium laticeps*
Cephaloscyllium laticeps 171, **198**, Plate 23
Cephaloscyllium nascione see *Cephaloscyllium* sp. A
Cephaloscyllium sp. A 170, **192**, Plate 23
Cephaloscyllium sp. B 170, **193**, Plate 24
Cephaloscyllium sp. C 171, **194**, Plate 24
Cephaloscyllium sp. D 170, **195**, Plate 23
Cephaloscyllium sp. E 171, **196**, Plate 23
Ceratoptera alfredi see *Manta birostris*
cerva, Raja 302, **347–8**, Plate 62

Cestracion heterodontus see *Heterodontus portusjacksoni*
Cetorhinidae 33, **159–60**
Cetorhinus maximus **159–60**, Plate 13
chaophraya, Himantura 380, **399–400**, Plate 69
Chiloscyllium furvum see *Brachaelurus waddi*
Chiloscyllium fuscum see *Brachaelurus waddi*
Chiloscyllium modestum see *Brachaelurus waddi*
Chiloscyllium punctatum 134, **135**, Plate 17
Chimaera sp. 1: May & Maxwell see *Chimaera* sp. A
Chimaera sp. A 468, **469–70**, Plate 83
Chimaera sp. B 468, **470–1**, Plate 83
Chimaera sp. C 468, **471–2**, Plate 83
Chimaera sp. D 467, **473**, Plate 83
Chimaera sp. E 467, **474**, Plate 84
Chimaeridae 35, **467–78**
Chlamydoselachidae 30, **36–7**
Chlamydoselachus anguineus **36–7**, Plate 1
circularis, Urolophus 418, **431**, Plate 73
cirratus, Pristiophorus 106, **109–10**, Plate 44
Cirrhigaleus barbifer 48, **68**, Plate 6
clavata, Pristis 361, **362–3**, Plate 43
coelolepis, Centroscymnus 53, **62–3**, Plate 10
colcloughi, Brachaelurus 122, **123**, Plate 17
collare, Parascyllium 117, **119**, Plate 16
cookei, Echinorhinus 44, **46**, Plate 11
crepidater, Centroscymnus 53, **64–5**, Plate 10
cruciatus, Urolophus 418, **432**, Plate 76
cuspidata, Anoxypristis 360, **361–2**, Plate 43
cuvier, Galeocerdo 7, 222, **258–9**, Plate 29

D

Dalatias licha 54, **69–70**, Plate 10
Dasyatididae 30, **380–415**
Dasyatis annotata 383, **386–7**, Plate 71
Dasyatis brevicaudata 384, **388–9**, Plate 72
Dasyatis fluviorum 384, **390–1**, Plate 72
Dasyatis guileri see *Dasyatis violacea*
Dasyatis kuhlii 383, **391–2**, Plate 71
Dasyatis leylandi 383, **393–4**, Plate 71
Dasyatis sp. A 384, **385–6**, Plate 73
Dasyatis sp.: Allen & Swainston see *Dasyatis leylandi*
Dasyatis thetidis 384, **394–5**, Plate 72
Dasyatis violacea 383, **396–7**, Plate 72
dasypogon, Eucrossorhinus 125, **127–8**, Plate 25
daviesi, Plesiobatis 416, **420–1**, Plate 73
Deania calcea 51, **70–1**, Plate 3
Deania quadrispinosa 51, **71–2**, Plate 3
djiddensis, Rhynchobatus 295, **297–8**, Plate 42
draco, Himantura 404
dussumieri, Carcharhinus 224, **239–40**, Plate 36

E

Echinorhinidae 31, **44–6**
Echinorhinus brucus 44, **45**, Plate 11
Echinorhinus cookei 44, **46**, Plate 11
Echinorhinus mccoyi see *Echinorhinus brucus*
elongata, Hemipristis 217, **219–20**, Plate 29
Emissola ganearum see *Mustelus antarcticus*
Emissola maugeana see *Mustelus antarcticus*
eregoodootenkee, Mobula 458, **460–1**, Plate 81
Etmopterus abernethyi see *Etmopterus lucifer*
Etmopterus baxteri see *Etmopterus granulosus*
Etmopterus brachyurus 51, **79–80**, Plate 7
Etmopterus granulosus 50, **80–1**, Plate 9
Etmopterus lucifer 51, **81–2**, Plate 7
Etmopterus molleri 51, **83–4**, Plate 7
Etmopterus pusillus 50, **84–5**, Plate 8
Etmopterus sp. A 50, **73**, Plate 8

Etmopterus sp. B 50, **74**, Plate 8
Etmopterus sp. C 50, **75**, Plate 8
Etmopterus sp. D 50, **76**, Plate 8
Etmopterus sp. E 51, **77**, Plate 7
Etmopterus sp. F 51, **78**, Plate 7
Etmopterus unicolor 74
Eucrossorhinus dasypogon 125, **127–8**, Plate 25
Eugomphodus taurus see *Carcharias taurus*
Eulamia ahenea see *Carcharhinus brachyurus*
Euprotomicrus bispinatus 54, **85–6**, Plate 9
Eusphyra blochii 270, **271–2**, Plate 37
expansus, Urolophus 420, **433**, Plate 74

F
fai, Himantura 382, **400–1**, Plate 69
fairchildi, Torpedo 371
falciformis, Carcharhinus 225, **241–2**, Plate 35
fasciata, Trygonorrhina 283, **293–3**, Plate 41
fasciatum, Cephaloscyllium 170, **197**, Plate 23
fasciatum, Stegastoma **138–9**, Plate 15
ferox, Odontaspis 144, **146–7**, Plate 15
ferrugineum, Parascyllium 117, **120**, Plate 16
ferrugineus, Nebrius 7, **140–1**, Plate 17
Figaro boardmani see *Galeus boardmani*
Figaro boardmani socius see *Galeus boardmani*
fitzroyensis, Carcharhinus 227, **242–3**, Plate 36
flavomosaicus, Urolophus 419, **434**, Plate 75
fluviorum, Dasyatis 384, **390–1**, Plate 72
Fur macki see *Furgaleus macki*
Fur ventralis see *Furgaleus macki*
Furgaleus macki 10, 205, **207**, Plate 27
Furgaleus ventralis see *Furgaleus macki*

G
galapagensis, Carcharhinus 226, **243–4**, Plate 35
galeatus, Heterodontus 112, **113**, Plate 2
Galeocerdo cuvier 7, 222, **258–9**, Plate 29
Galeocerdo rayneri see *Galeocerdo cuvier*
Galeolamna ahenea see *Carcharhinus brachyurus*
Galeolamna bogimba see *Carcharhinus leucas*
Galeolamna coongoola see *Carcharhinus amblyrhynchos*
Galeolamna dorsalis see *Carcharhinus plumbeus*
Galeolamna fitzroyensis see *Carcharhinus fitzroyensis*
Galeolamna fowleri see *Carcharhinus amblyrhynchos*
Galeolamna fowleri see *Carcharhinus brevipinna*
Galeolamna greyi cauta see *Carcharhinus cautus*
Galeolamna greyi mckaili see *Carcharhinus leucas*
Galeolamna greyi see *Carcharhinus brachyurus*
Galeolamna isobel see *Carcharhinus sorrah*
Galeolamna macrurus see *Carcharhinus obscurus*
Galeolamna mckaili see *Carcharhinus leucas*
Galeolamna stevensi see *Carcharhinus leucas*
Galeolamna tufiensis see *Carcharhinus amblyrhynchos*
Galeolamnoides isobel see *Carcharhinus sorrah*
Galeorhinus australis see *Galeorhinus galeus*
Galeorhinus galeus 205, **208**, Plate 27
Galeus australis see *Galeorhinus galeus*
Galeus boardmani 170, **201**, Plate 18
galeus, Galeorhinus 205, **208**, Plate 27
Galeus sp. A 169, **199**, Plate 18
Galeus sp. B 170, **200**, Plate 18
garricki, Iago 206, **212–13**, Plate 28
gigas, Urolophus 418, **435**, Plate 73
Gillisqualus amblyrhynchoides see *Carcharhinus amblyrhynchoides*
Ginglymostomatidae 32, **140–1**
glauca, Prionace 222, **263–4**, Plate 29
Glyphis glyphis 260

Glyphis sp. A 222, **259–60**, Plate 29
granulata, Himantura 381, **402–3**, Plate 69
granulosus, Centrophorus 52, **55–6**, Plate 4
granulosus, Etmopterus 50, **80–1**, Plate 9
griseus, Hexanchus 38, **40–1**, Plate 1
gudgeri, Raja 301, **349–50**, Plate 57
Gymnura australis **443–4**, Plate 79
Gymnuridae 29, **443–4**

H
haeckeli, Harriotta 479, **480–1**, Plate 82
Halaelurus analis see *Asymbolus analis*
Halaelurus boesemani 171, **203–4**, Plate 18
Halaelurus labiosus see *Aulohalaelurus labiosus*
Halaelurus sp. 1 see *Atelomycterus* sp. A
Halaelurus sp. 1 see *Halaelurus boesemani*
Halaelurus sp. A 171, **202**, Plate 18
Halaelurus vincenti see *Asymbolus vincenti*
Halsydrus maccoyi see *Cetorhinus maximus*
hamlyni, Myliobatis 447, **454–5**, Plate 80
Harriotta haeckeli 479, **480–1**, Plate 82
Harriotta raleighana 479, **481–2**, Plate 82
harrissoni, Centrophorus 52, **56–7**, Plate 4
Hemigaleidae 35, **217–20**
Hemigaleus microstoma 217, **218–19**, Plate 28
Hemipristis elongata 217, **219–20**, Plate 29
Hemiscylliidae 33, **134–7**
Hemiscyllium ocellatum 134, **136**, Plate 16
Hemiscyllium trispeculare 134, **137**, Plate 16
Hemitriakis sp. A 206, **209–10**, Plate 27
Hemitriakis sp. B 206, **210–11**, Plate 27
Heptranchias dakini see *Heptranchias perlo*
Heptranchias perlo 38, **39–40**, Plate 1
herklotsi, Apristurus 179
Heterodontidae 31, **112–16**
Heterodontus bonaespei see *Heterodontus portusjacksoni*
Heterodontus galeatus 112, **113**, Plate 2
Heterodontus portusjacksoni 112, **114–15**, Plate 2
Heterodontus zebra 112, **115–16**, Plate 2
Heteroscyllium colcloughi see *Brachaelurus colcloughi*
Hexanchidae 31, **38–43**
Hexanchus griseus 38, **40–1**, Plate 1
Hexanchus griseus australis see *Hexanchus griseus*
Hexanchus nakamurai 38, **41–2**, Plate 1
Hexanchus vitulus see *Hexanchus nakamurai*
Hexatrygon sp. A **445–6**, Plate 64
Hexatrygonidae 29, **445–6**
Himantura chaophraya 380, **399–400**, Plate 69
Himantura draco 404
Himantura fai 382, **400–1**, Plate 69
Himantura granulata 380, 381, **402–3**, Plate 69
Himantura jenkinsii 382, **403–4**, Plate 69
Himantura sp. 1: Gloerfelt-Tarp & Kailola see *Himantura uarnak*
Himantura sp. A 382, **397–8**, Plate 67
Himantura sp.: Compagno & Roberts see *Himantura chaophraya*
Himantura toshi 381, **405–6**, Plate 68
Himantura uarnak (spotted form) see *Himantura toshi*
Himantura uarnak 381, **406–7**, Plate 68
Himantura undulata 381, **408–9**, Plate 68
Hydrolagus lemures 468, **477**, Plate 84
Hydrolagus ogilbyi 468, **478**, Plate 84
Hydrolagus sp. A 468, **475**, Plate 84
Hydrolagus sp. B 468, **476**, Plate 84
Hydrolagus sp.: Last *et al.* see *Chimaera* sp. A
Hydrolagus waitei see *Hydrolagus ogilbyi*
hyperborea, Raja 300, **350–1**, Plate 55

Hypnidae 28, **372–3**
Hypnos monopterygium **372–3**, Plate 65
Hypnos subnigrum see *Hypnos monopterygium*
Hypogaleus hyugaensis 206, **211–12**, Plate 27
hyugaensis, Hypogaleus 206, **211–12**, Plate 27

I
Iago garricki 206, **212–13** Plate 28
Irolita sp. A 303, **307–8**, Plate 48
Irolita waitii 303, **308–9**, Plate 48
isabella, Cephaloscyllium 192
Isistius brasiliensis 54, **86–7**, Plate 3
Isuropsis mako see *Isurus oxyrinchus*
Isurus oxyrinchus 161, **163–4**, Plate 14
Isurus paucus 164

J
japanica, Mobula 458, **461–2**, Plate 81
javanica, Rhinoptera 457, Plate 80
jenkinsii, Himantura 382, **403–4**, Plate 69
Juncrus vincenti see *Asymbolus vincenti*

K
kamoharai, Centroscyllium 47, **61–2**, Plate 9
kamoharai, Pseudocarcharias **150–1**, Plate 14
kuhlii, Dasyatis 383, **391–2**, Plate 71

L
labiosus, Aulohalaelurus 168, **191**, Plate 24
Lamna nasus 161, **165–6**, Plate 14
Lamna whitleyi see *Lamna nasus*
Lamnidae 34, **161–6**
laticaudus, Squaliolus 91
laticeps, Cephaloscyllium 171, **198**, Plate 23
Leius ferox see *Isistius brasiliensis*
lemprieri, Raja 302, **352–3**, Plate 54
lemures, Hydrolagus 468, **477**, Plate 84
leucas, Carcharhinus 7, 226, **244–6**, Plate 34
lewini, Sphyrna 271, **272–3**, Plate 37
leylandi, Dasyatis 383, **393–4**, Plate 71
licha, Dalatias 54, **69–70**, Plate 10
limbatus, Carcharhinus 228, **246–7**, Plate 33
lobatus, Urolophus 417, **436**, Plate 76
longicephalus, Apristurus 168, **180**, Plate 19
longimanus, Carcharhinus 8, 224, **247–8**, Plate 31
Longmania calamaria see *Carcharhinus brevipinna*
Loxodon macrorhinus 223, **260–1**, Plate 30
lucifer, Etmopterus 51, **81–2**, Plate 7
lymma, Taeniura 382, **411–12**, Plate 70

M
macki, Furgaleus 10, 205, **207**, Plate 27
macleayi, Atelomycterus 167, **190**, Plate 22
macloti, Carcharhinus 223, **249–50**, Plate 36
macmillani, Parmaturus 204
macneilli, Torpedo 368, **370–1**, Plate 65
macrorhinus, Loxodon 223, **260–1**, Plate 30
maculatus, Orectolobus 126, **129–30**, Plate 26
Manta alfredi see *Manta birostris*
Manta birostris 458, **459–60**, Plate 81
maximus, Cetorhinus **159–60**, Plate 13
Megachasma pelagios **152–3**, Plate 13
Megachasmidae 34, **152–3**
megalops, Squalus 48, **99–100**, Plate 5
melaleuca, Trygonorrhina 283, **294**, Plate 41
melanopterus, Carcharhinus 227, **250–1**, Plate 32
meyeni, Taeniura 382, **412–13**, Plate 70

microcephalus, Somniosus 89
microdon, Pristis 361, **364–5**, Plate 42
microstoma, Hemigaleus 217, **218–19**, Plate 28
milii, Callorhinchus 10, **465–6**, Plate 82
mitosis, Urolophus 419, **437**, Plate 75
mitsukurii, Squalus 49, **101–2**, Plate 5
Mitsukurina owstoni **148–9**, Plate 13
Mitsukurinidae 32, **148–9**
Mobula eregoodootenkee 458, **460–1**, Plate 81
Mobula japanica 458, **461–2**, Plate 81
Mobula thurstoni 458, **463–4**, Plate 81
Mobulidae 30, **458–64**
mokarran, Sphyrna 271, **274–5**, Plate 37
molleri, Etmopterus 51, **83–4**, Plate 7
Molochophrys galeatus see *Heterodontus galeatus*
moluccensis, Centrophorus 52, **57–8**, Plate 4
monopterygium, Hypnos **372–3**, Plate 65
mucosa, Trygonoptera 417, **424**, Plate 78
Mustelus antarcticus 206, **215–16**, Plate 28
Mustelus manazo see *Mustelus* sp. B
Mustelus sp. A 206, **213–14**, Plate 28
Mustelus sp. B (eastern form) 206
Mustelus sp. B (western form) 206
Mustelus sp. B 206, **214–15**, Plate 28
Myliobatididae 30, **447–55**
Myliobatis aquila 455
Myliobatis australis 447, **453–4**, Plate 80
Myliobatis hamlyni 447, **454–5**, Plate 80
Myliobatis tenuicaudatus 454
Myliobatis tobijei 455
Mystidens inominatus see *Negaprion acutidens*

N
nakamurai, Hexanchus 38, **41–2**, Plate 1
Narcine sp. 1 see *Narcine* sp. B
Narcine sp. A 374, **375**, Plate 67
Narcine sp. B 374, **376**, Plate 66
Narcine sp. C 374, **377**, Plate 66
Narcine tasmaniensis 374, **378**, Plate 66
Narcine westraliensis 374, **379**, Plate 67
Narcinidae 28, **374–9**
narinari, Aetobatus **448–9**, Plate 79
nasus, Lamna 161, **165–6**, Plate 14
nasuta, Raja 340
Nebrius concolor see *Nebrius ferrugineus*
Nebrius ferrugineus 7, **140–1**, Plate 17
Nebrodes concolor ogilbyi see *Nebrius ferrugineus*
Negaprion acutidens 222, **262–3**, Plate 31
Negaprion queenslandicus see *Negaprion acutidens*
neglecta, Rhinoptera **456–7**, Plate 80
Negogaleus microstoma see *Hemigaleus microstoma*
nichofii, Aetomylaeus 448, **450–1**, Plate 79
nitida, Pavoraja 305, **322–3**, Plate 47
Notogaleus australis see *Galeorhinus galeus*
Notogaleus rhinophanes see *Galeorhinus galeus*
Notoraja sp. A 304, **309–10**, Plate 50
Notoraja sp. B 304, **311**, Plate 50
Notoraja sp. C 304, **312**, Plate 51
Notoraja sp. D 304, **313**, Plate 51
Notorynchus cepedianus 38, **42–3**, Plate 1
Notorynchus macdonaldi see *Notorynchus cepedianus*
Notostrape macneilli see *Torpedo macneilli*
nudipinnis, Pristiophorus 106, **111**, Plate 44

O

obesus, Triaenodon 221, **268–9**, Plate 31
obscurus, Carcharhinus 8, 226, **252–3**, Plate 35
ocellatum, Hemiscyllium 134, **136**, Plate 16
Odontaspididae 34, **144–7**
Odontaspis cinerea see *Carcharias taurus*
Odontaspis ferox 144, **146–7**, Plate 15
Odontaspis herbsti see *Odontaspis ferox*
ogilbyi, Hydrolagus 468, **478**, Plate 84
oligolinx, Rhizoprionodon 224, **266–7**, Plate 30
orarius, Urolophus 418, **438**, Plate 76
Orectolobidae 33, **125–33**
Orectolobus devisi see *Orectolobus ornatus*
Orectolobus maculatus 126, **129–30**, Plate 26
Orectolobus ogilbyi see *Eucrossorhinus dasypogon*
Orectolobus ornatus 126, **131**, Plate 26
Orectolobus ornatus halei see *Orectolobus ornatus*
Orectolobus sp. A 126, **128–9**, Plate 26
Orectolobus sp.: Hutchins & Swainston see *Orectolobus* sp. A
Orectolobus wardi 126, **132**, Plate 25
ornatus, Orectolobus 126, **131**, Plate 26
ovalis, Trygonoptera 417, **425**, Plate 77
owstoni, Centroscymnus 53, **65–6**, Plate 10
owstoni, Mitsukurina 148–9, Plate 13
Oxynotidae 31, **104–5**
Oxynotus bruniensis **104–5**, Plate 11
oxyrinchus, Isurus 161, **163–4**, Plate 14

P

pacifica, Rhinochimaera 479, **482–3**, Plate 82
pacificus, Somniosus 54, **89–90**, Plate 3
Parascylliidae 33, **117–21**
Parascyllium collare 117, **119**, Plate 16
Parascyllium ferrugineum 117, **120**, Plate 16
Parascyllium multimaculatum see *Parascyllium ferrugineum*
Parascyllium nuchalis see *Parascyllium variolatum*
Parascyllium sp. A 117, **118**, Plate 16
Parascyllium variolatum 117, **121**, Plate 16
Parmaturus macmillani 204
Parmaturus sp. A 169, **204**, Plate 20
Pastinachus sephen 383, **409–10**, Plate 70
Pastinachus sephen ater see *Pastinachus sephen*
paucimaculatus, Urolophus 419, **439**, Plate 74
paucus, Isurus 164
Pavoraja alleni 305, **320–1**, Plate 45
Pavoraja nitida 305, **322–3**, Plate 47
Pavoraja polyommata see *Raja polyommata*
Pavoraja sp. A 304, **314**, Plate 49
Pavoraja sp. B 304, **315**, Plate 49
Pavoraja sp. C 305, **316**, Plate 45
Pavoraja sp. D 305, **317**, Plate 46
Pavoraja sp. E 305, **318–19**, Plate 47
Pavoraja sp. F 305, **319–20**, Plate 46
pectinata, Pristis 361, **365–6**, Plate 43
pelagicus, Alopias 154, **155**, Plate 12
pelagios, Megachasma **152–3**, Plate 13
perlo, Heptranchias 38, **39–40**, Plate 1
perotteti, Pristis 365
personata, Trygonoptera 417, **426**, Plate 77
Physodon taylori see *Rhizoprionodon taylori*
Platypodon coatesi see *Carcharhinus dussumieri*
platyrhynchus, Apristurus 174
Plesiobatis daviesi 416, **420–1**, Plate 73
plumbeus, Carcharhinus 226, **253–4**, Plate 34
plunketi, Centroscymnus 53, **66–7**, Plate 10
polyommata, Raja 302, **353–4**, Plate 63
portusjacksoni, Heterodontus 112, **114–15**, Plate 2

Prionace glauca 222, **263–4**, Plate 29
Prionodon dussumieri see *Carcharhinus dussumieri*
Prionodon falciformis see *Carcharhinus falciformis*
Prionodon leucas see *Carcharhinus leucas*
Prionodon limbatus see *Carcharhinus limbatus*
Prionodon sorrah see *Carcharhinus sorrah*
Pristidae 27, **360–7**
Pristiophoridae 27, **106–11**
Pristiophorus cirratus 106, **109–10**, Plate 44
Pristiophorus nudipinnis 106, **111**, Plate 44
Pristiophorus owenii see *Pristiophorus nudipinnis*
Pristiophorus sp. A **107–8**, Plate 44
Pristiophorus sp. B 107, **108–9**, Plate 44
Pristiopsis leichhardti see *Pristis microdon*
Pristis clavata 361, **362–3**, Plate 43
Pristis leichhardti see *Pristis microdon*
Pristis microdon 361, **364–5**, Plate 42
Pristis pectinata 361, **365–6**, Plate 43
Pristis perotteti 365
Pristis pristis 365
Pristis zijsron 360, **366–7** Plate 43
Pristiurus boardmani see *Galeus boardmani*
Protozygaena taylori see *Rhizoprionodon taylori*
Psammobatis waitii see *Irolita waitii*
Pseudocarcharias kamoharai **150–1**, Plate 14
Pseudocarchariidae 34, **150–1**
Psychichthys waitei see *Hydrolagus ogilbyi*
Pteroplatytrygon 397
punctatum, Chiloscyllium 134, **135**, Plate 17
pusillus, Etmopterus 50, **84–5**, Plate 8

Q

quadrispinosa, Deania 51, **71–2**, Plate 3

R

Raia scabra see *Raja whitleyi*
Raja africana see *Urogymnus asperrimus*
Raja australis 303, **346–7**, Plate 61
Raja cerva 302, **347–8**, Plate 62
Raja dentata see *Raja lemprieri*
Raja gudgeri 301, **349–50**, Plate 57
Raja hyperborea 300, **350–1**, Plate 55
Raja lemprieri 302, **352–3**, Plate 54
Raja nasuta 340
Raja ogilbyi see *Raja whitleyi*
Raja polyommata 302, **353–4**, Plate 63
Raja sp.: Allen & Swainston see *Raja* sp. D
Raja sp. 1: Last *et al.* see *Raja gudgeri*
Raja sp. 1: Sainsbury *et al.* see *Raja* sp. D
Raja sp. 2: Last *et al.* see *Raja* sp. B
Raja sp. A 303, **323–4**, Plate 62
Raja sp. B 302, **325–6**, Plate 60
Raja sp. C 303, **326–7**, Plate 60
Raja sp. D 302, **328–9**, Plate 61
Raja sp. E 302, **329–30**, Plate 63
Raja sp. F 301, **330–1**, Plate 58
Raja sp. G 301, **332–3**, Plate 57
Raja sp. H 300, 301, **333–4**, Plate 56
Raja sp. I 300, **335–6**, Plate 58
Raja sp. J 301, **336–7**, Plate 59
Raja sp. K 301, **338–9**, Plate 59
Raja sp. L 301, **339–40**, Plate 55
Raja sp. M 302, **341–2**, Plate 54
Raja sp. N 300, **342–3**, Plate 56
Raja sp. O 300, **344**, Plate 52
Raja sp. P 300, **345–6**, Plate 52
Raja whitleyi 299, **355–6**, Plate 53

Rajidae 29, **299–356**
raleighana, *Harriotta* 479, **481–2**, Plate 82
Raya rostrata see *Raja whitleyi*
reticulatus, *Aetomylaeus* 452
Rhina ancylostoma 295, **296–7**, Plate 42
Rhincodon typus **142–3** Plate 15
Rhincodontidae 32, **142–3**
Rhiniodon typus see *Rhincodon typus*
Rhinobatidae 28, **283–94**
Rhinobatos batillum see *Rhinobatos typus*
Rhinobatos sp. 1: Gloerfelt-Tarp & Kailola see *Rhinobatos* sp. A
Rhinobatos sp. 1: Sainsbury *et al.* see *Rhinobatos* sp. A
Rhinobatos sp. 2 see *Rhinobatos* sp. A
Rhinobatos sp. A 284, **289**, Plate 39
Rhinobatos sp.: Allen & Swainston see *Rhinobatos* sp. A
Rhinobatos typus 284, **290–1**, Plate 39
Rhinobatos armatus see *Rhinobatos typus*
Rhinobatos banksii see *Aptychotrema rostrata*
Rhinobatos dumerilii see *Trygonorrhina fasciata*
Rhinobatos tuberculatus see *Aptychotrema rostrata*
Rhinochimaera africana 483
Rhinochimaera atlantica 483
Rhinochimaera pacifica 479, **482–3**, Plate 82
Rhinochimaeridae 35, **479–83**
Rhinoptera javanica 457, Plate 80
Rhinoptera neglecta **456–7**, Plate 80
Rhinopteridae 30, **456–7**
Rhizoprionodon acutus 223, **265–6**, Plate 30
Rhizoprionodon oligolinx 224, **266–7**, Plate 30
Rhizoprionodon taylori 224, **267–8**, Plate 30
Rhynchobatidae 28, **295–8**
Rhynchobatus djiddensis 295, **297–8**, Plate 42
Rhynchobatus djiddensis australiae see *Rhynchobatus djiddensis*
rostrata, *Aptychotrema* 284, **286–7**, Plate 40
Rubusqualus mccoyi see *Echinorhinus brucus*

S
Scapanorhynchus owstoni see *Mitsukurina owstoni*
Scoliodon affinis see *Loxodon macrorhinus*
Scoliodon jordani see *Loxodon macrorhinus*
Scoliodon longmani see *Rhizoprionodon acutus*
Scyliorhinidae 34, **167–204**
Scymnodalatias albicauda 54, **88–9**, Plate 9
Scymnodalatias sherwoodi 89
Scymnodon squamulosus see *Zameus squamulosus*
Scymnorhinus phillippsi see *Dalatias licha*
sephen, *Pastinachus* 383, **409–10**, Plate 70
sherwoodi, *Scymnodalatias* 89
sinensis, *Apristurus* 173
Somniosus antarcticus see *Somniosus pacificus*
Somniosus microcephalus 89
Somniosus pacificus 54, **89–90**, Plate 3
sorrah, *Carcharhinus* 10, 225, **255–6**, Plate 32
Sphyrna blochii see *Eusphyra blochii*
Sphyrna lewini 271, **272–3**, Plate 37
Sphyrna mokarran 271, **274–5**, Plate 37
Sphyrna zygaena 270, **275–6**, Plate 37
Sphyrnidae 31, **270–6**
Squalidae 31, **47–103**
Squaliolus aliae 54, **90–1**, Plate 9
Squaliolus laticaudus 91
Squalus acanthias 48, **98–9**, Plate 5
Squalus anisodon see *Pristiophorus cirratus*
Squalus appendiculatus see *Orectolobus maculatus*
Squalus barbatus see *Orectolobus maculatus*
Squalus blainvillei see *Squalus mitsukurii*
Squalus fernandinus see *Squalus acanthias*
Squalus jacksoni see *Heterodontus portusjacksoni*

Squalus kirki see *Squalus acanthias*
Squalus lobatus see *Orectolobus maculatus*
Squalus megalops 48, **99–100**, Plate 5
Squalus mitsukurii 49, **101–2**, Plate 5
Squalus oculatus see *Hemiscyllium ocellatum*
Squalus philippi see *Heterodontus portusjacksoni*
Squalus philippinus see *Heterodontus portusjacksoni*
Squalus sp. A 48, **91–2**, Plate 6
Squalus sp. B 49, **93**, Plate 6
Squalus sp. C 49, **94**, Plate 6
Squalus sp. D 48, **95**, Plate 6
Squalus sp. E 49, **96**, Plate 5
Squalus sp. F 49, **97**, Plate 5
Squalus tasmaniensis see *Squalus megalops*
Squalus tentaculatus see *Pristiophorus cirratus*
Squalus whitleyi see *Squalus acanthias*
squamosus, *Centrophorus* 52, **58–9**, Plate 4
squamulosus, *Zameus* 53, **102–3**, Plate 3
Squatina australis 277, **280–1**, Plate 39
Squatina sp. A 277, **278**, Plate 38
Squatina sp. B 277, **279–80**, Plate 38
Squatina tergocellata McCulloch 277, **281–2**, Plate 38
Squatina tergocellata Sainsbury *et al.* see *Squatina* sp. B
Squatinidae 27, **277–82**
Stegastoma fasciatum **138–9**, Plate 15
Stegastoma tigrinum naucum see *Stegastoma fasciatum*
Stegastomatidae 32, **138–9**
sufflavus, *Urolophus* 418, **440**, Plate 76
superciliosus, *Alopias* 154, **156**, Plate 12
Sutorectus tentaculatus 125, **133**, Plate 25
Sutorectus wardi see *Orectolobus wardi*

T
Taeniura lymma 382, **411–12**, Plate 70
Taeniura melanospila 413
Taeniura melanospilos see *Taeniura meyeni*
Taeniura meyeni 382, **412–13**, Plate 70
Taeniura mortoni see *Taeniura meyeni*
tasmaniensis, *Narcine* 374, **378**, Plate 66
taurus, *Carcharias* 144, **145–6**, Plate 15
taylori, *Rhizoprionodon* 224, **267–8**, Plate 30
tentaculatus, *Sutorectus* 125, **133**, Plate 25
tenuicaudatus, *Myliobatis* 454
tergocellata, *Squatina* 277, **281–2**, Plate 38
testacea, *Trygonoptera* 417, **427**, Plate 78
Tetroras maccoyi see *Cetorhinus maximus*
thetidis, *Dasyatis* 384, **394–5**, Plate 72
thurstoni, *Mobula* 458, **463–4**, Plate 81
tilstoni, *Carcharhinus* 10, 228, **256–7**, Plate 33
tobijei, *Myliobatis* 455
Torpedinidae 28, **368–71**
Torpedo fairchildi 371
Torpedo macneilli 368, **370–1**, Plate 65
Torpedo sp. A 368, **369**, Plate 65
toshi, *Himantura* 381, **405–6**, Plate 68
Triaenodon apicalis see *Triaenodon obesus*
Triaenodon obesus 221, **268–9**, Plate 31
Triakidae 34, **205–16**
trispeculare, *Hemiscyllium* 134, **137**, Plate 16
Trygonoptera australis see *Trygonoptera testacea*
Trygonoptera henlei see *Trygonoptera testacea*
Trygonoptera mucosa 417, **424**, Plate 78
Trygonoptera muelleri see *Trygonoptera testacea*
Trygonoptera ovalis 417, **425**, Plate 77
Trygonoptera personata 417, **426**, Plate 77
Trygonoptera sp. 1: Gloerfelt-Tarp & Kailola see *Urolophus flavomosaicus*
Trygonoptera sp. A 417, **422**, Plate 78

Trygonoptera sp. B 417, **423**, Plate 78
Trygonoptera testacea 417, **427**, Plate 78
Trygonorhina fasciata see *Trygonorrhina fasciata*
Trygonorhina guanerius see *Trygonorrhina fasciata*
Trygonorrhina fasciata 283, **292–3**, Plate 41
Trygonorrhina melaleuca 283, **294**, Plate 41
Trygonorrhina sp. A 283, **291–2**, Plate 41
typus, *Rhincodon* **142–3**, Plate 15
typus, *Rhinobatos* 284, **290–1**, Plate 39

U
uarnak, *Himantura* 381, **406–7**, Plate 68
undulata, *Himantura* 381, **408–9**, Plate 68
unicolor, *Etmopterus* 74
Uranga nasuta see *Carcharhinus brevipinna*
Uranganops fitzroyensis see *Carcharhinus fitzroyensis*
Urogymnus asperrimus 381, **414–15**, Plate 70
Urogymnus asperrimus solanderi see *Urogymnus asperrimus*
Urolophidae 30, 416–42
Urolophus bucculentus 419, **430**, Plate 75
Urolophus circularis 418, **431**, Plate 73
Urolophus cruciatus 418, **432**, Plate 76
Urolophus ephippiatus see *Urolophus cruciatus*
Urolophus expansus 420, **433**, Plate 74
Urolophus flavomosaicus 419, **434**, Plate 75
Urolophus gigas 418, **435**, Plate 73
Urolophus lobatus 417, **436**, Plate 76
Urolophus mitosis 419, **437**, Plate 75
Urolophus orarius 418, **438**, Plate 76
Urolophus paucimaculatus 419, **439**, Plate 74
Urolophus sp. 1: Hutchins & Thompson see *Trygonoptera ovalis*
Urolophus sp. 1: Sainsbury *et al.* see *Urolophus westraliensis*
Urolophus sp. 2: Gloerfelt-Tarp & Kailola see *Urolophus mitosis*
Urolophus sp. 2: Hutchins & Thompson see *Trygonoptera personata*
Urolophus sp. 2: Sainsbury *et al.* see *Urolophus flavomosaicus*
Urolophus sp. 3: Sainsbury *et al.* see *Urolophus mitosis*
Urolophus sp. A 419, **428**, Plate 74
Urolophus sp. B 419, **429**, Plate 77
Urolophus sp.: Hutchins & Swainston see *Trygonoptera ovalis*
Urolophus sp.: Hutchins & Swainston see *Trygonoptera personata*
Urolophus sp.: Hutchins & Swainston see *Urolophus gigas*
Urolophus sufflavus 418, **440**, Plate 76
Urolophus viridis 420, **441**, Plate 74
Urolophus westraliensis 419, **442**, Plate 75
Urotrygon sp. 1 see *Plesiobatis daviesi*
uyato, *Centrophorus* 52, **60–1**, Plate 4

V
variolatum, *Parascyllium* 117, **121**, Plate 16
vespertilio, *Aetomylaeus* 448, **451–2**, Plate 79
vincenti, *Asymbolus* 172, **188**, Plate 22
vincentiana, *Aptychotrema* 284, **287–8**, Plate 40
violacea, *Dasyatis* 383, **396–7**, Plate 72
viridis, *Urolophus* 420, **441**, Plate 74
vulpinus, *Alopias* 154, **157–8**, Plate 12

W
waddi, *Brachaelurus* 122, **124**, Plate 17
waitii, *Irolita* 303, **308–9**, Plate 48
wardi, *Orectolobus* 126, **132**, Plate 25
westraliensis, *Narcine* 374, **379**, Plate 67
westraliensis, *Urolophus* 419, **442**, Plate 75
wheeleri, *Carcharhinus* 233
whitleyi, *Raja* 299, **355–6**, Plate 53

Z
Zameus squamulosus 53, **102–3**, Plate 3
zebra, *Heterodontus* 112, **115–16**, Plate 2
zijsron, *Pristis* 360, **366–7**, Plate 43
zygaena, *Sphyrna* 270, **275–6**, Plate 37

INDEX
COMMON NAMES

Note: Recommended common names are in bold, page numbers in bold refer to major descriptions of species. Plates refer to the colour illustrations.

A
abyssal skate 306, Plate 53
Allens skate 320–1, Plate 45
angel sharks 27, **277–82**, Plates 38–9
angelshark 280
archbishop 281
Argus skate 353–4, Plate 63
Australian angel shark 280–1, Plate 39
Australian blacktip shark 10, **256–7**, Plate 33
Australian butterfly ray 443–4, Plate 79
Australian cownose ray 456–7, Plate 80
Australian devilray 459
Australian marbled catshark 190
Australian sawtail catshark 201
Australian sharpnose shark 267–8, Plate 30
Australian smooth hound 215
Australian spotted catshark 187
Australian swellshark 198

B
banana-tail ray 409
banded catshark 189, Plate 22
banded eagle ray 450–1, Plate 79
banded numbfish 379, Plate 67
banded shark 201
banded stingaree 432, Plate 76
banded wobbegong 131, Plate 26
banjo shark 291, 292
Banks shovelnose ray 286
bareskin dogfish 61–2, Plate 9
bartail spurdog 91–2, Plate 6
basking shark 33, **159–60**, Plate 13
bebil 424
bentfin devilray 463–4, Plate 81
bigeye sixgill shark 41–2, Plate 1
bigeye thresher 156–7, Plate 12
bigfin catshark 174, Plate 19
bighead catshark 178, Plate 20
Bight ghostshark 477
Bight skate 349–50, Plate 57
bignose shark 230–1, Plate 34
bigspine spookfish 481–2, Plate 82
birdbeak dogfish 70
black-blotched stingray 412
black ghostshark 475, Plate 84
black shark 69–70, Plate 10
black skate 394
black-spotted catshark 191, Plate 24
black-spotted stingray 412
black-spotted whipray 405–6, Plate 68
black stingaree 394
black stingray 394–5, Plate 72

black-vee whaler 232
black whaler 252
blackbelly lantern shark 81–2, Plate 7
blackfin ghostshark 477, Plate 84
blackmouth lantern shark 77, Plate 7
blacktip houndshark 211
blacktip reef shark 250–1, Plate 32
blacktip shark 250
blacktip skate 333–4, Plate 56
blacktip tope 211
blacktip whaler 246, 256
blind shark 124, 136, Plate 17
blind sharks 33, **122–4**, Plate 17
blotched catshark 182, Plate 22
blotched fantail ray 412–13, Plate 70
blotched skate 313, Plate 51
blotched stingaree 437
blue pointer 163
blue shark 263–4, Plate 29
blue skate 309–10, Plate 50
blue-spotted fantail ray 411–12, Plate 70
blue-spotted lagoon ray 411
blue-spotted maskray 391–2, Plate 71
blue-spotted ribbontail ray 411
blue-spotted stingaree 391
blue-spotted stingray 391
blue whaler 263
bluegrey carpet shark 123
blunthead shark 268
bluntnose sixgill shark 40–1, Plate 1
bonnet skate 448
boreal skate 350–1, Plate 55
bowmouth guitarfish 296
bramble shark 45, Plate 11
bramble sharks 31, **44–6**, Plate 11
brier shark 70–1, Plate 3
bristled lantern shark 74 Plate 8
broad-snout 42
broadnose sevengill shark 42–3, Plate 1
bronze whaler 8, **235–6**, 252, Plate 35
brown-banded bamboo shark 135
brown-banded catshark 135
brown catshark 124
brown-reticulate stingray 393
brown-spotted catshark 135
brown stingaree 442, Plate 75
brown stingray 386–7, 390
brown whipray 397–8, Plate 67
bull ray 453
bull shark 7, 8, 40, **244–6**, Plate 34
bulldog catshark 177, Plate 20
bullhead 114
bullhead shark 115
bumpytail ragged-tooth 146
butterfly rays 29, **443–4**, Plate 79

C

carpet shark 131
carpet sharks 33, **117–21, 134-7**, Plates 16–17
catsharks 34, **167–204**, Plates 18–24
Challenger skate 345–6, Plate 52
chimaeras Plates 83, 84
circular stingaree 431, Plate 73
coachwhip ray 405, 406
coastal stingaree 438, Plate 76
cobbler carpet shark 133
cobbler wobbegong 133, Plate 25
cocktail shark 235
coffin ray 28, **372–3**, Plate 65
Colcloughs shark 123, Plate 17
collared carpet shark 119, Plate 16
collared carpet sharks 33, **117–21**, Plate 16
collared catshark 119
common blacktip shark 246–7, Plate 33
common hammerhead 275
common sawshark 109–10, Plate 44
common shovelnose ray 290
common skate 346
common stingaree 427, Plate 78
cookie-cutter shark 86–7, Plate 3
Cooks bramble shark 46
copper shark 235
Coral Sea stingaree 429, Plate 77
cow ray 456
cowfish 453
cownose rays 30, **456–7**, Plate 80
cowshark 42
cowtail stingray 409–10, Plate 70
crampfish 372
creek whaler 242–3, Plate 36
crested bullhead shark 113
crested horn shark 113, Plate 2
crested Port Jackson shark 113
crocodile shark 34, **150–1**, Plate 14
crossback stingaree 432

D

darksnout hound shark 210–11, Plate 27
deepwater dogfish 64, 65
deepwater ghostshark 469
deepwater skate 336–7, Plate 59
deepwater spiny dogfish 58
deepwater stingray 420
denticulate skate 352
devilfish 459
devilrays 30, **458–64**, Plate 81
diamond fish 460
dindagubba 366
Dixons stingaree 439
dogfishes 7, 9, 31, **47–103**, Plates 3–10
dogshark 99
Dorian Grey 70
draughtboard shark 192, **198**, Plate 23
duckbill ray 448
dumb gulper shark 56
dumb shark 56
dusky catshark 202, Plate 18
dusky shark 8, 10, **252–3**, Plate 35
dusky skate 319–20, Plate 46
dwarf black stingray 385–6, Plate 73
dwarf catshark 181, Plate 22
dwarf sawfish 362–3, Plate 43

E

eagle rays 30, **447–55**, Plates 79–80
eastern angel shark 278, Plate 38
eastern fiddler ray 291–2, Plate 41
eastern highfin spurdog 93, Plate 6
eastern leg skate 359, Plate 64
eastern longnose spurdog 97, Plate 5
eastern looseskin skate 314, Plate 49
eastern numbfish 377, Plate 66
eastern sawshark 107–8, Plate 44
eastern shovelnose ray 286–7, Plate 40
eastern shovelnose stingaree 423, Plate 78
electric ray 370, 372, 378
electric rays 9, Plate 65
electric torpedo 378
elephant fish 10, **465–6**, Plate 82
elephant fishes 35, **465–6**, Plate 82
elephant shark 465
Endeavour dogfish 57–8, Plate 4
epaulette shark 136, Plate 16
eragoni 286
eregoodoo 460
estuary stingaree 390
estuary stingray 390–1, Plate 72
eye skate 328

F

false argus skate 328–9, Plate 61
false peacock skate 318–19, Plate 47
fantail ray 409
fantail rays **411–13**, Plate 70
fatspine spurdog 95, Plate 6
feathertail stingray 409
fiddler 292
fiddler ray 291, 292
fiddler rays **291–4**, Plate 41
fish shark 265
flake 215
flapray 456
fleshynose catshark 175, Plate 20
fossil shark 219–20, Plate 29
fox shark 157
freckled catshark 173, Plate 19
freshwater sawfish 364–5, Plate 42
freshwater whaler 244
freshwater whipray 399–400, Plate 69
frill shark 36,
frill-gilled shark 36
frilled shark 30, **36–7**, Plate 1

G

Galapagos shark 243–4, Plate 35
ghost shark 465
ghost skate 312, Plate 51
ghostshark 478
ghostsharks **475–8**, Plate 84
giant chimaera 473, Plate 83
giant guitarfish 297
giant lamniform sharks Plate 13
giant reef ray 412
giant shovelnose ray 290–1, Plate 39
giant stingaree 420–1, Plate 73
ginger carpet shark 118, Plate 16
goblin shark 32, **148–9**, Plate 13
golden dogfish 64–5, Plate 10
goldeneye shovelnose ray 289, Plate 39
graceful shark 231–2, Plate 33

graceful skate 322
Grahams skate 326–7, Plate 60
great blue shark 263
great hammerhead 274–5, Plate 37
great skate 355
great stingaree 430
great white shark 162
green sawfish 366–7, Plate 43
green skate 292
green-eyed dogfish 101
greenback skate 349
greenback stingaree 441, Plate 74
greeneye spurdog 101–2, Plate 5
Greenland shark 89
grey carpet shark 135, Plate 17
grey gummy shark 213–14, Plate 28
grey nurse shark 145–6, Plate 15
grey nurse sharks 8, 34, **144–7**, Plate 15
grey reef shark 232–3, Plate 32
grey sharpnose shark 266–7, Plate 30
grey skate 325–6, Plate 60
grey spiny dogfish 101
grey spotted catshark 187, Plate 21
ground shark 42
guergunna 409
Guilers stingray 396
guitarfishes **295–8**, Plate 42
gulf catshark 188, Plate 22
gulf wobbegong 131
guliman 250
gulper shark 55–6, Plate 4
gulper sharks Plate 4
gummy shark 7, 10, **215–16**, Plate 28
gummy sharks **213–16**, Plate 28

H
hammerhead sharks 8, 31, **270–6**, Plate 37
hardnose shark 249–50, Plate 36
Harrisons deepsea dogfish 56
Harrisons dogfish 56–7, Plate 4
Herbsts nurse shark 146
horn sharks 31, **112–16**, Plate 2
hound sharks 34, **205–16**, Plates 27–8

I
inkytail shark 237

J
Japanese devilray 461–2, Plate 81
Java shark 234
Javanese cownose ray 457, Plate 80
Jenkins whipray 403–4, Plate 69
Jordans blue dogshark 260

K
Kapala stingaree 428, Plate 74
kejetuck 424
kidney-headed shark 272
kitefin shark 69

L
lagoon ray 411
lantern sharks **73–85**, Plates 7–8
leafscale gulper shark 58–9, Plate 4
leg skates 29, **357–9**, Plate 64
Leichardts sawfish 364
lemon shark 262–3, Plate 31

Lemprieres skate 352
leopard shark 138
leopard whipray 408–9, Plate 68
lesser fantail ray 411
lesser soupfin shark 211
Leylands skate 330–1, Plate 58
lined lantern shark 78, Plate 7
little gulper shark 60
little numbfish 378
lobed stingaree 436, Plate 76
longfin mako 164
longlip spotted catshark 189
Longmans dogshark 265
longnose blacktail shark 232
longnose chimaeras 479
longnose grey shark 237
longnose hound shark 212–13, Plate 28
longnose rabbitfishes 479
longnose sawshark 109
longnose skate 323–4, Plate 62
longnose velvet dogfish 64
longsnout dogfish 71–2, Plate 3
longspine chimaera 471–2, Plate 83
longtail carpet sharks 33, **134–7**, Plate 17
longtail ray 406
longtail stingray 394
longtail torpedo ray 369, Plate 65
Lord Plunkets shark 66
lucifer shark 81
luminous shark 86

M
mackerel shark 163, 165
mackerel sharks 34, **161–6**, Plate 14
Macleays coachwhip ray 402
madame X 140
magpie fiddler ray 294, Plate 41
mako shark 163
mandarin shark 68, Plate 6
mangrove ray 402
mangrove whipray 402–3, Plate 69
manta 459
manta ray 459–60, Plate 81
marbled catshark 137, **190**, Plate 22
marbled ghostshark 476, Plate 84
masked stingaree 426, Plate 77
maskrays **386–7**, **391–4**, Plate 71
Maugean skate 339–40, Plate 55
megamouth shark 34, **152–3**, Plate 13
Melbourne skate 355–6, Plate 53
milk shark 265–6, Plate 30
mill ray 453
mitotic stingaree 437 Plate 75
Mollers deepsea shark 83
Mollers lantern shark 83–4, Plate 7
monkfish 280
mosaic skate 317, Plate 46
mud skate 296
munguna 459

N
narrow sawfish 361–2, Plate 43
narrowbar swell shark 195, Plate 23
narrownose chimaera 481, 482
narrowsnout sawfish 366
necklace carpet shark 121
nervous shark 238–9, Plate 32

New Zealand eagle ray 454
New Zealand lantern shark 80
New Zealand swellshark 192
Nilsons deepsea dogfish 58
northern draughtboard shark 194, Plate 24
northern sawtail shark 200, Plate 18
northern wobbegong 132, Plate 25
numbfish 372, 378
numbfishes 9, 28, **374–9**, Plates 66–7
numbie 372
nurse sharks 7, 32, **140–1**, Plate 17
nutcracker shark 198

O
oceanic whitetip shark 8, **247–8**, Plate 31
Ogilbys ghostshark 478, Plate 84
Ogilbys wobbegong 127
one-finned shark 39
orange spotted catshark 184, Plate 21
ornate angel shark 281–2, Plate 38
ornate eagle ray 451–2, Plate 79
ornate numbfish 375, Plate 67
ornate wobbegong 131
oscellate skate 329–30, Plate 63
Owstons dogfish 65–6, Plate 10
Owstons spiny dogfish 65
ox ray 460
oyster crusher 114

P
Pacific sleeper shark 89–90, Plate 3
Pacific spookfish 482–3, Plate 82
painted maskray 393–4, Plate 71
pale skate 311, Plate 50
pale spotted catshark 185, Plate 21
pale tropical skate 332–3, Plate 57
parrit 292
patchwork stingaree 434, Plate 75
peacock skate 322–3, Plate 47
pelagic stingray 396–7, Plate 72
pelagic thresher 155, Plate 12
pencil shark 211–12, Plate 27
perlon shark 39
pigeye shark 234–5, Plate 34
piked dogfish 98, 99
piked spurdog 99–100, Plate 5
pink lantern shark 76, Plate 8
pink whipray 400–1, Plate 69
pinocchio catshark 179, Plate 19
plain maskray 386–7, Plate 71
Plunkets dogfish 66–7, Plate 10
Plunkets shark 66
porbeagle 165–6 Plate 14
porcupine ray 414–15, Plate 70
Port Jackson shark 7, 9, **114–15**, Plate 2
Portuguese dogfish 62–3, Plate 10
prickly dogfish 104–5, Plate 11
prickly dogfishes 31, **104–5**, Plate 11
prickly shark 46, Plate 11
Prince Alfreds ray 459
purple eagle ray 454–5, Plate 80
pygmy devilray 460–1, Plate 81
pygmy lantern shark 75, Plate 8
pygmy shark 85–6, Plate 9
pygmy thornback skate 341–2, Plate 54

Q
Queensland deepwater skate 338–9, Plate 59
Queensland sawfish 362
Queensland shark 231

R
ratfishes 467
rat-tail ray 443
reef ray 411
reperepe 465
reticulate swell shark 197, Plate 23
reticulate whipray 406–7, Plate 68
river whaler 244
rough sharks 104
rough skate 355
roughback skate 322
roughskin catshark 176, Plate 20
roughskin dogfish 65
roughskin stingaree 414
round ribbontail ray 412
round skate 307, 308
rusty carpet shark 120, Plate 16
rusty catshark 120, 184

S
saddled catshark 183
saddled swell shark 193, Plate 24
sand tiger shark 145, **146–7**, Plate 15
sandbar shark 253–4, Plate 34
sandshark 297
sandy skate 316, Plate 45
sandyback stingaree 430, Plate 75
sawback skate 344, Plate 52
sawfish 366
sawfishes 27, **360–7**, Plates 42–3
sawsharks 27, **106–11**, Plate 44
sawtail shark 201, Plate 18
sawtail sharks **199–201**, Plate 18
scalloped hammerhead 272–3, Plate 37
school shark 10, **208–9**, 255, Plate 27
seal shark 69
seven-gilled shark 42
sevengill sharks 31, **38–43**, Plate 1
shark ray 296–7, Plate 42
sharkfin guitarfishes 28, **295–8**, Plate 42
sharpnose sevengill shark 39–40, Plate 1
sharpnose sharks 266–8, Plate 30
sharpsnout sevengill shark 39
sharptooth shark 262
shining skate 322
short-tail catshark 204, Plate 20
short-tail electric ray 372
short-tail lantern shark 79–80, Plate 7
short-tail stingray 388
short-tail torpedo ray 370–1, Plate 65
shortfin mako 163–4, Plate 14
shortlip spotted catshark 203
shortnose chimaeras 35, **467–78**, Plates 83–4
shortnose spiny dogfish 99
shortnose spurdog 99
shortspine chimaera 470–1, Plate 83
shortspine spurdog 101
shovelnose rays 28, **283–94**, Plates 39–41
shovelnose shark 286
shovelnose spiny dogfish 70
sicklefin hound shark 209–10, Plate 27
sicklefin weasel shark 218

silky shark 241–2, Plate 35
silvertip shark 228–9, Plate 31
Sinclairs stingaree 435
sixgill rays Plate 64
sixgill shark 40
sixgill sharks 31, **38–43**, Plate 1
sixgill stingray 445–6, Plate 64
sixgill stingrays 29, **445–6**, Plate 64
skate 443
skates 10, 29, **299–356**, Plates 45–63
skittle dog 99
sleepy joe 198
sleepy shark 140
slender dogshark 260
slender hammerhead 271
slender lantern shark 84–5, Plate 8
slender sawtail shark 199, Plate 18
slender sevengill shark 39
sliteye shark 260–1, Plate 30
smaller devilray 460
smalleye pygmy shark 90–1, Plate 9
smallfin gulper shark 57
smallspine spookfish 480–1, Plate 82
smalltooth sand tiger 146
smalltooth sawfish 364, 365
smooth hammerhead 275–6, Plate 37
smooth lantern shark 73, Plate 8
smooth stingray 388–9, Plate 72
smoothbelly catshark 180, Plate 19
smoothfang shark 237
snaggletooth shark 219–20
snapper shark 163, 208
Solanders ray 414
sorrah shark 255
soupfin shark 208
southern catshark 121
southern chimaera 469–70, Plate 83
southern dogfish 60–1, Plate 4
southern eagle ray 453–4, Plate 80
southern fiddler 292
southern fiddler ray 292–3, Plate 41
southern lantern shark 80–1, Plate 9
southern round skate 308–9, Plate 48
southern sawshark 111, Plate 44
southern shovelnose ray 287
sparsely-spotted stingaree 439, Plate 74
speartooth shark 259–60, Plate 29
speckled carpet shark 137, Plate 16
speckled catshark 137, **203–4**, Plate 18
speckled swell shark 196, Plate 23
spiked dogfish 99
spinner shark 237–8, Plate 33
spinous shark 45
spiny dogfish 98, 101
spitting shark 140
spookfish 478
spookfishes 35, **479–83**, Plate 82
spot-tail shark 10, **255–6**, Plate 32
spotted catshark 135, 184, 186
spotted catsharks **184–7**, Plate 21
spotted dogfish 187
spotted eagle ray 448
spotted ragged-tooth 145
spotted shovelnose ray 285–6 Plate 40
spotted spiny dogfish 98
spotted stingaree 435, Plate 73
spotted wobbegong 129–30, Plate 26

spurdog 98, 99
spurdogs **91–102**, Plates 5–6
stingaree 427
stingarees 7, 9, 30, **416–42**, Plates 73–8
stingrays 9, 30, **380–415**, Plates 67–73
striped stingaree 425, Plate 77
Swan River whaler 244
Sweet William 215
swell shark 192
swell sharks **192–8**, Plates 23–4
Sydney skate 346–7, Plate 61

T
tabbigaw 114
Tasmanian dogfish 99
Tasmanian numbfish 378, Plate 66
Tasmanian tiger shark 42
tasselled wobbegong 127–8, Plate 25
tawny nurse shark 140
tawny shark 140–1, Plate 17
Taylors shark 267
thickskin shark 253
thintail skate 342–3, Plate 56
thintail thresher 157
Thompsons deepsea dogfish 70
Thompsons shark 70
thornback skate 341, **352–3**, Plate 54
thorntail stingray 394
thorny ray 414
thresher shark 157–8, Plate 12
thresher sharks 31, **154–8**, Plate 12
tiger shark 7, 8, **258–9**, Plate 29
tope 208
torpedo 370, 372
torpedo rays 9, 28, **368–71**, Plate 65
Toshs longtail ray 405
tropical sawshark 108–9, Plate 44

V
varied carpet shark 121, Plate 16
varied catshark 121
variegated catshark 183, Plate 21
velvet dogfish 102–3, Plate 3
Victorian spotted dogfish 98
violet stingray 396

W
Waites deepsea dogfish 66
Wards wobbegong 132
weasel shark 218–19, Plate 28
weasel sharks 7, 35, **217–20**, Plates 28–9
wedgenose skate 355
Wengs skate 335–6, Plate 58
weralli 409
western angel shark 279–80, Plate 38
western highfin spurdog 94, Plate 6
western leg skate 358, Plate 64
western longnose spurdog 96, Plate 5
western looseskin skate 315, Plate 49
western numbfish 376, Plate 66
western round skate 307–8, Plate 48
western school shark 211
western shovelnose ray 287–8, Plate 40
western shovelnose stingaree 424, Plate 78
western spotted catshark 186, Plate 21
western stingaree 424
western wobbegong 128–9, Plate 26

whale shark 32, **142–3**, Plate 15
whaler sharks 7, 35, **221–69**, Plates 29–36
whip ray 453
whiprays 397–409, Plates 68–9
whiptail ray 453
whiskery shark 10, **207**, Plate 27
white death 162
white-eye shark 265
white pointer 162
white shark 8, **162–3**, Plate 14
white-spotted dogfish 98
white-spotted eagle ray 448–9, Plate 79
white-spotted guitarfish 297–8, Plate 42
white-spotted gummy shark 214–15, Plate 28
white-spotted skate 347–8, Plate 62
white-spotted spurdog 98–9, Plate 5
white-spotted stingaree 439
whitecheek shark 239–40, Plate 36
whitefin chimaera 474, Plate 84
whitefin swell shark 192, Plate 23
whitefish 465, 478
whitespot ray 297
whitespot shovelnose ray 297
whitetail dogfish 88–9, Plate 9
whitetip reef shark 268–9, Plate 31
whitetip shark 268
whitetip whaler 247
Whitleys skate 355
wide sawfish 365–6, Plate 43
wide stingaree 433, Plate 74
widemouth blackspot shark 239
winghead shark 271–2, Plate 37
wobbegong 129
wobbegongs 8, 33, **125–33**, Plates 25–6
wulura 405

Y
yellow shovelnose ray 287
yellow shovelnose stingaree 422, Plate 78
yellowback stingaree 440, Plate 76

Z
zebra bullhead shark 115
zebra horn shark 115–16, Plate 2
zebra Port Jackson shark 115
zebra shark 32, **138–9**, Plate 15

COLOUR PLATES

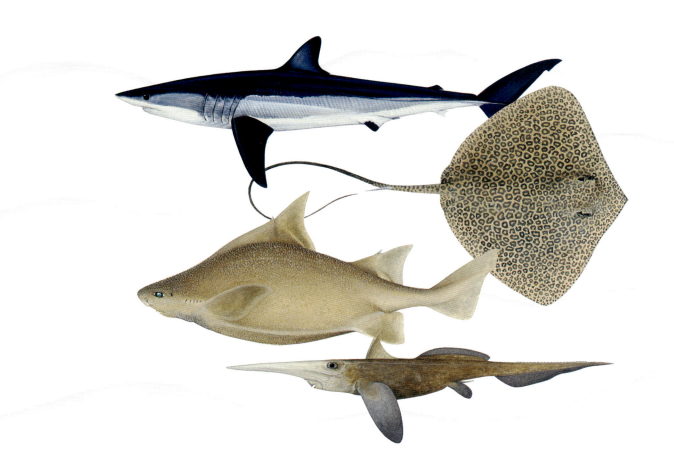

List of plates

Plate 1 — Frill, Sixgill and Sevengill Sharks
Plate 2 — Horn Sharks
Plate 3 — Dogfishes
Plate 4 — Dogfishes (gulper sharks)
Plate 5 — Dogfishes (spurdogs)
Plate 6 — Dogfishes (spurdogs)
Plate 7 — Dogfishes (lantern sharks)
Plate 8 — Dogfishes (lantern sharks)
Plate 9 — Dogfishes
Plate 10 — Dogfishes
Plate 11 — Prickly Dogfish and Bramble Sharks
Plate 12 — Thresher Sharks
Plate 13 — Giant Lamniform Sharks
Plate 14 — Mackerel and Crocodile Sharks
Plate 15 — Grey Nurse, Whale and Zebra Sharks
Plate 16 — Epaulette and Collared Carpet Sharks
Plate 17 — Blind, Nurse and Longtail Carpet Sharks
Plate 18 — Catsharks
Plate 19 — Catsharks
Plate 20 — Catsharks
Plate 21 — Catsharks
Plate 22 — Catsharks
Plate 23 — Catsharks (swell sharks)
Plate 24 — Catsharks (swell sharks)
Plate 25 — Wobbegongs
Plate 26 — Wobbegongs
Plate 27 — Hound Sharks
Plate 28 — Hound and Weasel Sharks
Plate 29 — Weasel and Whaler Sharks
Plate 30 — Whaler Sharks
Plate 31 — Whaler Sharks
Plate 32 — Whaler Sharks
Plate 33 — Whaler Sharks
Plate 34 — Whaler Sharks
Plate 35 — Whaler Sharks
Plate 36 — Whaler Sharks
Plate 37 — Hammerhead Sharks
Plate 38 — Angel Sharks
Plate 39 — Angel Sharks and Shovelnose Rays
Plate 40 — Shovelnose Rays
Plate 41 — Shovelnose Rays (fiddler rays)
Plate 42 — Guitar and Sawfishes
Plate 43 — Sawfishes
Plate 44 — Sawsharks
Plate 45 — Skates
Plate 46 — Skates
Plate 47 — Skates
Plate 48 — Skates
Plate 49 — Skates
Plate 50 — Skates
Plate 51 — Skates
Plate 52 — Skates
Plate 53 — Skates
Plate 54 — Skates
Plate 55 — Skates
Plate 56 — Skates
Plate 57 — Skates
Plate 58 — Skates
Plate 59 — Skates
Plate 60 — Skates
Plate 61 — Skates
Plate 62 — Skates
Plate 63 — Skates
Plate 64 — Leg Skates and Sixgill Ray
Plate 65 — Electric Rays
Plate 66 — Numbfishes
Plate 67 — Numbfishes and Stingray
Plate 68 — Stingrays (whiprays)
Plate 69 — Stingrays (whiprays)
Plate 70 — Stingrays
Plate 71 — Stingrays (maskrays)
Plate 72 — Stingrays
Plate 73 — Stingray and Stingarees
Plate 74 — Stingarees
Plate 75 — Stingarees
Plate 76 — Stingarees
Plate 77 — Stingarees
Plate 78 — Stingarees
Plate 79 — Butterfly and Eagle Rays
Plate 80 — Eagle and Cownose Rays
Plate 81 — Devilrays
Plate 82 — Elephant and Spookfishes
Plate 83 — Chimaeras
Plate 84 — Ghostsharks

Plate 1 — **Frilled, Sixgill and Sevengill Sharks**

5.1 *Chlamydoselachus anguineus* (frilled shark) female; **6.2** *Hexanchus griseus* (bluntnose sixgill shark) juvenile; **6.3** *Hexanchus nakamurai* (bigeye sixgill shark) female; **6.1** *Heptranchias perlo* (sharpnose sevengill shark) female; **6.4** *Notorynchus cepedianus* (broadnose sevengill shark) female.

Plate 2 — **Horn Sharks**

11.1 *Heterodontus galeatus* (crested horn shark) female; **11.2** *Heterodontus portusjacksoni* (Port Jackson shark) female; **11.3** *Heterodontus zebra* (zebra horn shark) female.

Plate 3 — **Dogfishes**

8.13 *Deania calcea* (brier shark) male; **8.14** *Deania quadrispinosa* (longsnout dogfish) female; **8.27** *Isistius brasiliensis* (cookie-cutter shark) male; **8.40** *Zameus squamulosus* (velvet dogfish) female; **8.29** *Somniosus pacificus* (Pacific sleeper shark) juvenile.

Plate 4 — **Dogfishes (gulper sharks)**

8.4 *Centrophorus squamosus* (leafscale gulper shark) female; **8.1** *Centrophorus granulosus* (gulper shark) female;
8.5 *Centrophorus uyato* (southern dogfish) female; **8.2** *Centrophorus harrissoni* (Harrisons dogfish) male;
8.3 *Centrophorus moluccensis* (Endeavour dogfish) female.

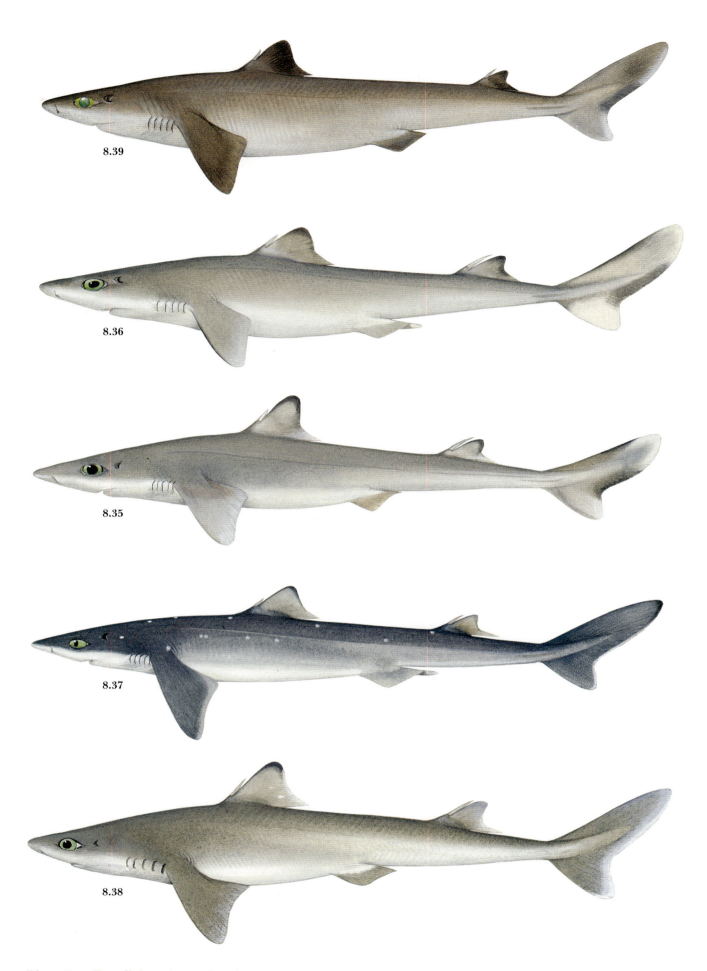

Plate 5 — **Dogfishes (spurdogs)**

8.39 *Squalus mitsukurii* (greeneye spurdog) female; **8.36** *Squalus* sp. F (eastern longnose spurdog) male; **8.35** *Squalus* sp. E (western longnose spurdog) female; **8.37** *Squalus acanthias* (white-spotted spurdog) male; **8.38** *Squalus megalops* (piked spurdog) female.

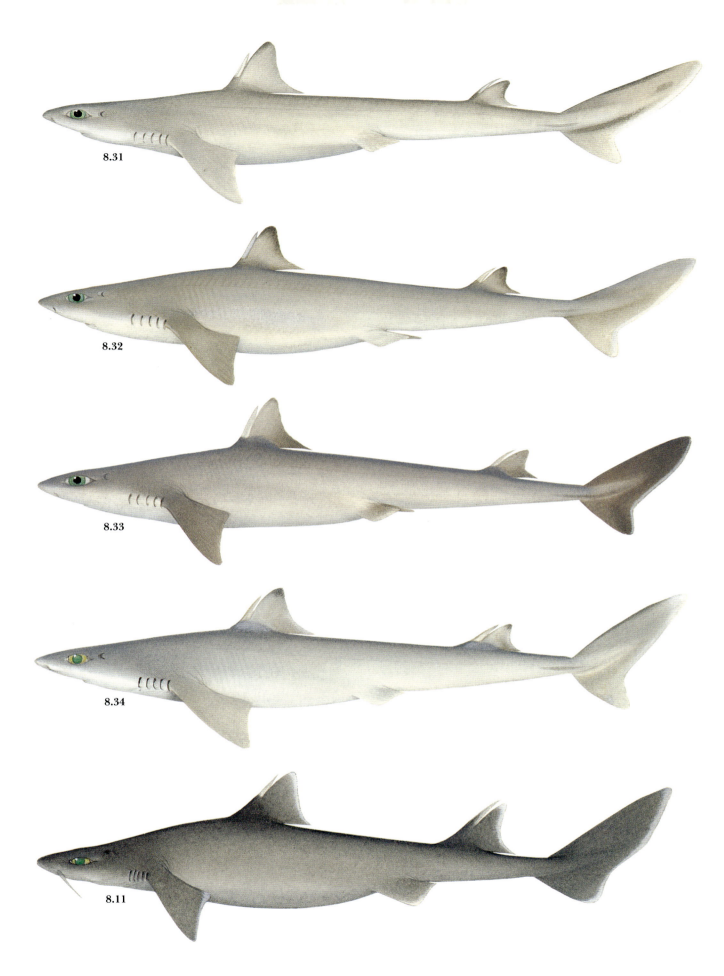

Plate 6 — **Dogfishes (spurdogs)**

8.31 *Squalus* sp. A (bartail spurdog) female; **8.32** *Squalus* sp. B (eastern highfin spurdog) male; **8.33** *Squalus* sp. C (western highfin spurdog) male; **8.34** *Squalus* sp. D (fatspine spurdog) female; **8.11** *Cirrhigaleus barbifer* (mandarin shark) female.

Plate 7 — **Dogfishes (lantern sharks)**

8.21 *Etmopterus brachyurus* (short-tail lantern shark) female; **8.23** *Etmopterus lucifer* (blackbelly lantern shark) female; **8.24** *Etmopterus molleri* (Mollers lantern shark) female; **8.20** *Etmopterus* sp. F (lined lantern shark) female; **8.19** *Etmopterus* sp. E (blackmouth lantern shark) female.

Plate 8 — **Dogfishes (lantern sharks)**

8.18 *Etmopterus* sp. D (pink lantern shark) female; **8.17** *Etmopterus* sp. C (pygmy lantern shark) male; **8.25** *Etmopterus pusillus* (slender lantern shark) male; **8.15** *Etmopterus* sp. A (smooth lantern shark) male; **8.16** *Etmopterus* sp. B (bristled lantern shark) male.

Plate 9 — **Dogfishes**

8.22 *Etmopterus granulosus* (southern lantern shark) male; **8.6** *Centroscyllium kamoharai* (bareskin dogfish) female; **8.28** *Scymnodalatias albicauda* (whitetail dogfish) female; **8.26** *Euprotomicrus bispinatus* (pygmy shark) female; **8.30** *Squaliolus aliae* (smalleye pygmy shark) female.

Plate 10 — **Dogfishes**

8.7 *Centroscymnus coelolepis* (Portuguese dogfish) male; **8.9** *Centroscymnus owstoni* (Owstons dogfish) female;
8.8 *Centroscymnus crepidater* (golden dogfish) female; **8.10** *Centroscymnus plunketi* (Plunkets dogfish) immature;
8.12 *Dalatias licha* (black shark) female.

Plate 11 — **Prickly Dogfish and Bramble Sharks**

9.1 *Oxynotus bruniensis* (prickly dogfish) immature; **7.2** *Echinorhinus cookei* (prickly shark) female;
7.1 *Echinorhinus brucus* (bramble shark) immature.

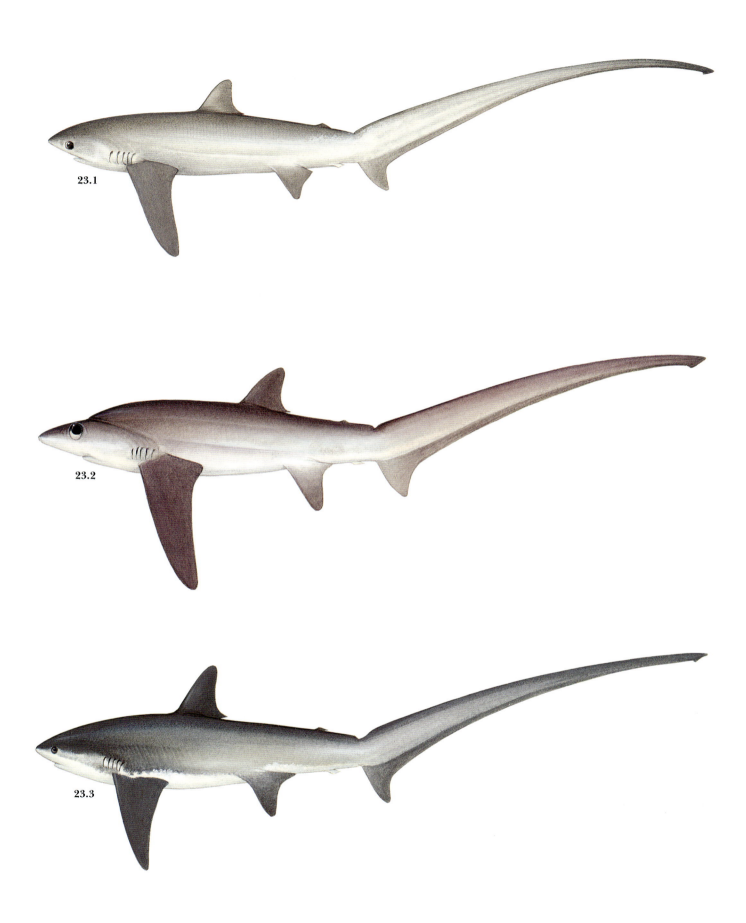

Plate 12 — **Thresher Sharks**

23.1 *Alopias pelagicus* (pelagic thresher) immature; **23.2** *Alopias superciliosus* (bigeye thresher) immature; **23.3** *Alopias vulpinus* (thresher shark) female.

Plate 13 — **Giant Lamniform Sharks**

24.1 *Cetorhinus maximus* (basking shark) female; **22.1** *Megachasma pelagios* (megamouth shark) female; **20.1** *Mitsukurina owstoni* (goblin shark) female.

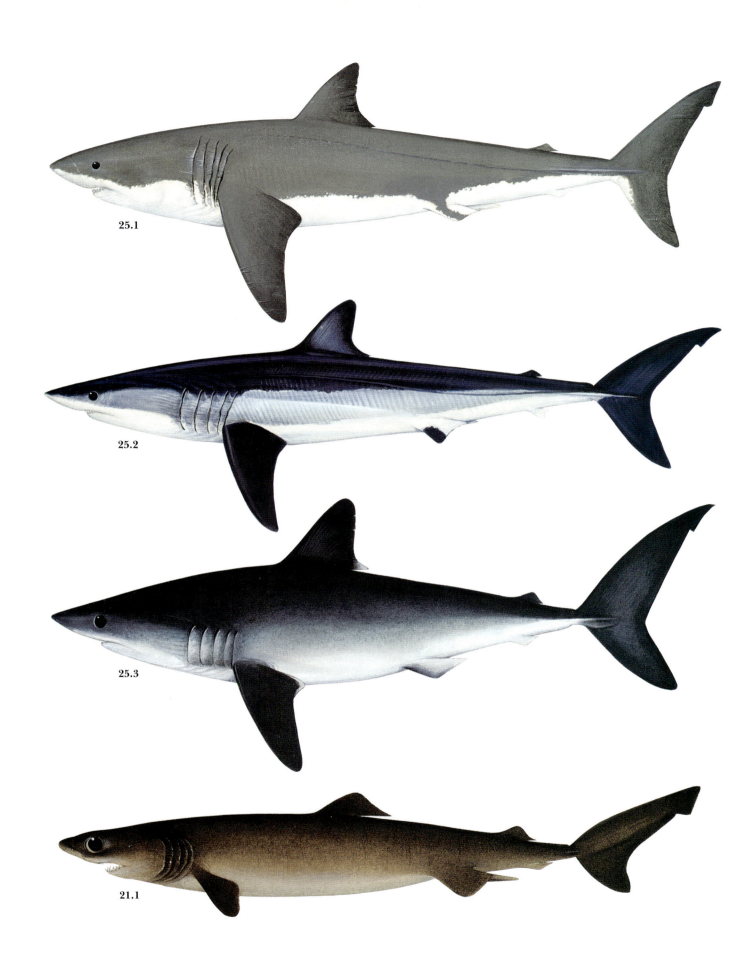

Plate 14 — **Mackerel and Crocodile Sharks**

25.1 *Carcharodon carcharias* (white shark) male; **25.2** *Isurus oxyrinchus* (shortfin mako) immature; **25.3** *Lamna nasus* (porbeagle) immature; **21.1** *Pseudocarcharias kamoharai* (crocodile shark) male.

Plate 15 — **Grey Nurse, Whale and Zebra Sharks**

19.1 *Carcharias taurus* (grey nurse shark) immature; **19.2** *Odontaspis ferox* (sand tiger shark) female;
18.1 *Rhincodon typus* (whale shark) female; **16.1a** *Stegastoma fasciatum* (zebra shark) adult;
16.1b *Stegastoma fasciatum* (zebra shark) juvenile.

Plate 16 — **Epaulette and Collared Carpet Sharks**

12.2 *Parascyllium collare* (collared carpet shark) male; **12.3** *Parascyllium ferrugineum* (rusty carpet shark) female; **12.4** *Parascyllium variolatum* (varied carpet shark) female; **12.1** *Parascyllium* sp. A (ginger carpet shark) female; **15.2** *Hemiscyllium ocellatum* (epaulette shark) female; **15.3** *Hemiscyllium trispeculare* (speckled carpet shark) female.

Plate 17 — **Blind, Nurse and Longtail Carpet Sharks**

15.1 *Chiloscyllium punctatum* (grey carpet shark) female; **13.1** *Brachaelurus colcloughi* (Colcloughs shark) female; **13.2** *Brachaelurus waddi* (blind shark) male; **17.1** *Nebrius ferrugineus* (tawny shark) immature.

Plate 18 — **Catsharks**

26.27 *Galeus* sp. A (slender sawtail shark) female; **26.28** *Galeus* sp. B (northern sawtail shark) male; **26.29** *Galeus boardmani* (sawtail shark) female; **26.30** *Halaelurus* sp. A (dusky catshark) immature; **26.31** *Halaelurus boesemani* (speckled catshark) female.

Plate 19 — **Catsharks**

26.7a *Apristurus* sp. G (pinocchio catshark — northern form) female; **26.7b** *Apristurus* sp. G (pinocchio catshark — southern form) male; **26.8** *Apristurus longicephalus* (smoothbelly catshark) male; **26.1** *Apristurus* sp. A (freckled catshark) female; **26.2** *Apristurus* sp. B (bigfin catshark) female.

Plate 20 — **Catsharks**

26.3 *Apristurus* sp. C (fleshynose catshark) male; **26.5** *Apristurus* sp. E (bulldog catshark) male; **26.6** *Apristurus* sp. F (bighead catshark) male; **26.4** *Apristurus* sp. D (roughskin catshark) male; **26.32** *Parmaturus* sp. A (short-tail catshark) female.

Plate 21 — **Catsharks**

26.12 *Asymbolus* sp. D (orange spotted catshark) male; **26.13** *Asymbolus* sp. E (pale spotted catshark) female; **26.14a** *Asymbolus* sp. F (western spotted catshark) male; **26.11** *Asymbolus* sp. C (variegated catshark) female; **26.15** *Asymbolus analis* (grey spotted catshark) male; **26.14b** *Asymbolus* sp. F (western spotted catshark) juvenile.

Plate 22 — **Catsharks**

26.10 *Asymbolus* sp. B (blotched catshark) female; **26.16** *Asymbolus vincenti* (gulf catshark) female; **26.9** *Asymbolus* sp. A (dwarf catshark) male; **26.17a** *Atelomycterus* sp. A (banded catshark) female; **26.17b** *Atelomycterus* sp. A (banded catshark — Torres Strait form) female; **26.18** *Atelomycterus macleayi* (marbled catshark) male.

Plate 23 — **Catsharks (swell sharks)**

26.24 *Cephaloscyllium* sp. E (speckled swell shark) immature; **26.23** *Cephaloscyllium* sp. D (narrowbar swell shark) female; **26.25** *Cephaloscyllium fasciatum* (reticulate swell shark) male; **26.26** *Cephaloscyllium laticeps* (draughtboard shark) male; **26.20** *Cephaloscyllium* sp. A (whitefin swell shark) immature.

Plate 24 — **Catsharks (swell sharks)**

26.21 *Cephaloscyllium* sp. B (saddled swell shark) male; **26.22a** *Cephaloscyllium* sp. C (northern draughtboard shark) female; **26.22b** *Cephaloscyllium* sp. C (northern draughtboard shark) juvenile; **26.19** *Aulohalaelurus labiosus* (black-spotted catshark) male.

Plate 25 — **Wobbegongs**

14.1 *Eucrossorhinus dasypogon* (tasselled wobbegong) immature; **14.6** *Sutorectus tentaculatus* (cobbler wobbegong) male; **14.5** *Orectolobus wardi* (northern wobbegong) female.

Plate 26 — **Wobbegongs**

14.3 *Orectolobus maculatus* (spotted wobbegong) immature; **14.4** *Orectolobus ornatus* (banded wobbegong) female; **14.2** *Orectolobus* sp. A (western wobbegong) female.

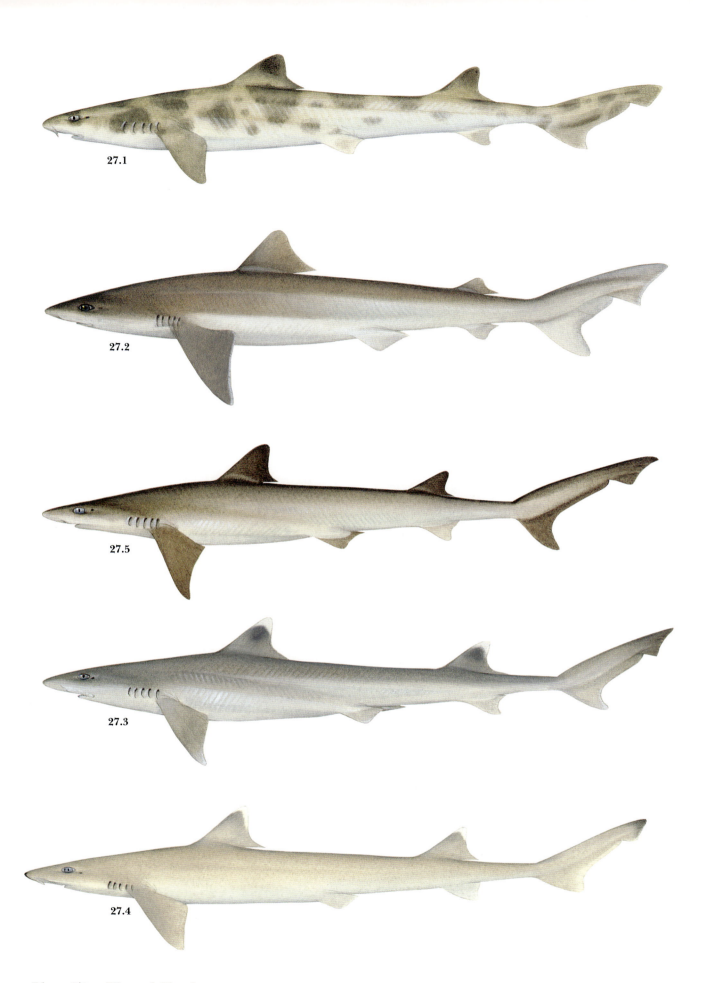

Plate 27 — **Hound Sharks**

27.1 *Furgaleus macki* (whiskery shark) juvenile; **27.2** *Galeorhinus galeus* (school shark) female;
27.5 *Hypogaleus hyugaensis* (pencil shark) female; **27.3** *Hemitriakis* sp. A (sicklefin hound shark) male;
27.4 *Hemitriakis* sp. B (darksnout hound shark) female.

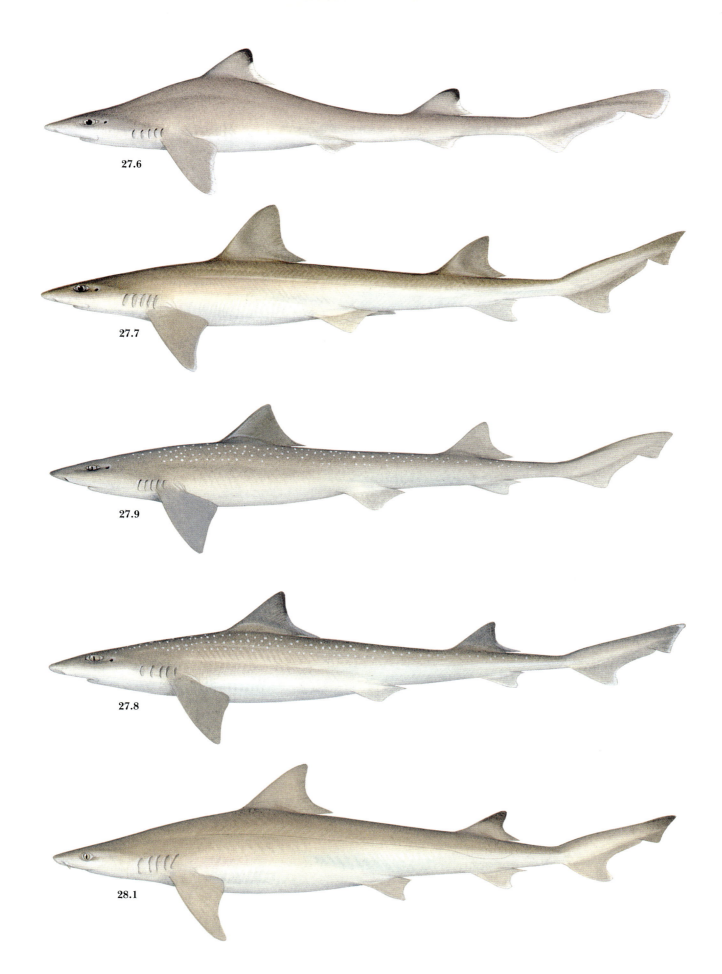

Plate 28 — **Hound and Weasel Sharks**

27.6 *Iago garricki* (longnose hound shark) female; **27.7** *Mustelus* sp. A (grey gummy shark) male; **27.9** *Mustelus antarcticus* (gummy shark) female; **27.8** *Mustelus* sp. B (white-spotted gummy shark) female; **28.1** *Hemigaleus microstoma* (weasel shark) female.

Plate 29 — **Weasel and Whaler Sharks**

28.2 *Hemipristis elongata* (fossil shark) female; **29.22** *Galeocerdo cuvier* (tiger shark) male; **29.23** *Glyphis* sp. A (speartooth shark) immature; **29.26** *Prionace glauca* (blue shark) male.

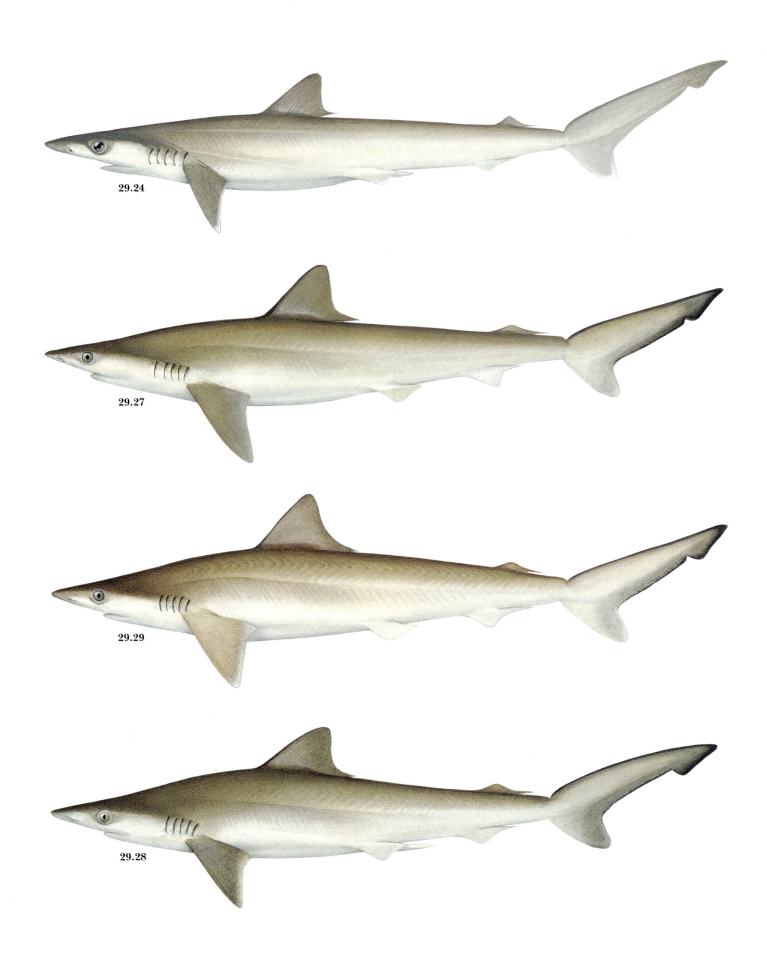

Plate 30 — **Whaler Sharks**

29.24 *Loxodon macrorhinus* (sliteye shark) male; **29.27** *Rhizoprionodon acutus* (milk shark) immature; **29.29** *Rhizoprionodon taylori* (Australian sharpnose shark) female; **29.28** *Rhizoprionodon oligolinx* (grey sharpnose shark) male.

Plate 31 — **Whaler Sharks**

29.25 *Negaprion acutidens* (lemon shark) female; **29.30** *Triaenodon obesus* (whitetip reef shark) immature; **29.1** *Carcharhinus albimarginatus* (silvertip shark) immature; **29.15** *Carcharhinus longimanus* (oceanic whitetip shark) female.

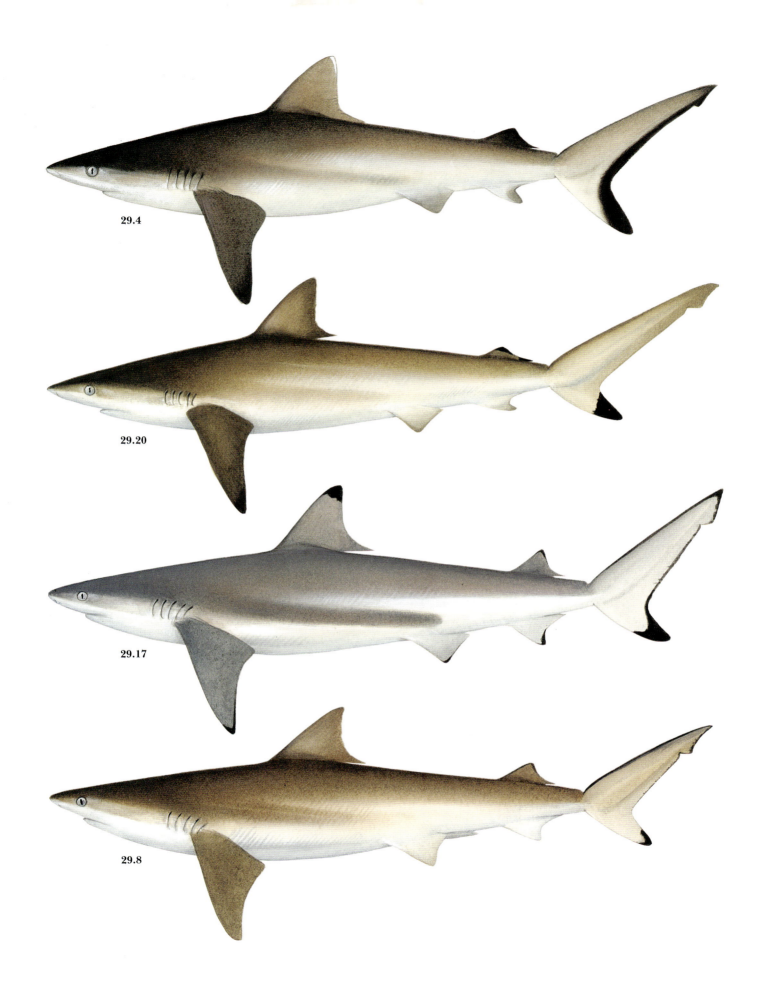

Plate 32 — **Whaler Sharks**

29.4 *Carcharhinus amblyrhynchos* (grey reef shark) female; **29.20** *Carcharhinus sorrah* (spot-tail shark) female; **29.17** *Carcharhinus melanopterus* (blacktip reef shark) female; **29.8** *Carcharhinus cautus* (nervous shark) male.

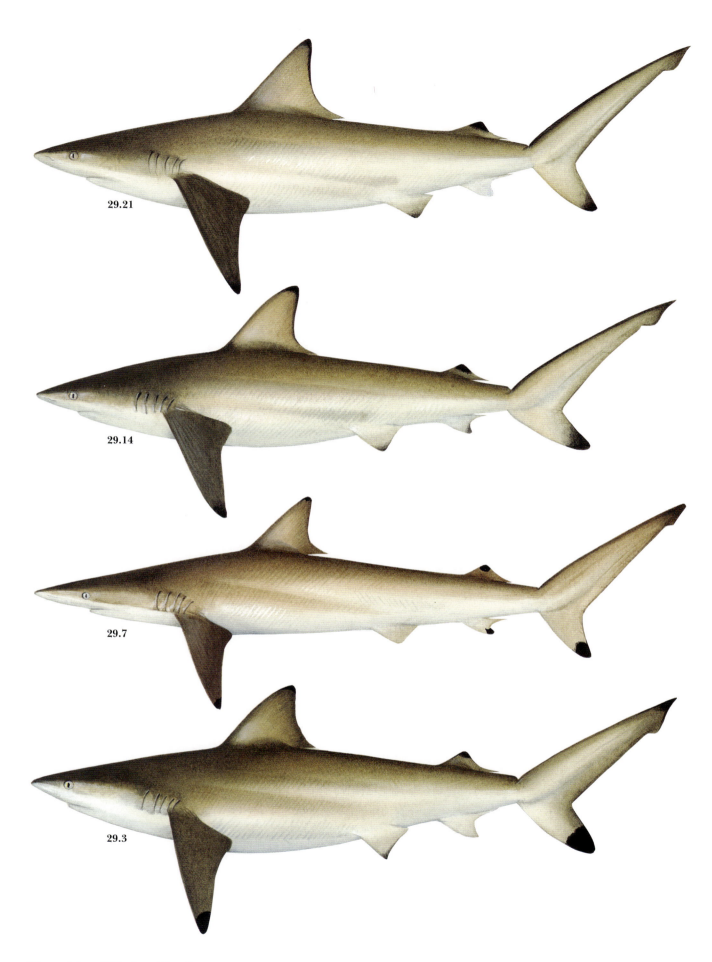

Plate 33 — **Whaler Sharks**

29.21 *Carcharhinus tilstoni* (Australian blacktip shark) female; **29.14** *Carcharhinus limbatus* (common blacktip shark) immature; **29.7** *Carcharhinus brevipinna* (spinner shark) immature; **29.3** *Carcharhinus amblyrhynchoides* (graceful shark) immature.

Plate 34 — **Whaler Sharks**

29.5 *Carcharhinus amboinensis* (pigeye shark) female; **29.13** *Carcharhinus leucas* (bull shark) juvenile;
29.2 *Carcharhinus altimus* (bignose shark) immature; **29.19** *Carcharhinus plumbeus* (sandbar shark) male.

Plate 35 — **Whaler Sharks**

29.18 *Carcharhinus obscurus* (dusky shark) juvenile; **29.12** *Carcharhinus galapagensis* (Galapagos shark) immature; **29.6** *Carcharhinus brachyurus* (bronze whaler) immature; **29.10** *Carcharhinus falciformis* (silky shark) immature.

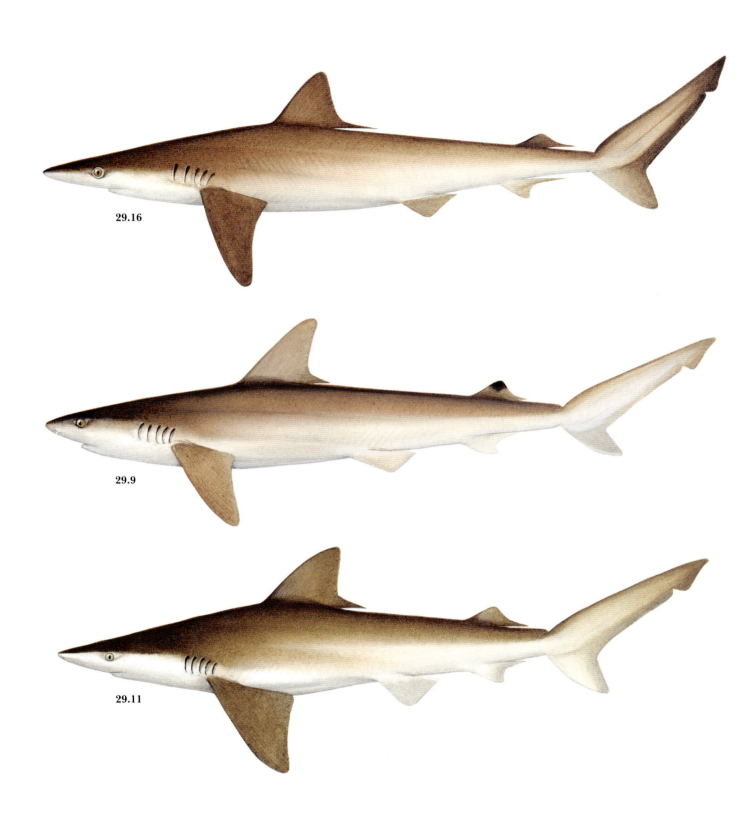

Plate 36 — **Whaler Sharks**

29.16 *Carcharhinus macloti* (hardnose shark) female; **29.9** *Carcharhinus dussumieri* (whitecheek shark) female; **29.11** *Carcharhinus fitzroyensis* (creek whaler) immature.

Plate 37 — **Hammerhead Sharks**

30.1 *Eusphyra blochii* (winghead shark) female; **30.3** *Sphyrna mokarran* (great hammerhead) immature;
30.2 *Sphyrna lewini* (scalloped hammerhead) female; **30.4** *Sphyrna zygaena* (smooth hammerhead) juvenile.

Plate 38 — **Angel Sharks**

31.1 *Squatina* sp. A (eastern angel shark) female; **31.2** *Squatina* sp. B (western angel shark) female; **31.4** *Squatina tergocellata* (ornate angel shark) female.

Plate 39 — **Angel Sharks and Shovelnose Rays**

31.3 *Squatina australis* (Australian angel shark) female; **32.5** *Rhinobatos typus* (giant shovelnose ray) immature;
32.4 *Rhinobatos* sp. A (goldeneye shovelnose ray) female.

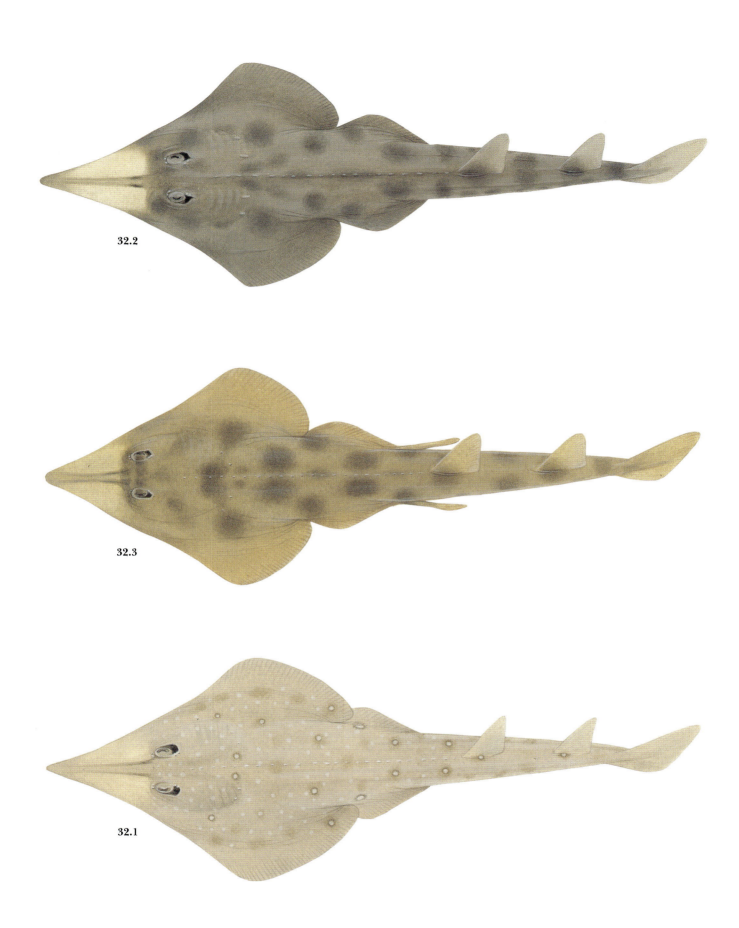

Plate 40 — **Shovelnose Rays**

32.2 *Aptychotrema rostrata* (eastern shovelnose ray) female; **32.3** *Aptychotrema vincentiana* (western shovelnose ray) male; **32.1** *Aptychotrema* sp. A (spotted shovelnose ray) female.

Plate 41 — **Shovelnose Rays (fiddler rays)**

32.6 *Trygonorrhina* sp. A (eastern fiddler ray) female; **32.7** *Trygonorrhina fasciata* (southern fiddler ray) female; **32.8** *Trygonorrhina melaleuca* (magpie fiddler ray) male.

Plate 42 — **Guitar and Sawfishes**

33.2 *Rhynchobatus djiddensis* (white-spotted guitarfish) immature; **33.1** *Rhina ancylostoma* (shark ray) immature; **36.3** *Pristis microdon* (freshwater sawfish) immature.

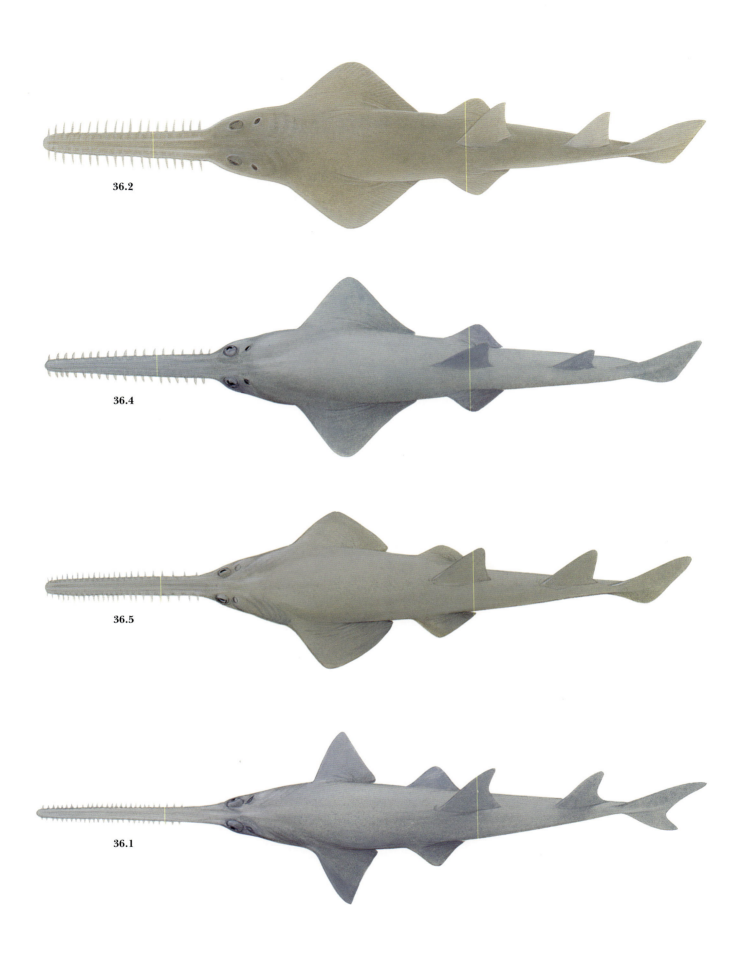

Plate 43 — **Sawfishes**

36.2 *Pristis clavata* (dwarf sawfish) immature; **36.4** *Pristis pectinata* (wide sawfish) female; **36.5** *Pristis zijsron* (green sawfish) female; **36.1** *Anoxypristis cuspidata* (narrow sawfish) female.

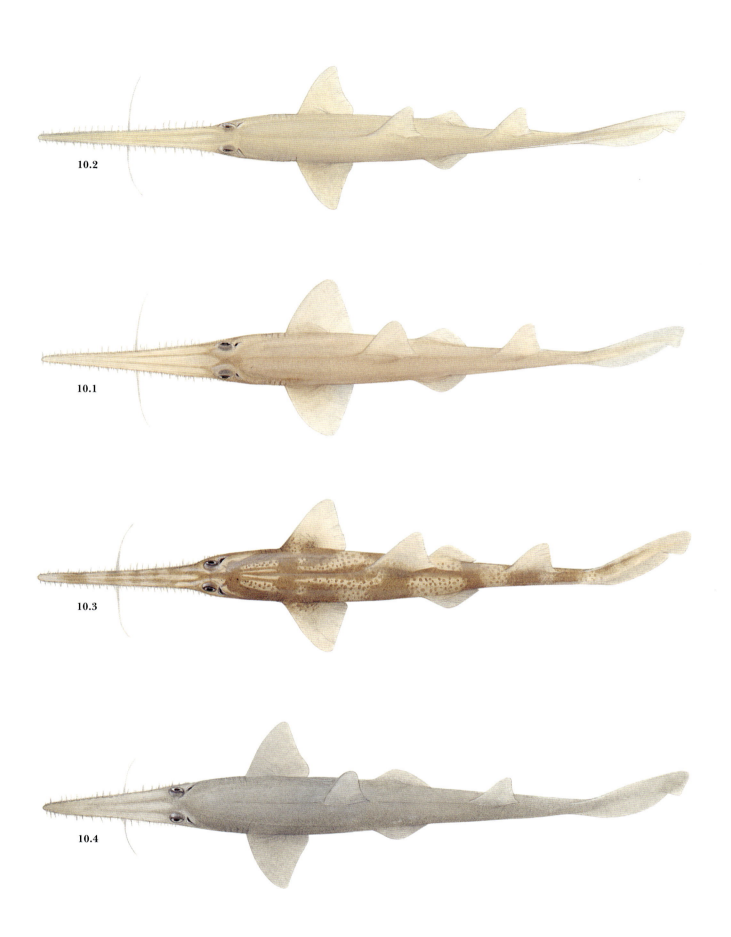

Plate 44 — **Sawsharks**

10.2 *Pristiophorus* sp. B (tropical sawshark) female; **10.1** *Pristiophorus* sp. A (eastern sawshark) female;
10.3 *Pristiophorus cirratus* (common sawshark) female; **10.4** *Pristiophorus nudipinnis* (southern sawshark) female.

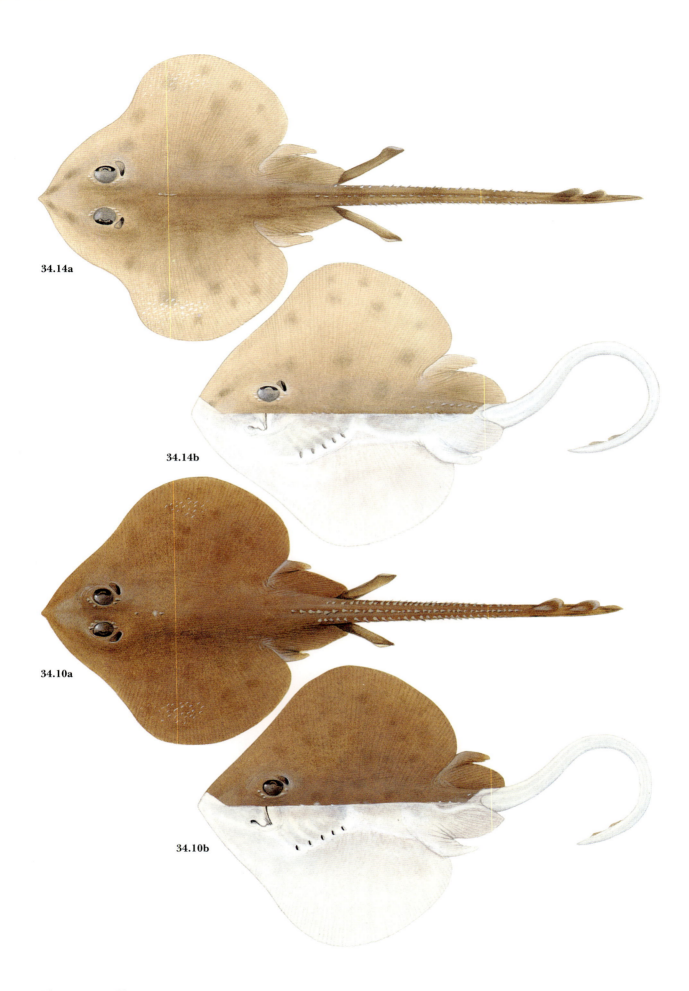

Plate 45 — **Skates**

34.14a *Pavoraja alleni* (Allens skate) male; **34.14b** *Pavoraja alleni* (Allens skate) female (both surfaces);
34.10a *Pavoraja* sp. C (sandy skate) male; **34.10b** *Pavoraja* sp. C (sandy skate) female (both surfaces).

Plate 46 — **Skates**

34.13a *Pavoraja* sp. F (dusky skate) male; **34.13b** *Pavoraja* sp. F (dusky skate) female (both surfaces);
34.11a *Pavoraja* sp. D (mosaic skate) male; **34.11b** *Pavoraja* sp. D (mosaic skate) female (both surfaces).

Plate 47 — **Skates**

34.15a *Pavoraja nitida* (peacock skate) male; **34.15b** *Pavoraja nitida* (peacock skate) female (both surfaces);
34.12a *Pavoraja* sp. E (false peacock skate) male; **34.12b** *Pavoraja* sp. E (false peacock skate) female (both surfaces).

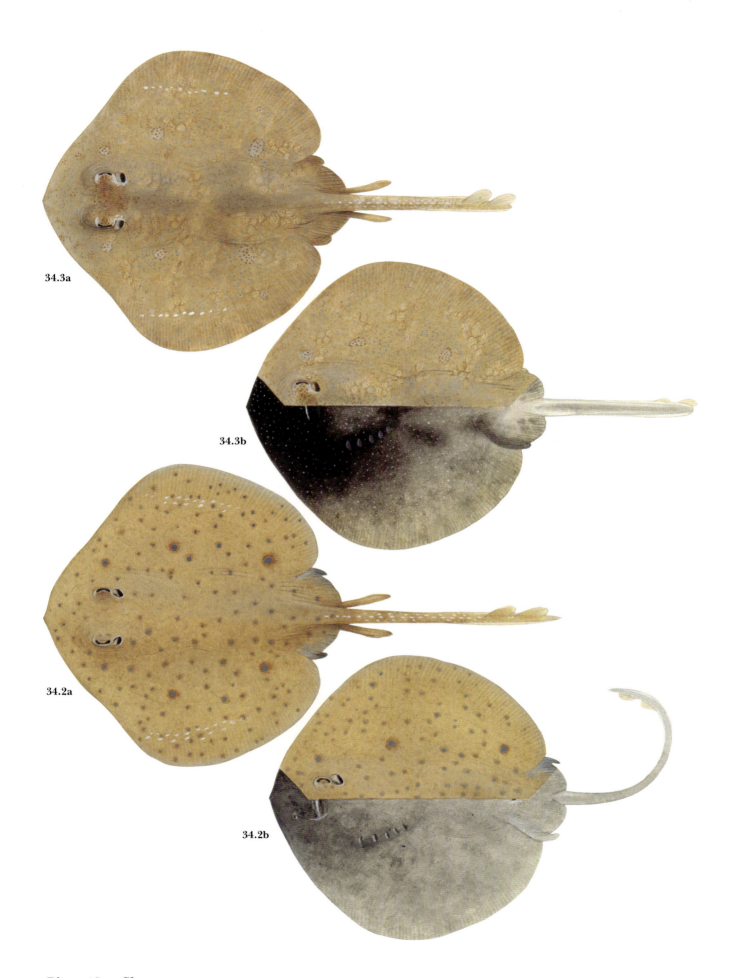

Plate 48 — **Skates**

34.3a *Irolita waitii* (southern round skate) male; **34.3b** *Irolita waitii* (southern round skate) female (both surfaces);
34.2a *Irolita* sp. A (western round skate) male; **34.2b** *Irolita* sp. A (western round skate) female (both surfaces).

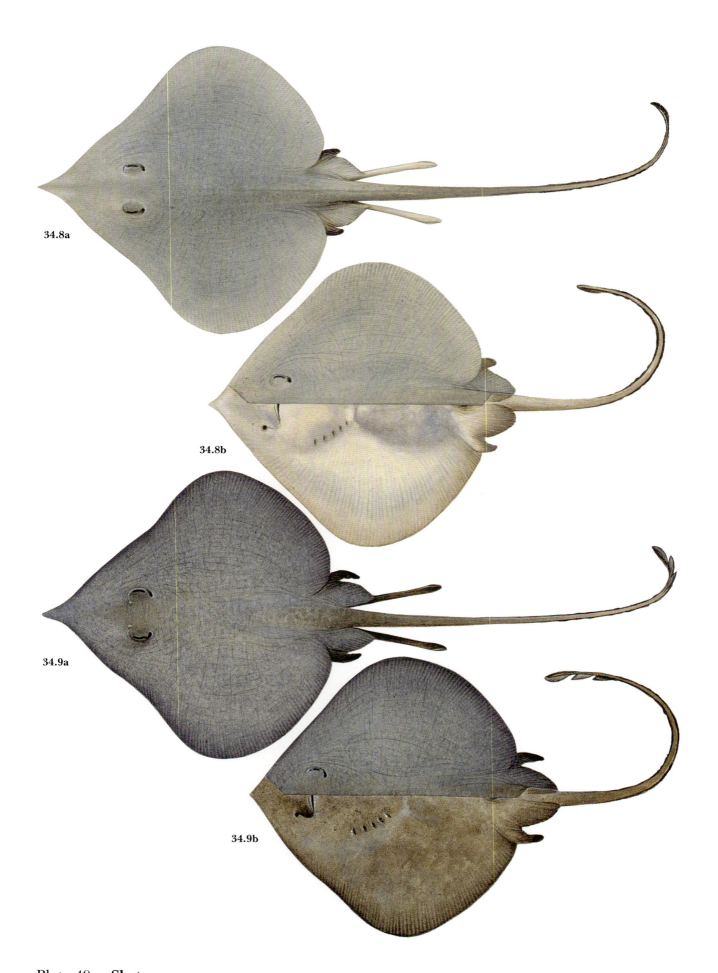

Plate 49 — **Skates**

34.8a *Pavoraja* sp. A (eastern looseskin skate) male; **34.8b** *Pavoraja* sp. A (eastern looseskin skate) female (both surfaces); **34.9a** *Pavoraja* sp. B (western looseskin skate) male; **34.9b** *Pavoraja* sp. B (western looseskin skate) female (both surfaces).

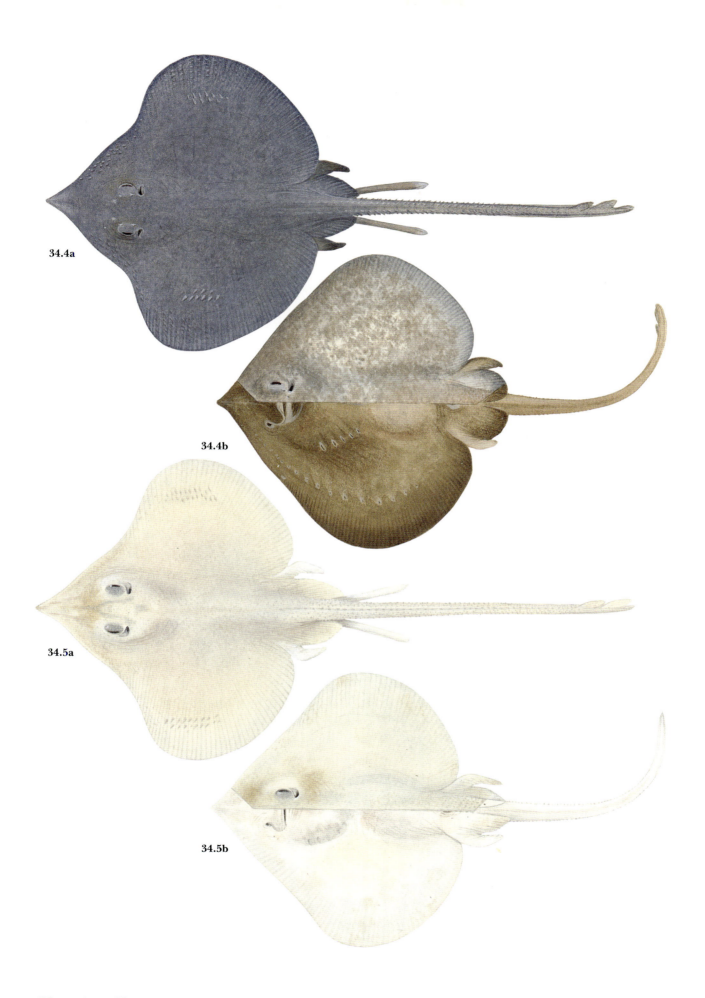

Plate 50 — **Skates**

34.4a *Notoraja* sp. A (blue skate) male; **34.4b** *Notoraja* sp. A (blue skate) female (mottled form); **34.5a** *Notoraja* sp. B (pale skate) male; **34.5b** *Notoraja* sp. B (pale skate) female (both surfaces).

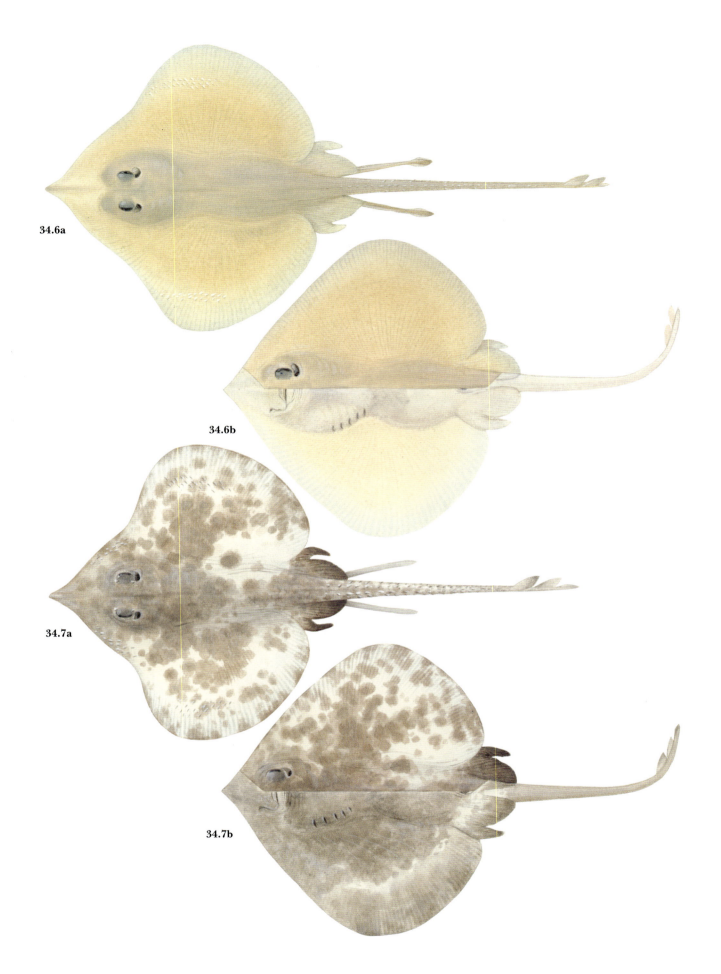

Plate 51 — **Skates**

34.6a *Notoraja* sp. C (ghost skate) male; **34.6b** *Notoraja* sp. C (ghost skate) female (both surfaces); **34.7a** *Notoraja* sp. D (blotched skate) male; **34.7b** *Notoraja* sp. D (blotched skate) female (both surfaces).

Plate 52 — **Skates**

34.31a *Raja* sp. P (Challenger skate) male; **34.31b** *Raja* sp. P (Challenger skate) female (both surfaces);
34.30a *Raja* sp. O (sawback skate) male; **34.30b** *Raja* sp. O (sawback skate) female (both surfaces).

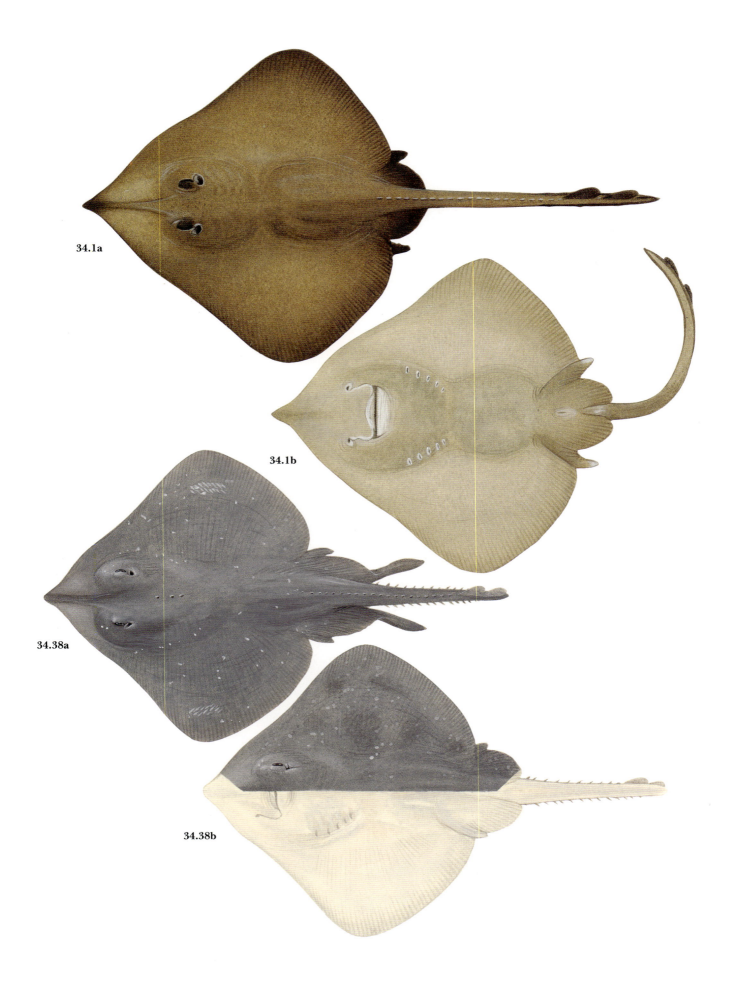

Plate 53 — **Skates**

34.1a *Bathyraja* sp. A (abyssal skate) female (dorsal surface); **34.1b** *Bathyraja* sp. A (abyssal skate) female (ventral surface); **34.38a** *Raja whitleyi* (Melbourne skate) male; **34.38b** *Raja whitleyi* (Melbourne skate) female (both surfaces).

Plate 54 — **Skates**

34.36a *Raja lemprieri* (thornback skate) male; **34.36b** *Raja lemprieri* (thornback skate) female (both surfaces);
34.28a *Raja* sp. M (pygmy thornback skate) male; **34.28b** *Raja* sp. M (pygmy thornback skate) female (both surfaces).

Plate 55 — **Skates**

34.35a *Raja hyperborea* (boreal skate) male (dorsal surface); **34.35b** *Raja hyperborea* (boreal skate) male (ventral surface); **34.27a** *Raja* sp. L (maugean skate) male (dorsal surface); **34.27b** *Raja* sp. L (maugean skate) male (ventral surface).

Plate 56 — **Skates**

34.29a *Raja* sp. N (thintail skate) male; **34.29b** *Raja* sp. N (thintail skate) female (both surfaces); **34.23a** *Raja* sp. H (blacktip skate) male; **34.23b** *Raja* sp. H (blacktip skate) female (both surfaces).

Plate 57 — **Skates**

34.34a *Raja gudgeri* (bight skate) male; **34.34b** *Raja gudgeri* (bight skate) female (both surfaces);
34.22a *Raja* sp. G (pale tropical skate) male; **34.22b** *Raja* sp. G (pale tropical skate) female (both surfaces).

Plate 58 — **Skates**

34.24a *Raja* sp. I (Wengs skate) male; **34.24b** *Raja* sp. I (Wengs skate) female (both surfaces); **34.21a** *Raja* sp. F (Leylands skate) male; **34.21b** *Raja* sp. F (Leylands skate) female (both surfaces).

Plate 59 — **Skates**

34.25a *Raja* sp. J (deepwater skate) male; **34.25b** *Raja* sp. J (deepwater skate) female (both surfaces);
34.26a *Raja* sp. K (Queensland deepwater skate) male; **34.26b** *Raja* sp. K (Queensland deepwater skate) female (both surfaces).

Plate 60 — **Skates**

34.17a *Raja* sp. B (grey skate) male; **34.17b** *Raja* sp. B (grey skate) female (both surfaces); **34.18a** *Raja* sp. C (Grahams skate) male; **34.18b** *Raja* sp. C (Grahams skate) female (both surfaces).

Plate 61 — **Skates**

34.32a *Raja australis* (Sydney skate) male; **34.32b** *Raja australis* (Sydney skate) female (both surfaces); **34.19a** *Raja* sp. D (false argus skate) male; **34.19b** *Raja* sp. D (false argus skate) female (both surfaces).

Plate 62 — **Skates**

34.33a *Raja cerva* (white-spotted skate) male; **34.33b** *Raja cerva* (white-spotted skate) female (both surfaces); **34.16a** *Raja* sp. A (longnose skate) male; **34.16b** *Raja* sp. A (longnose skate) female (both surfaces).

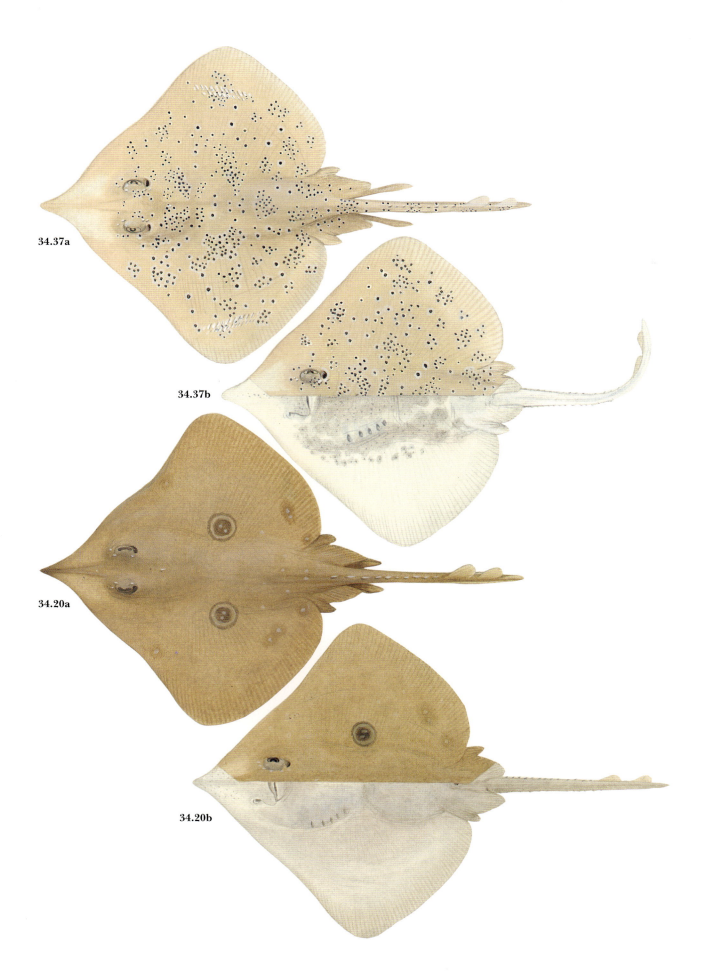

Plate 63 — **Skates**

34.37a *Raja polyommata* (argus skate) male; **34.37b** *Raja polyommata* (argus skate) female (both surfaces);
34.20a *Raja* sp. E (oscellate skate) male; **34.20b** *Raja* sp. E (oscellate skate) female (both surfaces).

Plate 64 — **Leg Skates and Sixgill Ray**

35.2a *Anacanthobatis* sp. B (eastern leg skate) male; **35.2b** *Anacanthobatis* sp. B (eastern leg skate) female (both surfaces); **35.1** *Anacanthobatis* sp. A (western leg skate) female; **43.1** *Hexatrygon* sp. A (sixgill stingray) female.

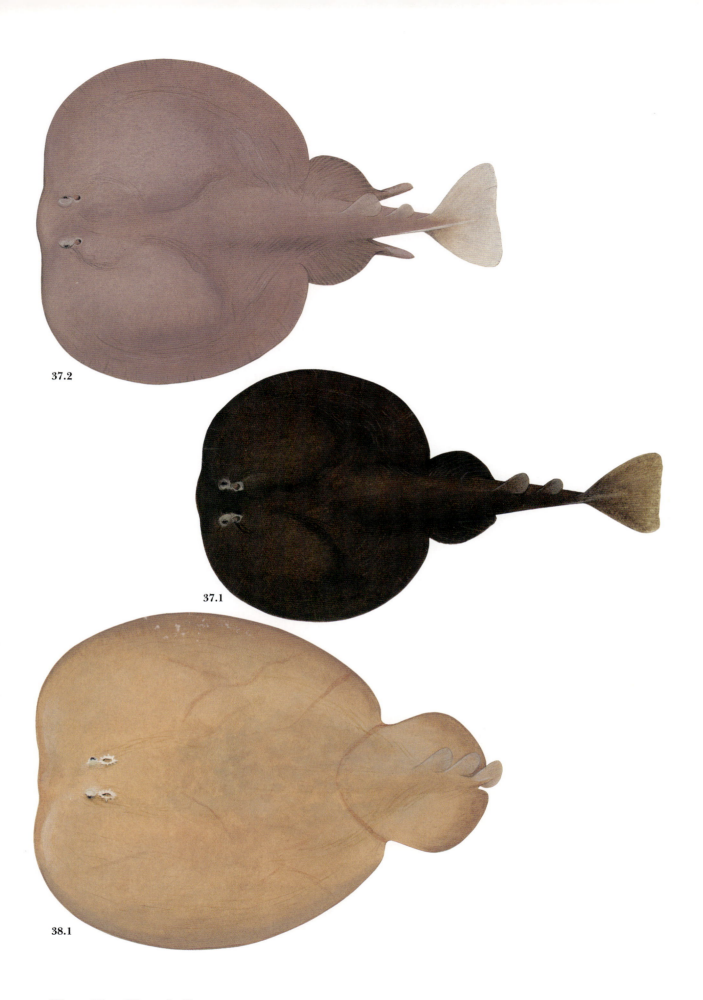

Plate 65 — **Electric Rays**

37.2 *Torpedo macneilli* (short-tail torpedo ray) male; **37.1** *Torpedo* sp. A (longtail torpedo ray) female;
38.1 *Hypnos monopterygium* (coffin ray) female.

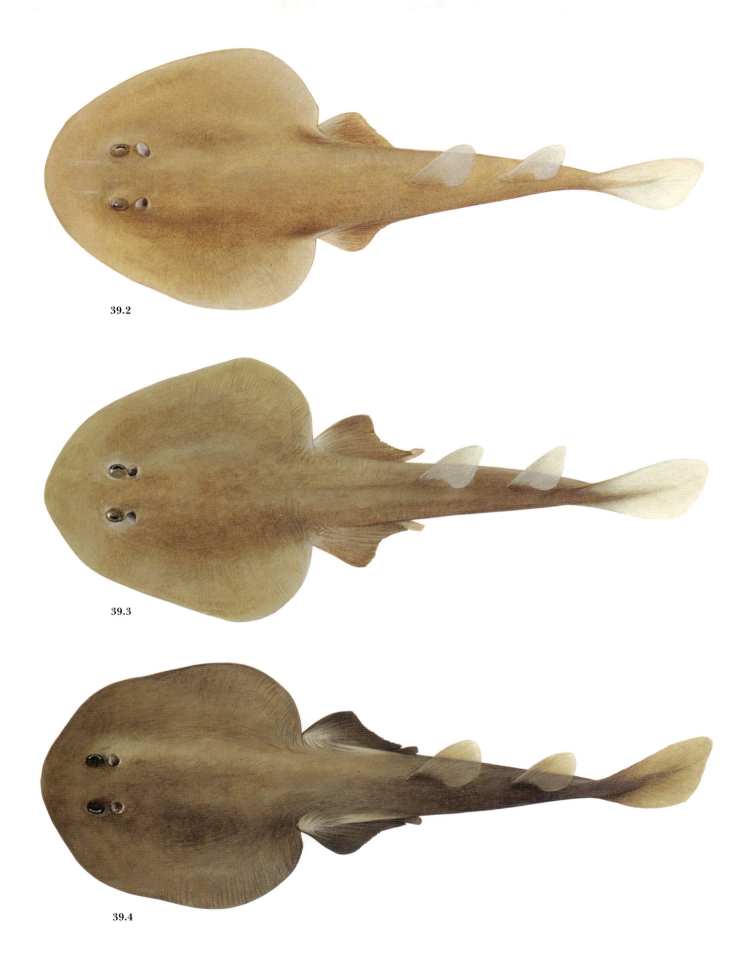

Plate 66 — **Numbfishes**

39.2 *Narcine* sp. B (western numbfish) female; **39.3** *Narcine* sp. C (eastern numbfish) female; **39.4** *Narcine tasmaniensis* (Tasmanian numbfish) female.

Plate 67 — **Numbfishes and Stingray**

39.5 *Narcine westraliensis* (banded numbfish) female; **39.1** *Narcine* sp. A (ornate numbfish) male;
40.9 *Himantura* sp. A (brown whipray) female.

Plate 68 — **Stingrays (whiprays)**

40.14a *Himantura toshi* (black-spotted whipray — rare white-spotted form) male; **40.14b** *Himantura toshi* (black-spotted whipray) female; **40.16** *Himantura undulata* (leopard whipray) female; **40.15** *Himantura uarnak* (reticulate whipray) female.

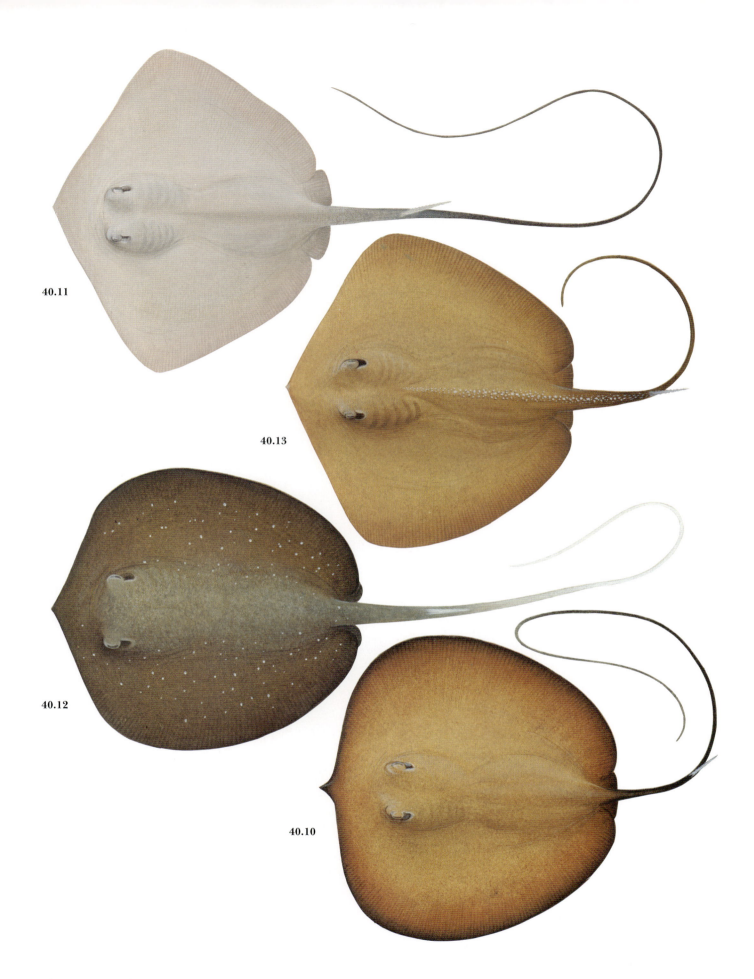

Plate 69 — **Stingrays (whiprays)**

40.11 *Himantura fai* (pink whipray) female; **40.13** *Himantura jenkinsii* (Jenkins whipray) female; **40.12** *Himantura granulata* (mangrove whipray) immature; **40.10** *Himantura chaophraya* (freshwater whipray) immature.

Plate 70 — **Stingrays**

40.17 *Pastinachus sephen* (cowtail stingray) immature; **40.18** *Taeniura lymma* (blue-spotted fantail ray) immature; **40.19** *Taeniura meyeni* (blotched fantail ray) female; **40.20** *Urogymnus asperrimus* (porcupine ray) male.

Plate 71 — **Stingrays (maskrays)**

40.2 *Dasyatis annotata* (plain maskray) female; **40.5** *Dasyatis kuhlii* (blue-spotted maskray) female; **40.6a** *Dasyatis leylandi* (painted maskray — western form) female; **40.6b** *Dasyatis leylandi* (painted maskray — eastern form) female.

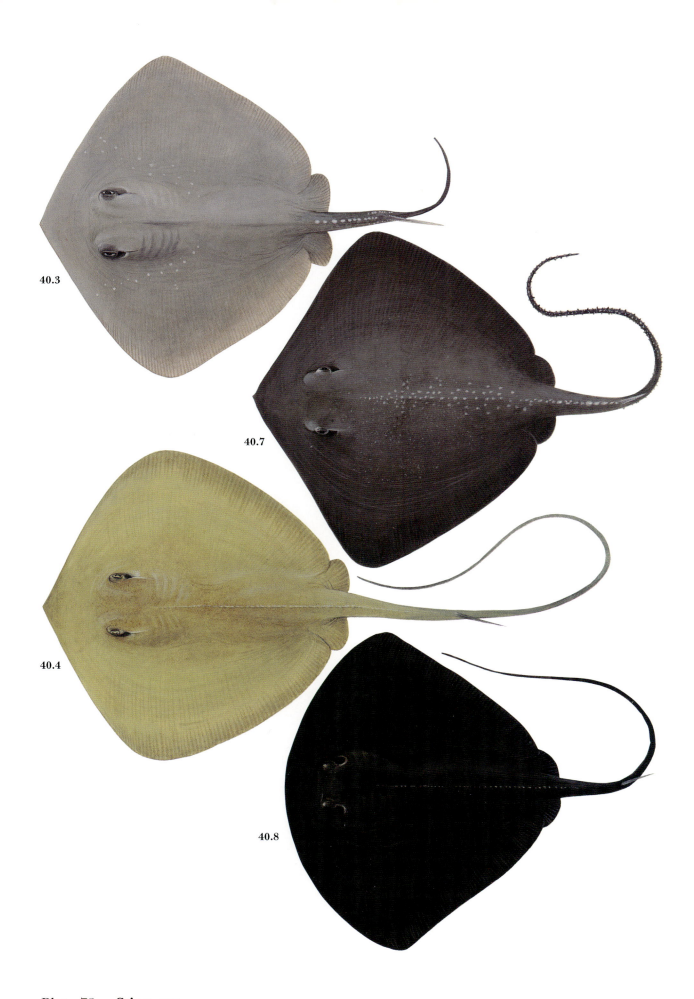

Plate 72 — **Stingrays**

40.3 *Dasyatis brevicaudata* (smooth stingray) female; **40.7** *Dasyatis thetidis* (black stingray) female; **40.4** *Dasyatis fluviorum* (estuary stingray) female; **40.8** *Dasyatis violacea* (pelagic stingray) female.

Plate 73 — **Stingray and Stingarees**

40.1 *Dasyatis* sp. A (dwarf black stingray) male; **41.1** *Plesiobatis daviesi* (giant stingaree) male;
41.11 *Urolophus circularis* (circular stingaree) female; **41.15** *Urolophus gigas* (spotted stingaree) female.

Plate 74 — **Stingarees**

41.21 *Urolophus viridis* (greenback stingaree) female; **41.13** *Urolophus expansus* (wide stingaree) male; **41.8** *Urolophus* sp. A (Kapala stingaree) male; **41.19** *Urolophus paucimaculatus* (sparsely-spotted stingaree) male.

Plate 75 — **Stingarees**

41.17 *Urolophus mitosis* (mitotic stingaree); **41.22** *Urolophus westraliensis* (brown stingaree); **41.10** *Urolophus bucculentus* (sandyback stingaree); **41.14** *Urolophus flavomosaicus* (patchwork stingaree).

Plate 76 — **Stingarees**

41.20 *Urolophus sufflavus* (yellowback stingaree) female; **41.18** *Urolophus orarius* (coastal stingaree) male; **41.16** *Urolophus lobatus* (lobed stingaree) male; **41.12** *Urolophus cruciatus* (banded stingaree) female.

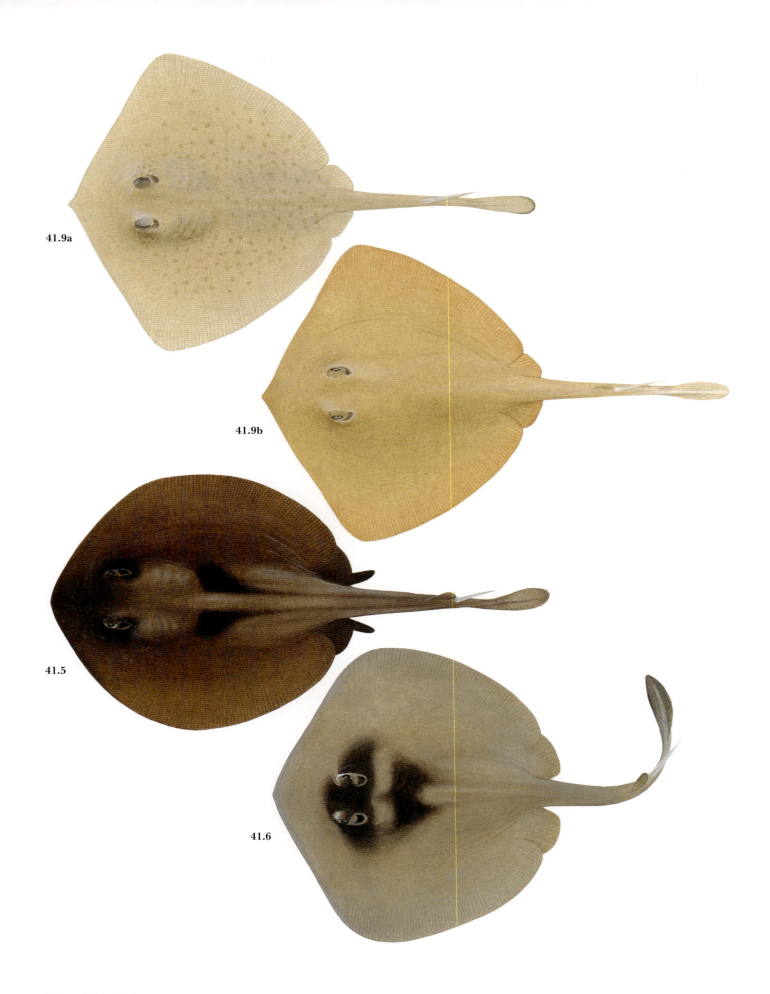

Plate 77 — **Stingarees**

41.9a *Urolophus* sp. B (Coral Sea stingaree) female; **41.9b** *Urolophus* sp. B (Coral Sea stingaree — brown form) female; **41.5** *Trygonoptera ovalis* (striped stingaree) male; **41.6** *Trygonoptera personata* (masked stingaree) female.

Plate 78 — **Stingarees**

41.2 *Trygonoptera* sp. A (yellow shovelnose stingaree) male; **41.4** *Trygonoptera mucosa* (western shovelnose stingaree) female; **41.3** *Trygonoptera* sp. B (eastern shovelnose stingaree) female; **41.7** *Trygonoptera testacea* (common stingaree) female.

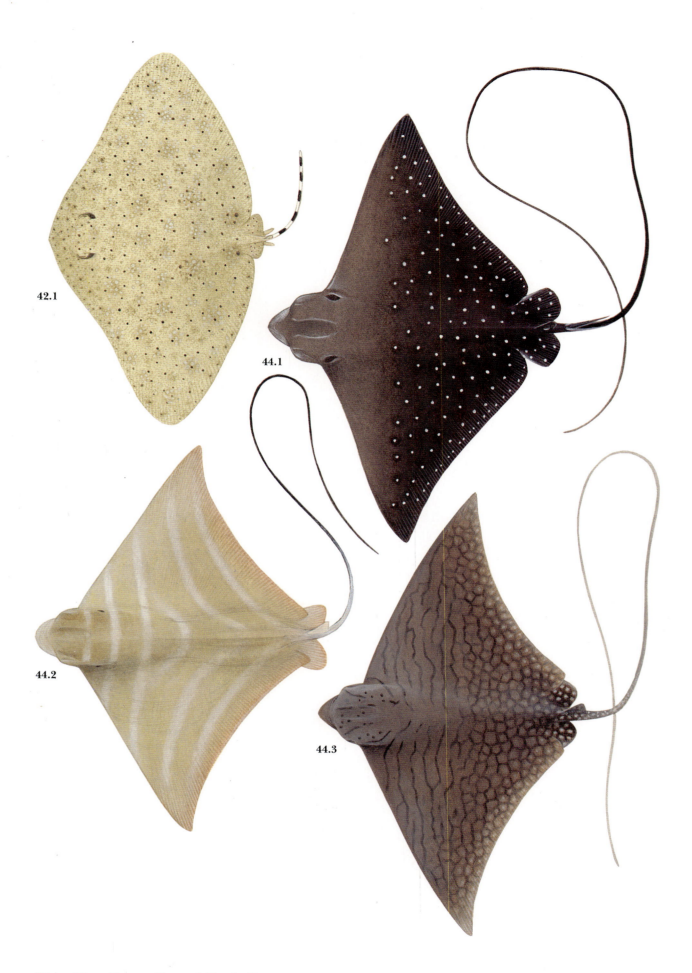

Plate 79 — **Butterfly and Eagle Rays**

42.1 *Gymnura australis* (Australian butterfly ray) male; **44.1** *Aetobatus narinari* (white-spotted ray) female; **44.2** *Aetomylaeus nichofii* (banded eagle ray) female; **44.3** *Aetomylaeus vespertilio* (ornate eagle ray) female.

Plate 80 — **Eagle and Cownose Rays**

44.4 *Myliobatis australis* (southern eagle ray) female; **44.5** *Myliobatis hamlyni* (purple eagle ray) female; **45.1a** *Rhinoptera neglecta* (Australian cownose ray) immature; **45.1b** *Rhinoptera javanica* (Javanese cownose ray) immature.

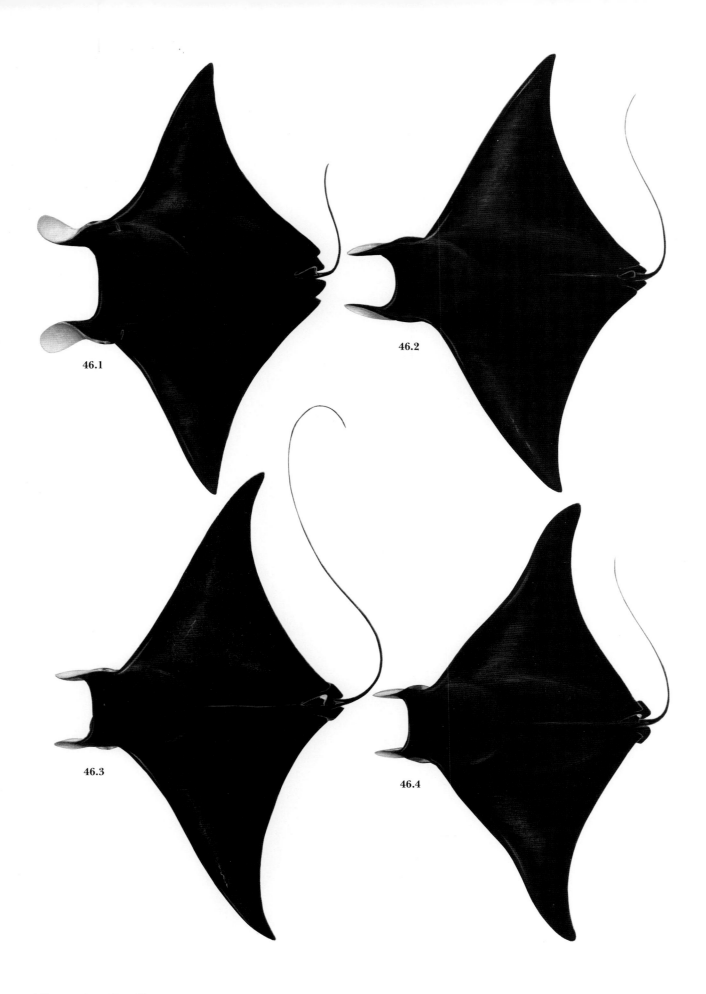

Plate 81 — **Devilrays**

46.1 *Manta birostris* (manta ray) female; **46.2** *Mobula eregoodootenkee* (pygmy devilray) female; **46.3** *Mobula japanica* (Japanese devilray) female; **46.4** *Mobula thurstoni* (bentfin devilray) female.

Plate 82 — **Elephant and Spookfishes**

47.1 *Callorhinchus milii* (elephant fish) female; **49.2** *Harriotta raleighana* (bigspine spookfish) female; **49.1** *Harriotta haeckeli* (smallspine spookfish) female; **49.3** *Rhinochimaera pacifica* (Pacific spookfish) male.

Plate 83 — **Chimaeras**

48.1 *Chimaera* sp. A (southern chimaera) female; **48.2** *Chimaera* sp. B (shortspine chimaera) female; **48.3** *Chimaera* sp. C (longspine chimaera) female; **48.4** *Chimaera* sp. D (giant chimaera) male.

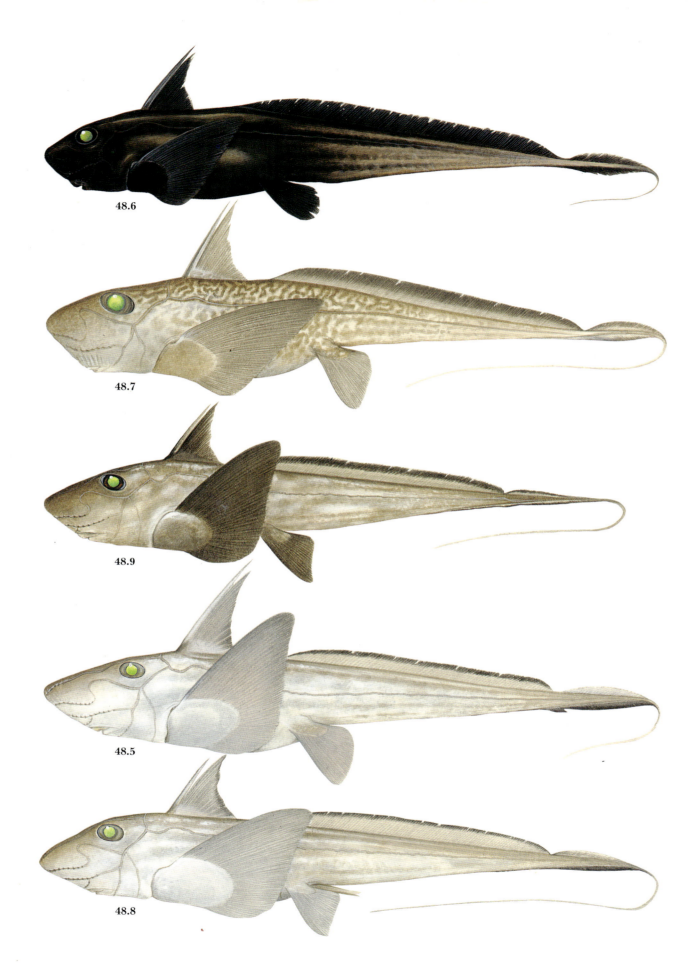

Plate 84 — **Ghostsharks**

48.6 *Hydrolagus* sp. A (black ghostshark) female; **48.7** *Hydrolagus* sp. B (marbled ghostshark) female; **48.9** *Hydrolagus ogilbyi* (Ogilbys ghostshark) female; **48.5** *Chimaera* sp. E (whitefin chimaera) female; **48.8** *Hydrolagus lemures* (blackfin ghostshark) male.

2